AF327320

CONTROLLED
THERMONUCLEAR REACTIONS

CONTROLLED THERMONUCLEAR REACTIONS

An Introduction to Theory and Experiment

by

SAMUEL GLASSTONE

Consultant to the United States Atomic Energy Commission

AND

RALPH H. LOVBERG

University of California
Los Alamos Scientific Laboratory

PREPARED UNDER AUSPICES OF
THE OFFICE OF TECHNICAL INFORMATION,
UNITED STATES ATOMIC ENERGY COMMISSION

ROBERT E. KRIEGER PUBLISHING COMPANY
MALABAR, FLORIDA

Original Edition 1960
Reprint 1975

Printed and Published by
ROBERT E. KRIEGER PUBLISHING CO., INC
KRIEGER DRIVE
MALABAR, FLORIDA 32950

Copyright © 1960
LITTON EDUCATIONAL PUBLISHING, INC.
Reprinted by Arrangement
Van Nostrand Reinhold Company

Printed in the United States of America

Library of Congress Cataloging in Publication Data

Glasstone, Samuel, 1897-
   Controlled thermonulcear reactions.

   "Prepared under auspices of the Office of Technical Information, United States Atomic Energy Commission."
   Reprint of the ed. published by Van Nostrand Reinhold Co., New York.
   Includes bibliographies.
   1. Controlled fusion. I. Lovberg, Ralph Harvey, joint author. II. Title.
[QC791.73.G49 1975]   539.7'64   75-11911
ISBN 0-88275-326-6

FOREWORD

by

Arthur E. Ruark, Chief, Controlled Thermonuclear Branch
United States Atomic Energy Commission

At the 1958 Geneva Conference on the Peaceful Uses of Atomic Energy, research on fusion power was declassified by the United Kingdom, the United States, and the USSR. The main results obtained by hundreds of scientists over a period of more than six years were presented at the conference. It was clear that a flood of papers would soon be published and that steps should be taken to provide a comprehensive book which would weave together the results and help to point the way to further progress. The matter was all the more urgent because of growing interest in plasma physics outside the controlled fusion fraternity. The physics and technology of controlled fusion have strong interactions with research in astrophysics, ionospherics, accelerator design, the electromagnetic method of isotope separation, direct conversion of heat to power, and space propulsion.

The Atomic Energy Commission has long been aware of the need for excellent books in new and growing fields while they are still in the process of growth. The Commission has relied heavily on Dr. Samuel Glasstone for preparation of such material in the field of fission science and technology, so it was natural to turn to him for similar service in the field of fusion. There are now several excellent books in English on plasma physics, and some brief accounts of controlled thermonuclear research. Dr. Glasstone, ably assisted by Dr. Ralph H. Lovberg, has written, to the best of our knowledge, the first reasonably comprehensive treatise on fusion-power research presented on a practical scale—not too short, and not too long. The authors have had the benefit of many helpful comments from colleagues in American laboratories, but the book is substantially their own. I anticipate that it will command the respect and gratitude of a world-wide audience.

PREFACE

Thermonuclear fusion reactions are known to be the energy source in two widely different situations. At one extreme, relatively slow reactions of this type produce the energy of the sun and the stars, whereas at the other extreme, rapid thermonuclear reactions are responsible for the explosive power of the hydrogen bomb. Somewhere between these two extremes, it should be possible to bring about thermonuclear reactions under conditions which will permit energy to be released at a controllable rate. Controlled thermonuclear reactions, in which there is a net gain of energy, are possible in principle, and so could serve as a source of useful power. Since the raw materials would be water and, perhaps, lithium minerals, which are available all over the world, the achievement of economical controlled thermonuclear power would have highly significant consequences, both economic and political. For this impelling reason, and also because of the exceptionally challenging nature of the scientific and technical problems involved, the study of thermonuclear reactions, which is being actively pursued in many countries, will inevitably attract increasing interest.

The main purpose of this book is to provide the necessary background for physicists and engineers who are planning to enter the field of research into controlled thermonuclear reactions. The treatment is directed largely to experimentalists, but sufficient theory is introduced to enable the reader to appreciate the principles underlying the observed phenomena. The theoretical derivations given are often simplified, since they are intended to provide a physical picture rather than to represent a rigorous mathematical approach. Nevertheless, it is hoped that theorists will find the book a useful guide to the many experimental results which require interpretation.

In general, the standard of the book is that of the first-year graduate level in either physics or engineering. Apart from a basic knowledge of physics and mathematics appropriate to this level, all that is needed for a satisfactory comprehension of most of the book's contents is an understanding of the elements of electromagnetism. Because this book is directed at both engineers and physicists, some concern has been felt over the matter of units. However, the result of inquiries made of a number of scientists presently engaged in thermonuclear research indicated a decided preference for Gaussian units for electrical and magnetic quantities, because they are best suited to the description of microscopic phenomena in plasmas. Consequently, these units have been generally used, although in certain instances, where macroscopic current and voltage measurements are treated, mks units are employed. Such a compromise is not completely satisfactory, but it is felt that it is the best that can be done, for in the study of thermonuclear plasmas extreme points of view merge.

An effort has been made to cover the now extensive literature of thermonuclear research by means of specific references collected at the end of each chapter, with a more general list given at the end of the book. It should be emphasized, however, that there is no intention, either in the choice of references or in their order of presentation, to assign priority or special credit in connection with any particular development or discovery. This task is rendered difficult, if not impossible, by the fact that similar research on controlled thermonuclear reactions has been proceeding simultaneously but quite independently in different laboratories.

Los Alamos, New Mexico
July 1960

Samuel Glasstone
Ralph H. Lovberg

ACKNOWLEDGMENT

In producing this book, we received considerable help from a large number of individuals to all of whom we wish to record our sincere appreciation of their efforts. Some provided information that was not readily accessible and many, who are engaged in thermonuclear research, reviewed portions of the draft manuscript and made valuable comments and suggestions. It is not possible to name all those who contributed in one way or another, but we wish to thank the following in particular: L. Spitzer, Jr., and the members of the Project Matterhorn (Princeton University) staff, including D. J. Grove, M. A. Heald (now at Swarthmore College), and R. M. Sinclair; C. M. Van Atta and the staff of Project Sherwood at the Lawrence Radiation Laboratory, University of California, Berkeley and Livermore, including J. Berkowitz (from New York University), N. C. Christofilos, S. A. Colgate, H. P. Furth, W. B. Kunkel, and R. F. Post; J. L. Tuck, Jr., and his staff at the Los Alamos Scientific Laboratory, University of California, including J. A. Phillips, E. J. Stovall, and T. F. Stratton; W. R. Faust and A. C. Kolb, U. S. Naval Research Laboratory, Washington, D.C.; M. A. Biondi, Westinghouse Research Laboratory; J. W. Butler, Argonne National Laboratory; H. L. Davis, *Nucleonics*, McGraw-Hill Publishing Co.; H. Grad, Institute of Mathematical Sciences, New York University; H. Hurwitz, General Electric Research Laboratory; D. W. Kerst and M. N. Rosenbluth, General Atomic Research Laboratory; J. R. Lamarsh, Cornell University; A. H. Snell and A. Simon, Oak Ridge National Laboratory; and E. S. Weibel, Space Technology Laboratories. We are also grateful to P. C. Thonemann, Chief of the thermonuclear research project at the Atomic Energy Research Establishment, Harwell, England, who reviewed parts of the manuscript while he was a guest at Project Matterhorn.

Our special thanks are due to B. R. Suydam, Los Alamos Scientific Laboratory, for undertaking the difficult task of preparing a draft chapter on Plasma Stability Theory. We feel that he has been successful in explaining how mathematical physicists are approaching this highly complex problem.

We must express our indebtedness for support, encouragement, and assistance received from members of the U.S. Atomic Energy Commission, Washington, D.C., including A. E. Ruark, Chief, Controlled Thermonuclear Branch, Division of Research, and his staff; M. S. Day, Director, Office of Technical Information, E. J. Brunenkant, Chief, Industrial Information Branch, OTI, and, especially, J. W. Young, Chief, Technical Book Section, J. G. Gratton, project officer for this book, and C. B. Holmes, Chief, Publishing Branch, OTI Extension, Oak Ridge.

Finally, we wish to thank the Administration of the Los Alamos Scientific

Laboratory, N. E. Bradbury, Director, D. K. Froman, Technical Associate Director, P. F. Belcher, W. H. Crew, and H. R. Hoyt, Assistant Directors, and J. M. B. Kellogg, Physics Division Leader, for their courtesy and co-operation in making available to us the facilities of the Laboratory.

SAMUEL GLASSTONE
RALPH H. LOVBERG

TABLE OF CONTENTS

CONTENTS

Chapter 1

INTRODUCTION

FUSION OF HYDROGEN ISOTOPES

1.1. In both the lightest and the heaviest nuclei, the mean binding energy per nuclear particle is less than it is in those of intermediate mass number. Consequently, it should be possible, in principle, to obtain nuclear energy either by fission of the nuclei of the heaviest elements or by the fusion (or combination) of the lightest ones, the products in each case being nuclei in the intermediate mass range. As is well known, the fission process has been extensively investigated in recent years, and it is already being used as a source of useful power. Serious attention is now being given to the alternative, but much more difficult, problem of obtaining energy by nuclear fusion.

1.2. There are many reactions involving the combination of two light nuclei which are accompanied by the release of energy. However, before the positively charged nuclei can be brought close enough together for fusion to take place, sufficient energy must be supplied to overcome the force of electrostatic repulsion between them. This force is proportional to the product of the nuclear charges, and so it increases as the atomic numbers of the interacting nuclei increase. Hence, for nuclei carrying a given amount of energy, the rates of the fusion reactions decrease, in general, with increasing atomic number. It is only with the very lightest nuclei, e.g., those of hydrogen and helium isotopes, that conditions appear to be attainable under which the fusion reactions become fast enough to have potential practical value.

1.3. From measurements of the cross sections (or probabilities) of reactions involving light nuclei, it has been concluded that, except for a possible combination of deuterium and the rare isotope helium-3, the only fusion processes which offer any promise of producing more energy than is expended in bringing them about are those involving the two heavier isotopes of hydrogen, namely, deuterium and tritium. One of the difficulties associated with making nuclear fusion self-sustaining, at least, so that just as much energy is released as is consumed or lost while the reactions proceed, is that very large amounts of energy escape in the form of radiation. The fact that these losses increase with increasing atomic number of the nuclei is another reason why, as far as can

1

be seen at present, the only fusion processes that might lead to a net gain of energy are those in which the nuclei of hydrogen isotopes take part. In many stars the fusion of the lightest isotope of hydrogen is the main source of energy, whereas in others fusion reactions involving heavier elements play the major role. The situation is too complex for discussion here, but it may be stated that the reactions taking place are too slow to be of any practical use on the earth. It is only because of the enormous mass and size of the sun, for example, that it produces energy at such a high rate in spite of the slowness of the fusion reactions.

1.4. Information concerning nuclear reactions between the isotopes of hydrogen has been obtained from two very different experimental approaches. In the first place, reactions involving deuterons, i.e., deuterium nuclei, alone or deuterons and tritons, i.e., tritium nuclei, have been studied and their cross sections determined by laboratory measurements. For example, deuterons can be accelerated in a cyclotron (or similar machine) and allowed to impinge on a solid target containing deuterium or tritium. Fusion reactions accompanied by the liberation of energy occur, but, for reasons which will be apparent later, the energy produced in this manner is considerably less than that expended in accelerating the deuterons.

1.5. The other way in which nuclear fusion has been achieved on the earth is in the device commonly known as the hydrogen bomb. At the high temperatures produced in a fission reaction, nuclei of isotopes of hydrogen undergo fusion with the liberation of energy. In these circumstances, the nuclear reactions are propagated so rapidly that the energy is released in an uncontrolled (or explosive) manner.

1.6. It seems reasonable to hope that, somewhere between the trivial production of fusion energy achieved by means of accelerated particles, on the one hand, and its release in an explosive manner, on the other hand, controlled nuclear fusion will be possible. In essence, a controlled fusion reactor would be a device in which appropriate isotopes of hydrogen combine, the end result being the production and extraction, in a manner that can be regulated at will, of useful quantities of energy in excess of the amount required to operate the device.

ENERGY RELEASE IN FUSION REACTIONS

1.7. It is probable that the basic "fuel" material for fusion power will be deuterium, the isotope of hydrogen which is present to the extent of 1 atom to 6500 atoms of ordinary hydrogen in water. In spite of this apparently small proportion, it is estimated that the oceans and other surface waters contain a total of more than 10^{14} tons of deuterium. If this could be utilized, in well-known nuclear fusion reactions, for the release of energy, the total available would be at least 10^{21} kilowatt-years. The world's present energy consumption is at the rate of approximately 4×10^9 kilowatts. Even if this were increased by a large factor, to allow for increases in both the earth's population and the per capita power demand, the available deuterium would be sufficient for many million years.

1.8. Another way of emphasizing the vast amounts of energy that might be realized by controlled fusion is to state that 1 gram of deuterium could yield a maximum of something like 8×10^{10} calories. To produe this quantity of deuterium requires about 8 gallons of water, so that 1 gallon of water has a fusion energy equivalent of 10^{10} calories. The combustion of 1 gallon of gasoline yields a little more than 3×10^7 calories. Hence, the nuclear fusion energy that could, in theory, be obtained from 1 gallon of water would be equivalent to over 300 gallons of gasoline.

1.9. In spite of the fact that the nuclear energy potential of a gram of deuterium is equivalent to the combustion energy of more than 2500 gallons of gasoline or to the explosive energy of some 80 tons of TNT, a fusion reactor may be expected to be completely safe. It appears that there would be absolutely no danger of a destructive runaway if there were a loss of control due to an accident or to earthquake, lightning, or other natural phenomenon. The reason is that the reactor would operate at a very low gas density, and so the total energy density would not exceed 20 calories/cm^3. The disruption of a reactor having a volume of 1500 liters would thus release no more energy than is produced by the combustion of a gallon of gasoline. It is probable, therefore, that there would be no serious danger of explosion in the operation of a controlled fusion reactor.

1.10. In connection with the hazard problem, it may be noted that, in contrast to the fission process, no appreciable amount of radioactive material would result from the reactions occurring in a fusion reactor. There would thus be no problem of the disposal of radioactive waste and no danger of the contamination of the surrounding area in the event of an accident leading to disruption of the reactor. However, like a fission reactor, a fusion device would require considerable shielding to prevent the escape of neutrons and various harmful radiations.

OPERATIONAL ASPECTS OF FUSION REACTORS

1.11. Ideally, a fusion reactor would function on deuterium alone as the feed material, but there are indications that the required operating conditions would be extremely difficult to attain. It will undoubtedly be easier, although less desirable economically, to employ a mixture of deuterium and tritium. Unfortunately, tritium exists in nature in the merest traces only, and to obtain it in quantity it must be made by the interaction of neutrons with lithium. For the start-up of the reactor, it would be necessary to provide a supply of tritium, but once the operating conditions have been attained, it is anticipated that the required tritium could be regenerated from the neutrons produced in the fusion reactions. Apart from the initial tritium supply, the feed materials would then be deuterium and lithium. Since there are ample lithium supplies in nature, the fusion power potential would not be seriously affected.

1.12. No attempt will be made here to consider the economics of fusion power because of the many uncertainties involved. It can be stated, however, that fuel costs will not be an important consideration. At the present time the price of deuterium, which is available in the form of deuterium oxide (or heavy water) extracted from ordinary water, is equivalent to about \$140 per pound. Since this could, in principle, produce some 4×10^7 kw-hr of energy by fusion reactions, the fuel cost would be roughly 0.00035 cent per kw-hr. This may be compared with about 0.08 cent, including inventory charges, for nuclear fission fuels and 0.2 to 0.3 cent for fossil fuels. However, fuel cost will probably not be an important consideration, since it will represent only a small fraction of the cost of producing fusion power. What is significant is that the raw materials, namely, water and lithium minerals, are widely distributed and readily available all over the world.

1.13. An interesting aspect of controlled fusion reactors is the possibility that, in certain proposed devices, a portion of the energy released could be extracted directly as electrical power. While this would partly eliminate the requirement for turbines and boilers, other equipment would be necessary, and it is not certain that the efficiency for conversion of the nuclear energy into electricity would be higher than can be obtained from a conventional thermal cycle. In any event, a proportion of the nuclear fusion energy would be released as heat, so that the usual methods of recovery would have to be applied. However, it is not impossible that, by the time fusion power has become feasible, efficient methods for the direct conversion of heat into electrical energy will have been developed.

THERMONUCLEAR REACTIONS

1.14. As pointed out above, in order to bring about fusion reactions it is necessary that the interacting nuclei collide with sufficient energy to overcome the forces of electrostatic repulsion which tend to keep them apart. It is known that this energy can be supplied, under laboratory conditions, by means of a charged-particle accelerator. Alternatively, the interacting nuclei may acquire the necessary energy if they are part of a system at sufficiently high temperatures, as is the case in the sun and stars. The two situations are, however, different in an important respect. Accelerated particles all have essentially the same energy and move in the same direction, but when the energy is acquired as a result of raising the temperature the particles exhibit random motion combined with a wide distribution of energies. In the latter case, the nuclear fusion reactions are referred to as *thermonuclear reactions.**

---

* Strictly speaking, the term "thermonuclear" should be applied to the reacting system only when the velocity distribution of the particles is Maxwellian in character, so that a true kinetic temperature may be defined (see Chapter 2).

1.15. Both acceleration and heating methods have been proposed for conferring energy upon deuterium nuclei in order to bring about controlled fusion reactions. It seems fairly certain, however, for reasons which will be apparent in later chapters, that, when acceleration is used, it would be advantageous if the directed motion of the accelerated nuclei were converted into random motion with a distribution of energies. This means that, if controlled fusion can be achieved on a sufficient scale to be of practical value, it will probably be by way of reactions which are essentially thermonuclear in nature.

1.16. It will be shown in the next chapter that in a thermonuclear reactor capable of producing net power the energy distribution among the reacting nuclei will correspond to a temperature of 100,000,000°C or more. Such temperatures are considerably higher than those in the interior of the sun. The nuclei must be confined somehow at these fantastically high temperatures for a sufficiently long time to permit them to undergo fusion and release an amount of energy greater than the total required to heat and confine the fuel and to compensate for losses.

1.17. It is because of the truly formidable problems involved that achievement of economical thermonuclear power represents an exceptional scientific and technical challenge. The difficulties are so great that it is by no means completely certain that this objective will ever be attained. However, a reactor for the controlled release of power by nuclear fusion seems to be possible in principle and its realization would make available a virtually inexhaustible supply of energy. Consequently, scientists and engineers in many countries are becoming interested in the study of thermonuclear reactions. It is as an introduction to both the theoretical and experimental aspects of this study that the succeeding chapters are presented.

Chapter 2

PHYSICAL CONDITIONS FOR THERMONUCLEAR REACTIONS

MECHANISM OF THERMONUCLEAR REACTIONS

THE COULOMB BARRIER

2.1. Many reactions involving the combination of nuclei of low atomic number, e.g., isotopes of hydrogen, helium, and lithium, are accompanied by a liberation of energy. Such reactions are therefore of possible interest in connection with the production of fusion power.

2.2. Since atomic nuclei are positively charged, when two nuclei are brought together as a preliminary to combination (or fusion), there is an increasing force of electrostatic (or Coulomb) repulsion of their positive charges. At a certain distance apart, however, the short-range nuclear attractive forces just exceed the long-range forces of repulsion. At this point, fusion of the nuclei becomes possible. The variation in the potential energy of the system of two nuclei, with their distance apart, is shown in Fig. 2.1; a negative slope of the potential energy curve indicates net repulsion whereas a positive slope implies net attraction.

2.3. According to classical theory, the energy which must be supplied to the nuclei to surmount the Coulomb barrier, i.e., the energy required to overcome the electrostatic repulsion so that fusion can occur, is given by

$$\text{Energy to surmount Coulomb barrier} = \frac{Z_1 Z_2 e^2}{R_0}, \tag{2.1}$$

where Z_1 and Z_2 are the respective charges (or atomic numbers) of the interacting nuclei, e is the unit (or electronic) charge, and R_0 is the distance between the centers of the nuclei at which the attractive forces become dominant (see Fig. 2.1). For light nuclei, which are of interest for controlled thermonuclear reactions, R_0 may be taken as approximately equal to a nuclear diameter, i.e., 5×10^{-13} cm; and since e is 4.80×10^{-10} esu (statcoulomb), it follows from equation (2.1) that

6

$$\text{Energy to surmount Coulomb barrier} = 4.6 \times 10^{-7} Z_1 Z_2 \text{ erg}$$
$$= 0.28 Z_1 Z_2 \text{ Mev}, \qquad (2.2)$$

where 1 Mev (million electron volts) is equivalent to 1.60×10^{-6} erg.

2.4. It is seen from equation (2.2) that the energy which the nuclei must acquire before they can combine increases with the atomic numbers, Z_1 and Z_2, and even for reactions among the isotopes of hydrogen, for which $Z_1 = Z_2 = 1$, the minimum energy, according to classical theory, is about 0.28 Mev. Even larger energies should be required for reactions involving nuclei of higher atomic number because of the increased electrostatic repulsion.

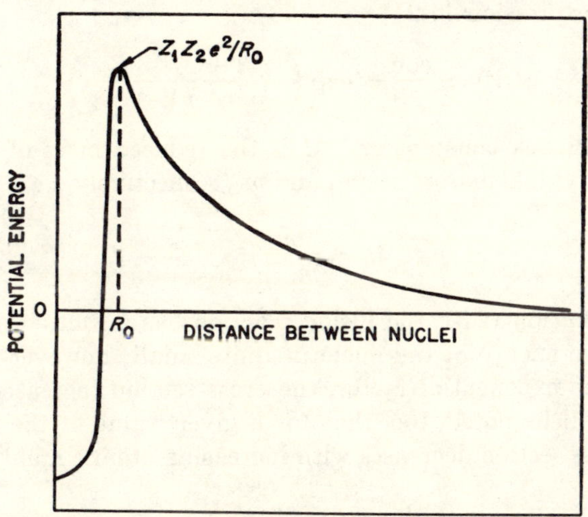

Fig. 2.1.   Variation of Coulomb potential energy with distance between nuclei.

2.5. Although energies of the order of magnitude indicated by equation (2.2) must be supplied to nuclei to cause them to combine fairly rapidly, experiments made with accelerated nuclei have shown that nuclear reactions can take place at detectable rates even when the energies are considerably below those corresponding to the top of the Coulomb barrier. In other words, there is no threshold energy, determined by the maximum electrostatic repulsion of the interacting nuclei, below which the fusion reaction will not occur. Such behavior, which cannot be explained in terms of classical mechanics, can be interpreted by means of wave mechanics.

2.6. It can be shown that there is a certain probability that two nuclei will combine even though they do not have sufficient energy to surmount the Coulomb barrier. This effect is commonly referred to as *barrier penetration*.* The penetration probability, and hence the cross section for a given nuclear

* It is sometimes called the wave mechanical *tunnel effect*.

fusion reaction, is high when the particle energy is large and approaches the value at the top of the barrier. The reaction cross section decreases with decreasing energy, but it is nevertheless finite even at quite low energies. This means that thermonuclear reactions will occur at energies much below those to be expected from classical considerations.

2.7. It has been found theoretically that, provided the energy of the interacting particles is well below the top of the Coulomb barrier, the cross section for the combination of two nuclei (atomic numbers Z_1 and Z_2) can be represented, to a good approximation, as a function of the relative particle energy W, i.e., the total kinetic energy of the two nuclei in the center-of-mass system, by the expression [1,2]*

$$\sigma(W) \approx \frac{\text{const.}}{W} \exp\left(-\frac{2^{3/2}\pi^2 M^{1/2} Z_1 Z_2 e^2}{h W^{1/2}}\right), \tag{2.3}$$

where h is the Planck constant and M is the reduced mass of the interacting nuclei whose individual masses are m_1 and m_2, respectively, i.e.,

$$M = \frac{m_1 m_2}{m_1 + m_2}. \tag{2.4}$$

According to equation (2.3), the fusion reaction has a finite cross section even when the relative energy of the nuclei is quite small; however, because of the dominance of the exponential factor, the cross section increases rapidly as W increases. It will be noted, too, that for a given value of the relative energy the reaction cross section decreases with increasing atomic number of the interacting nuclei.

THERMONUCLEAR REACTIONS

2.8. In a thermonuclear reaction system the energies of the reacting nuclei would have a Maxwellian distribution; for the present problem, this distribution can be represented by

$$dn = \text{const.} \times \frac{W^{1/2}}{T^{3/2}} \exp\left(-\frac{W}{kT}\right) dW, \tag{2.5}$$

where dn is the number of nuclei per unit volume whose energies, in the frame of the system, lie in the range from W to $W + dW$; k is the Boltzmann constant, 1.38×10^{-16} erg/°K, and T is the *kinetic temperature*.† The kinetic temperature of a system of particles is defined as the temperature appropriate to the Maxwellian distribution assumed by the particles upon equipartition of energy among the three translational degrees of freedom. The mean particle energy is then $\frac{3}{2}kT$.

---

*The numbers in square brackets apply to the references at the end of the chapter.

† The reason for distinguishing between the "temperature" and the "kinetic temperature" is discussed in §2.57.

2.9. The contribution to the overall reaction rate (per unit energy interval) made by nuclei with relative energy in the range from W to $W + dW$ in a thermonuclear system, at the kinetic temperature T, is proportional to the product of $\sigma(W)$ and of dn/dW for that particular temperature. This contribution may be represented by $R(W)$, so that, from equations (2.3) and (2.5),

$$R(W) \approx \frac{C}{W^{\frac{1}{2}}T^{\frac{3}{2}}} \exp\left(-\frac{2^{\frac{3}{2}}\pi^2 M^{\frac{1}{2}}Z_1 Z_2 e^2}{hW^{\frac{1}{2}}} - \frac{W}{kT}\right), \qquad (2.6)$$

where C is a constant.

2.10. The significance of equation (2.6) may be most readily seen by means of a graphical representation. The curve marked dn/dW in Fig. 2.2 depicts a

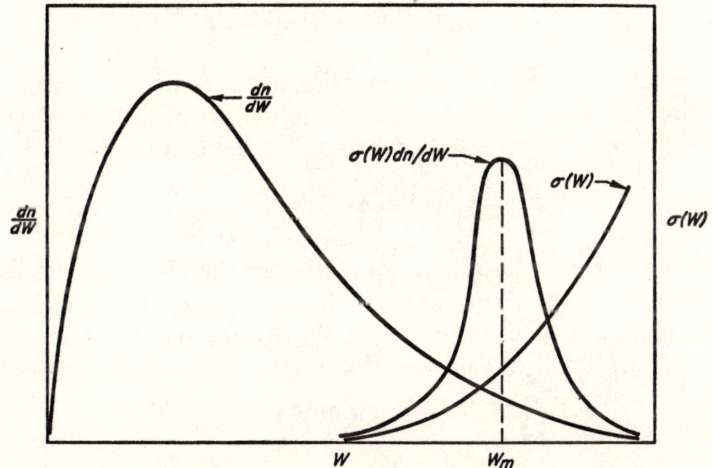

Fig. 2.2. Effect of Maxwellian energy distribution on nuclear reaction rate.

typical Maxwellian distribution of the relative particle energies for a specified kinetic temperature. The variation of the cross section for the nuclear fusion reaction with the relative energy, as determined from equation (2.3), is shown by the curve $\sigma(W)$. The dependence of the contribution to the reaction rate made by particles of relative energy W, obtained by multiplying the ordinates of the other two curves, is indicated by the curve in the center of Fig. 2.2. It is seen that this curve has a distinct maximum corresponding to the relative energy W_m, so that nuclei having this amount of relative energy make the maximum contribution to the total fusion reaction rate. The average energy of the nuclei is in the vicinity of the maximum of the dn/dW curve; hence it is evident that W_m is larger than the average energy for the given kinetic temperature [1].

2.11. The total reaction rate is determined by the area under the $\sigma(W)$ dn/dW curve, over the entire range of energies. Consequently, it is apparent, most of the observed thermonuclear reaction will be due to a relatively small fraction of the nuclear collisions in which the relative energies are greatly in excess of the average. This explains why it is advantageous in performing a nuclear fusion reaction, e.g., with uniformly accelerated particles, to permit the nuclei to become "thermalized," that is, to attain a Maxwellian distribution of energies, as a result of collisions.

2.12. Provided the kinetic temperature is not too high, the variation of $R(W)$ with W is determined almost entirely by the exponential factor in equation (2.6); hence, a good approximation to the value of W_m in Fig. 2.2 can be obtained by calculating the energy for which this factor is a maximum. Upon differentiating with respect to W and setting the result equal to zero, it is found that

$$W_m \approx \left[\frac{(2M)^{\frac{1}{2}}\pi^2 Z_1 Z_2 e^2 kT}{h}\right]^{\frac{2}{3}}.$$

This expression gives the relative energy in nuclear collisions making the maximum contribution to the reaction rate at the not too high kinetic temperature T. Upon dividing through both sides by kT, the result is

$$\frac{W_m}{kT} \approx \left[\frac{(2M)^{\frac{1}{2}}\pi^2 Z_1 Z_2 e^2}{h}\right]^{\frac{2}{3}} \frac{1}{(kT)^{\frac{1}{3}}}. \tag{2.7}$$

2.13. It has become the common practice in thermonuclear studies to express kinetic temperatures in terms of the corresponding energy kT in kilo-electron volts, i.e., in kev units. Since the Boltzmann constant k is 1.38×10^{-16} erg/°K and 1 kev is equivalent to 1.60×10^{-9} erg, it follows that

$$k = 8.6 \times 10^{-8} \text{ kev/°K}$$

or

$$1 \text{ kev} \equiv 1.16 \times 10^7 \text{ °K}.$$

Thus a temperature of T kev is equivalent to $1.16 \times 10^7 T$°K.

2.14. For a reaction between two deuterium nuclei, for example, Z_1 and Z_2 are both unity and M is 1.67×10^{-24} gram; taking e as 4.80×10^{-10} esu (statcoulomb) and expressing the temperatures or, more exactly kT, in kilo-electron volts, equation (2.7) takes the form

$$\frac{W_m}{T} = \frac{6.3}{T^{\frac{1}{3}}}.$$

Hence, at a kinetic temperature of 1 kev the greatest contribution to the rate of a thermonuclear reaction comes from nuclear collisions between deuterons with a relative energy of about 6.3 kev. These deuterons constitute a small fraction only of the total present.

RATES OF THERMONUCLEAR REACTIONS

2.15. Consider a binary reaction in a system containing n_1 nuclei/cm³ of one reacting species and n_2 nuclei/cm³ of the other. To determine the rate at which the two nuclear species interact, it may be supposed that the nuclei of the first kind form a stationary lattice within which the nuclei of the second kind move at random with a constant velocity v cm/sec, equal to the relative velocity of the nuclei. Let σ cm^2 be the cross section for the given reaction corresponding to this relative velocity. The total cross section for all the stationary nuclei in 1 cm^3 is then $n_1\sigma$ nuclei/cm. This gives the number of nuclei of the first kind with which each nucleus of the second kind will react while traveling a distance of 1 cm. The total distance traversed in 1 sec by all the nuclei of the latter type present in 1 cm^3 is equal to $n_2 v$ nuclei/(cm^2) (sec). Hence, the nuclear reaction rate R_{12} is equal to the product of $n_1\sigma$ and $n_2 v$; thus

$$R_{12} = n_1 n_2 \sigma v \text{ interactions/(cm}^3)(\text{sec}). \tag{2.8}$$

If the reaction occurs between two nuclei of the same kind, e.g., two deuterons, so that n_1 and n_2 are equal, the expression for the nuclear reaction rate, represented by R_{11}, becomes

$$R_{11} = \tfrac{1}{2} n^2 \sigma v \text{ interactions/(cm}^3)(\text{sec}), \tag{2.9}$$

where n is the number of reactant nuclei/cm^3. In order that each interaction between identical nuclei should not be counted twice, the factor of $\tfrac{1}{2}$ is introduced into equation (2.9).

2.16. The foregoing equations (2.8) and (2.9) are applicable when the relative velocity of the interacting nuclei is constant, as is true, approximately at least, for high-energy particles from an accelerator. For thermonuclear reactions, however, there will be a distribution of velocities (and energies) over a wide range. Since the reaction cross section is dependent on the energy (or velocity), it follows that, for the general case, the product σv in equations (2.8) and (2.9) should be replaced by a value $\overline{\sigma v}$ averaged over the whole range of relative velocities. Thus, equation (2.8) should be written as

$$R_{12} = n_1 n_2 \overline{\sigma v} \text{ interactions/(cm}^3)(\text{sec}), \tag{2.10}$$

and equation (2.9) becomes

$$R_{11} = \tfrac{1}{2} n^2 \overline{\sigma v} \text{ interactions/(cm}^3)(\text{sec}). \tag{2.11}$$

2.17. If the velocity distribution is Maxwellian, then the equation for the distribution in terms of the relative velocity is obtained upon substituting the reduced mass M of the interacting nuclei for the individual masses in the familiar expression [3]; thus,

$$dn = n \left(\frac{M}{2\pi kT} \right)^{3/2} \exp \left(-\frac{Mv^2}{2kT} \right) v^2 \, dv,$$

where dn is the number of particles whose velocities relative to that of a given particle lie in the range from v to $v + dv$. Hence, it follows that

$$\overline{\sigma v} = \frac{\int_0^\infty \sigma v \, dn}{\int_0^\infty dn}$$

$$= \frac{\int_0^\infty \sigma v \left[\exp\left(-\frac{Mv^2}{2kT}\right) v^2 \, dv \right]}{\int_0^\infty \exp\left(-\frac{Mv^2}{2kT}\right) v^2 \, dv}. \tag{2.12}$$

The integral in the denominator of equation (2.12) is equal to $(2kT/M)^{3/2}$ $\pi^{1/2}/4$, and so this equation becomes

$$\overline{\sigma v} = \frac{4}{\pi^{1/2}} \left(\frac{M}{2kT}\right)^{3/2} \int_0^\infty \sigma \exp\left(-\frac{Mv^2}{2kT}\right) v^3 \, dv. \tag{2.13}$$

2.18. The integral in equation (2.13) can be evaluated by changing the variable. Since nuclear cross sections are always determined and expressed as a function of the energy of the bombarding particle, the bombarded particle being essentially at rest in the target, the actual velocity of the bombarding nucleus is also its relative velocity. Hence, if W is the actual energy, in the laboratory system, of the bombarding nucleus of mass m, then

$$W = \tfrac{1}{2}mv^2,$$

so that

$$v = \left(\frac{2W}{m}\right)^{1/2}$$

and

$$v^3 \, dv = \frac{2W}{m^2} \, dW.$$

Substitution of this result into equation (2.13) yields

$$\overline{\sigma v} = \frac{8}{\pi^{1/2}} \left(\frac{M}{2kT}\right)^{3/2} \frac{1}{m^2} \int_0^\infty \sigma \exp\left(-\frac{MW}{mkT}\right) W \, dW, \tag{2.14}$$

where σ in the integrand is the cross section for a bombarding nucleus of mass m and energy W.

2.19. If the temperature T in equation (2.14) is expressed in kilo-electron volts, and the values of W are in the same units, it is convenient to rewrite the equation in the form

$$\overline{\sigma v} = \left(\frac{8}{\pi T}\right)^{1/2} \frac{M^{3/2}}{m^2} \int_0^\infty \sigma \exp\left(-\frac{MW}{mT}\right) \frac{W}{T} \, dW, \tag{2.15}$$

where the quantity W/T is dimensionless. If σ, determined experimentally, can be expressed as a relatively simple function of W, as is sometimes the case, the integration in equation (2.15) may be performed analytically. Alternatively, numerical methods, for example, Simpson's rule, may be employed.

In any event, the values of $\overline{\sigma v}$ for various kinetic temperatures can be derived from equation (2.15), based on a Maxwellian distribution of energies (or velocities), and the results can be inserted in equation (2.10) or (2.11) to give the rate of a thermonuclear reaction at a specified temperature [4, 5].

NUCLEAR FUSION REACTIONS

2.20. Because of the increased height of the Coulomb energy barrier with increasing atomic number, it is generally true that, at a given temperature, reactions involving the nuclei of hydrogen isotopes take place more readily than do analogous reactions with heavier nuclei. In view of the great abundance of the lightest isotope of hydrogen, with mass number 1, it is natural to see if nuclear fusion reactions involving this isotope could be used for the release of energy. It is unfortunate, however, that the three possible reactions between H nuclei alone or with deuterium (D) or tritium (T) nuclei, i.e.,

$$_1H^1 + {}_1H^1 \rightarrow {}_1D^2 + {}_1e^0$$

$$_1H^1 + {}_1D^2 \rightarrow {}_2He^3 + \gamma$$

$$_1H^1 + {}_1T^3 \rightarrow {}_2He^4 + \gamma,$$

are known to have cross sections that are too small to permit a net gain of energy at temperatures which may be regarded as attainable.

2.21.   Consequently, recourse must be had to the next most abundant isotope, i.e., deuterium, and here two reactions, which occur at approximately the same rate over a considerable range of energies, are of interest; these are the D-D reactions

$$_1D^2 + {}_1D^2 \rightarrow {}_2He^3 + {}_0n^1 + 3.27 \text{ Mev}$$

and

$$_1D^2 + {}_1D^2 \rightarrow {}_1T^3 + {}_1H^1 + 4.03 \text{ Mev},$$

called the "neutron branch" and the "proton branch," respectively. The tritium produced in the proton branch or obtained in another way, as explained below, can then react, at a considerably faster rate, with deuterium nuclei in the D-T reaction

$$_1D^2 + {}_1T^3 \rightarrow {}_2He^4 + {}_0n^1 + 17.6 \text{ Mev}.$$

The He^3 formed in the first D-D reaction can also react with deuterium; thus,

$$_1D^2 + {}_2He^3 \rightarrow {}_2He^4 + {}_1H^1 + 18.3 \text{ Mev}.$$

This reaction is of interest because, as in the D-T reaction, there is a large energy release; the D-He^3 reaction is, however, slower than the others at low thermonuclear temperatures, but its rate approaches that of the D-D reactions at 100 kev (see Fig. 2.3).

2.22.   In the methods currently being considered for the production of useful thermonuclear power, the fast neutrons produced in the neutron branch of the D-D reaction and also in the D-T reaction would probably escape from the immediate reaction environment. It is to be expected that these neutrons will be slowed down in a suitable moderator, e.g., water, lithium, or beryllium, with the liberation of their kinetic energy as heat which can be utilized. The slow neutrons can then be captured in lithium-6, which constitutes 7.5 atomic per cent of natural lithium, by the reaction

$$_3\text{Li}^6 + {}_0n^1 \rightarrow {}_2\text{He}^4 + {}_1\text{T}^3 + 4.6 \text{ Mev},$$

leading to the production of tritium. The energy released can be used as heat, and the tritium can, in principle, be transferred to the thermonuclear system to react with deuterium.

2.23. At sufficiently high temperatures, the four nuclear reactions given in §2.21 would occur, and the two neutrons produced would subsequently be captured by lithium-6. The total energy release due to the consumption of six deuterons would thus be 52.4 Mev, i.e., 8.7 Mev per deuteron. However, for kinetic reaction temperatures below about 50 kev, the contribution of the D-He^3 reaction is not significant. If this is neglected, it is seen that 34.1 Mev would be released for every five deuterons consumed, i.e., roughly 7 Mev per deuteron. It was on this latter result that the energy estimates given in Chapter 1 were based. For some purposes it is of interest to know how much energy is liberated in a thermonuclear D-D system, apart from that accompanying capture of the neutrons produced and disregarding the D-He^3 reaction. This is seen to be the sum of 3.27 and 4.03 Mev, for the two D-D reactions, and 17.6 Mev for the accompanying D-T reaction, making a total of 24.9 Mev. This means an average of nearly 12.5 Mev for every D-D interaction or roughly 5 Mev for each of the five deuterons involved.

CHARGED PARTICLE ENERGIES

2.24. For reasons which will be apparent shortly, it is necessary to distinguish between the energy carried by the charged particles and by neutrons, respectively, in the reactions occurring in a thermonuclear system. In any nuclear interaction, momentum must be conserved and so, assuming that the reacting nuclei are essentially at rest,

$$mv = m'v', \tag{2.16}$$

where m and m' are the masses of the particles produced in the reaction and v and v' are their respective velocities. Upon taking the square of both sides of equation (2.16), it is seen that

$$m(mv^2) = m'(m'v'^2)$$

or

$$mq = m'q',$$

where q and q' are the energies carried by the particles of masses m and m', respectively. If the total energy release of the nuclear reaction is Q^*, then

$$Q = q + q'$$

and, therefore,

$$q = \frac{m'Q}{m + m'}$$

and

$$q' = \frac{mQ}{m + m'}.$$

In other words, the total energy is distributed between the two product particles in inverse proportion to their masses.

2.25. Utilizing this result, the reactions in §2.21 may be rewritten in the following form, the amount of energy carried off by each particle being indicated:

$$D + D \rightarrow \underset{(0.82 \text{ Mev})}{He^3} + \underset{(2.45 \text{ Mev})}{n}$$

$$D + D \rightarrow \underset{(1.01 \text{ Mev})}{T} + \underset{(3.02 \text{ Mev})}{H}$$

$$D + T \rightarrow \underset{(3.5 \text{ Mev})}{He^4} + \underset{(14.1 \text{ Mev})}{n}$$

$$D + He^3 \rightarrow \underset{(3.6 \text{ Mev})}{He^4} + \underset{(14.7 \text{ Mev})}{H}.$$

2.26. There are two aspects of the distribution of the energy between charged and uncharged particles which are of importance. First, in all the forms of controlled thermonuclear reactors at present envisaged, the neutrons will escape from the reacting system and deposit their energy elsewhere. Only the energy of the charged particles will be retained within the reaction region. Hence, only this energy, at most, will be available internally to compensate for energy losses and to sustain the thermonuclear reactions. Second, it is only the energy of the charged particles that might be used to generate electricity directly, in accordance with the possibility mentioned in §1.13.

2.27. In a thermonuclear reacting system consisting of deuterium alone, the two D-D reactions would take place at essentially equal rates, and the tritium formed in the proton branch would then react, relatively rapidly, with some of the deuterium. Ignoring the D-He³ reaction for the present because of its slow rate, it is seen that, in a deuterium system, 8.3 Mev out of a total of 24.9 Mev, i.e., about 33 per cent, would be carried by the charged particles. Thus, for every two D-D reactions, followed by a D-T reaction, 8.3 Mev is deposited within the reacting system and is available internally. This is also the maximum amount which could be converted directly into electrical

---

*The quantity Q does not include the energy, if any, emitted as gamma radiation.

energy. The other 16.6 Mev would be recoverable only as heat. In a system consisting of 50 atomic per cent each of deuterium and tritium, 3.5 Mev out of 17.6 Mev total, i.e., about 20 per cent, would be carried by the charged particles per reaction.

2.28. It is because of the possible advantages accruing from the formation of charged particles in fusion reactions that the D-He^3 process is of special interest. In this case, both the porducts are charged particles, i.e., atomic nuclei. Thus, all the energy produced, which is larger than for the three reactions considered above, would be available internally and could, if desired, be converted directly into electricity. If it develops, as is not at all impossible, that thermonuclear reactors will have to be operated at temperatures as high as 100 kev, then the contribution of the D-He^3 reaction could be quite significant. In this event, it might prove advantageous to prepare tritium by the capture of neutrons in lithium-6, and then let the tritium undergo radioactive decay (half life 12.26 years) to produce helium-3; thus,

$$_1\text{H}^3 \rightarrow {_2\text{He}^3} + {_{-1}e^0}.$$

The helium-3 would then be used as the feed material, with deuterium, in a thermonuclear reacting system.

THERMONUCLEAR REACTION CROSS SECTIONS

2.29. By means of the quantum-mechanical theory of Coulomb barrier penetration, it is often possible to make order-of-magnitude estimates, at least, of the rates (or cross sections) of thermonuclear reactions [6]. It is much more satisfactory, however, to make use of cross sections obtained experimentally. By bombarding targets containing deuterium, tritium, or helium-3 with deuterons of known energies, obtained by means of a Cockcroft-Walton or other charged-particle accelerator, cross sections for the D-D, D-T, and D-He^3 reactions have been determined. The results, expressed in barns (1 barn = 10^{-24} cm^2), are shown in Fig. 2.3; the curve marked D-D gives the sum of the cross sections for the two D-D reactions. The value for each individual reaction may be taken as half the total, at least up to deuteron energies of 120 kev [7, 8].

2.30. It will be observed that the D-T curve exhibits a maximum at an energy of 110 kev; this is an example of the resonance phenomenon which often occurs in nuclear reactions. Below about 100 kev, the D-T cross sections are roughly 100 times as large as those for the sum of the two D-D reactions. In the same energy range, the D-He^3 cross sections are appreciably less than the D-D and D-T values, but the curve is rising so rapidly with increasing deuteron energy that it crosses the D-D curve at about 120 kev. Incidentally, the appreciable cross sections for energies well below the top of the Coulomb barrier for each of the reactions studied provide an experimental demonstration of the reality of the barrier penetration effect.

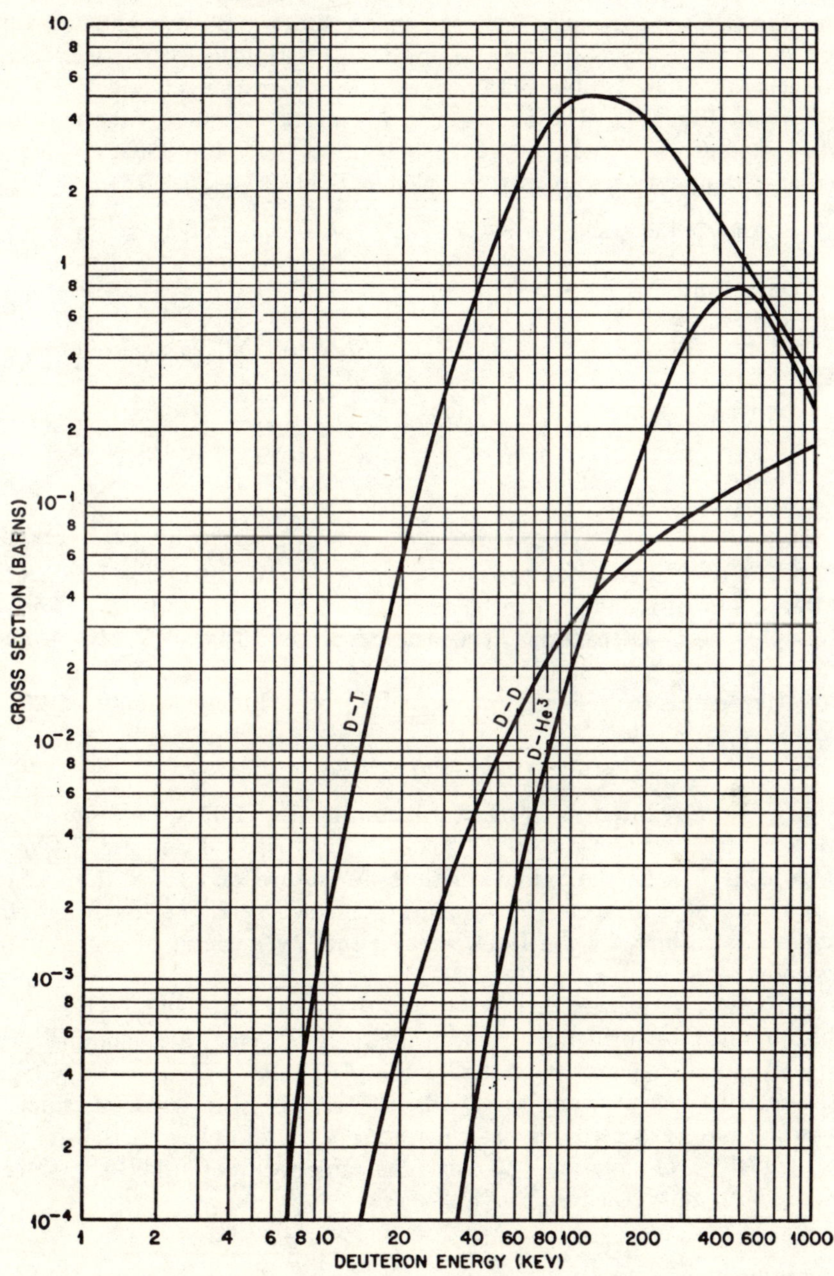

FIG 2.3.   Cross sections for D-T, D-D (total), and D-He³ reactions.

2.31. For use in connection with thermonuclear reaction calculations, the data in Fig. 2.3, for particular deuteron energies, are applied to the determination of the average $\overline{\sigma v}$, assuming a Maxwellian distribution of particle energies (or velocities). The results obtained by the integration of equation (2.15) are shown in Fig. 2.4. The curves give $\overline{\sigma v}$ in cm³/sec as a function of the kinetic temperature of the reaction system in kilo-electron volts. Values for a number of temperatures are also recorded in Table 2.1 [4, 5, 9-11].

TABLE 2.1. VALUES OF $\overline{\sigma v}$ AT SPECIFIED KINETIC TEMPERATURES

Temperature (kev)	D-D (cm³/sec)	D-T (cm³/sec)	D-He³ (cm³/sec)
1.0	2×10^{-22}	7×10^{-21}	6×10^{-26}
2.0	5×10^{-21}	3×10^{-19}	2×10^{-23}
5.0	1.5×10^{-19}	1.4×10^{-17}	1×10^{-20}
10.0	8.6×10^{-19}	1.1×10^{-16}	2.4×10^{-19}
20.0	3.6×10^{-18}	4.3×10^{-16}	3.2×10^{-18}
60.0	1.6×10^{-17}	8.7×10^{-16}	7×10^{-17}
100.0	3.0×10^{-17}	8.1×10^{-16}	1.7×10^{-16}

2.32. The overall effect on the nuclear reaction rates of taking into account the Maxwellian distribution can be found from Figs. 2.3 and 2.4. Consider, for example, the D-D reactions at a temperature of 20 kev. If all the nuclei had the same energy, σ would be about 6×10^{-4} barn, i.e., 6×10^{-28} cm². The relative deuteron velocity corresponding to a kinetic energy of 20 kev is given by

$$\tfrac{1}{2}M_{\mathrm{D}}v^2 = 20\,\mathrm{kev} = 20 \times 1.60 \times 10^{-9}\,\mathrm{erg},$$

where M_{D} is the reduced mass of two deuterons, i.e., 1.67×10^{-24} g. Hence, v is about 2×10^8 cm/sec and σv is 1.2×10^{-19} cm³/sec. From Fig. 2.4, on the other hand, it is seen that at a kinetic temperature of 20 kev the value of $\overline{\sigma v}$ is 3.6×10^{-18} cm³/sec. Consequently, the net result of the Maxwellian distribution is to increase the D-D reaction rate by a factor of roughly 30 at the specified temperature.

2.33. Analytical expressions for σ and $\overline{\sigma v}$ for the D-D and D-T reactions can be obtained by utilizing equation (2.3) in a somewhat modified form. The relative kinetic energy W of the interacting nuclei is equal to $\tfrac{1}{2}Mv^2$, where v is the relative velocity, and the deuteron energy W_{D}, in terms of which the cross sections are expressed, is $\tfrac{1}{2}m_{\mathrm{D}}v^2$, where m_{D} is the mass of the deuteron. Hence, $(M/W)^{1/2}$ in equation (2.3) may be replaced by $(m_{\mathrm{D}}/W_{\mathrm{D}})^{1/2}$; since Z_1 and Z_2 are both unity, the result is

$$\sigma(W_{\mathrm{D}}) = \frac{\text{const.}}{W_{\mathrm{D}}} \exp\left(-\frac{2^{3/2}\pi^2 m_{\mathrm{D}}^{1/2}e^2}{hW_{\mathrm{D}}^{1/2}}\right)$$

$$= \frac{\text{const.}}{W_{\mathrm{D}}} \exp\left(-\frac{44.24}{W_{\mathrm{D}}^{1/2}}\right),$$

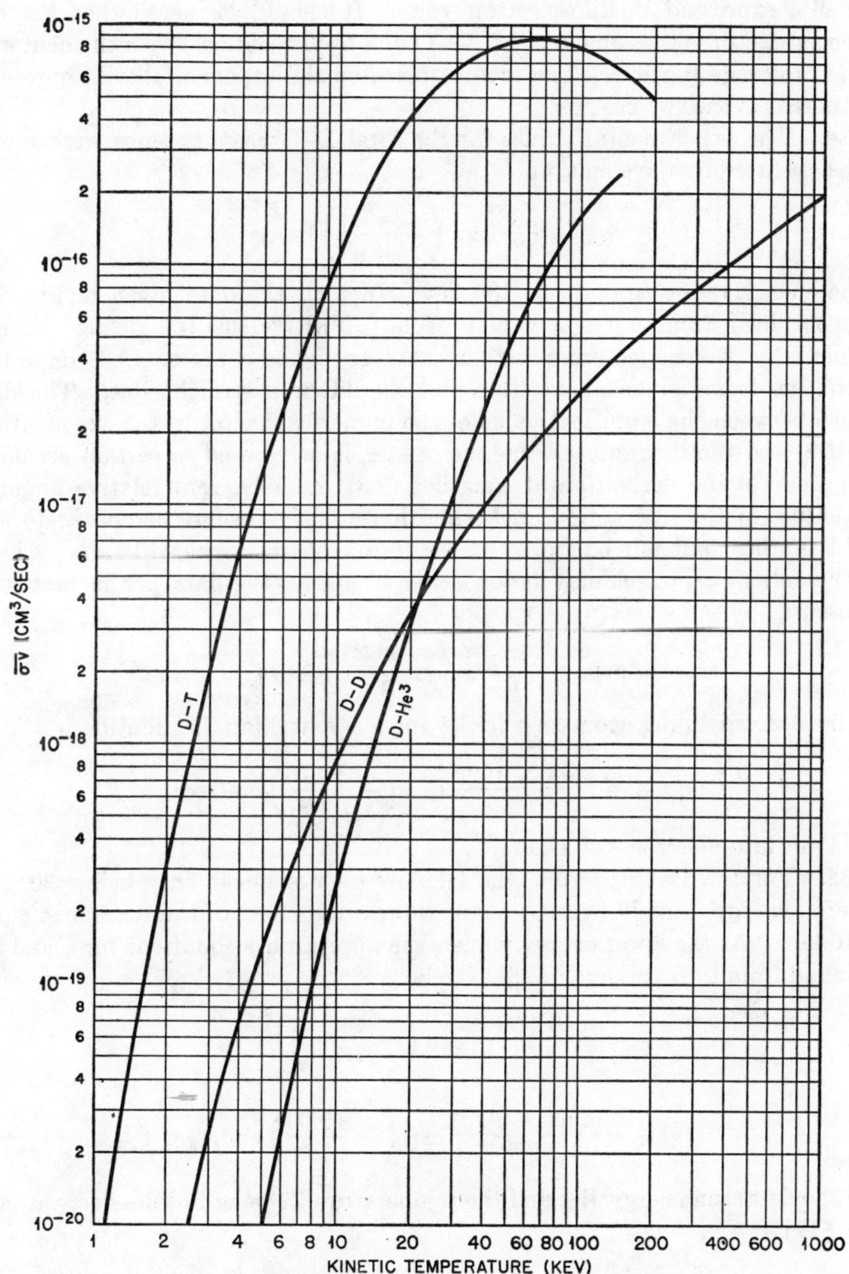

FIG. 2.4. Values of $\overline{\sigma v}$ based on Maxwellian distribution for D-T, D-D (total), and D-He³ reactions.

with W_D expressed in kilo-electron volts. It should be noted that the exponential factor is the same for both D-D and D-T reactions, with the deuteron as the projectile particle. The factor preceding the exponential will, however, be different in the two cases.

2.34. The experimental values for the total D-D cross sections were found to be best fitted by the equation

$$\sigma_{DD} = \frac{288}{W_D} \exp\left(-\frac{45.8}{W_D^{1/2}}\right) \text{ barns}$$

in the energy range from 10 to 100 kev. In fact, the data given in Fig. 2.3 were obtained from a Gamow plot of $\ln(\sigma W_D)$ versus W_D, using the experimentally determined cross sections. According to equation (2.3) or to the alternative forms given above, this plot should be a straight line. The difference between the empirical value of the numerical factor in the exponential, i.e., 45.8, and the theoretical value, i.e., 44.24, is attributed to certain assumptions made in the derivation of equation (2.3), namely, zero relative angular momentum of the interacting nuclei and low relative energy compared to the height of the Coulomb barrier. For deuteron energies below 13 kev, where the theoretical exponent may be expected to apply, the data are in fact well represented by

$$\sigma_{DD} = \frac{182}{W_D} \exp\left(-\frac{44.24}{W_D^{1/2}}\right) \text{ barns,}$$

and the corresponding expression for $\overline{\sigma v}$ for a Maxwellian distribution is

$$(\overline{\sigma v})_{DD} = \frac{2.33 \times 10^{-14}}{T^{2/3}} \exp\left(-\frac{18.76}{T^{1/3}}\right) \text{ cm}^3/\text{sec,}$$

with T in kilo-electron volts [12].

2.35. For the D-T reaction, the Gamow plot is linear only below 20 kev deuteron energy, largely because of the interfering effect of the resonance close to 110 kev. At the lower energy values, the appropriate equations for σ and $\overline{\sigma v}$ are found from the experimental data to be

$$\sigma_{DT} = \frac{2.19 \times 10^4}{W_D} \exp\left(-\frac{44.24}{W_D^{1/2}}\right) \text{ barns}$$

and

$$(\overline{\sigma v})_{DT} = \frac{3.68 \times 10^{-12}}{T^{2/3}} \exp\left(-\frac{19.94}{T^{1/3}}\right) \text{ cm}^3/\text{sec,}$$

both the deuteron energy W_D and the temperature T being in kilo-electron volt units [12].

REACTION MEAN FREE PATH

2.36. In a system containing n nuclei/cm^3 of a particular reacting species, the reaction mean free path λ, i.e., the average distance traveled by a nucleus before it undergoes reaction, is equal to $1/n\sigma$, where σ is the cross section for the given reaction [13].   If a Maxwellian distribution is assumed, then σ must be replaced by $\bar{\sigma}$, averaged over all energies from zero to infinity, at a given kinetic temperature.   The value of the mean free path of a deuteron in centimeters is plotted in Fig. 2.5 as a function of the deuteron particle density n, in nuclei/cm³, for the D-D and D-T reactions at two kinetic temperatures, 10 and 100 kev, in each case.   As will be seen in §2.68, tem-

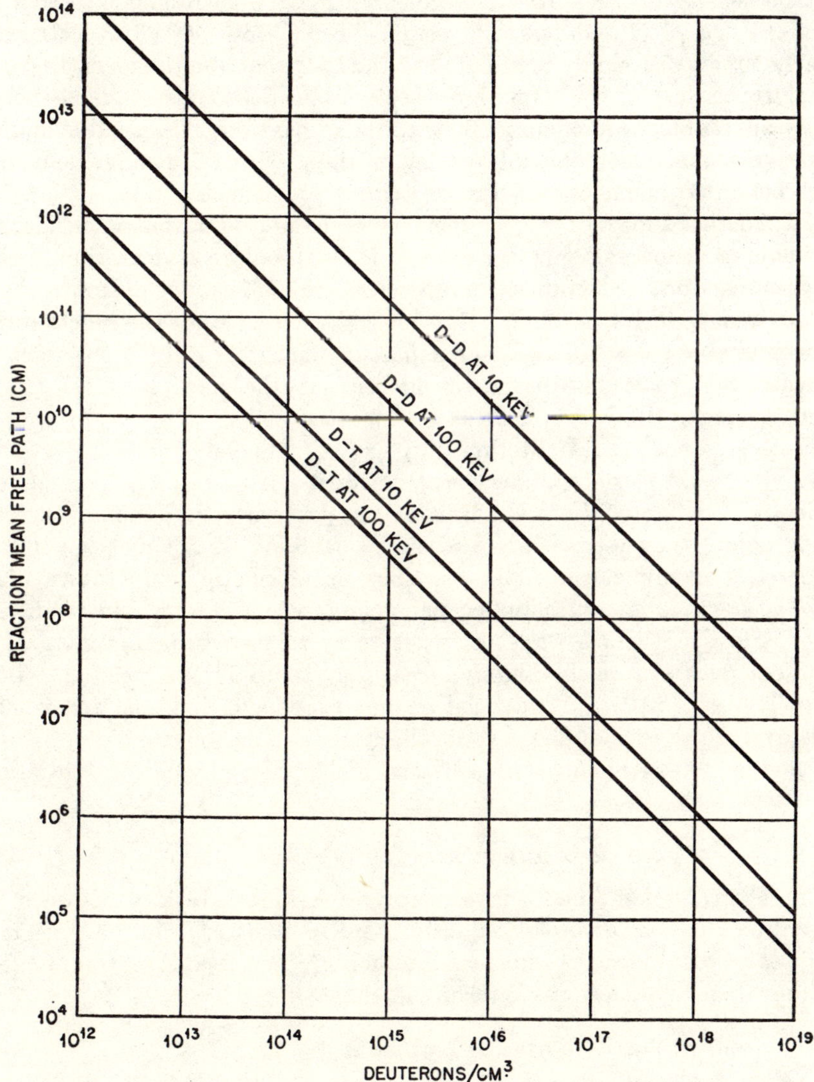

FIG. 2.5.   Mean free paths for D-T and D-D (total) thermonuclear reactions.

peratures of these orders of magnitude would be required in a controlled thermonuclear reactor [14].

2.37. For a particle density of about 10^{15} deuterons/cm³, which is one that might be of practical interest for reasons which will be apparent later, the mean free path at 100 kev for the D-D reaction is about 2×10^{10} cm. This means that, at the specified temperature and particle density, a deuteron would travel on the average a distance of some 120,000 miles before reacting. Even at a particle density of 2.7×10^{19} deuterons/cm³, which corresponds to that of a gas at STP, i.e., 0°C and 1 atm pressure, the reaction mean free path would be nearly 10^6 cm or roughly 6 miles. For the D-T reaction the mean free paths are shorter, because of the larger cross sections for deuterons of given energies, but they are still large in comparison with the dimensions of normal equipment. These results have an important bearing on the problem of confinement of the particles in a thermonuclear reacting system.

2.38. Another aspect of the confinement problem arises from a consideration of the average time elapsing (or *mean lifetime*) before a deuterium nucleus undergoes reaction. At a kinetic temperature of 100 kev, for example, the average deuteron velocity is roughly 5×10^8 cm/sec, and since the reaction mean free path is about 2×10^{10} cm, for a particle density of 10^{15} deuterons/cm³, the mean D-D reaction lifetime is 40 sec. If the particle density is increased, the mean lifetime is decreased proportionately. Increase of temperature also decreases the mean lifetime of the reacting nuclei.

2.39. It should be noted, however, that for a self-sustaining thermonuclear reactor, the average confinement time of the deuterons, i.e., the average time elapsing before they escape, need be only a fraction of the reaction lifetime. Consider, for example, the D-D reactions at a kinetic temperature of 100 kev. The average energy supplied per deuteron is 150 kev, and if an equal amount is lost in various ways, it is necessary to recover a minimum of 300 kev of energy. For each deuteron reacting, there is produced a total of roughly 7 Mev (§2.23); hence, the energy expended could be recovered in a little over 4 per cent of the reaction lifetime, i.e., in about 1.6 sec. For the D-T reaction, with its shorter lifetime and higher energy yield, the situation is even better.

THERMONUCLEAR REACTION POWER DENSITY

2.40. The rate of thermonuclear energy production is readily obtained from the reaction rate, as given by equation (2.10) or (2.11). If Q ergs is the average energy produced per nuclear interaction, then it follows from equation (2.11), for the D-D reactions, for example, that

$$\text{Rate of energy release} = \tfrac{1}{2} n_D{}^2 \overline{\sigma v}\, Q \text{ ergs } /(\text{cm}^3)(\text{sec}). \tag{2.17}$$

Using the symbol P_{DD} to represent the power density in watts/cm³, i.e., 10^7 ergs/(cm³)(sec), then

$$P_{\text{DD}} = \tfrac{1}{2}n_{\text{D}}^2\overline{\sigma v}Q \times 10^{-7} \text{ watts/cm}^3, \qquad (2.18)$$

with n_{D} in deuterons/cm³, $\overline{\sigma v}$ in cm³/sec, and Q in ergs. It was shown in §2.27 that, for every two D-D interactions, an average of 8.3 Mev of energy is deposited within the reacting system. The energy Q per interaction is thus $\tfrac{1}{2} \times 8.3 \times 1.60 \times 10^{-6} = 6.6 \times 10^{-6}$ erg, and upon substitution into equation (2.17), the result is

$$P_{\text{DD}} = 3.3 \times 10^{-13}n_{\text{D}}^2\overline{\sigma v} \text{ watts/cm}^3. \qquad (2.19)$$

Since $\overline{\sigma v}$ can be obtained from Fig. 2.4 or from Table 2.1 for a given kinetic temperature, the power density can be evaluated for any specified particle density. At 10 kev, for example, $\overline{\sigma v}$ for the D-D reactions is 8.6×10^{-19} cm³/sec, and

$$P_{\text{DD}} \text{ (10 kev)} = 2.8 \times 10^{-31}n_{\text{D}}^2 \text{ watts/cm}^3,$$

and at 100 kev, where $\overline{\sigma v}$ is 3.0×10^{-17} cm³/sec,

$$P_{\text{DD}} \text{ (100 kev)} = 10^{-29}n_{\text{D}}^2 \text{ watts/cm}^3.$$

2.41. For the D-T reaction, the energy remaining in the system per interaction is 3.5 Mev, i.e., $3.5 \times 1.6 \times 10^{-6} = 5.6 \times 10^{-6}$ erg. The reaction rate is given by equation (2.10), and so the thermonuclear reactor power density is

$$P_{\text{DT}} = n_{\text{D}}n_{\text{T}}\overline{\sigma v}Q \times 10^{-7} \text{ watts/cm}^3, \qquad (2.20)$$

where Q is 5.6×10^{-6} erg; hence,

$$P_{\text{DT}} = 5.6 \times 10^{-13}\overline{\sigma v}n_{\text{D}}n_{\text{T}} \text{ watts/cm}^3.$$

Utilizing the values of $\overline{\sigma v}$ from Table 2.1, it follows that

$$P_{\text{DT}} \text{ (10 kev)} = 6.2 \times 10^{-29}n_{\text{D}}n_{\text{T}} \text{ watts/cm}^3$$

and

$$P_{\text{DT}} \text{ (100 kev)} = 4.5 \times 10^{-28}n_{\text{D}}n_{\text{T}} \text{ watts/cm}^3.$$

2.42. Power density values, calculated from these equations, are shown in Fig. 2.6 as a function of the deuteron particle density [14]. For the D-T reaction, the particle densities n_{D} and n_{T} are taken as equal, a 50 atomic per cent mixture of deuterium and tritium being assumed. It is of interest to note that the power density of a nuclear fission reactor is of the order of 10^2 watts/cm³. Hence, power densities similar to those in fission reactors would mean operating a thermonuclear reactor at particle densities in the region of 10^{15} or 10^{16} nuclei/cm³. At ordinary temperatures this corresponds to the values at pressures as low as 10^{-4} or 10^{-3} atm, i.e., in a fairly good vacuum. However, at temperatures of thermonuclear interest these very small particle densities would generate pressures of many atmospheres (cf. Fig. 2.7).

2.43. There is not, of course, an exact parallel between the conditions, e.g., heat transfer and operating pressures, which limit the power density of a fission reactor and those which might apply to a thermonuclear fusion reactor.

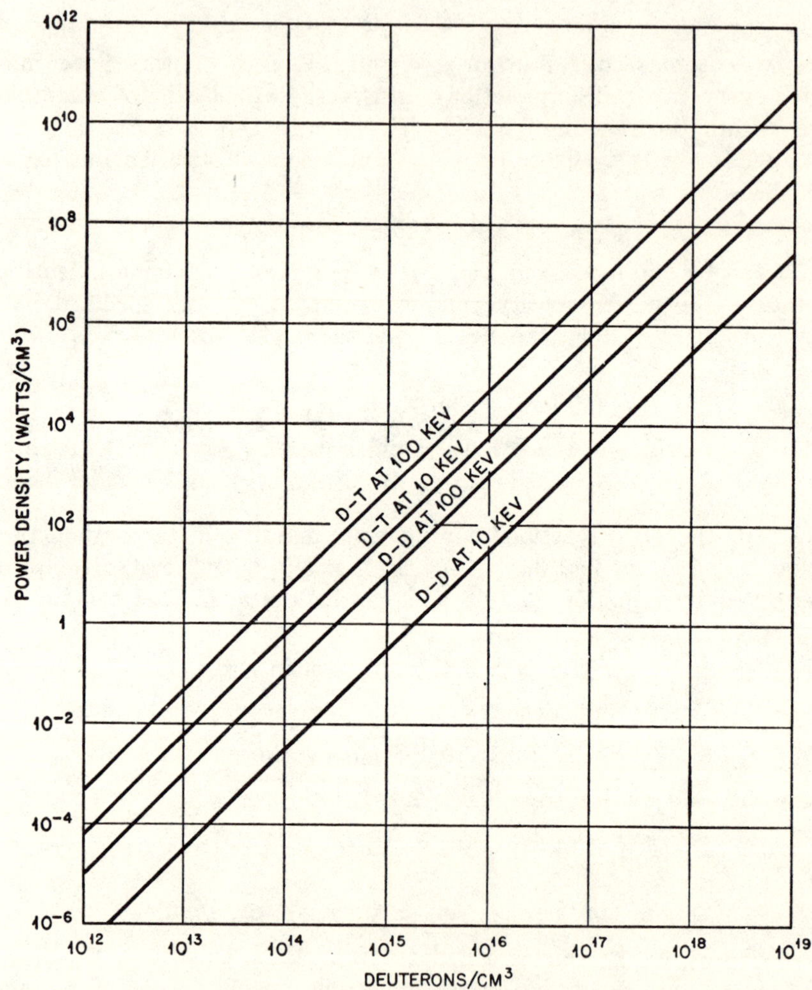

Fɪɢ. 2.6. Power densities for D-T and D-D (total) thermonuclear reactions.

Nevertheless, there must be similar limitations upon power transfer in a continuously operating thermonuclear reactor as in other electrical power systems. One conclusion appears to be inevitable: operation at particle densities similar to those under normal atmospheric conditions, viz, about 10^{19} nuclei/cm³, seems impossible.   A large steam-powered electrical generating plant has a power of about 500 megawatts, i.e., 5×10^8 watts. With deuteron particle densities equivalent to those at standard temperature and pressure, i.e., 2.7×10^{19} deuterons/cm³, it appears from Fig. 2.6 that at 100 kev in a D-D reactor a power of 5×10^8 watts would be obtained in a reacting volume of only 0.03

cm^3. At the same time the gas kinetic pressure exerted by the thermonuclear fuel would be about 10^7 atm or 1.5×10^8 psi. Since the mean reaction lifetime is only a few milliseconds under the conditions specified, it is obvious that the situation would be completely impractical.

2.44. At the other extreme are the consequences of operating at too low a particle density. Several proposed schemes for producing thermonuclear power fail economically, although not in principle, merely because the particle densities that can be achieved are not sufficiently high. When the density is down to 10^{12} nuclei/cm^3, for example, the curve in Fig. 2.6 shows that the D-D power density at 100 kev has fallen to 10^{-5} watt/cm^3. This value is far too low to be economically interesting. For a D-T reactor the power density would be more than a hundred times greater, but even this would be too small to be of any value.

2.45. The reason for the rapid change of power density with particle density arises from the fact, apparent from equation (2.17), that the power density is proportional to the *square* of the deuteron density. For the D-T reaction, the situation is equivalent, since P_{DT} is proportional to the product of n_D and n_T, and this is the same as $n_D{}^2$ in the case of a 50 atomic per cent mixture.

2.46. From the foregoing considerations it may be concluded that, as far as can be seen at present, the particle density in a practical thermonuclear reactor must be in the vicinity of 10^{15} nuclei/cm^3. There is another reason, associated with the problem of confinement, why the density cannot be much larger, as will be explained in §3.21. It is because of this low particle density and, consequently, small mass density, e.g., about 7×10^{-8} g/cm^3, that the heat content of a thermonuclear reactor would not be large, even at the extremely high operating temperatures. The explosion hazard associated with such a reactor would thus appear to be negligible.

RADIATION LOSSES FROM A PLASMA

BREMSSTRAHLUNG

2.47. The discussion so far has referred only to the energy (or power) that might be produced in a thermonuclear reactor. This energy production must compete with inevitable losses, and the role of the processes which result in such losses is crucial in determining the operating temperature of a thermonuclear reactor. Some energy losses can be minimized by a suitable choice of certain design parameters (see Chapter 12), but others are inherent in the reacting system; these will be considered here.

2.48. At kinetic temperatures in the region of 1 kev or more, substances of low mass number are not only wholly vaporized and dissociated into atoms, but the latter are entirely stripped of their orbital electrons. In other words, matter is in a state of complete ionization; it consists of a gas composed of positively charged nuclei and an equivalent number of negative electrons, with no neutral particles. An ionized gaseous system consisting of equivalent numbers of positive ions and electrons, irrespective of whether neutral particles are present or not, is referred to as a *plasma* [15]. At sufficiently high temperatures, when there are no neutral particles and the ions consist of bare nuclei only, with no orbital electrons, the plasma may be said to be completely (or fully) ionized.

2.49. If the ions in a plasma are not completely stripped, emission of energy will occur in the form of optical (or excitation) radiation. An electron attached to such an ion can absorb energy, e.g., as the result of a collision with a free electron, and thus be raised to an excited state. When the electron returns to a lower quantum level, the excitation energy is emitted in the form of radiation. This represents a possible source of energy loss from the plasma in a thermonuclear reaction system. Hydrogen isotope atoms have only a single electron and are completely stripped at a temperature of about 0.05 kev, so that there is no excitation radiation above this temperature. However, if impurities of higher atomic number are present, energy losses in the form of excitation radiation can become very significant, especially at the lower temperatures, while the plasma is being heated, and even at temperatures as high as 10 kev. Since, in principle, the emission of excitation radiation from impurities is an extraneous rather than an intrinsic cause of energy loss, it will be discussed more fully in Chapter 12.

2.50. Ignoring impurities for the present, it may be stated that the plasma in a thermonuclear reactor will consist of completely stripped nuclei of hydrogen isotopes together with an equal number of electrons. From such a plasma, energy will inevitably be lost in the form of *bremsstrahlung*, that is, continuous radiation emitted by charged particles, mainly the electrons, as a result of deflection by the Coulomb fields of other charged particles. Except possibly at temperatures above 50 kev (see §2.70), the bremsstrahlung from a plasma arises almost entirely from electron-ion interactions. Since the electron is free before its encounter with an ion and remains free subsequently, the transitions are often described as "free-free."

2.51. In theory, the losses due to bremsstrahlung could be decreased if the dimensions of the system were larger than the mean free path for absorption of the radiation photons under the existing conditions. This would mean that the system would be impossibly large, e.g., with dimensions of 10^8 cm (600 miles) or more, even at very high plasma densities. In a system of this impractical size, a thermonuclear reaction involving deuterium could become self-sustaining without the application of energy from outside sources (§2.54). In other words, a sufficiently large mass of deuterium could attain a critical size, by the propagation of a thermal chain reaction, just as does a suitable mass of fissionable material as the result of a neutron chain reaction.

2.52. Consider a mass of deuterium so large that it behaves as an optically thick (or opaque) body as far as bremsstrahlung is concerned, and these radiations are essentially completely absorbed within the system. The energy loss will then be given by the black-body radiation corresponding to the existing temperature. Even at ordinary temperatures, some D-D reactions will occur, although at an extremely slow rate. The energy produced in these reactions will be absorbed and will serve to raise the temperature in the interior, so that the nuclear reaction rate will increase owing to the increasing cross sections. This will result in a more rapid temperature rise accompanied by a further gain in the reaction rate, and so on. If the system were large enough, a temperature would ultimately be attained at which the rate of energy loss by black-body radiation is equal to the rate of energy produced by nuclear fusion. This would represent a self-sustaining thermonuclear reactor from which energy could be drawn off continuously as radiation, provided the deuterium is replaced as fast as it is used up.

2.53.   The size of such a critical reactor of deuterium can be calculated approximately in the following manner. If the system behaves as a black body, in which there is a state of radiation equilibrium between the particles present and the radiation field, the radiation flux from the surface is expressed by the Stefan-Boltzmann law $I = \alpha T'^4$, where α is 5.7×10^{-5} erg/(cm^2) (sec) (°K^4) and T is expressed in °K*. If the critical system is a sphere of radius r, the area is $4\pi r^2$ and so the radiation flux is $4\pi r^2 \alpha T'^4$. The rate of energy production in the sphere of volume $\frac{4}{3}\pi r^3$ is obtained upon multiplying by $\frac{1}{2}n_D{}^2\overline{\sigma v}Q$, in accordance with equation (2.17). It follows, therefore, that in the critical system

$$4\pi r^2 \alpha T^4 = \tfrac{2}{3}\pi r^3 n_D{}^2 \overline{\sigma v} Q,$$

so that the critical radius is given by

$$r = \frac{6\alpha T^4}{n_D{}^2 \overline{\sigma v} Q}. \tag{2.21}$$

2.54.   Since the sphere will probably be large enough to retain the neutrons, all the energy produced in the D-D and D-T reactions will be deposited in the reacting region. As seen in §2.23, neglecting the D-He^3 reaction, this will amount to 24.9 Mev for every two D-D interactions, or roughly 12.5 Mev per interaction; this is the value of Q to be used in equation (2.21). In choosing a possible operating temperature, an optimum value is selected, namely, the one at which $T^4/\overline{\sigma v}$ is a minimum for the D-D reactions; this is about 10 kev or 1.16×10^8 °K, when $\overline{\sigma v}$ is 8.6×10^{-19} cm^3/sec. Finally, a maximum value of 3×10^{21} nuclei/cm^3 may be postulated for n_D, since this would represent a pressure of about 100 atm at ordinary temperatures. Upon inserting these figures into equation (2.21), it is found that the critical radius of the deuterium sphere would be at least 10^8 cm, i.e., over 600 miles [16].

---

*The Stefan-Boltzmann constant is generally represented by the symbol σ; however, α is used here to avoid confusion with the cross section.

2.55.   A further examination of the situation in a controlled thermonuclear system shows that conditions must be sought in which equilibrium with black-body radiation does not exist.   From the Stefan-Boltzmann equation it is found that at a temperature of 10 kev the radiation flux would be 10^{21} watts/cm². In view of the conclusion reached earlier that a practical power density would be in the vicinity of 100 watts/cm³, a radiation flux of the rate indicated would be intolerable except, as shown above, for a system of impractical size.

2.56.   Incidentally, when a system is in black-body radiation equilibrium, the radiation pressure is equal to aT^4/c, where c is the velocity of light.   For a temperature of 10 kev, i.e., 1.16×10^8 °K, this would be the order of 10^{11} atm.   In stars, such high pressures are balanced by gravitational forces due to the enormous masses.   No practical controlled thermonuclear reactor could withstand the pressures resulting from equilibrium with the radiation at extremely high temperatures.

2.57.   The solution to this problem is to be found in the use of the very low particle densities required by other considerations.   A system of this type is optically "thin" and transparent to essentially all the bremsstrahlung from a hot plasma; it is a poor absorber, and hence also a poor emitter, of this radiation.   The radiation field with which the particles may be in equilibrium is then very much weaker than black-body radiation.   In other words, the equivalent radiation temperature is much lower than the kinetic temperature, which is related to the energy distribution among the particles (§2.8).   It is for this reason that the term kinetic temperature, rather than temperature without qualification, has been frequently used in the preceding discussion.   Strictly speaking, "temperature" implies thermodynamic equilibrium, which means both kinetic and black-body radiation equilibrium [17].

2.58.   A similar departure from thermodynamic equilibrium is known to occur frequently in gaseous nebulae and in stellar atmospheres.   The distribution of velocities is Maxwellian, so that a definite kinetic temperature can be ascribed to the system.   The radiation field is, however, much weaker than would be expected from black-body equilibrium at that temperature.

2.59.   For systems with dimensions that are small compared to the mean free path for absorption of photons at an energy corresponding to the kinetic temperature, a rough rule is that the actual radiation from the medium will be of the order of the black-body value diminished by the ratio of the dimensions of the system to the mean free path.   For the absorption of a photon in a hydrogen isotope plasma, as the result of a free-free transition, the mean free path is represented approximately [18] by

$$\lambda_\nu \approx 7 \times 10^{-5} \frac{T^{1/2} \nu^3}{n^2} \text{ cm,}$$

where T is the kinetic temperature in kilo-electron volts, n is the number density of either electrons or ions, and ν sec^{-1} is the radiation frequency.   At a kinetic temperature of 10 kev, for example, the average wave length of the bremsstrahlung may be taken to be about 3 angstroms (see Fig. 2.8). The corresponding average frequency is then 10^{18} sec^{-1}, and so for a density of 10^{15} ions (or electrons)/cm^3, the absorption mean free path would be more than 10^{20} cm. Hence, in a system having dimensions of the order of 10^3 cm or so, the radiation flux will be less than the black-body value for the same temperature by a factor of about 10^{17}.

2.60. Although bremsstrahlung losses from a plasma at temperatures of thermonuclear significance are high, as will be seen below, they are, nevertheless, not necessarily disastrous. Fortunately, the rate of energy loss due to bremsstrahlung increases as $T^{1/2}$ for a Maxwellian distribution, whereas the rate of thermonuclear energy production varies very roughly as $T^{5/2}$ in the temperature region of interest. Hence, with increasing temperature, the ratio of the power produced to that leaving the system as bremsstrahlung, which is less than unity at low and moderately high temperatures, becomes greater than unity at very high kinetic temperatures. A thermonuclear reactor which produces more energy within the reacting system than escapes owing to unavoidable losses can then be realized in principle. However, such temperatures, as will be seen shortly, are in the range of 10 to 100 kev, and it is entirely due to the loss of energy as bremsstrahlung that these ultrahigh temperatures are necessary. If a method could be discovered for preventing the loss of energy as bremsstrahlung without making the system impossibly large, thermonuclear reactors operating at lower temperatures would be feasible; but no such prospect is in sight.

RATE OF BREMSSTRAHLUNG EMISSION

2.61. An expression for the rate of electron-ion bremsstrahlung energy emission of the correct form [19, 20], but differing by a small numerical factor from the result obtained by a more rigorous procedure, can be derived from the classical expression for the rate P_e at which energy is radiated by an accelerated electron, namely,

$$P_e = \frac{2e^2}{3c^3} a^2, \tag{2.22}$$

where e is the electronic charge, c is the velocity of light, and a is the acceleration. Suppose an electron moves past a relatively stationary ion of charge Ze with an impact parameter b^*, as indicated in Fig. 2.7. The Coulomb force between the charged particles is then Ze^2/b^2. Let m_e be the electron rest mass; then its acceleration is Ze^2/b^2m_e, and the rate of energy loss as radiation is given by equation (2.22) as

$$P_e \approx \frac{2e^6Z^2}{3m_e^2c^3b^4}.$$

---

* For a more complete discussion of the significance of the term "impact parameter" see §4.48.

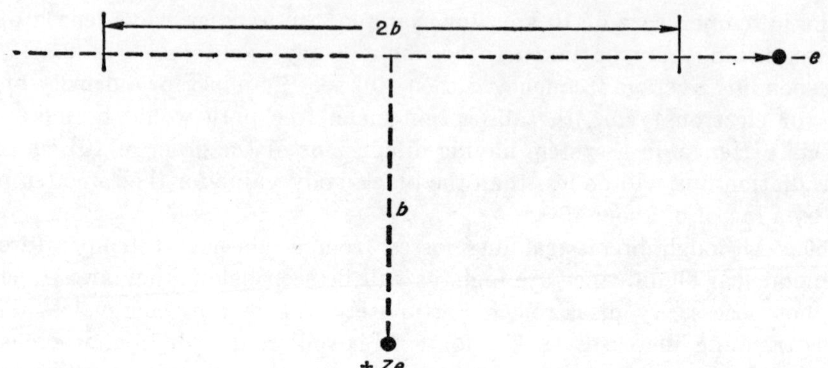

Fig. 2.7. Coulomb interaction of electron with a nucleus.

2.62. The electron path length over which the Coulomb force is effective is $2b$ (see §4.54 footnote), and if the velocity is v, the time during which acceleration occurs is $2b/v$. If the acceleration is assumed to be constant during this time, the total energy E_e radiated as the electron moves past an ion with an impact parameter b is then

$$E_e \approx \frac{4e^6 Z^2}{3 m_e^2 c^3 b^3 v}.$$

If this is multiplied by n_e and n_i, the numbers of electrons and ions, respectively, per unit volume, and also by v, the result is the rate of energy loss P_a per unit impact area for all ion-electron collisions occurring in unit volume at an impact parameter b; hence,

$$P_a \approx \frac{4e^6 n_e n_i Z^2}{3 m_e^2 c^3 b^3}.$$

The total power P_{br} radiated as bremsstrahlung per unit volume is obtained upon multiplying by $2\pi b\, db$ and integrating over all values of b from $b_{\min}$, the distance of closest approach of an electron to an ion, to infinity; thus,

$$P_{\text{br}} \approx \frac{8\pi e^6 n_e n_i Z^2}{3 m_e^2 c^3} \int_{b_{\min}}^{\infty} \frac{db}{b^2}$$

$$= \frac{8\pi e^6 n_e n_i Z^2}{3 m_e^2 c^3 b_{\min}}. \tag{2.23}$$

2.63. An estimate of the minimum value of the impact parameter can be made by utilizing the uncertainty principle relationship, i.e.,

$$\Delta x\, \Delta p \approx \frac{h}{2\pi},$$

when Δx and Δp are the uncertainties in position and momentum, respectively, of a particle and h is Planck's constant. The uncertainty in the momentum may be set equal to the momentum $m_e v$ of the electron and Δx may then be identified with $b_{\min}$, so that

$$b_{\min} \approx \frac{h}{2\pi m_e v}.$$

Furthermore, assuming a Maxwellian distribution of velocity among the electrons, it is possible to write

$$\tfrac{1}{2}m_e v^2 = \tfrac{3}{2}kT_e,$$

where T_e is the kinetic temperature of the electrons; hence,

$$b_{\min} \approx \frac{h}{2\pi (3kTm_e)^{\frac{1}{2}}}.$$

Upon inserting this result into equation (2.23), it is found that

$$P_{\mathrm{br}} \approx \frac{16\pi^2}{3^{\frac{1}{2}}} \cdot \frac{(kT_e)^{\frac{1}{2}}e^6}{m_e^{\frac{3}{2}}c^3 h}\, n_e n_i Z^2.$$

The foregoing derivation has referred to a system containing a single ionic species of charge Z. In the case of a mixture of ions (or nuclei), it can be readily seen that the quantity $n_i Z^2$ should be replaced by $\Sigma(n_i Z^2)$, where the summation is taken over all the ions present.*

2.64. A more precise treatment, assuming a Maxwellian distribution of electron velocities, gives for the rate of bremsstrahlung energy emission per unit volume

$$P_{\mathrm{br}} = g\,\frac{32\pi}{3^{\frac{3}{2}}} \cdot \frac{(2\pi kT)^{\frac{1}{2}}e^6}{m_e^{\frac{3}{2}}c^3 h}\, n_e \sum (n_i Z^2), \tag{2.24}$$

where g is the Gaunt factor which corrects the classical expression for the requirements of quantum mechanics [21-28]. At high temperatures, the correction factor approaches a limiting value of $2 \times 3^{\frac{1}{2}}/\pi$ and utilizing this result, together with the known values of k in ergs/°K, e in statcoulombs, and m_e, c, and h in cgs units, equation (2.24) becomes

$$P_{\mathrm{br}} = 1.57 \times 10^{-27} n_e \sum (n_i Z^2)T_e^{\frac{1}{2}} \text{ ergs}/(\mathrm{cm}^3)(\mathrm{sec}), \tag{2.25}$$

where T_e is the electron temperature in °K, or making use of the conversion factor given in §2.13,

$$P_{\mathrm{br}} = 5.35 \times 10^{-24} n_e \sum (n_i Z^2)T_e^{\frac{1}{2}} \text{ ergs}/(\mathrm{cm}^3)(\mathrm{sec})$$

$$= 5.35 \times 10^{-31} n_e \sum (n_i Z^2)T_e^{\frac{1}{2}} \text{ watts}/\mathrm{cm}^3,$$

where T_e is now in kilo-electron volts.

2.65. For a plasma consisting only of hydrogen isotopes, Z is 1, and n_i and n_e are equal, so that the factor $n_e\Sigma(n_i Z^2)$ may be replaced by n^2, where n is

---

*The factor $n_e\Sigma n_i Z^2$ is sometimes written in the form $n_e^2(\Sigma n_e Z^2/\Sigma n_i Z)$, since n_e is equal to $\Sigma n_i Z$.

the particle density of either electrons or nuclei. Hence, for a deuterium system, for example,

$$P_{\text{DD(br)}} = 5.35 \times 10^{-31} n_{\text{D}}^2 T_e^{\frac{1}{2}} \text{ watts/cm}^3. \tag{2.26}$$

For a 50 atomic per cent mixture of deuterium and tritium, n^2 is equal to $4 n_{\text{D}} n_{\text{T}}$; in this case, therefore,

$$P_{\text{DT(br)}} = 2.14 \times 10^{-30} n_{\text{D}} n_{\text{T}} T_e^{\frac{1}{2}} \text{ watts/cm}^3. \tag{2.27}$$

Upon comparison of these expressions with those in §§2.40 and 2.41 for the thermonuclear power density, it is found that the rate of energy release for the D-D reactions will exceed the rate of loss as bremsstrahlung at a temperature between 10 and 100 kev. For the D-T reaction, the corresponding temperature is somewhat less than 10 kev.

2.66. The foregoing equations give the total rate of emission of energy as bremsstrahlung over all wave lengths. It is of interest, however, for various purposes, to consider the energy distribution as a function of the wave length of the radiation. The classical expression for the rate of bremsstrahlung emission per unit volume per unit frequency interval in the frequency range from ν to $\nu + d\nu$ is

$$dP_\nu = g \frac{32\pi}{3^{\frac{3}{2}}} \left(\frac{2\pi}{kT}\right)^{\frac{1}{2}} \frac{e^6}{m_e^{\frac{3}{2}} c^3} n_e \sum (n_i Z^2) \exp(-h\nu/kT)\, d\nu. \tag{2.28}$$

Upon integration over all frequencies, this expression leads to equation (2.24). For the present purpose, it is more convenient to express equation (2.28) in terms of wave length; it is then found that the rate of energy emission per unit wave length in the interval from λ to $\lambda + d\lambda$ is

$$dP_\lambda = 6.01 \times 10^{-30} g n_e \sum (n_i Z^2) T_e^{-\frac{1}{2}} \lambda^{-2} \exp(-12.40/\lambda T_e)\, d\lambda$$
$$\text{watts/(cm}^3\text{)(angstrom)}, \tag{2.29}$$

where the temperature is in kilo-electron volts and the wave lengths are in angstroms. Assuming the Gaunt factor to remain constant, as is not strictly correct, the relative values of $dP_\lambda/d\lambda$ obtained from equation (2.29), for arbitrary electron and ion densities, have been plotted as a function of wave length in Fig. 2.8 for electron temperatures of 1, 10, and 100 kev. It is seen that each curve passes through a maximum at a wave length which differentiation of equation (2.29) shows to be equal to $6.20/T_e$ angstroms. To the left of the maximum the energy emission as bremsstrahlung is dominated by the exponential term and decreases rapidly with decreasing wave length. To the right of the maximum, however, the variation approaches a dependence upon $1/\lambda^2$ and the energy emission falls off more slowly with increasing wave length of the radiation.

IDEAL IGNITION TEMPERATURE

2.67. The minimum operating temperature for a self-sustaining thermonuclear reactor is that at which the energy deposited by nuclear fusion within the reacting system just exceeds that lost from the system as a result of bremsstrahlung emission.   To determine its value it is necessary to calculate the rates of thermonuclear energy production at a number of temperatures, utilizing equations (2.19) and (2.20) together with Fig. 2.4, *for charged-particle*

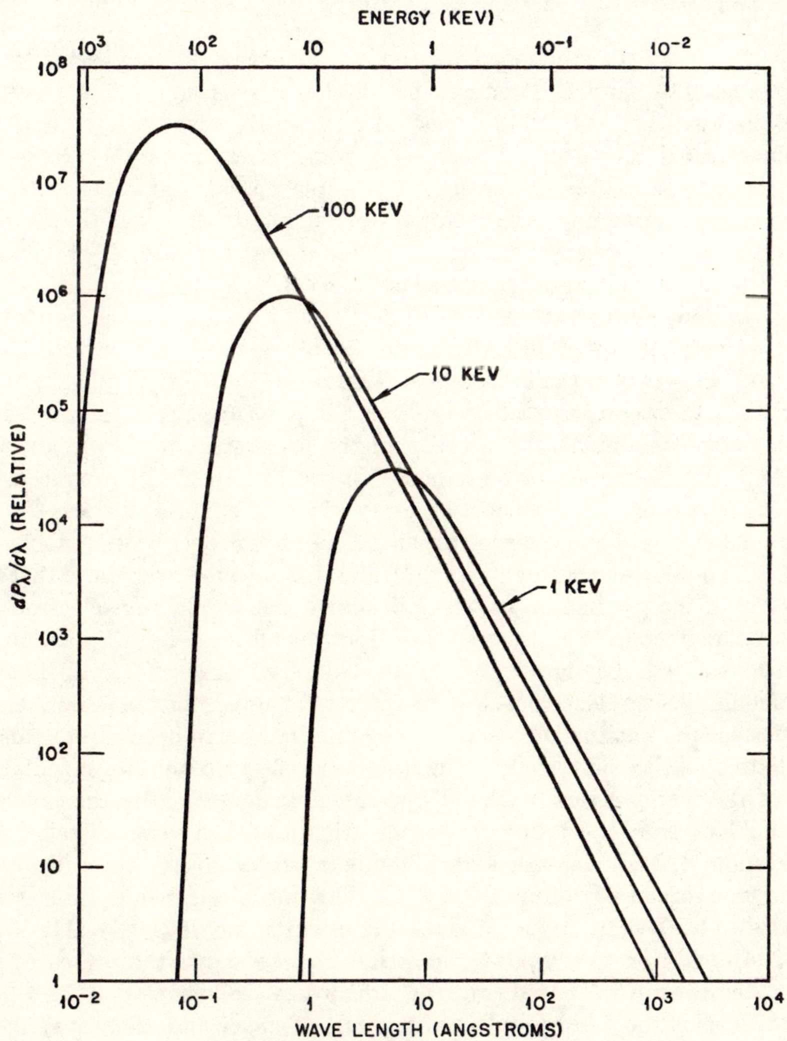

FIG. 2.8. Bremsstrahlung power distribution at kinetic temperatures of 1, 10, and 100 kev.

products only, and to compare the results with the rates of energy loss as bremsstrahlung derived from equations (2.26) and (2.27). The assumption is made that, in the given plasma, the kinetic ion (nuclear) temperature and the electron temperature are the same. For purposes of illustration, n_D is taken as 10^{15} nuclei/cm^3 for the D-D reactions, whereas n_D and n_T are each 0.5×10^{15} nuclei/cm^3 for the D-T reaction; this makes the bremsstrahlung losses the same for the two cases. The results of the calculations are shown in Fig. 2.9; the energy rates are expressed in terms of the respective power densities, i.e., energy produced or lost per unit time per unit volume of reacting system [29].

2.68. It is seen that the curve for the rate of energy loss as bremsstrahlung intersects the D-T and D-D energy production curves at the temperatures of 4 and 36 kev, i.e., 4.6×10^7 and 4.1×10^8 °K, respectively. They are sometimes called the *ideal ignition temperatures*. If equation (2.24) is accurate, then these represent the lowest possible operating temperatures for a self-sustaining thermonuclear reactor. For temperatures lower than the ideal ignition values, the bremsstrahlung loss would exceed the rate of thermonuclear energy deposition by charged particles in the reacting system.

2.69. In addition to various losses apart from bremsstrahlung which can be minimized but not completely eliminated in a practical reactor (see Chapter 12), there are two factors which will require the actual plasma kinetic temperature to exceed the ideal ignition temperature given above. In the first place, no account has been taken of the bremsstrahlung emission arising from Coulomb interaction of electrons with the helium nuclei produced in the thermonuclear reactions. Since they carry two unit charges, the loss of energy will be greater than for the same concentration of hydrogen isotope ions.

2.70. Secondly, at the high temperatures that would exist in a thermonuclear reactor, the production of bremsstrahlung due to electron-electron interactions, as distinct from those resulting from the electron-ion interactions considered above, will be significant. Provided relativistic effects do not arise, there should be no electron-electron bremsstrahlung, but at high electron velocities such is not the case and appreciable losses can occur from this form of radiation. Although some attempts have been made to calculate the extent of the energy emission, there is no complete agreement in the conclusions reached. However, the following results will provide a general indication of the situation. At an electron kinetic temperature of 25 kev the ratio of electron-electron bremsstrahlung energy to that for electron-ion interaction is estimated to be 0.06, at 50 kev it is 0.13, and at 100 kev it is 0.34 [11, 30].

2.71. In addition to the power densities, Fig. 2.9 shows the pressures at the various temperatures, based on the ideal gas equation $p = (n_i + n_e)kT$, where $n_i + n_e$ is the total number of particles (nuclei and electrons) per cm^3 and T is the kinetic temperature in °K. In the present case $n_i = n_e = 10^{15}$ particles/cm^3, so that $n_i + n_e = 2 \times 10^{15}$. With k in ergs/°K, the values are

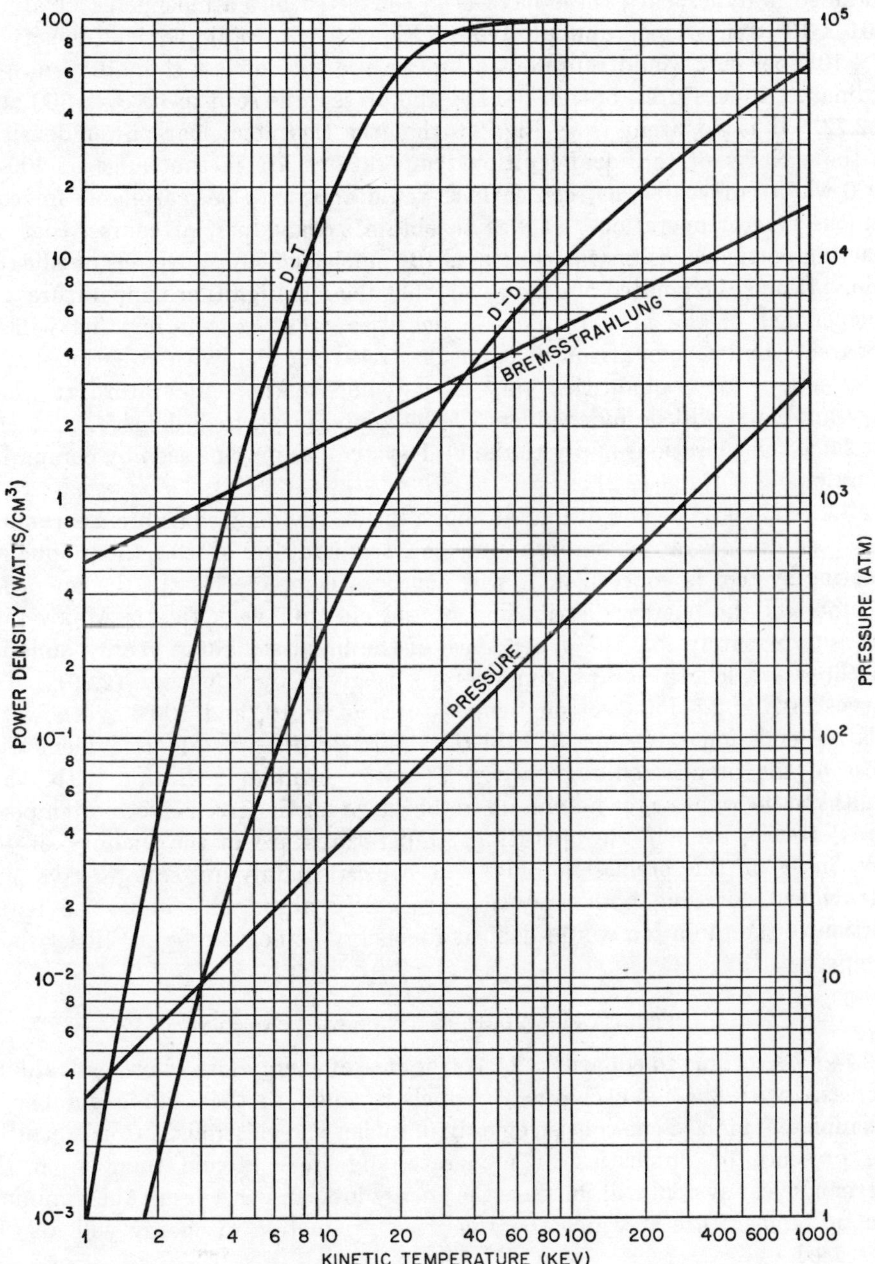

Fig. 2.9. Characteristics of thermonuclear reactions and the ideal ignition temperature.

obtained in dynes/cm^2; these have been converted into atmospheres (1 atm = 1.01×10^6 dynes/cm^2) and plotted in Fig. 2.9. A total particle density of 2×10^{15} per cm^3 would represent a kinetic pressure of less than 10^{-4} atm at ordinary temperatures, but at 100 kev the pressure is seen to be over 300 atm.

2.72. It is apparent from Fig. 2.9 that the thermonuclear power densities in the vicinity of the ideal ignition temperatures are in the range of 100 to 1000 watts/cm^3 which, as seen earlier, would appear to be reasonable for continuous reactor operation. It was to obtain such values, of course, that the reacting particle density was chosen as 10^{15} nuclei/cm^3 for purposes of illustration. It must be pointed out, however, that the ideal ignition temperatures are independent of the nuclear particle densities. The reason for this will be apparent from a comparison of equation (2.19) for P_{DD} with equation (2.26) for $P_{DD(br)}$. Both quantities vary as $n_D{}^2$, and so the temperature at which they are equal will be independent of the deuteron particle density. Exactly the same considerations apply to the D-T system, as may be seen by comparing equations (2.20) and (2.27).

2.73. It should be noted that, although the energy emitted as bremsstrahlung may be lost as far as maintaining the temperature of the thermonuclear reacting system is concerned, it will not be a complete loss in the operating reactor. If the energy distribution of the electron velocities is Maxwellian (or approximately so), the dependence of the bremsstrahlung energy emission on the wave length (or photon energy) is given by equation (2.29). The curves in Fig. 2.8 for electron temperatures of 1, 10, and 100 kev show the relative emission rates for an arbitrary particle density. It is evident that most of the bremsstrahlung energy is in the form of radiation with wave lengths to the right of the maximum in the curve for the given electron temperature. Hence, even if the operating temperatures are in the vicinity of 100 kev, most of the bremsstrahlung will consist mainly of soft X-rays and ultraviolet radiation, both of which are easily absorbed. Thus, the major portion of the radiation would deposit its energy as heat and very little would escape.

EFFECT OF IMPURITIES ON RADIATION LOSS

2.74. According to equation (2.24), the rate of energy loss as bremsstrahlung increases with the ionic charge Z, which is equal to the atomic number in a completely ionized gas consisting only of nuclei and electrons. Consequently, the presence of impurities of moderate and high atomic number in the thermonuclear system will increase the energy loss, and as a result the minimum kinetic temperature at which there is a net production of energy will also be increased.

2.75. Consider a completely ionized mixture containing n_1 nuclei/cm^3 of hydrogen isotopes ($Z = 1$) and n_z nuclei/cm^3 of an impurity of atomic number Z. The electron density n_e is then $n_1 + n_z Z$ per cm^3. The factor $n_e \Sigma (n_i Z^2)$ in equation (2.23) is consequently equal to $(n_1 + n_z Z)(n_1 + n_z Z^2)$. In the absence of the impurity, however, the corresponding factor would be $n_1{}^2$. It follows, therefore, from equation (2.24) that

$$\frac{\text{Power loss in presence of impurity}}{\text{Power loss in absence of impurity}} = \frac{(n_1 + n_z Z)(n_1 + n_z Z^2)}{n_1{}^2}$$

$$= 1 + f^2 Z^3 + fZ(Z + 1).$$

where $f = n_z/n_1$, i.e., the fraction of impurity atoms.

2.76. Suppose, for example, that the impurity is oxygen $(Z = 8)$, and that it is present to the extent of 1 atomic per cent, so that $f = 0.01$. In this case,

$$\frac{\text{Power loss in presence of impurity}}{\text{Power loss in absence of impurity}} = 1.77$$

In other words, the presence of only 1 atomic per cent of oxygen impurity will increase the rate of energy loss as bremsstrahlung by 77 per cent. In the case of the D-D system, Fig. 2.9 shows that this would raise the ideal ignition temperature from 36 to 80 kev; for the D-T reaction the corresponding increase would be from 4 to 5.5 kev. For impurities of higher atomic number, the increase in the already extremely high ideal ignition temperature would be even greater. It thus appears to be an important requirement of a thermonuclear reactor that even traces of impurities, especially those of moderate and high atomic number, should be rigorously excluded from the reacting plasma.*

CYCLOTRON RADIATION

2.77. In addition to the loss of energy as bremsstrahlung from a high-temperature plasma, there is another possible way in which energy may be radiated from a thermonuclear reaction system. It will be seen in Chapter 3 that, in general, the approaches being considered at the present time to the problem of confining a plasma at extremely high temperatures are based on the use of magnetic fields. In such fields, the ions and electrons spiral about the lines of force at definite frequencies, called the gyromagnetic (or cyclotron) frequencies (§4.13). The centripetal acceleration of the charged gyrating particles is accompanied by the emission of what has been called *cyclotron* (or *synchrotron*) *radiation* [31-35]. Because of their relatively large mass, the ions, e.g., deuterons, in a plasma move moderately slowly and the associated emission of cyclotron radiation is insignificant under conditions of immediate interest. For electrons, however, the rate of energy loss may be so large as greatly to exceed that resulting from the emission of bremsstrahlung from the given plasma.

2.78. Apart from their relative magnitude, there is another important difference between bremsstrahlung and cyclotron radiation. The former are

* In Chapter 10, reference will be made to a possible exception to this rule.

mostly in the X-ray and ultraviolet regions of the spectrum (cf. Fig. 2.8), but the cyclotron radiation is expected to be mainly in the infrared and conventional microwave regions. Consequently, whereas the escape of bremsstrahlung from a plasma cannot be prevented, it is probable that the cyclotron radiation will be partially absorbed in the plasma, so that some of the energy, at least, remains in the reaction system. For the preliminary discussion, however, it will be assumed that the cyclotron radiation, like the bremsstrahlung, escapes from the plasma.

2.79. If the electrons have a Maxwellian distribution of velocities, the power of the cyclotron radiation per unit volume is

$$P_{cy} \approx \frac{4e^4k}{3m_e{}^3c^5} B^2 n_e T_e, \tag{2.29}$$

where B is the strength of the magnetic field in which the electrons are gyrating, and the other symbols have the same significance as before. For electrons of high (relativistic) energies, an additional factor must be included, but this may be neglected here. It will be shown in §3.18 that a magnetic field exerts a pressure equal to $B^2/8\pi$. For the purpose of the following discussion it will be assumed that the particle (or kinetic) pressure of the plasma, i.e., $n_i k T_i + n_e k T_e$, is equal to the magnetic pressure; thus,

$$\frac{B^2}{8\pi} = n_i k T_i + n_e k T_e.$$

In a hydrogen isotope plasma, n_i and n_e are equal and may be replaced by n, which is the particle density of either ions or electrons. Furthermore, if T_e and T_i are not very different, it follows that

$$B^2 \approx 16\pi n k T_e,$$

and upon substituting this value for B^2 in equation (2.29), the result is

$$P_{cy} \approx \frac{64\pi e^4 k^2}{3m_e{}^3c^5} n^2 T_e{}^2$$

$$\approx 5.0 \times 10^{-32} n^2 T_e{}^2 \text{ watts/cm}^3, \tag{2.30}$$

where T is now in kilo-electron volts. For purposes of comparison, the rate of energy loss as bremsstrahlung from a hydrogen isotope plasma may be written in the analogous form

$$P_{br} \approx 5.3 \times 10^{-31} n^2 T_e{}^{1/2} \text{ watts/cm}^3.$$

2.80. It will be observed that, whereas the bremsstrahlung loss increases as $T^{1/2}$, the cyclotron radiation varies as T^2. At temperatures below about 5 kev, the cyclotron radiation is less than that due to bremsstrahlung, but at higher temperatures the energy emission rate of the former radiation increases very rapidly with temperature and soon greatly exceeds that of the latter. If, as tentatively postulated above, the cyclotron radiation escapes completely from the plasma, then the rate of energy loss would be so large that a self-sustaining thermonuclear reaction system involving deuterium only would become impossible.   In other words, the rate of energy release from the two D-D reactions, followed by a D-T interaction, can never exceed the rate at which energy is emitted as cyclotron radiation, according to equation (2.30).   On the other hand, because the rate of the D-T reaction is already large in the vicinity of 5 kev and increases faster than T^2, the cyclotron radiation loss would result in only a slight increase in the ideal ignition temperature, perhaps to about 7 kev.

2.81.   There are two possible ways in which the apparently serious situation outlined above for a D-D reaction system might possibly be alleviated.   It was assumed, for example, that the electron and ion temperatures are equal.   In an operating thermonuclear reactor, the energy released will be taken up first by the ions and will then be transferred to the electrons in collisions; the latter will then emit the appropriate electromagnetic radiation.   If the rate of energy transfer by collision is less than the radiation loss at the same temperature, a steady state will be attained in which the electron temperature is lower than the ion temperature (cf. §4.83).   In these circumstances, the rate of energy loss as cyclotron radiation would be less than given by equation (2.30) and the conditions might be more favorable for the realization of a self-sustaining thermonuclear system.   However, it turns out that the lower electron temperature is not of itself sufficient to improve the prospects for a D-D reactor.

2.82.   Another possibility is that, for a sufficiently thick plasma, the emitted cyclotron radiation may be largely reabsorbed, thus reducing the total loss rate significantly below that computed for the case where all the radiation escapes.   The cyclotron radiation is emitted in discrete spectral lines, all harmonics of the fundamental gyromagnetic (or cyclotron) frequency of the electrons.   Theory indicates that, while the plasma absorption coefficient is large for the fundamental and low harmonics, it is small for the higher harmonics.   Furthermore, the proportion of the energy contained in the higher harmonics increases with the plasma temperature.   The effect of the reabsorption of the radiation can be shown by calculating the radius of a body of plasma which would absorb sufficient radiation to permit the operation of a D-D reactor under specified conditions of magnetic field strength, plasma density, and temperature.

2.83.   In one particular case, the magnetic field strength in the plasma was taken to be 10,000 gauss, the kinetic particle pressure and the magnetic pressure were assumed to be equal, as postulated above, and allowance was made for the difference in electron and ion temperatures.   At an ion temperature of 100 kev, the electron temperature in the steady state was estimated to be 65 kev; the minimum size of the plasma for a self-sustaining D-D reactor was found to be 13 meters. For an ion temperature of 200 kev, the electron temperature would be 86 kev and the reactor would have the impracticable dimension of 95 meters [31].

2.84. A later calculation, however, indicates a very much more hopeful situation. This takes specific account of the angular distribution of the radiation, which is emitted preferentially in a narrow cone whose axis lies along the direction of motion of the electron, and also of the polarization of the cyclotron radiation. As a result the critical (or minimum) size of a D-D reactor, under the same conditions as were given above, at an ion temperature of 235 kev and an electron temperature of 80 kev, is only 38 cm [35]. Other considerations would require larger dimensions than this, in any event. Unfortunately, the same calculations show that the minimum size is markedly increased by a decrease in the ratio of the kinetic pressure to the magnetic pressure in the plasma. Thus, if this ratio is 0.1, which is probably a not unreasonable value in an operating reactor, the minimum size becomes 670 meters!

2.85. An effective means for reducing the critical size appears to be the use of reflectors. Since most of the cyclotron radiation power loss is in the range of wave lengths from a few millimeters to the far infrared, ordinary metals, of which the reaction chambers would be made, will serve as excellent reflectors. It has been shown theoretically that, if R is the reflectivity of the wall, the minimum size of the D-D reactor will be reduced by a factor of $1/(1 - R)$, which could become very large. It seems therefore, that conditions can, in principle, be realized under which the cyclotron radiation loss would not rule out the possibility of a thermonuclear reactor operating on deuterium alone.

2.86. Experimentally, cyclotron radiation is difficult to produce and detect since electron temperatures of several kilo-electron volts must be attained before it reaches noticeable intensities. Only when the temperatures reached in controlled thermonuclear research approach those for an operating reactor will cyclotron radiation loss become a matter of practical concern. This development is not yet in sight, but the possible serious effects of such loss at sufficiently high temperatures must not be overlooked.

2.87. It should be mentioned, in conclusion, that the foregoing discussion has been based on the supposition that there is a magnetic field within the plasma. If the plasma were bounded by a magnetic field, but contained none within it, it should not suffer loss of energy in the form of cyclotron radiation, except in the form of a surface effect at the boundary between the plasma and the field.

OPERATIONAL CONDITIONS

CONTINUOUS OR PULSED OPERATION

2.88. Thermonuclear reaction systems at present under consideration fall into two groups depending on how they would have to operate, whether continuously or in pulses. In a continuously operating (or steady-state) reactor, the fuel would be fed steadily into the reaction chamber maintained at an essentially constant thermonuclear temperature. Part of the energy produced would be used up in increasing the temperature of the reacting species and compensating for bremsstrahlung and other losses, and the remainder would be available as heat for conversion into electricity by means of a turbogenerator or by a thermocouple or analogous system. Direct conversion of the thermonuclear energy carried by the charged products of the reaction into electricity would be difficult. For a reactor of this type, it would appear to be advantageous if most of the thermonuclear energy were carried by neutrons, which readily escape from the reaction chamber and deposit their energy as heat, rather than by charged particles. It is evident, therefore, that the preferred fuel would be an equimolecular mixture of deuterium and tritium.

2.89. For a reactor operating in pulses, direct conversion of part of the thermonuclear energy into electricity is feasible, at least in principle. A quantity of fuel is injected into the chamber and is compressed and heated rapidly to permit thermonuclear reaction to occur. Expansion of the charged-particle products in a magnetic field would then produce electrical energy, at the expense of part of the energy released. In this case, it would be desirable to use deuterium, perhaps with a small proportion of tritium, as the fuel, rather than an equimolecular mixture of the two isotopes. For the latter situation, only 20 per cent of the total energy yield is carried by the charged particles.

OPERATING LIMITATIONS

2.90. There are certain limitations, apart from that of temperature, placed upon the conditions of operation of a thermonuclear reactor which can be derived by considering an ideal cycle in a pulsed system [36]. Suppose that in the reactor the fuel is heated very rapidly to the kinetic temperature T; this temperature is maintained for a time t to permit thermonuclear reaction to take place. The plasma is then allowed to cool to a relatively low temperature to complete the cycle (or pulse). If the fuel gas density is n atoms/cm^3, the completely ionized hydrogen isotope plasma will contain $2n$ particles/cm^3. Assuming a Maxwellian distribution, the average energy per particle will be $\frac{3}{2} kT$, and so the energy required to raise the temperature of the fuel is close to $3nkT$ per cm^3, since the initial temperature will be very small in comparison with the thermonuclear temperature T.

2.91. Let P_{th} and P_{br} represent the thermonuclear power produced and the bremsstrahlung power loss, respectively, per cm^3; in time t, the corresponding energies will be tP_{th} and tP_{br}. If the bremsstrahlung energy is absorbed, as mentioned in §2.73, the total energy available after the thermonuclear reaction pulse will be $tP_{th} + tP_{br} + 3nkT$ per cm^3. Suppose the overall efficiency for the conversion of this energy into useful power, either indirectly, through heat, or partly in a direct manner, is taken to be one third [37]. The minimum condition for the practical operation of the reactor is then that the power thus produced shall be sufficient to raise the temperature of the fuel and to compensate for bremsstrahlung losses; that is,

$$\tfrac{1}{3}(tP_{\mathrm{th}} + tP_{\mathrm{br}} + 3nkT) > tP_{\mathrm{br}} + 3nkT.$$

Upon rearrangement, this condition is found to be equivalent to

$$R \equiv \frac{P_{\mathrm{th}}/3n^2kT}{(P_{\mathrm{br}}/3n^2kT) + 1/nt} > 2.$$

2.92.   Since P_{th} and P_{br} both vary as n^2, it is evident that the parameters which determine the value of the fraction on the left side of this inequality, designated by R, are T and nt only.   Obviously, T must exceed the ideal ignition temperature, and the minimum value of nt can be found by plotting R, as calculated from the equations for P_{th} and P_{br} given earlier, as a function of T for various values of nt.   It has been found for the D-D reaction system that nt must exceed 10^{16} and for the D-T reaction it must be greater than 10^{14} (atoms) (sec)/cm^3 for the operation of a pulsed thermonuclear reactor to produce more energy than is supplied.   The minimum values quoted are, of course, dependent upon the efficiency with which the available energy can be converted into useful power.   If this efficiency exceeds the one third postulated here, the minimum values of nt will be somewhat less than those given above; but it is doubtful whether a reduction by as much as a factor of 10 would be possible.

2.93.   Although the foregoing discussion has applied in particular to pulsed operation, with t equal to the length of the reaction period, similar considerations must apply to a steady-state thermonuclear device.   In this case, t would represent the average confinement time of the reacting nuclei.

UTILIZATION OF THERMONUCLEAR NEUTRONS

2.94.   In addition to the heat generated by the slowing down of the escaping neutrons produced in one of the D-D reactions (2.45 Mev energy) and in the D-T reaction (14.1 Mev energy), there are at least two practical applications for the neutrons themselves.   If controlled thermonuclear power proves feasible while there are still ample supplies of uranium-238 and thorium-232 available, the neutrons, after being slowed down, could be used to convert these fertile materials into the fissionable species, plutonium-239 and uranium-233, respectively.   This might make breeder reactors unnecessary and would greatly simplify the problem of the design of fission power reactors to permit the efficient utilization of the available resources of possible nuclear fuels for such reactors.

2.95.   The second application, which has been mentioned in §2.22, is to capture the slow neutrons in lithium-6, thus producing tritium for addition to the deuterium to be used as the thermonuclear fuel.   A mixture of deuterium and tritium has an important advantage over deuterium alone because of the significant decrease in the ideal ignition temperature. On the other hand, the addition of tritium to deuterium results in a decrease in the proportion of the total thermonuclear energy which could be converted directly into electrical energy in the manner described above. Whether the benefit of the lower operating temperature would more than compensate for the loss in direct electric power production would depend to a great extent on the design of the thermonuclear system.

2.96. In any event, it is of interest to see how the efficiency of the conversion of neutrons into tritium would determine the deuterium-tritium ratio that might be realized in the thermonuclear reactor fuel [38]. Information on this matter can be obtained from the reaction rate equations given earlier. Since the D-D reactions have almost identical cross sections, neutrons are produced in half of the D-D reactions and in all the D-T reactions. It follows, therefore, from equations (2.10) and (2.11) that

$$\text{Rate of neutron formation} = \tfrac{1}{4}n_D^2\overline{\sigma v}_{DD} + n_D n_T \overline{\sigma v}_{DT} \text{ per cm}^3 \text{ per sec.}$$

Let n be the number of tritium nuclei produced, on the average, by capture in lithium-6, for every neutron released in the thermonuclear reactions; the value of n cannot exceed unity, except possibly in special circumstances to be mentioned below, and it will probably be less since some neutrons will inevitably escape or be captured by nuclei other than those of lithium-6. Hence,

$$\text{Rate of T recovery} = n n_D(\tfrac{1}{4}n_D\overline{\sigma v}_{DD} + n_T\overline{\sigma v}_{DT}) \text{ per cm}^3 \text{ per sec.} \quad (2.31)$$

Since a tritium nucleus is consumed in each D-T reaction and one is produced in half of the D-D reactions, it is seen that

$$\begin{aligned}
\text{Rate of T consumption} &= n_D n_T \overline{\sigma v}_{DT} - \tfrac{1}{4}n_D^2\overline{\sigma v}_{DD} \\
&= n_D(n_T\overline{\sigma v}_{DT} - \tfrac{1}{4}n_D\overline{\sigma v}_{DD}) \text{ per cm}^3 \text{ per sec.} \quad (2.32)
\end{aligned}$$

2.97. If all the tritium recovered from the capture of neutrons in lithium-6 is returned to the thermonuclear fuel gas and none is added from outside sources, then after a short time for attainment of an equilibrium state, the rates given by equations (2.31) and (2.32) will be equal; hence,

$$n(\tfrac{1}{4}n_D\overline{\sigma v}_{DD} + n_T\overline{\sigma v}_{DT}) = n_T\overline{\sigma v}_{DT} - \tfrac{1}{4}n_D\overline{\sigma v}_{DD}.$$

Upon rearrangement, it is found that

$$\frac{n_T}{n_D} = \frac{1+n}{1-n} \cdot \frac{\overline{\sigma v}_{DD}}{4\overline{\sigma v}_{DT}}. \quad (2.33)$$

2.98. At a temperature of 25 kev, as is seen from Fig. 2.4, $\overline{\sigma v}_{DD}$ is close to 5×10^{-18} and $\overline{\sigma v}_{DT}$ is about 5×10^{-16} cm³/sec, so that $\overline{\sigma v}_{DD}/\overline{\sigma v}_{DT}$ is 0.01. Consequently, equation (2.33) reduces to

$$\frac{n_T}{n_D} = 2.5 \times 10^{-3} \frac{1+n}{1-n}.$$

It is thus a simple matter to calculate the equilibrium ratio n_T/n_D for various postulated values of the efficiency of conversion of neutrons into tritium; the results are given in Table 2.2. The figures in the last column are expressed as atomic per cent of tritium in the equilibrium fuel gas, i.e., $100 \times n_T/(n_D + n_T)$.

TABLE 2.2. TRITIUM RECOVERY EFFICIENCY AND THERMONUCLEAR FUEL COMPOSITION AT 25 KEV

n	n_T/n_D	Atomic % T
0.50	0.0083	0.83
0.75	0.0175	1.72
0.90	0.0475	4.5
0.99	0.50	16.7
0.995	~1.0	~50

2.99. It is evident that, if it is required to maintain a composition of 50 atomic per cent of tritium, it would be necessary to achieve a **99.5 per cent** conversion of the thermonuclear neutrons into tritium nuclei. If the efficiency dropped to 99 per cent, the equilibrium fuel gas would contain only 16.7 per cent of tritium. Although these values apply, strictly, to operation at a kinetic thermonuclear reaction temperature of 25 kev, the results are much the same at other temperatures; in fact, they are even less favorable at lower temperatures, such as might be contemplated on the supposition that appreciable concentrations of tritium could be achieved in the thermonuclear fuel mixture.

2.100. If it should transpire that, in order to avoid operation at too high temperatures, a large proportion of tritium is necessary in the thermonuclear fuel, advantage may be taken of the neutron multiplication resulting from the $(n,2n)$ reactions, e.g., with beryllium-9, bismuth-209, or lead-207, of the neutrons produced in the D-D and D-T reactions. Another possibility is to permit the thermonuclear neutrons to cause fission in uranium-238 or thorium-232, thereby yielding fission energy as well as neutron multiplication. Although some of the thermonuclear reaction energy will be lost in the $(n, 2n)$ reactions it will be offset by the 4.6 Mev released in the (n, t) reaction with lithium-6.

REFERENCES FOR CHAPTER 2

1. R. d'E. Atkinson and F. Houtermans, *Z. Physik,* **54,** 656 (1928).
2. G. Gamow, *Phys. Rev.,* **53,** 598 (1938).
3. L. D. Landau and E. M. Lifschitz, *Statistical Physics.* Trans. by E. and R. F. Peierls, Pergamon Press, Inc., 1958, p. 115.
4. J. L. Tuck, USAEC Report LA-1190 (1955).
5. W. B. Thompson, UKAEA Report AERE T/M-138 (1956); *Proc. Phys. Soc.* (*London*), **B70,** 1 (1957).
6. G. Gamow and C. L. Critchfield, *Theory of Atomic Nucleus and Nuclear Energy Sources,* Oxford University Press, 1949, Chap. X.

7. W. R. Arnold, J. A. Phillips, G. A. Sawyer, E. J. Stovall, and J. L. Tuck, *Phys. Rev.*, **93**, 483 (1954).
8. USAEC Report LA-2014 (1957), N. Jarmie and J. D. Seagrave, Eds.
9. J. L. Tuck, USAEC Report LAMS-1640 (1954).
10. G. Boulègue, P. Chanson, R. Combe, M. Feix, and P. Strasma, *Proc. Second U.N. Conf. on Peaceful Uses of Atomic Energy*, **32**, 409 (1958).   The values for the D-D reactions in Table 1 are too large by a factor of 10.
11. C. F. Wandel, T. H. Jensen, and O. K. Hansen, *Nuclear Instr.*, **4**, 249 (1959).
12. E. J. Stovall, unpublished.
13. S. Glasstone and M. C. Edlund, *Elements of Nuclear Reactor Theory*, D. Van Nostrand Co. Inc., 1952, pp. 45 *et seq.*
14. R. F. Post, *Rev. Mod. Phys.*, **28**, 338 (1956).
15. I. Langmuir, *Phys. Rev.*, **33**, 954 (1929).
16. A. Simon, *An Introduction to Thermonuclear Research*, Pergamon Press, Inc., 1959, p. 8.
17. L. H. Aller, *Astrophysics*, Ronald Press Co., 1953, Vol. I, p. 76.
18. L. Spitzer, *Physics of Fully Ionized Gases*, Interscience Publishers, Inc., 1956, p. 89.
19. L. Spitzer, USAEC Report NYO-6049 (1954), p. 9.
20. Reference 16, p. 16.
21. H. A. Kramers, *Phil. Mag.*, **46**, 836 (1923).
22. G. Cillié, *Monthly Notices Roy. Astron. Soc.*, **92**, 820 (1932).
23. L. Spitzer, *Astrophys. J.*, **95**, 329 (1942).
24. W. B. Thompson, UKAEA Report AERE T/M-73 (1954).
25. G. Elwert, *Z. Naturforsch.*, **9a**, 637 (1954).
26. J. E. Allen and W. R. Hindmarsh, UKAEA Report AERE GP/R-1761 (1055).
27. W. I. Kogan and A. B. Migdal, *Plasma Physics and the Problem of Controlled Thermonuclear Reactions*, Pergamon Press, Inc., Vol. I, 1960.
28. V. V. Babikov, *Plasma Physics and the Problem of Controlled Thermonuclear Reactions*, Pergamon Press, Inc., Vol. II, 1959, p. 309.
29. J. L. Tuck, USAEC Report WASH-115 (1952), p. 39.
30. R. F. Post, *Ann. Rev. Nuclear Sci.*, **9**, 367 (1959).
31. B. A. Trubnikov and V. S. Kudryavtsev, *Proc. Second U.N. Conf. on Peaceful Uses of Atomic Energy*, **31**, 93 (1958).
32. B. A. Trubnikov and A. E. Bazhnova, *Plasma Physics and the Problem of Controlled Thermonuclear Reactions*, Pergamon Press, Inc., Vol. III, 1959, p. 141.
33. D. B. Beard, *Phys. Rev. Letters*, **2**, 81 (1959); *Phys. Fluids*, **2**, 379 (1959).
34. T. E. Allibone and D. R. Chick, *Proc. Inst. Elec. Eng.* (*London*), **106A**, Suppl. No. 2, 44 (1959).
35. J. E. Drummond and M. Rosenbluth, *Phys. Fluids*, **3**, 45, 491 (1960).
36. J. D. Lawson, *Proc. Phys. Soc.* (*London*), **B70**, 6 (1957).
37. J. D. Jukes, *Proc. Inst. Elec. Eng.* (*London*), **106A**, Suppl. No. 2, 173 (1959).
38. L. Spitzer, D. J. Grove, W. E. Johnson, L. Tonks, and W. F. Westendorp, USAEC Report NYO-6047 (1954), Chapter X.

Chapter 3

POSSIBLE APPROACHES TO CONTROLLED FUSION

CONFINEMENT OF PLASMA

INTRODUCTION

3.1. It is generally felt that if a plasma could be confined in some manner for a sufficient time, so as to satisfy at least the minimum conditions derived in §2.92, the other problems associated with the attainment of a thermonuclear reaction system, e.g., production and heating of the plasma, could be solved. The required kinetic temperature might be attained either directly within the reaction vessel by the use of electrical and magnetic fields in various ways, or particles of high energy might be injected from an external source and then permitted to thermalize before reacting. These important matters will be treated in Chapter 5. The purpose of the present chapter is to outline some of the proposals which have been considered for the realization of controlled thermonuclear reactions, so as to provide a background for the subsequent discussion [1]. Special emphasis is placed on the possible methods for confining a plasma under conditions appropriate to a thermonuclear reactor. More detailed descriptions of the approaches to which the main attention is being devoted at present will be found in Chapters 7 through 11.

3.2. At the particle densities, e.g., about 10^{15} nuclei/cm^3, and the temperatures which would exist in a thermonuclear reactor, the mean free path for collisions between particles in the plasma would be very long, many times the dimensions of the containing vessel. If collisions represent the only means of preventing the escape of the particles from the high-temperature region, the tendency for ions and electrons to escape from this region will be very great. In order to prevent the complete dispersal of the particles, some form of confinement is necessary. Ordinary materials are useless for this purpose, not only because of the intense local heating which would result, but also because collisions of the hot particles with the material walls of the container would result in a transfer of energy to the walls. The cooled particles would then

46

return to the reacting region and serve to lower the temperature of the remaining particles.

3.3.   Consider, for example, a plasma with a particle density of 10^{15} deuterons/cm^3 at a temperature of 100 kev.   The mean free path for the D-D reaction is seen from Fig. 2.5 to be just over 10^{10} cm, provided the energy distribution is Maxwellian.   In a reaction vessel having a diameter of 100 cm, a deuteron will, therefore, collide with the walls roughly 10^8 times before reacting with another deuteron.   Suppose that at each collision it loses only 1 kev, but as a result of each reaction roughly 10 Mev, i.e., 10^4 kev, is liberated. The ratio of energy loss to the walls to the energy released by thermonuclear reaction is approximately $10^8/10^4$, i.e., 10^4.   It is obvious that the decrease in temperature of the plasma due to wall collisions would be intolerable.   For the D-T reaction at the same particle density and a reaction temperature of 10 kev, the ratio of wall loss to the reaction energy would be less, but it would still be too large.

3.4.   Although the heating of the walls would not appear to be serious at first sight, the indirect results could be disastrous.   At a kinetic temperature of 100 kev, the average energy per particle in a Maxwellian system is 150 kev. The heat content of a reactor in which the number densities of nuclei and electrons are each 10^{15} particles/cm^3 would thus be 3×10^{17} kev/cm^3 or roughly 12 cal/cm^3.   Suppose that, in a reactor vessel in which the volume in cubic centimeters was 50 times the surface area in square centimeters, all the particles suddenly collided with the walls and gave up their energy.   Each square centimeter of wall would then receive 600 calories, and the construction of a vessel capable of absorbing this quantity of heat without damage would appear to present no insuperable problem.   However, the important point is that if the energy release occurred rapidly, most would be absorbed in a thin layer of the wall of the containing vessel.   Here the temperature could become so high that there would be considerable evaporation and sputtering of the wall materials into the plasma [2, 3].   The introduction of these impurities would be accompanied by a large increase in the energy losses due to bremsstrahlung at the higher temperatures and to excitation radiation at the lower temperatures, as explained in Chapter 2.

ELECTRIC AND MAGNETIC FIELDS

3.5.   The solution to the problem of confinement of a plasma at the very high temperatures and pressures that would exist in a thermonuclear reactor clearly cannot lie in the use of material walls, and so some other method must be sought.   The fact that, at temperatures above about 50 ev, i.e., 5.8×10^5 °K, deuterium and tritium will be in the form of a completely ionized plasma of charged particles suggests the possibility that electric or magnetic fields might be employed for confinement for a sufficient time, at least, to permit thermonuclear reaction to occur to a useful extent.   Both electric and magnetic fields can effectively exert a pressure on the charged particles and so restrain them from leaving the reaction region and reaching the walls of the reactor (or containing) vessel.*

3.6. Before considering the matter of the confinement of a plasma by electric or magnetic fields, a brief digression is necessary to discuss the forms of the Maxwell equations that are appropriate to the present situation and the units used to express the various electrical and magnetic quantities.

3.7. Although mks units are widely employed in connection with macroscopic electrical quantities, the common practice in plasma studies is to use the Gaussian system, since it is more suited to the treatment of the microscopic behavior of charged particles. In this system, which will be used except where macroscopic measurements are involved, electrical quantities are given in terms of electrostatic (cgs) units and magnetic quantities in electromagnetic (cgs) units. Thus, potential difference and current are in statvolts and statamperes, respectively, where 1 statvolt is equal to 300 volts and 1 statampere is equivalent to $10/c$ ampere, where c is the velocity of light, i.e., 3.00×10^{10} cm/sec. Electrical charge is expressed in statcoulombs, so that the electronic charge has the familiar value of 4.803×10^{-10} esu. Magnetic quantities are stated in terms of the oersted and the gauss; for reasons to be given below, only the latter will be used in this book.

3.8. For material media, Maxwell's equations of electromagnetism are generally written, in Gaussian units, as

$$\nabla \cdot \mathbf{D} = 4\pi\sigma \tag{3.1}$$

$$\nabla \cdot \mathbf{B} = 0 \tag{3.2}$$

$$\nabla \times \mathbf{E} = -\frac{1}{c} \cdot \frac{\partial \mathbf{B}}{\partial t} \tag{3.3}$$

$$\nabla \times \mathbf{H} = \frac{1}{c}\left(4\pi\mathbf{j} + \frac{\partial \mathbf{D}}{\partial t}\right), \tag{3.4}$$

where $\mathbf{E}$ is the electric field strength (or electric intensity), $\mathbf{D}$ is the electric flux density (or electric induction or displacement), $\mathbf{H}$ is the magnetic field strength (or magnetic intensity), $\mathbf{B}$ is the magnetic flux density (or magnetic induction), σ is the electric charge density, and $\mathbf{j}$ is the current density; t is the time and c the velocity of light.

3.9. In this form of the Maxwell equations, σ represents the free charge density and $\mathbf{j}$ is the free current density, as distinguished from the bound

---

* Although the terms "confinement" and "containment" are used somewhat interchangeably in connection with plasmas, the former term will be employed here for the restraint placed upon the charged particles, e.g., by a magnetic field, whereas the latter will refer to the vessel or container in which the plasma is held or enclosed. For example, it may be stated that "a plasma contained in a toroidal tube is confined by a magnetic field."

charges and currents. The latter currents are then supposed to give rise to a magnetization **M**, such that

$$\mathbf{B} = \mathbf{H} + 4\pi\mathbf{M}.$$

In plasma physics, however, it is more convenient to treat all current densities explicitly, that is, to include in **j** both the "free" current and that which may produce diamagnetism. It is correct, therefore, to use **B** rather than **H**, and the term $\nabla \times \mathbf{B}$ should then be substituted for $\nabla \times \mathbf{H}$ in equation (3.4). ·

3.10. The specification of σ as the free charge implies the existence in general material media of bound charges. In dielectrics, such charges are indeed present, and the difference between **D** and **E** is proportional to the density of dipole moments, or polarization **P**, produced by the distorting effect of the imposed field **E** on the charge configuration; thus,

$$\mathbf{D} = \mathbf{E} + 4\pi\mathbf{P}.$$

In a fully ionized and stripped plasma, there are no bound charges and hence no polarization under a steady applied field. The dielectric constant K, defined by $\mathbf{D} = K\mathbf{E}$, is then unity, and **D** in equations (3.1) and (3.4) may be replaced by **E**.

3.11. In applying Maxwell's equations to a plasma, therefore, they will be written in the form

$$\nabla \cdot \mathbf{E} = 4\pi\sigma \tag{3.5}$$

$$\nabla \cdot \mathbf{B} = 0 \tag{3.6}$$

$$\nabla \times \mathbf{E} = -\frac{1}{c} \cdot \frac{\partial \mathbf{B}}{\partial t} \tag{3.7}$$

$$\nabla \times \mathbf{B} = \frac{1}{c}\left(4\pi\mathbf{j} + \frac{\partial \mathbf{E}}{\partial t}\right), \tag{3.8}$$

where **B** is generally referred to as the magnetic field strength and is expressed in gauss. Basically, these equations imply that, in their application to a plasma, **E, D, B,** and **H** are the microscopic vacuum field values, so that **E** and **D** are equal and so also are **B** and **H,** and all charges and currents are treated explicitly.

CONFINEMENT BY ELECTRIC FIELD

3.12. It does not seem that confinement of a plasma by an electrostatic field will be possible for several reasons. Suppose that an electrostatic "cage" is established by means of a set of charged electrodes and that the field is able to prevent the plasma from making contact with the material of the electrodes. If such a system is capable of confining electrically charged particles of one sign, e.g., ions, then clearly those of the opposite sign, i.e., electrons, present in the plasma cannot be confined, since the electric field has oppositely directed effects on the positive and negative particles. Furthermore, according to Earnshaw's theorem of classical electrostatics, there is no position of stable equilibrium for a charged particle in an electrostatic field, no matter how complex its structure [1]. A consequence of this theorem is that no configuration of charges, such as a plasma, can exist in stable equilibrium under the influence of purely electrostatic forces [4].

3.13. In addition to the foregoing qualitative arguments, there is a quantitative reason why confinement of a plasma by means of an electrostatic field is not likely to be practical. An essentially static electric field can exert an effective pressure on a system of charged particles which is limited to the energy density of the field, given by $E^2/8\pi$.* If E is the field strength in statvolts/cm, then the energy density (or pressure) will be obtained in ergs/cm^3 (or dynes/cm^2). The pressure of the plasma, treated as an ideal gas, is nkT, where n is the total number of particles per cubic centimeter, i.e., the total particle density of the plasma. If the number densities of ions and electrons are each 10^{15} particles/cm^3, then n is 2×10^{15} particles/cm^3; and suppose T (or rather kT) is 100 kev. The minimum value of E required to contain the plasma is found by equating $E^2/8\pi$ to nkT; thus,

$$\frac{E^2}{8\pi} \geqslant 2 \times 10^{15} \times 100 \times 1.6 \times 10^{-9} = 3.2 \times 10^8 \text{ ergs/cm}^3,$$

where 1.6×10^{-9} is the factor for converting kilo-electron volts into ergs. It is seen that E must be nearly 9×10^4 statvolts/cm or about 2.7×10^7 volts/cm. Thus, a stationary electrostatic field of impossibly large magnitude would be required to confine a plasma of reasonable particle density such as might be used in a thermonuclear reactor.

CONFINEMENT BY MAGNETIC FIELD

3.14. It will be shown in Chapter 4 that in a magnetic field charged particles gyrate about the lines of force, the positive particles in one direction and the negative particles in the opposite direction. Hence, apart from the effect of collisions, in a uniform magnetic field the ions and electrons remain tied to the field lines. Although they can move freely along (or parallel) to these lines, in either direction, they cannot cross the lines if there are no collisions among the particles. Hence, if the ions and electrons in a plasma can in some manner be prevented from escaping at the ends of the containing vessel, e.g., by means of an endless tube of toroidal form or in other ways, the use of a magnetic field appears to offer promise for confinement of a plasma. It remains to be seen, however, if the strength of the field required would be reasonable. For this purpose it is necessary to determine the relationship between the field strength and the effective pressure which the field could exert on the plasma. A simplified derivation of this relationship is given below.

* The equivalent expression for a magnetic field is derived below (§3.20).

3.15. The plasma will be regarded as consisting of singly charged positive and negative particles, i.e., hydrogen isotope nuclei and electrons, moving independently, so that collision forces can be neglected. If, in general, both electric and magnetic fields are present, the force in dynes acting on a single particle of charge $\pm e$ statcoulombs due to the electric field $\mathbf{E}$ statvolts/cm is $\pm eE$; and that due to the magnetic field of $\mathbf{B}$ gauss is $e(\mathbf{v} \times \mathbf{B})/c$, where $\mathbf{v}$ is the particle velocity in cm/sec and c is the velocity of light in the same units.

3.16. In the steady state, the force on all the particles in unit volume, i.e., the force density, is just balanced by the rate of momentum transfer which is equal to the gradient of the pressure. It follows, therefore, that for a plasma containing n_i ions/cm³ having an average velocity $\mathbf{v}_i$,

$$n_i e \left[\mathbf{E} + \frac{1}{c}\, (\mathbf{v}_i \times \mathbf{B}) \right] = \nabla p_i \tag{3.9}$$

and for the n_e electrons/cm³ with an average velocity $\mathbf{v}_e$,

$$-n_e e \left[\mathbf{E} + \frac{1}{c}\, (\mathbf{v}_e \times \mathbf{B}) \right] = \nabla p_e \tag{3.10}$$

where ∇p_i and ∇p_e represent the pressure gradients due to ions and electrons, respectively. In a hydrogen isotope plasma, n_i and n_e are equal, and so addition of equations (3.9) and (3.10) gives

$$\frac{ne}{c} \left[(\mathbf{v}_i - \mathbf{v}_e) \times \mathbf{B} \right] = \nabla p, \tag{3.11}$$

where n is equal to both n_e and n_i, and ∇p is the total pressure gradient in the plasma. The quantity $ne\,(\mathbf{v}_i - \mathbf{v}_e)$ is the net rate of movement of charge and so is equal to the current density $\mathbf{j}$; it follows, therefore, from equation (3.11) that

$$\frac{1}{c}\, (\mathbf{j} \times \mathbf{B}) = \nabla p. \tag{3.12}$$

3.17. For an electric field which does not vary with time, Maxwell's equation (3.8) reduces to

$$\nabla \times \mathbf{B} = \frac{1}{c}\, (4\pi \mathbf{j}), \tag{3.13}$$

and combination with equation (3.12) gives

$$\frac{1}{4\pi}\, (\nabla \times \mathbf{B}) \times \mathbf{B} = \nabla p. \tag{3.14}$$

In general,

$$(\nabla \times \mathbf{B}) \times \mathbf{B} = -\tfrac{1}{2}\nabla B^2 + (\mathbf{B} \cdot \nabla)\mathbf{B},$$

and in a magnetic field in which the lines of force are straight and parallel, the last term on the right is zero; upon making this substitution and rearranging, equation (3.14) becomes

$$\nabla \left(p + \frac{B^2}{8\pi} \right) = 0,$$

so that

$$p + \frac{B^2}{8\pi} = \text{constant.} \tag{3.15}$$

3.18. The quantity $B^2/8\pi$, which has the units of energy/volume, is the *energy density* of the magnetic field. However, since energy/volume is equivalent to force/area, $B^2/8\pi$ is also regarded as the *magnetic pressure* of the field. It follows, therefore, from equation (3.15) that the sum of the kinetic and magnetic pressures is constant in a plasma in a straight magnetic field. If the plasma is completely confined by an external magnetic field of strength B_0, the pressure at the outside must fall to zero, and so it is seen that the constant in equation (3.15) is equal to $B_0^2/8\pi$; that is,

$$p + \frac{B^2}{8\pi} = \frac{B_0^2}{8\pi}. \tag{3.16}$$

Hence, an external magnetic field B_0 can confine a plasma having a kinetic pressure p and a contained field B, where B_0, B, and p are related by equation (3.16). It will be noted that in the steady state the magnetic field within a plasma having a finite kinetic pressure must always be less than the external field. Consequently, a plasma confined by a magnetic field tends to be diamagnetic.

3.19. In the foregoing treatment, ∇p has been treated as the gradient of an isotropic scalar plasma pressure. A more rigorous derivation of equation (3.12) shows, however, that ∇p is really $\nabla \cdot \mathbf{p}$, the divergence of a stress tensor [5].* The pressure (or stress) tensor may then be described in terms of two equal scalar components $p_\perp$ at right angles to the field lines and the component $p_{\parallel}$ parallel to these lines. The p term in equation (3.16) is then strictly $p_\perp$; this means that the magnetic field can support a plasma (scalar) pressure only in the direction perpendicular to the field lines. Since the plasma exerts its pressure in all directions, escape of the particles along the field lines is possible, as indicated above, unless steps are taken to prevent it.

3.20. It will be seen in subsequent chapters that it is often convenient to express the kinetic pressure of the particles in a plasma in terms of its ratio to the external magnetic pressure (or energy density). The dimensionless ratio β is then defined by†

---

* In addition, the term $\rho \mathbf{v} \nabla \cdot \mathbf{v}$ should appear on the right of equation (3.12), but this is usually taken to be small in a quiescent plasma (cf. §13.47).

† A few writers have used the symbol β to represent the ratio of the kinetic pressure at a given point to the magnetic energy density *at the same point*.

$$\beta \equiv \frac{p}{B_0^2/8\pi}, \tag{3.17}$$

so that equation (3.16) may be written as

$$\beta = 1 - \frac{B^2}{B_0^2}. \tag{3.18}$$

Since the minimum value of B is zero, the ratio β has a maximum value of unity; this would represent the ideal case of a perfectly diamagnetic plasma from which the magnetic field was completely excluded. In this event, equation (3.16) may be written as

$$p_{\max} = \frac{B_0^2}{8\pi}, \tag{3.19}$$

where $p_{\max}$ is the maximum kinetic pressure of a plasma that can be confined, in a steady state, by an external magnetic field of strength B_0.

3.21. Although stability requirements frequently restrict β in a magnetically confined plasma to values appreciably less than unity, equation (3.19) may be used to indicate the order of magnitude of the field that would be required to confine a plasma under conceivable conditions in a thermonuclear reactor. As was seen above, the kinetic pressure may be set equal to nkT, where n is the total particle density. Taking n as 2×10^{15} particles/cm³ and the temperature (or kT) as 100 kev, then

$$\frac{B_0^2}{8\pi} \approx 1.6 \times 10^8 \text{ ergs/cm}^3 \text{ (or dynes/cm}^2\text{)}$$

and, consequently,

$$B_0 \approx 9 \times 10^4 \text{ gauss.}$$

An external field of about 90,000 gauss would thus be required to confine the specified plasma. A field of this order is by no means outside the realm of practicality, provided the dimensions of the containing vessel are not too large. It is seen, therefore, that confinement of the plasma in a thermonuclear reactor by means of a magnetic field should be a definite possibility. Attention may be called to the fact that the strength of available magnetic fields sets an upper limit of 10^{15} or 10^{16} particles/cm³ for the plasma density. It is a fortunate circumstance that this corresponds to a situation in which the power density has a reasonable value (§2.42).

3.22. It will be seen in the next chapter (§4.14) that the radius of gyration r_g cm of a charged particle around the lines of force of a magnetic field, in a hydrogen isotope plasma, is given by

$$r_g = \frac{m v_\perp c}{eB}, \tag{3.20}$$

where m grams is the mass of the particle, $v_\perp$ cm/sec is its velocity of gyration perpendicular to the field direction, B gauss is the magnetic field strength, e statcoulombs is the electronic charge, and c cm/sec is the velocity of light. Since there are two degrees of freedom perpendicular to the field lines, the energy of gyration, $\frac{1}{2}mv_\perp^2$, is equal to kT for a Maxwellian distribution; hence, equation (3.20) may be written as

$$r_g = \frac{c(2mkT)^{1/2}}{eB}. \tag{3.21}$$

At a temperature of 100 kev, i.e., $kT = 100$ kev $= 1.6 \times 10^{-7}$ erg, and a field strength of 100,000 gauss, the gyration radius would be 0.010 cm for an electron, 0.60 cm for a deuteron, and 0.72 cm for a triton.*

3.23. As a result of inhomogeneities in the magnetic field, of collisions among the particles, and of the effects of electric fields, the particles in a plasma will not rigorously follow the magnetic lines of force, but will tend to drift across them to some extent. Nevertheless, to a first approximation, the radii of gyration calculated above indicate the distances from a particular line of force within which the respective charged particles remain when confined by a magnetic field. Assuming that the lines of force do not intersect the material walls of the vessel containing the plasma, so that the particles are not actually led into them, it seems that in a vessel of reasonable dimensions, e.g., at least several centimeters, the number of charged particles reaching the walls, as a result of gyration about the field lines, will be relatively small. At lower temperatures and higher field strengths, the values of r_g are decreased, and so also will be the wall losses from the source under consideration.

3.24. A number of methods for utilizing a magnetic field to confine a plasma at high temperatures have been proposed. Four of these in particular have been the subject of experimental study, and their basic principles will be outlined here. Further details, as well as a discussion of methods of producing and heating the plasma, will be given in later chapters. At the present time it is impossible to say which, if any, of the confinement methods to be described will prove successful. One of the main difficulties is concerned with the fact that, except in special circumstances, a plasma confined by a magnetic field is subject to instabilities which can result in the energetic particles reaching the walls of the containing vessel. As was seen above, this would lead to a general lowering of the temperature of the reacting system. The problem, which is as yet far from being solved, is to overcome the instabilities for a sufficient time for the temperature of the plasma to be raised above the thermonuclear ignition temperature, and also to satisfy the requirement that the product of the particle density and the reaction time exceed a certain minimum value.

---

* For a particle of charge Ze and mass m, the value of r_g is proportional to $m^{1/2}/Z$; hence the gyration radius of a He^4 nucleus, which is one of the products of the D-T reaction, will be less than that of a deuteron. The value for a He^3 nucleus will be smaller still.

METHODS OF PLASMA CONFINEMENT *

THE PINCH EFFECT

3.25. One method of confinement, which was proposed independently by a number of investigators in various parts of the world, makes use of what is called the *pinch effect;* it is the self-constriction that occurs in a plasma as a result of the passage of a unidirectional current (see Chapter 7). Such a current produces an azimuthal self-magnetic field (Fig. 3.1) that tends to

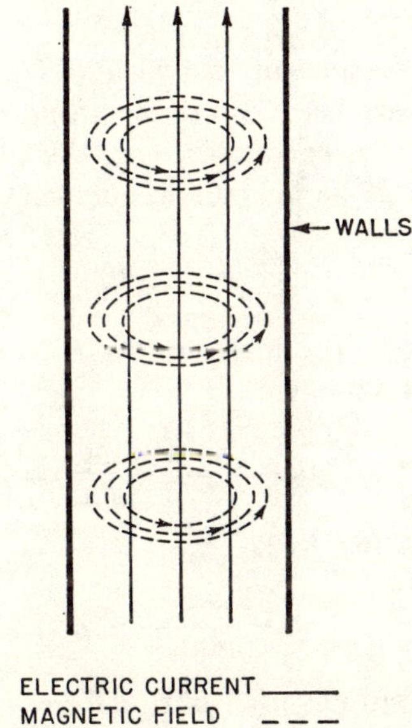

Fig. 3.1. Confinement of plasma by azimuthal self-magnetic field.

constrict (or pinch) the plasma. The phenomena is equivalent to the familiar one in which parallel circuits carrying current in the same direction attract each other.

3.26. If the discharge current is large enough, the constricting effect of the self-magnetic field can pull the plasma away from the walls of the containing vessel, so that magnetic confinement, in the sense discussed earlier, can be achieved. The fact that the discharge becomes pinched is direct evidence that the current carriers, i.e., the charged particles, are being prevented by the azimuthal self-magnetic field from moving in a radial direction.

3.27. An expression for the magnitude of the plasma current required to produce a pinched discharge will be derived in Chapter 7, but its physical

* The material in this section is intended to be introductory only, and so appropriate references are deferred to later chapters where the topics are discussed in detail.

basis will be apparent from the following simplified treatment. If there is no magnetic field trapped within the plasma in a pinched discharge, equation (3.19) takes the form

$$p = \frac{B_\theta^2}{8\pi},\tag{3.22}$$

where the kinetic pressure p of the particles in the plasma is balanced, in the steady state, by the pressure (or energy density) of the azimuthal self-magnetic field B_θ produced by the discharge. Actually, the field strength decreases with distance from the discharge, and the value considered here is that just outside the pinched plasma.

3.28. To express the relationship between the discharge current and the magnetic field, mks units are used; thus, according to the Biot-Savart law, a current of I amp flowing in a conductor, e.g., the plasma, of radius r meters produces an azimuthal magnetic field B_θ webers/meter^2 given by

$$B_\theta = \frac{\mu_0 I}{2\pi r} = 2 \times 10^{-7} \frac{I}{r},\tag{3.23}$$

since $\mu_0/4\pi$ is equal to 10^{-7}. In pinch studies, B_θ is generally stated in gauss, i.e., 10^{-4} webers/meter^2, and r in cm, so that

$$B_\theta \text{ (gauss)} = \frac{I \text{ (amp)}}{5r \text{ (cm)}},\tag{3.24}$$

and combination with equation (3.22) gives

$$p = \frac{I^2}{200\pi r^2}.$$

Expressing the kinetic pressure as nkT, with kT in ergs, it follows that

$$I^2 = 200NkT,\tag{3.25}$$

where N, equal to $\pi r^2 n$, is the linear particle density, i.e., the total number of particles per cm length of the discharge.

3.29. Some indication of the current that might be required to raise the temperature of a pinched plasma may be obtained from equation (3.25). Suppose the total particle density, i.e., hydrogen isotope nuclei and electrons, is 2×10^{15} particles/cm^3 and the cross-sectional area of the containing vessel is 1000 cm^2, so that N is 2×10^{18} particles/cm. If the temperature in the discharge is to reach 100 kev, the required current in amperes is given by equation (3.25) as

$$I^2 = 200 \times 2 \times 10^{18} \times 100 \times 1.60 \times 10^{-9}$$

$$= 64 \times 10^{12}$$

so that

$$I = 8 \times 10^6 \text{ amp.}$$

Thus, currents of the order of millions of amperes would be required to produce a pinched discharge under the conditions specified. For lower temperatures and particle densities, smaller currents would of course be adequate.

3.30. A pinched discharge can be most easily produced by applying a high voltage between two electrodes placed at the ends of a straight tube containing a gas at low pressure. Although linear discharges of this type have been commonly employed in the study of pinch phenomena, there might be two drawbacks in a thermonuclear reactor operating at very high temperatures. In the first place, since the electrodes are in contact with the plasma at all times, they may remove energy from the charged particles and thus lower the temperature. Furthermore, the sputtering from the electrodes could introduce elements of high atomic number into the plasma and so increase the energy lost as bremsstrahlung. Possibly because of the lower temperatures involved, neither of these effects has hitherto been serious in experimental work on linear pinched discharges.

3.31. The problems arising from the presence of electrodes can be overcome by the use of a toroidal tube to contain the gas. The latter acts as the secondary of a transformer in which an electric field is induced from an external primary circuit that partly or wholly surrounds the torus. The discharge within the gas then flows in a closed loop and no electrodes are needed. With a sufficiently large induced current in the plasma, the discharge is constricted, and can pull away from the walls, just as in a straight tube with electrodes within the gas.

3.32. One consequence of the use of an induced discharge is that the current flow cannot be steady. However, the intermittent (or pulsed) nature of the discharge has the advantage that, if rapid thermonuclear reaction can be achieved, some of the energy produced can be converted directly into electrical power. As the kinetic pressure of the plasma particles increases owing to the increase of temperature resulting from the release of nuclear fusion energy, the pinch can no longer be sustained; the constricted plasma will start to expand. The expanding plasma behaves like a mechanically driven armature in a conventional electric generator; in doing work against an external stationary magnetic field an emf is produced.

3.33. The outstanding difficulty in utilizing the pinch effect as a means for plasma confinement lies in the instability of the discharge. Several types of instability have been observed, and to these various names have been applied. There are, for example, the "sausage type" instability, or "necking off," in which the plasma becomes highly constricted at certain points so that it tends to break up into sausagelike links, and "kink" instability, and "wriggling," which are self-descriptive terms.

3.34. It appears from both theoretical arguments and experimental observations that the period of duration of the pinch can be extended, i.e., somewhat increased stability can be achieved, by (*a*) making the torus walls of a conducting material (or with a conducting coating); and (*b*) by including an axial (or longitudinal) magnetic field in the pinched discharge. This longitudinal field, with lines parallel to the direction of current flow in the plasma, as distinct from the azimuthal self-magnetic field, is supplied by a solenoidal winding around the torus. The trapping of the longitudinal magnetic lines of force within the constricted plasma adds to the stability of the latter. However, even with these modifications, the pinched discharge has been found to have a relatively short life, and various other developments are being considered, as will be explained in Chapter **7**.

THE STELLARATOR

3.35.   In principle, it should be possible to confine a plasma by means of an external magnetic field of sufficient strength produced by means of a solenoid. The situation in the case of a cylindrical tube is shown in Fig. 3.2, the magnetic

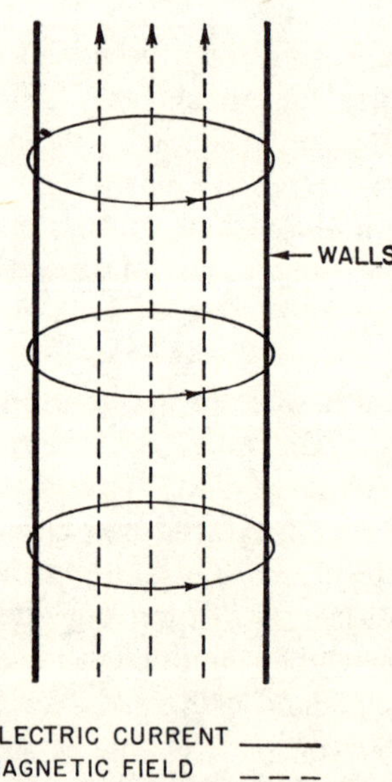

FIG. 3.2. Confinement of plasma by axial magnetic field.

lines of force being parallel to the cylinder axis. The minimum strength of the magnetic field required at the walls is given by equation (3.19) for the necessary particle density and temperature. Since the lines of force lead into the ends of the tube, the particles would lose considerable amounts of energy, and so the use of a torus in this case is again indicated. However, a fundamental difficulty now arises: the strength of the magnetic field in the plasma due to the current in the external solenoid decreases from the inner to the outer major radius of the torus. It will be shown in §4.36 that, in a nonuniform (or inhomogeneous) magnetic field, charged particles tend to drift perpendicularly to the field lines, the electrons in one direction and the ions in the opposite direction. The resulting separation of charges produces an electric field which, in conjunction with the magnetic field, forces the plasma toward the walls. The accompanying energy loss will then be very large.

3.36. A partial solution to this difficulty is to twist the torus into a shape like a figure eight (Fig. 3.3), since the particles would tend to drift in opposite

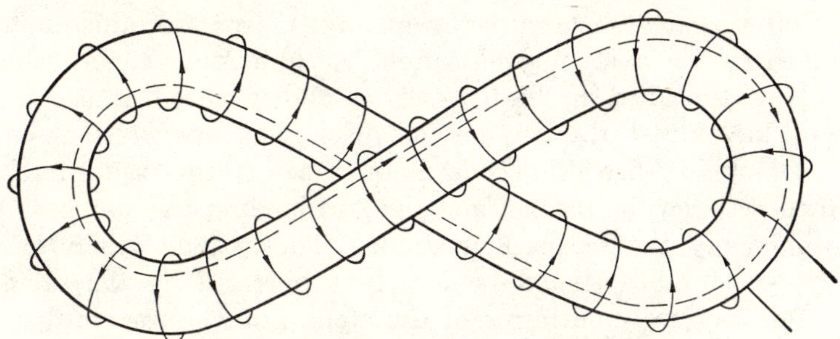

Fig. 3.3. Figure-eight form of stellarator.

directions in the two bends. This was the original form of the device called a *stellarator* (see Chapter 8), with the confining (axial) magnetic field produced by a current passing through a solenoid wound around the tube.

3.37. Basically, the difference in behavior of a plasma confined by an axial magnetic field in a planar torus and in one bent into a figure eight lies in a characteristic of the lines of force. If a line of force is followed around a planar torus, it should close upon itself. In the figure-eight configuration, however, this is not the case; after making a single circuit, the line of force is somewhat displaced from its original position. As a consequence of this displacement, a path becomes available for the charged particles to move upward or downward without crossing the lines of force. Hence, the charge separation which tends to occur, due to the variation in the magnetic field strength across the tube, can be partly neutralized. The drift of the plasma to the walls can thus be decreased.

3.38. The same result can be achieved, in principle, by any distortion of a planar torus or even in a planar torus if the lines of force are twisted by means of a helical magnetic field superimposed on the confining field. The name stellarator has been applied to systems in which the lines of force of an axial magnetic field are displaced in each circuit of a toroidal tube, by means of one of the methods mentioned above.

3.39. The plasma in a stellarator, as in several other systems using magnetic confinement, is subject to various instabilities causing a movement of the plasma to the walls and a consequent loss of energy. One of these instabilities, which is similar to the kink instability of a pinched discharge (§3.33), can arise when the current flowing around the stellarator, e.g., for heating the plasma, exceeds a certain value. Such a current produces an azimuthal magnetic field similar to that responsible for the pinch effect described above but usually of a much lower strength. This field tends to oppose the displacement effect on the lines of force produced by distortion of the torus in a stellarator system. When the current is equal to or greater than that value which would reduce the displacement in a single circuit to zero, kinklike perturbations in the plasma can grow until the plasma touches the walls of the container.

3.40. Another type of instability to which the plasma contained in a twisted torus confined by an axial magnetic field is subject is the interchange instability. It is also called the "flute type" instability since it is associated with the development of perturbations on the plasma surface which resemble the flutes of a column. When this occurs in such a way that the interchange of lines of force between the plasma and the surrounding vacuum field leads to a decrease in energy, the plasma is unstable. Theory indicates, however, that the interchange stability should be largely suppressed by the same helical magnetic field that, in some forms of the stellarator, is used to displace the field lines.

3.41. While the charged particles move around the stellarator, the gradient of the magnetic field in the curved ends causes them to drift normal to the field direction and to the field gradient. The tendency is thus for the particles of a given sign to drift upward in one loop and downward in the other loop of the stellarator. The tendency toward charge separation is counteracted by currents flowing along the lines of force, as indicated above. These currents produce a secondary magnetic field in a plane at right angles to the axis of the straight sections of the stellarator which is proportional to the density of the plasma. Some of the lines of force thus intersect the walls of the tube and lead to a decrease in confinement of the plasma.

3.42. As a consequence of this effect, the kinetic pressure of the particles which can be confined by the applied field is considerably less than the maximum value of $B_0^2/8\pi$ given by equation (3.19). In other words, a much stronger magnetic field is required to confine a plasma having a given particle density than would otherwise have been the case. Since the power of a thermonuclear reactor is proportional to the square of the particle density, this is a serious drawback. A possible method for overcoming the effect of the secondary currents, which involves special shaping of the stellarator tube, will be described in Chapter 8.

3.43. An important aspect of the stellarator is that it is possible and would be desirable to operate it continuously, instead of intermittently as appears to be necessary for pinched-plasma systems. The preference for continuous operation arises from the fact that in the stellarator there is no appreciable contraction of the plasma and so it is not easy to convert the thermonuclear energy directly into electric power by expansion of the plasma in a magnetic field. Consequently, there is nothing to be gained from a pulsed operation, whereas there is much to be lost because the thermonuclear reactions do not occur in the intervals between pulses. Thus, for the stellarator, continuous operation, probably with an equimolar mixture of deuterium and tritium as fuel (§2.88), would be advantageous.

MAGNETIC MIRROR SYSTEMS

3.44. The third proposed method of magnetic confinement of a plasma to be described here avoids the problems of drift in an endless tube by employing a straight tube. A longitudinal magnetic field, for confinement, is applied externally by means of a solenoid, but instead of being uniform, the field strength is increased at the ends of the tube (Fig. 3.4). The region of

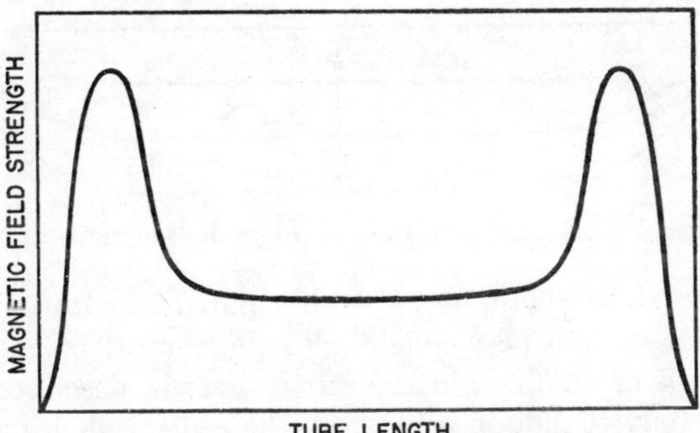

FIG. 3.4. Variation of magnetic field in magnetic mirror system.

enhanced magnetic field is referred to as a *magnetic mirror* (§9.1 *et seq.*); under suitable conditions charged particles, moving from the region of lower to that of higher field strength, will be reflected back into the former region. The magnetic field shown in Fig. 3.4 thus acts as a sort of potential well which inhibits the escape of many of the charged particles (and loss of energy) at the ends of a cylindrical tube.

3.45. The condition for reflection to occur is that in the region between the mirrors the ratio of $W_\parallel/W_\perp$, where $W_\parallel$ and $W_\perp$ are the components of the particle energy parallel and perpendicular, respectively, to the magnetic lines in the central field, should be equal to or less than $R - 1$, where R is the mirror ratio, i.e., the ratio of strengths of the stronger and weaker fields. The physical interpretation of this result may be understood with the aid of Fig. 3.5, which shows some of the magnetic lines of force in a straight tube with mirrors at the ends. Where the field is stronger, i.e., at the mirrors, the lines crowd closer together, so that they change direction in the region in which the field strength is increasing. The force exerted by a magnetic field on a charged particle is always in the direction perpendicular to the field lines, as shown by the arrows in the figure. It is seen, therefore, that, as a result of the curvature of the lines of force in the mirror region, the direction of the magnetic force is such that there is a backward component tending to push the particle back into the region between the mirrors. In other words, the effect of the mirror field is to decrease the longitudinal component of the particle velocity, i.e., the velocity component parallel to the direction of the field in the central region. If the original velocity vector is such that this component is reduced to zero somewhere in the mirror region, the particle will be reflected and will not escape.

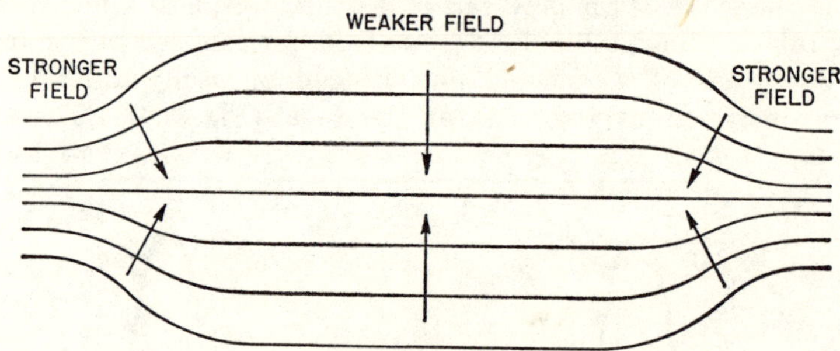

FIG. 3.5. Field lines and direction of force in magnetic mirror system.

3.46. It is apparent, therefore, that only particles with kinetic energy due primarily to their motion perpendicular to the central magnetic field, i.e., to the tube axis, will be confined in the mirror system described above. If the energy is due mainly to motion parallel to the central field lines, they will not be reflected at the ends and will escape through the mirrors. Even if all the charged particles initially in the plasma satisfied the condition for reflection, scattering collisions among the particles will inevitably occur. The partial randomization of the motion will thus mean that some of the particles will then not be reflected and will consequently be lost when they enter the mirror region.

3.47. Several schemes have been proposed for minimizing the losses due to escape from the magnetic field. One, for example, is to increase the value of the mirror ratio. Another possibility is to operate at temperatures greatly in excess of the ideal ignition temperature (§2.68). By increasing the temperature, the reaction cross sections are increased, but the scattering cross sections are decreased. The probability of energy gain due to thermonuclear reaction is thus increased relative to that of loss at the ends of the tube.

3.48. In order to supply energy to the plasma, it may be necessary to compress it adiabatically (§5.18). Axial (or longitudinal) compression can be achieved by moving the magnetic mirrors closer together, either mechanically or electrically. To obtain radial compression, the strength of the magnetic field between the mirrors is increased by increasing the current through the solenoid. Preferably both the mirror and central fields should be increased proportionately, so that the mirror ratio remains constant. When the thermonuclear reaction occurs sufficiently rapidly, so that more energy is being produced than is lost in one way or another, the plasma can expand against the confining magnetic field and electricity could be generated directly in the external coils.

3.49. Although a plasma confined in a mirror system is subject to certain instabilities, e.g., the flute-type instability, they appear, at least at low plasma densities, to be much less serious than those occurring in a pinched discharge or in a stellarator. However, there are other problems which are not encountered in the latter devices. One of these, to be considered more fully in later chapters, is concerned with the production and injection of the plasma between the mirrors. Another difficulty may arise from the need for operating at temperatures considerably in excess of the minimum theoretical value. The particle density of the plasma which can be contained by the available magnetic fields will then be relatively low, resulting in a small power output. If the temperatures required prove to be too high to be practical with deuterium alone, it may be necessary to use a deuterium-tritium mixture.

THE ASTRON SYSTEM

3.50. Another proposed method for confining a plasma by means of a magnetic field is based on the *Astron concept* which is described more fully in Chapter 10. An axial magnetic field is produced in a long evacuated cylindrical chamber by means of a solenoidal coil, in the usual manner; then a beam of high-energy electrons, probably 30 to 50 Mev energy in an actual thermonuclear reactor, is injected at one end. As a result of the action of the magnetic field, the electrons are expected to form a circulating layer of current about the central axis. When the current in this layer, called the E-layer, reaches a certain value, the configuration of the magnetic field will change so as to form a pattern of closed magnetic lines of force, as shown in Fig. 3.6.

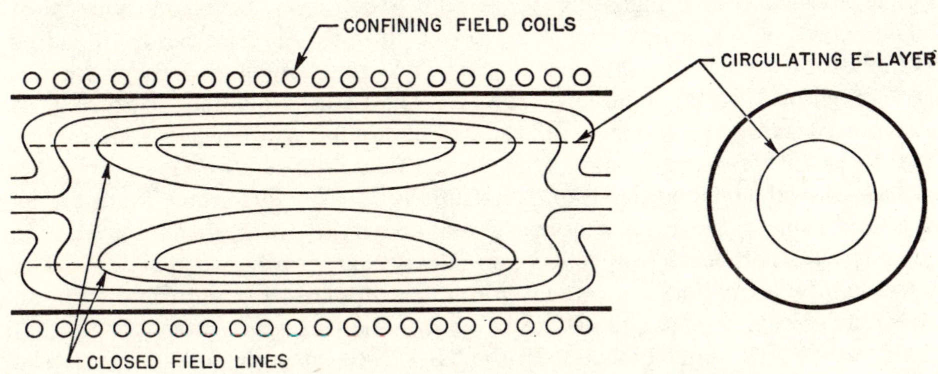

FIG. 3.6. Magnetic field configuration in Astron system.

This field can provide confinement of a plasma since the charged particles can escape only by the slow process of diffusion across the field lines.

3.51. If cold deuterium gas or, preferably, a mixture of deuterium and tritium is now injected, it will be immediately ionized by the high-energy electrons of the E-layer. The resultant plasma will be trapped by the closed system of magnetic field lines formed by the combination of the current in the external field coils and the E-layer current. The charged particles in the plasma will gain energy by collisions with the energetic electrons, so that temperatures may be attained at which thermonuclear D-T reactions take place at a useful rate.

3.52. A theoretical analysis indicates that the plasma in an Astron system should be stable against the more catastrophic perturbations. However, the concept involves such novel ideas, concerning which there has been little practical experience, that only the experimental investigations now proceeding can provide answers to the problems involved.

FUSION REACTIONS WITH ACCELERATED PARTICLES

BOMBARDMENT OF A SOLID TARGET

3.53. Apart from the proposal to inject deuterons of high energy, e.g., 300 kev or more, produced by a conventional accelerator, into a magnetic mirror system where they assume an approximately Maxwellian energy distribution before combining, a number of other suggestions have been made for initiating fusion reactions by the use of accelerated particles. Some of these proposals will be examined below [1].

3.54. An obvious method for bringing about a D-D or D-T reaction is to bombard a solid target containing deuterium or tritium with a beam of high-energy deuterons. In this case, most of the energy of the accelerated particles is expended uselessly in ionization of the target atoms. It will be shown in Chapter 4 that the Coulomb scattering cross section for the close-range interaction of a high-energy (or "hot") deuteron, having an energy of 100 kev, with a low-energy (or "cold") electron in the target exceeds 10^7 barns (§4.52) compared with 0.03 barn for the D-D reactions and 5 barns for the D-T reaction. It is apparent, therefore, that the Coulomb scattering of electrons, leading to ionization of the target atoms, will occur far more frequently than fusion reactions leading to the release of energy [1, 6].

3.55. For the bombardment of deuterium by 100-kev deuterons, for example, a comparison of the cross sections given above shows that there would be roughly 3.3×10^8 scattering collisions for every one leading to reaction. The fraction of energy lost by a deuteron in each scattering collision with an electron is roughly equal to the ratio of the mass of the electron to that of the deuteron, i.e., about 1/3660. In the 3.3×10^8 scattering collisions between 100-kev deuterons and cold electrons, the energy loss is thus $3.3 \times 10^8 \times 100 \times$

$1/3660 \approx 9 \times 10^6$ kev or 9×10^3 Mev. The average energy produced in the one reactive D-D collision is about 12.5 Mev (§2.23), so that far more energy is used in accelerating the particles than is recovered by the fusion reactions. For a tritium target bombarded by deuterons, the situation is somewhat more favorable, since in a D-T mixture reaction there are roughly 2×10^6 collisions for each one leading to reaction. In this case, however, it still is necessary to expend about 50 Mev to recover 14 Mev of fusion energy.

3.56. In the foregoing calculation, no allowance has been made for the fact that the scattered incident deuteron may still have enough energy to interact with a deuterium (or tritium) nucleus in the target. However, it is known that, as the energy of the deuteron decreases, the scattering cross section increases whereas the reaction cross section decreases. Another factor which has not been taken into account is multiple small-angle scatterings of the deuteron. Although this may not necessarily lead to ionization by ejection of electrons, it will result in considerable loss of energy by the deuteron.

BOMBARDMENT OF A PLASMA TARGET

3.57. A possible modification of the procedure described above, and one which would represent a decided improvement, is to use a plasma from an ordinary gas discharge as the target. In this way the problem of energy losses due to ionization of the target atoms would be largely eliminated. However, an ordinary gas discharge, in which the electron temperatures are of the order of 10 ev, offers little hope of achieving a net gain of energy as a result of fusion reactions. Since the 10-ev electrons are cold in comparison with the hot incoming ions, with energies of 100 kev or more, it appears that in these circumstances, as will be shown below, the transfer of energy from the hot deuteron to the cold electron is much more probable than reaction of the deuteron with a deuterium or tritium nucleus in the gas. Furthermore, as a result of this energy transfer the deuteron energy is reduced to a value at which the D-D or D-T reaction cross section is very small.

3.58. At a particle density of 10^{15} electrons/cm³ and an electron temperature of 10 ev, the time required for 100-kev deuterons to transfer most of their energy to the electrons is about 10^{-6} sec (§4.82). This is very short in comparison with the D-D mean reaction time of approximately 40 sec or even with the 1 sec for the D-T reaction at the given energy and a density of 10^{15} deuterons/ cm³ (§2.38). Thus, most of the energy of the bombarding deuterons would be utilized in raising the temperature of the electrons.

3.59. If the initial temperature of the electrons in the plasma were considerably higher than the 10 ev assumed above, the situation might be very different [7]. Calculations indicate that bombardment of a deuterium plasma in which the electron temperature was 10 kev, and possibly as low as 1 kev, by accelerated tritons having energy in the range of 100 to 300 kev could lead to a considerable net gain of energy. The temperature of the deuterons in the plasma target apparently does not need to exceed a few hundred electron volts. The deuterium plasma would of course have to be confined by a magnetic field, e.g., in a stellarator-type system. Initial preferential heating of the electrons could be readily achieved in a conventional manner by passage of an electrical discharge through the ionized gas (§5.9).

HEATING BY ACCELERATED PARTICLES

3.60. When an accelerated particle strikes a cold target and loses most of its energy in causing ionization, much of the lost energy will shortly appear in the form of heat. For example, the ionized atoms (or nuclei) in the target will recapture the electrons removed in the ionization process and emit energy as radiation. In a solid target, especially, the radiation will be readily absorbed and converted into heat. It may be wondered, therefore, if thermonuclear reactions might not be brought about by utilizing the heat generated locally by the accelerated particles. However, the following calculation, based on the most optimistic assumptions, shows that the temperature attained would be far below the minimum value of 10 kev, at least, required for the production of more energy by thermonuclear fusion than is lost by radiation as bremsstrahlung.

3.61. Suppose a beam of 1-Mev deuterons impinges on a solid target. It will be assumed that the beam current density is of the order of 100 amp/cm^2, which is probably higher than is actually attainable because of space-charge effects. Since 1 amp is 3×10^9 statamp and the charge carried by a deuteron is 4.8×10^{-10} statcoulomb, the particle flux in the beam is $(100 \times 3 \times 10^9)/(4.8 \times 10^{-10}) \approx 6 \times 10^{20}$ deuterons/(cm^2)(sec). The deuteron energy is 1 Mev, i.e., 10^6 ev, and so the energy flux of the beam is 6×10^{26} ev/(cm^2)(sec).

3.62. The range of a 1-Mev deuteron in matter is about 1.5 mg/cm^2. Hence, the mass of deuterium exposed to the accelerated deuterons in an area of 1 cm^2 is 1.5×10^{-3} g. Since 2 g of deuterium contain 6.0×10^{23} (the Avogadro number) atoms, the total number of particles, both nuclei and electrons, in the target volume is given by

$$\text{Number of particles in the target volume} = 2 \times \tfrac{1}{2} \times 6.0 \times 10^{23} \times 1.5 \times 10^{-3}$$
$$= 9 \times 10^{20}.$$

Suppose the time an accelerated deuteron and a target particle spend in the vicinity of each other is 10^{-6} sec; this is probably longer than the actual time, but it will be used here, since it leads to results which are more favorable than can be expected. During this time the energy falling on 1 cm^2 is $6 \times 10^{26} \times 10^{-6} = 6 \times 10^{20}$ ev. If this is shared equally among the target particles, it follows that

$$\text{Energy per target particle} = \frac{6 \times 10^{20}}{9 \times 10^{20}}$$
$$\approx 0.7 \text{ ev}.$$

This energy is of course much too low to satisfy the requirements for obtaining net power from nuclear fusion. No reasonable changes in the conditions postulated above for the purpose of the calculation could lead to an energy per particle in the required kilo-electron volt range [6].

3.63. Since, in the procedure just outlined, the accelerated deuterons are merely used to raise the temperature of the target nuclei and need not take part in any reaction, other accelerated ions could be used for this purpose. With nuclei of higher atomic number, more energy could be transferred to the target particles. Nevertheless, it is not apparent how conditions for useful nuclear fusion could be attained in this manner.

COLLIDING BEAMS OF ACCELERATED PARTICLES

3.64. Another method for utilizing accelerated deuterons which has been proposed from time to time is that two beams of deuterons (or one of deuterons and one of tritons) be directed at each other, thus obtaining reactions as a result of mutual collisions. This suggestion has the merit of apparently solving the problem of increasing the relative collision energy and thus increasing the effective cross section for the reaction. Furthermore, by confining all the reacting nuclei to beams, heat losses, e.g., to the walls of the vessel, are reduced, temporarily at least.

3.65. Like some of the other schemes for utilizing beams of accelerated particles proposed above, this one can also be shown to fail when considered quantitatively. There are two main reasons for this situation: in the first place, the cross section for large-angle scattering of deuterons by deuterons is several thousand times as great as that for the D-D reaction (§4.62). Consequently, most of the accelerated particles would be scattered out of the beam without reacting. The fusion energy released by the relatively few deuterons that did succeed in reacting would be only a small proportion of that utilized to accelerate them.

3.66. The second factor making the method of colliding beams unattractive for the production of useful power is the very low maximum particle density which can be attained and the consequent small power density [1]. Suppose the energy of the deuterons in each beam is 50 kev, so that 100 kev is the relative energy in a collision. The ion current density may be taken as having a value of 100 amp/cm^2, corresponding to 6×10^{20} deuterons/(cm^2)(sec). This is probably much larger than is actually attainable. The mean velocity of a 50-kev deuteron is about 2×10^8 cm/sec, and so the particle density in the accelerated beam is 3×10^{12} deuterons/cm^3. Utilizing the cross section for the D-D reactions at 100 kev, from Fig. 2.3, it is found that the power density in the colliding beams would be about 10^{-4} watt/cm^3. For the D-T reaction, the corresponding power density would be approximately 2×10^{-2} watt/cm^3. These are too small to be of any practical value. Increasing the acceleration energy to 500 kev, so that 1 Mev is available per collision, would increase the power density for the D-D reaction to 7×10^{-4} watt/cm³, whereas that for the D-T reaction would be less than at 100 kev because of the decrease in the cross section (see Fig. 2.3).

REFERENCES FOR CHAPTER 3

1. R. F. Post, *Rev. Mod. Phys.*, **28**, 338 (1956).
2. J. L. Craston, R. Hancox, A. E. Robson, S. Kaufman, H. T. Miles, A. A. Ware, and J. A. Wesson, *Proc. Second U.N. Conf. on Peaceful Uses of Atomic Energy*, **32**, 414 (1958).
3. A. E. Robson and R. Hancox, *Proc. Inst. Elec. Eng. (London)*, **106A,** Suppl. No. 2, 47 (1959).
4. G. P. Harnwell, *Principles of Electricity and Magnetism*, McGraw-Hill Book Co., Inc., 1949, p. 63, footnote.
5. R. F. Post, *Ann. Rev. Nuclear Sci.*, **9**, 367 (1959).
6. J. D. Lawson, UKAEA Report AERE GP/M-185 (1955).
7. W. I. Linlor, Hughes Aircraft Co. Research Report 128 (1959).

Chapter 4

ELEMENTARY PLASMA THEORY

INTRODUCTION

PLASMA PROPERTIES

4.1. It was seen in Chapter 3 that the hopes for confining a plasma consisting almost entirely of hydrogen isotope nuclei and electrons at high temperatures rest mainly in the use of magnetic fields. To understand and overcome the many problems arising in the implementation of the ideas reviewed in the preceding chapter, a detailed knowledge, both theoretical and experimental, is required of the behavior of plasmas in electromagnetic fields [1-5].

4.2. The gross behavior of a plasma is governed by interactions among the charged particles themselves and between the particles and external electromagnetic fields. Strictly speaking these interactions are not independent, since the plasma may profoundly affect the character of an external applied field through collective effects. Nevertheless, by treating the processes separately, it is often possible to derive a reasonable approximation to the actual behavior of a plasma.

4.3. In the present chapter it is proposed only to examine some of the simpler theoretical aspects of the properties and behavior of plasmas such as are necessary for a general understanding of the topics to be considered in later chapters. The discussion here is divided into three main parts: (1) the motion of individual charged particles in electromagnetic fields; (2) the interactions among the particles in a plasma; and (3) certain cooperative (or collective) properties of the plasma as a whole.

THEORETICAL PLASMA MODELS

4.4. The theoretical treatment of a plasma may be undertaken from several points of view, that is, by using different models of the plasma. The model used in any particular circumstance will depend on the property being examined and on the physical conditions assumed for the problem being considered. Two of the most important problems in the study of controlled thermonuclear reactions are confinement of the plasma by magnetic fields and the stability of various plasma-field configurations. It is convenient in these cases to use either of two simplified pictures of the ionized gas. The first is called the *hydromagnetic model,* in which the plasma is treated as a compressible conducting fluid that is subjected to the action of electromagnetic forces. In the second, known as the *single-particle model,* the system behavior is obtained by assuming that each charged particle is acted upon individually by the externally applied field and that collision effects, which would lead to interaction among the particles, are minor.

4.5. The hydromagnetic model is of greatest utility in studying the stability of the plasma in magnetic fields of various configurations and of the closely related problem of hydromagnetic waves in a plasma. The calculations require only a knowledge of the magnetic field configuration and of the density and possibly the conductivity of the plasma, which is treated as an ideal hydrodynamic fluid (see Chapter 13).

4.6. For the investigation of confinement by a magnetic field, both the hydromagnetic and single-particle models may be used to provide results of interest. A simple example of the hydromagnetic method is the treatment in Chapter 3 which led to the derivation of equation (3.15) relating the strength of a uniform magnetic field to the maximum pressure of a confined plasma in a steady-state system. In the present chapter, the single-particle approach will be used to provide information concerning the motion of charged particles in electromagnetic fields since this has a bearing on the problem of confinement.

4.7. The rigorous starting point, from which the two models described above appear as approximations, is the Boltzmann equation for the distribution of particles in phase space as a function of position and velocity (see §13.40). As applied to a plasma, the force term in the ordinary form of the Boltzmann equation is set equal to the electromagnetic force on a charged particle. By taking successive moments of the Boltzmann equation, i.e., upon multiplying by various functions of the velocity and integrating over velocity space, equations of particle conservation, momentum transfer, and energy transfer are obtained. The momentum transfer equation may then be subjected to various approximations in order to obtain the equations of motion for the hydromagnetic and single-particle models of a plasma.

4.8. The treatment of intensive properties of a plasma, e.g., diffusion coefficient, thermal and electrical conductivities, radiation effects, and energy transfer between two particles, ordinarily requires detailed analysis of the collisions in an electron-ion gas from the standpoint of kinetic theory modified to take into account Coulomb forces. Some of these properties will be treated in this chapter (§4.48 *et seq.*) and others will be discussed in Chapter 12. If external electromagnetic fields are imposed on the system, they are usually treated as perturbations on the properties which are determined essentially by collisions.

MOTION OF CHARGED PARTICLES IN ELECTRIC AND MAGNETIC FIELDS

INTRODUCTION

4.9. In principle, the behavior of a plasma in an electromagnetic field can be determined by finding a solution for the motion of each plasma particle in the field produced by all its neighbors combined with that applied externally. This represents, essentially, a self-consistent application of both single-particle and hydromagnetic models of the plasma. Such an approach is generally too complicated for satisfactory analysis, but in the limit of relatively strong magnetic fields and low plasma densities, such as would probably exist in a thermonuclear reactor, useful information can be obtained simply by considering the motion of a single particle in the applied electric and magnetic fields [3, 4, 6-9].

4.10. It has been mentioned earlier that in a magnetic field a charged particle gyrates about the field lines; the center of gyration at any instant is called the *guiding center* of the particle. As the gyrating particle moves along a line of force in a uniform field, it will follow a helical path, but its guiding center will remain on the field line. However, if the magnetic field is not homogeneous, if there is an applied electric field, or if gravitational forces are significant, various drift motions, which can be expressed as motions of the guiding center, will be superimposed on the normal helical path of the charged particle.

4.11. In determining the motion of the guiding center, the first step is to derive a general expression for the force acting on a particle as a result of its motion in electric and magnetic fields. The equation of motion (or force) for a single charged particle of mass m and charge $\pm e$, moving with a velocity v and exposed to the action of an electric field $\mathbf{E}$ and a magnetic field $\mathbf{B}$, can be derived in a manner exactly similar to that employed in §3.15. If the forces acting on the charged particle are equated to its rate of change of momentum, the result, expressed in Gaussian-cgs units, is

$$m \frac{d\mathbf{v}}{dt} = e\left[\mathbf{E} + \frac{1}{c}\left(\mathbf{v} \times \mathbf{B}\right)\right]. \tag{4.1}$$

This equation implies that, if the electric field has a component parallel to the motion of the charged particle, the latter moves with a constant acceleration $e\mathbf{E}/m$, so that its kinetic energy increases continuously. The magnetic field, on the other hand, exerts its force in a direction perpendicular to both the direction of motion and that of the field lines; there is then no force along the direction of motion. Consequently, the magnetic field produces a curvature in the path of the particle, but there is no change in its scalar velocity, and hence its kinetic energy is unaffected by the field. Equation (4.1) has simple solutions in a number of special cases which will now be considered.

MOTION IN A UNIFORM MAGNETIC FIELD

4.12.   For the case in which there is no electric field and $\mathbf{B}$ is constant in space and time, equation (4.1) reduces to

$$m \frac{d\mathbf{v}}{dt} = \frac{e}{c} \, (\mathbf{v} \times \mathbf{B}). \tag{4.2}$$

Consequently, the acceleration $d\mathbf{v}/dt$ is perpendicular to the velocity; this means that the velocity does not change in magnitude but only in direction. The particle thus moves in a circular path with a constant velocity v while being acted upon by a force evB/c.   If this force, due to the magnetic field, is equated numerically to the centrifugal force mv^2/r_g, where r_g is the radius of the circle, i.e.,

$$\frac{evB}{c} = \frac{mv^2}{r_g},$$

it is found that the angular frequency ω_g, which is equal to v/r_g, is given by

$$\omega_g = \frac{eB}{mc}. \tag{4.3}$$

This is called the *gyromagnetic frequency* of the particle.   The radius of the circle of gyration, known as the *gyromagnetic radius*, is

$$r_g = \frac{mvc}{eB}. \tag{4.4}$$

If the particle under consideration is an electron or a hydrogen isotope ion, e is equal to the electronic charge, i.e., 4.80×10^{-10} esu.

4.13.   The gyration frequency derived above is sometimes referred to as the *gyrofrequency*, and also as the *cyclotron frequency* because it is identical with the expression for the angular frequency of gyration of particles in a cyclotron. The adjective "gyromagnetic" is used in this book, however, as it is more completely descriptive of the situation than either of the others.   Unfortunately, the terms Larmor frequency and Larmor radius have become widely, but incorrectly, used in plasma physics for the gyromagnetic frequency and radius, respectively.   The Larmor angular frequency for the precession of an orbital electron in a magnetic field is $eB/2mc$, which is half the value of the frequency given by equation (4.3).

4.14.   If the direction of motion of the particle is not initially perpendicular to the magnetic field, v in equation (4.4) must be replaced by $v_\perp$, the component perpendicular to $\mathbf{B}$, so that in this case

$$r_g = \frac{mv_\perp c}{eB}. \tag{4.5}$$

The component of v parallel to $\mathbf{B}$, i.e., $v_{||}$, will not be affected by the magnetic field. By superposing the motions perpendicular and parallel to field, it is seen that the particle will move with a constant velocity component along a magnetic line of force while gyrating around it at the angular frequency eB/mc. In other words, the orbit described by the charged particle is a helix of constant pitch, its axis being one of the field lines. Consequently, as stated earlier, in a uniform magnetic field the guiding center of the particle moves along a line of force.

4.15. According to equation (4.1), the direction of the force exerted by a magnetic field on a charged particle depends on the sign of the charge. Consequently, the positive ions (nuclei) and negative electrons in a plasma gyrate in opposite directions about the magnetic field lines. The gyromagnetic frequency of the ions in a given magnetic field is less than that of the electrons, because, as equation (4.3) shows, this frequency is inversely proportional to the mass, i.e., ω_i/ω_e is equal to m_e/m_i, where the subscripts e and i represent electrons and ions, respectively. For a deuteron, equation (4.3) gives the gyromagnetic frequency as

$$\omega_g = 4.8 \times 10^3 B \text{ radians/sec,}$$

with B in gauss; the value for an electron is correspondingly greater by the factor of m_i/m_e, which is close to 3660 in this case.

4.16. The ratio of the gyromagnetic radii of ions and electrons depends on the velocities (or kinetic energies) of the particles perpendicular to the field lines. If the perpendicular component $W_\perp$ of the kinetic energy is the same for both ions and electrons, as would be the case if they had the same kinetic temperature, then $\frac{1}{2}mv^2_\perp$ is the same for both kinds of particles. It follows then from equation (4.5) that the ions execute orbits which are greater than the electron orbits by a factor of $(m_i/m_e)^{1/2}$. The product of magnetic field strength and the radius of gyration, a useful quantity experimentally, is given by equation (4.4) as

$$Br_g = \frac{mv_\perp c}{e}$$

$$= \frac{(2mW_\perp)^{1/2}c}{e}, \tag{4.6}$$

which is equivalent to equation (3.21) for a Maxwellian distribution, since $W_\perp$ is then equal to kT. For a deuteron, equation (4.6) becomes

$$Br_g = 6.4 \times 10^3 W_\perp^{1/2} \text{ gauss-cm,}$$

with $W_\perp$ in kilo-electron volts. The value of Br_g for an electron having the same value of $W_\perp$ is obtained upon dividing by $(m_i/m_e)^{1/2}$, which is approximately 60.

4.17. The time required for a particle to make a single turn in its gyration about the magnetic field lines is called the *gyromagnetic period* or *gyration* *time*, τ_g. It is equal to the length of a turn, i.e., $2\pi r_g$, divided by the velocity $v_\perp$ as given by equation (4.5); thus,

$$\tau_g = \frac{2\pi r_g}{v_\perp}$$

$$= \frac{2\pi mc}{eB}. \tag{4.7}$$

If Gaussian-cgs units are used, equation (4.7) gives τ_g in seconds.

MOTION IN ELECTRIC AND MAGNETIC FIELDS

4.18. When an electric field having a component perpendicular to the magnetic field is present, the particle path consists of a helical motion with a superposed drift at constant velocity in a direction perpendicular both to the magnetic field and to the component of the electric field. The actual motion may be regarded as arising from a transverse drift of the guiding center of the particle.

4.19. The mechanism which results in this drift of a charged particle across a magnetic field can be readily understood from a qualitative point of view. Consider a positively charged particle rotating about a line of force in a magnetic field, directed upward and perpendicular to the plane of the page, as

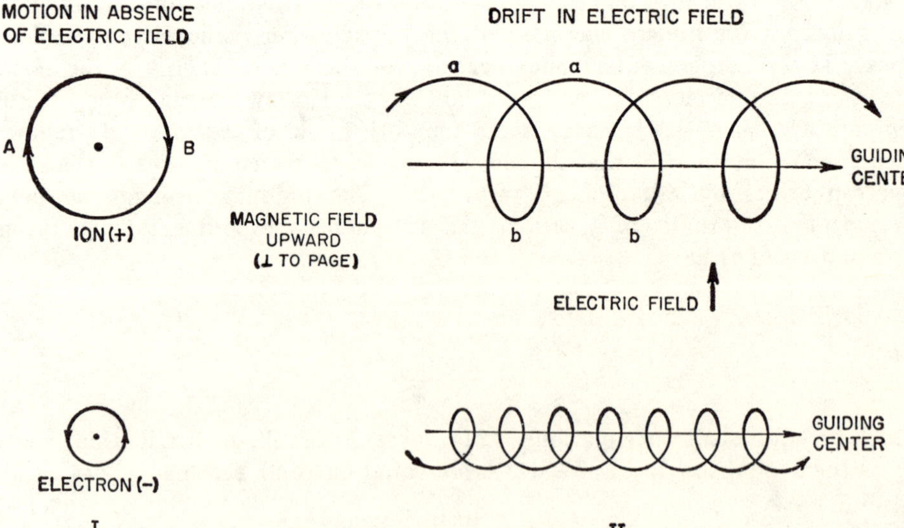

Fig. 4.1. Drift of charged particles in crossed electric and magnetic fields.

indicated in Fig. 4.1, I. If an electric field is applied, in the direction shown, i.e., from the bottom toward the top of the page, as seen in Fig. 4.1, II, so that it is perpendicular to the magnetic field, the positive ion will now be accelerated as it moves in the region A and decelerated at B. According to equation (4.5), the radius of curvature of the path is proportional to the velocity. Consequently the radius is increased at a and decreased at b, in Fig. 4.1, II. The net result is that the ion follows a cycloidal path with the guiding center drifting to the right. For a negatively charged particle, e.g., an electron, the rotation in the same magnetic field is in the opposite direction to that of an ion, but it can be readily seen that the drift will also be to the right, as shown in the figure. Thus, the direction of drift is independent of the charge of the particle. Since the radius r_g of the helix is proportional to $mv_\perp$, by equation (4.5), it will be smaller for an electron than for an ion having the same rotational energy, i.e., perpendicular to the magnetic field direction.

4.20.   In order to calculate the drift velocity of a charged particle, i.e., of its guiding center, in magnetic and electric fields at right angles, it will be supposed that the actual velocity $\mathbf{v}$ of the particle may be expressed by

$$\mathbf{v} = \mathbf{v}_0 + c\,\frac{\mathbf{E} \times \mathbf{B}}{B^2}, \tag{4.8}$$

where the significance of the two terms on the right will be seen below. Suppose that both $\mathbf{E}$ and $\mathbf{B}$ are constant in space and time; substitution of equation (4.8) into (4.1) then leads to

$$m\,\frac{d\mathbf{v}_0}{dt} = e\left[\mathbf{E} + \frac{1}{c}\,(\mathbf{v}_0 \times \mathbf{B}) + \frac{1}{B^2}\,(\mathbf{E} \times \mathbf{B}) \times \mathbf{B}\right]. \tag{4.9}$$

If, further, it is postulated that $\mathbf{E}$ is perpendicular to $\mathbf{B}$, so that $\mathbf{E}\cdot\mathbf{B}$ is zero, it is found that

$$(\mathbf{E} \times \mathbf{B}) \times \mathbf{B} = -B^2\mathbf{E},$$

and insertion of this result into the last term of equation (4.9) gives

$$m\,\frac{d\mathbf{v}_0}{dt} = \frac{e}{c}\,(\mathbf{v}_0 \times \mathbf{B}). \tag{4.10}$$

This expression, which is similar to equation (4.2), shows that $\mathbf{v}_0$, as defined by equation (4.8), is independent of the electric field and consists of a gyration about the lines of force at the frequency ω_g.

4.21.   If the movement of the particle may be regarded as consisting of a helical motion combined with a drift, it is evident that $\mathbf{v}_0$ represents the former. Hence, the drift velocity v_d is given by the last term on the right of equation (4.8) ; thus,

$$v_d = c\,\frac{\mathbf{E} \times \mathbf{B}}{B^2}$$

or, since it has been postulated that $\mathbf{E}$ and $\mathbf{B}$ are perpendicular,

$$v_d = c\,\frac{E}{B}. \tag{4.11}$$

By expressing E in volts/cm, instead of in statvolts/cm, the result is

$$v_d \text{ (cm/sec)} = \frac{10^8 E \text{ (volt/cm)}}{B \text{ (gauss)}}. \tag{4.12}$$

4.22. It is seen that the drift velocity is independent of the mass and initial velocity of the charged particle. As was stated above, the direction of drift is also independent of the sign of the charge. It should be mentioned that equation (4.11) is applicable provided the calculated value of the drift velocity v_d is less than the velocity of light. For a magnetic field of 10^5 gauss, for example, the results would be valid provided the electric field strength is less than the very large value of 3×10^7 volts/cm.

MOTION IN A TIME-VARYING MAGNETIC FIELD

4.23. An important application of equation (4.11) for the drift velocity is in connection with a magnetic field which varies with time. Suppose that a uniform magnetic field, provided by a long solenoid, is increasing with time. In these circumstances, an electric field appears, consisting of circular lines of electric force centered on the axis of the solenoid. The magnitude E of this field, in the laboratory frame of reference, can be determind by means of the electromagnetic induction equation

$$\oint E dl = -\frac{1}{c} \int \left(\frac{dB}{dt}\right) dA,$$

where dl is an element of length around the boundary of the contour of an element of area dA. For the case of a circular contour of radius r centered on the solenoid axis,

$$2\pi r E = -\frac{\pi r^2}{c} \cdot \frac{dB}{dt},$$

so that

$$E = -\frac{r}{2c} \cdot \frac{dB}{dt}. \tag{4.13}$$

4.24. For a plasma of low density (or high temperature), especially in a strong magnetic field, the gyromagnetic frequency is much greater than the frequency of collisions which tend to shift the guiding centers of the particles across the field lines.* The charged particles may then be regarded as being tightly bound to the lines of force. In that event the particle drift velocity in a time-varying magnetic field is represented by dr/dt; hence, by equation (4.11), this may be set equal to cE/B, and it then follows from equation (4.13) that

$$\frac{2}{r} \cdot \frac{dr}{dt} = -\frac{1}{B} \cdot \frac{dB}{dt}.$$

* When collisions are relatively frequent, the particles are able to cross the magnetic field lines and then diffusion can occur; this is discussed in Chapter 12.

Upon integration this leads to the result

$$r^2B = \text{constant.} \tag{4.14}$$

4.25. The conclusion to be drawn from equation (4.14) is that, in the radial motion of any particle with reference to the solenoid axis, in a time-varying magnetic field, the total flux between the moving point and the axis remains constant. In other words, the guiding centers of particles moving at the local drift velocity remain on the surface of some collapsing (or expanding) flux tube of the changing magnetic field. Another way of stating this conclusion is that the charged particles tend to stick to the lines of force.

MOTION IN AN INHOMOGENEOUS MAGNETIC FIELD

4.26. In the presence of a magnetic field with strength varying in space, so that magnetic gradients exist, the motion of a charged particle is somewhat complex. When the gradients are small, the motion can be represented by simple relationships. The case of a magnetic field with a gradient perpendicular to the local direction of the field lines is of particular interest.* The effects on a positive ion and an electron are shown qualitatively in Fig. 4.2; the magnetic field is directed upward as in Fig. 4.1, and the gradient is

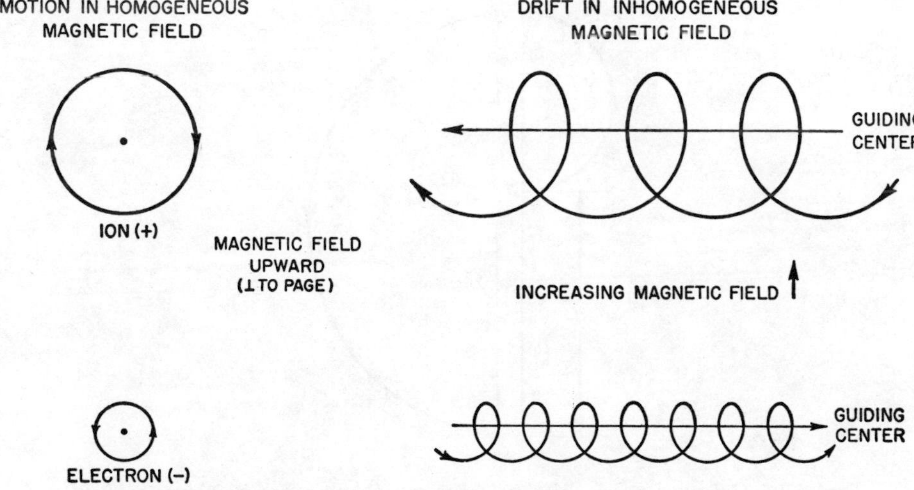

FIG. 4.2. Drift of charged particles in inhomogeneous magnetic field.

in a perpendicular direction, i.e., in the plane of the page. Where the field is stronger, the radius of curvature of the path is smaller than average, in accordance with equation (4.5); on the other hand, where the field is weaker, the radius is greater. The result is that the charged particles follow a cycloidal

---

* When the magnetic field gradient is parallel to the field lines, the situation corresponds to a "magnetic mirror," which will be discussed in Chapter 9.

path, but the guiding centers of the positive and negative particles now drift in opposite directions perpendicular both to the magnetic field and the gradient.

4.27. In a plasma located in an inhomogeneous magnetic field, therefore, the tendency for the ions and the electrons to drift in opposite directions would result in a charge separation. This, in turn, would produce a local electrostatic (space-charge) field of large magnitude. Such a field acting in conjunction with the magnetic field at right angles causes the ions and electrons, i.e., the plasma as a whole, to drift perpendicular to both fields, in the direction of decreasing magnetic field.

4.28. The derivation of the drift velocity of an individual particle in an inhomogeneous magnetic field is complicated [10]; however, the following approximate method, based on a simple physical model, leads to a substantially correct result [11]. Suppose the magnetic field acts in a direction perpendicular to and out of the plane of the page, and that the charged particle moves in this plane. Instead of a gradual variation in the strength of the inhomogeneous magnetic field, the simple case will be considered in which the field strength changes sharply from B_1, at the left of the vertical line in Fig. 4.3, to B_2, at the right of the line. If the particle starts at a and makes half

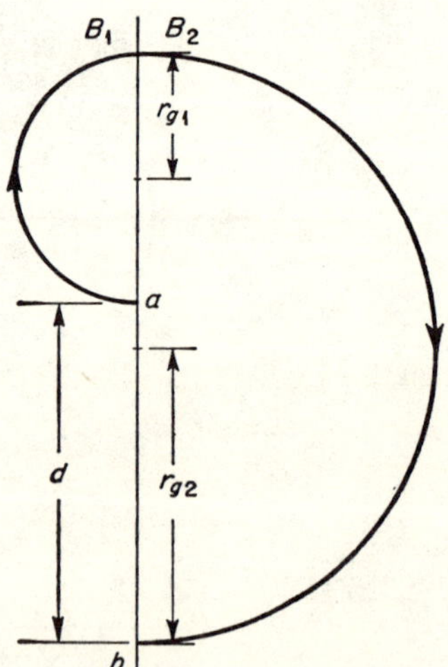

Fig. 4.3. Calculation of drift velocity in inhomogeneous magnetic field.

a gyration in the field B_1 and the other half in B_2, it will finish up at b. Hence, the net distance drifted will be d, which is equal to $2(r_{g2} - r_{g1})$, where r_{g1} and r_{g2} are the respective gyromagnetic radii in the two fields. The time required to traverse this distance is $\frac{1}{2}(\tau_1 + \tau_2)$, where τ_1 and τ_2 are the respective gyromagnetic periods. The drift velocity v_d can then be represented, approximately, by

$$v_d \approx \frac{2(r_{g2} - r_{g1})}{\frac{1}{2}(\tau_1 + \tau_2)}.$$

Upon substituting the appropriate expressions for r_g and τ, using equations (4.5) and (4.7), respectively, it is found that

$$v_d \approx \frac{2v_\perp}{\pi} \cdot \frac{B_1 - B_2}{B_1 + B_2}. \tag{4.15}$$

4.29. The average distance of a semicircular path from the diameter is equal to $\frac{1}{4}\pi$ times the radius. Hence, in the present case, the gradient of the magnitude of the field in the plane perpendicular to the field direction may be represented by

$$\nabla_\perp B \approx \frac{B_1 - B_2}{\frac{1}{4}\pi(r_{g2} + r_{g1})},$$

so that

$$B_1 - B_2 \approx \frac{1}{4}\pi(r_{g2} + r_{g1})\,\nabla_\perp B$$

$$= \frac{\pi}{4} \cdot \frac{mv_\perp c}{e} \frac{B_1 B_2}{B_1 + B_2}\,\nabla_\perp B.$$

Furthermore, if B is the average field strength,

$$B_1 + B_2 = 2B \quad \text{and} \quad B_1 B_2 \approx B^2.$$

Upon making the appropriate substitutions into equation (4.15), it is found that

$$v_d \approx \frac{\frac{1}{2}mv_\perp^2 c}{eB^2}\,\nabla_\perp B = \frac{W_\perp c}{eB^2}\,\nabla_\perp B, \tag{4.16}$$

where $W_\perp$ is the kinetic energy component perpendicular to the field lines. A more rigorous treatment shows that equation (4.16) is valid only if $v_d \ll v_\perp$, i.e., if $\frac{1}{2}r_g(\nabla_\perp B/B) \ll 1$; this condition implies that the fractional change in magnetic field strength across an orbit must be small.

EFFECT OF GRAVITATIONAL AND CENTRIFUGAL FORCES

4.30. When a charged particle in a homogeneous magnetic field is subjected to a gravitational force which has a component $mg_\perp$ perpendicular to **B**, a drift will result. The situation is similar to that arising from an applied electric field **E**, except that the gravitational force $mg_\perp$ replaces the force **E**e. The drift velocity in the gravitational field is then found by substituting $mg_\perp/e$ for E in equation (4.11); hence,

$$v_d = \frac{mg_\perp c}{eB}. \tag{4.17}$$

If equation (4.17) is combined with equation (4.3) for the gyromagnetic frequency ω_g of the particle, the result is

$$v_d = \frac{g_\perp}{\omega_g}. \qquad (4.18)$$

The direction of drift is perpendicular to both the magnetic and gravitational fields. Contrary to the behavior in an electric field, ions and electrons now drift in opposite directions. This is because they gyrate in opposite directions and the effect of the gravitational force is independent of the sign of the charge. Since, for deuterons (§4.15), ω_g is equal to $4.8 \times 10^3 B$, with B in gauss, it is evident that $g_\perp/\omega_g$ is very small for magnetic fields in the kilogauss range, and so the gravitational drift velocity is negligible in situations of interest.

4.31. If a particle is gyrating about a line of force which is curved, a centrifugal acceleration will arise which results in a drift similar to that due to a gravitational force. The drift velocity can in fact be obtained upon replacing $g_\perp$ in equation (4.17) by the centrifugal acceleration. Suppose the particle is executing a helical path, with the guiding center moving along a curved line of force, with a velocity component of $v_\parallel$ along this line. Then, if r is the radius of curvature of the magnetic field, the centrifugal acceleration is $v_\parallel^2/r$ and

$$v_d = \frac{mv_\parallel^2 c}{eBr}$$

$$= \frac{2W_\parallel c}{eBr}, \qquad (4.19)$$

where $W_\parallel$ is the kinetic energy component of the particle in the direction of the field line.

4.32. In contrast to gravitational drifts, which are generally small, the drift velocity due to curvature of the magnetic lines of force may be substantial. However, as in gravitational and inhomogeneous magnetic fields, the ions and electrons tend to drift in opposite directions. This may consequently give rise to charge separation and the development of an appreciable electrostatic (space-charge) field. The result, as mentioned above in connection with inhomogeneous magnetic fields, is that the plasma will move as a whole in the direction of the weaker magnetic field.

PARTICLE DRIFT IN A TOROIDAL MAGNETIC FIELD

4.33. A consequence of fundamental significance in connection with the achievement of controlled thermonuclear reactions can be derived from the results presented above. As stated in §3.14, it has been suggested that energy losses which would normally occur at the ends of a tube containing a plasma confined by a simple, axial magnetic field could be avoided by bringing the ends together to form a closed torus. Unfortunately, this is not a satisfactory solution to the problem because the magnetic field in the torus is both curved and nonuniform and as a result the plasma would drift to the walls.

4.34. Consider a torus containing a plasma in which a magnetic field is produced by passage of current through a solenoid wound around the torus (Fig. 4.4). In the absence of an electric field (or in the presence of an electric field that is constant in time) the Maxwell equation (3.8) for the curl of the magnetic field strength may be written as

$$\mathbf{\nabla} \times \mathbf{B} = \frac{4\pi}{c}\,\mathbf{j}, \tag{4.20}$$

where $\mathbf{j}$ is the current density in the solenoid. Integrating both sides of this equation over the area enclosed by a circle of the type shown by the dotted

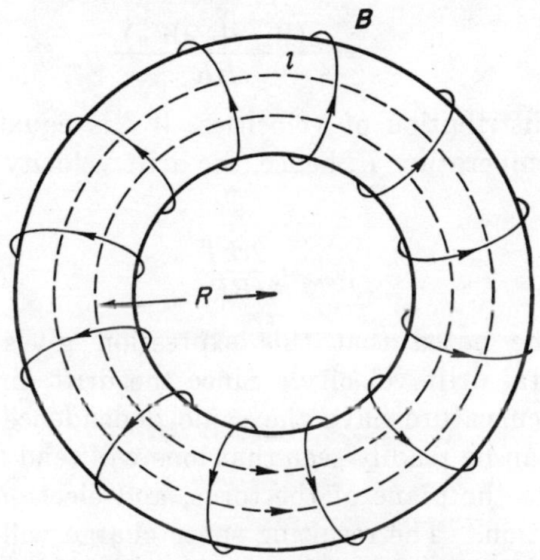

FIG. 4.4. Inhomogeneous axial magnetic field in a torus.

(field) lines in the Fig. 4.4, not necessarily in the median plane of the torus, it follows that, if dA is an element of area,

$$\int (\mathbf{\nabla} \times \mathbf{B})\cdot dA = \frac{4\pi}{c}\int \mathbf{j}\cdot dA = \text{constant},$$

since the integral over $\mathbf{j}\cdot dA$ is constant. The first integral may be written in an alternative form as the integral around the torus of $\mathbf{B}\cdot dl$, where dl is an element of length of the circular path; thus,

$$\int (\mathbf{\nabla} \times \mathbf{B})\cdot dA = \oint \mathbf{B}\cdot dl = \text{constant}.$$

If R is the radius of the path, so that $l = 2\pi R$,

$$\oint \mathbf{B}\cdot dl = 2\pi R B = \text{constant},$$

and hence

$$B = \frac{\text{constant}}{R}. \tag{4.21}$$

The magnetic field strength is thus inhomogeneous, since its value at any point is inversely proportional to the distance R of that point from the major axis of the torus.

4.35. The total drift velocity of a charged particle, perpendicular to the field lines, is the sum of that due to inhomogeneity of the magnetic field and that due to the curvature of the lines of force. The former contribution is given by equation (4.16), with $\nabla_\perp B/B$ equal to $1/R$, since BR is constant, and the latter by equation (4.19), so that

$$|v_d| = \frac{W_\perp c}{eBR} + \frac{2W_{||}c}{eBR}$$

$$= \frac{c(W_\perp + 2W_{||})}{eBR}.$$

For a Maxwellian distribution of velocities, $W_\perp$ is equal to kT and $W_{||}$ to $\frac{1}{2}kT$ at the kinetic temperature T; hence, the drift velocity may be represented by

$$|v_d| = \frac{2ckT}{eBR}. \tag{4.22}$$

4.36. It should be noted that this expression gives only the absolute magnitude of the total drift velocity. Since the drift directions due to both inhomogeneity and curvature have the same dependence on the sign of the charged particle, it can be readily seen that ions will tend to drift in one direction, perpendicular to the plane of the torus, and electrons will tend to drift in the opposite direction. The resulting space charge will produce an electric field which, acting in conjunction with the magnetic field, will cause the plasma as a whole to drift in a direction perpendicular to both fields, as in §4.19. There is no simple relationship between the drift velocity of the plasma as a whole and the individual particle drift velocities, given by equation (4.22). However, since the magnitude of the space-charge electric field must be related to v_d, it would appear that the plasma drift in a toroidal magnetic field can be reduced by increasing the magnetic field strength and the major radius of the torus.

COLLISION PHENOMENA IN A PLASMA

INTRODUCTION

4.37. In the preceding discussion it has been supposed that collisions among the charged particles are rare; that is to say, the collision mean free path is long compared with the dimensions of the confining field, so that the single-particle picture of a plasma is valid. In the present section various aspects of the collision behavior of charged particles will be examined. Estimates will be made of the mean free path for collisions and the exchange of energy accompanying collisions between charged particles. Scattering collisions and energy changes may be regarded as having a perturbing effect on the single-particle behavior in electromagnetic fields.

4.38. The collision between charged particles differs in a highly important respect from that between neutral particles or between a neutral and a charged particle. In the latter cases, there is a fairly definite collision diameter; whenever two particles are within this distance from each other a collision will have occurred. Any approach of the two particles at distances greater than the collision diameter will not result in any interaction. With two charged particles, on the other hand, the situation is very different because the effective range of the Coulomb force, upon which scattering collisions depend, is infinite. In considering charged-particle collisions, it is necessary first to define exactly what type of encounter is to be regarded as a collision. This is generally taken as the interaction which will lead to a deflection (or scattering) through a large angle, namely, 90° or more.

4.39. For simplicity of treatment, the collisions are divided into two categories, although no such distinction actually exists in a plasma. In the first category are the *short-range encounters* (or close collisions) which lead to a scattering angle of 90° or more in a single interaction between a pair of charged particles. The second type is that of *long-range encounters* (or distant collisions); these represent the multiple interactions of a single particle with many other particles such that the net effect is to give a large-angle scattering, i.e., about 90°. In principle, these long-range encounters can extend over the whole distance over which the Coulomb forces are effective, i.e., the whole of the plasma. However, in order to make possible the calculation of the cross section (or equivalent mean free path) for distant collisions of the type just defined, it is necessary to choose a characteristic distance, called the Debye shielding length, within which interaction of a given charged particle with other charged particles may be supposed to occur. Beyond this distance, the plasma may be regarded as being electrically neutral, both macroscopically and microscopically, so that the particle under consideration is not affected by Coulomb forces [2, 7, 8, 12].

ELECTRICAL NEUTRALITY OF A PLASMA

4.40. A basic property of a plasma, which is a consequence of long-range collective interactions among the charged particles, is the tendency toward electrical neutrality. If, over a relatively large volume of the plasma, the density of electrons should differ appreciably from the positive ion density, large electrostatic forces will come into play. As a result, the charged particles will move rapidly in such a manner as to approach a condition of charge equality.

4.41. Some indication of the order of magnitude of the electrostatic fields that would result from a departure from electrical neutrality over an appreciable volume may be obtained by considering a hydrogen isotope plasma in which the density of both ions and electrons is 10^{15} particles/cm³. Suppose that in some manner all the electrons present in a sphere of plasma of r cm radius were suddenly removed; the strength of the resulting electrostatic field E, as derived from Gauss's law, would be

$$E = \frac{Q}{r^2},\tag{4.23}$$

where Q is the value of the charge removed. If n_e is the electron number density of the plasma, then

$$Q = \tfrac{4}{3}\pi r^3 n_e e,$$

when e is the electronic charge. Upon substituting this result into equation (4.23), it follows that

$$E = \tfrac{4}{3}\pi r n_e e.$$

4.42. If the radius of the sphere is taken as 1 cm, then since n_e is 10^{15} electrons/cm³ and e is 4.80×10^{-10} statcoulomb, it follows that

$$E = \tfrac{4}{3}\pi \times 10^{15} \times 4.80 \times 10^{-10}$$
$$= 2 \times 10^6 \text{ statvolts/cm}$$
$$= 6 \times 10^8 \text{ volts/cm},$$

so that the field strength has an enormous value. A departure of only 1 part in a million from charge equality would give rise to a field of 600 volts/cm near a sphere of radius 1 cm. Since E increases in proportion to r, at a given plasma density, the field strength would increase with the radius of the sphere of plasma [7].

THE SHIELDING DISTANCE

4.43. Although on a macroscopic scale the distribution of positive and negative charges in a plasma must be the same, there are, in a sense, microscopic deviations from neutrality. Consider, for example, a small volume element in the vicinity of a positive ion. As a result of the thermal motion of the charged particles there will sometimes be an excess of positive charges and sometimes an excess of negative charges in this volume element. However, if a time average is taken, it will be found, as a consequence of the electrostatic field, that the negative charge density due to electrons will exceed the positive charge density of the ions. The reverse situation will of course exist in a volume element near to an electron. The difference between the positive and negative charge densities will obviously be greater in the immediate vicinity of any given charged particle and it will fall off with increasing distance.

4.44. The foregoing arguments lead to the conclusion that every charged particle may be regarded as being surrounded by an "atmosphere" having a net charge of opposite sign. The uniform distribution of these microscopic atmospheres throughout the plasma then leads to the macroscopic neutrality described above. The effective radius of the oppositely charged atmosphere surrounding an ion can be obtained by a procedure analogous to that employed in the study of the solutions of electrolytes [13, 14]. By postulating that the charged particles in a field of varying potential energy have a Boltzmann distribution and assuming that the gradient of the electrical potential can be expressed, in the usual manner, by Poisson's equation, based on Coulomb's law, it can be shown that the radius λ_d of the atmosphere surrounding a positive ion is

$$\lambda_d = \left(\frac{kT}{4\pi n_e e^2}\right)^{\frac{1}{2}}, \tag{4.24}$$

where T is the kinetic temperature of both electrons and positive ions. If T is expressed in terms of kilo-electron volts (§2.13), this equation becomes

$$\lambda_d = 2.35 \times 10^4 \left(\frac{T}{n_e}\right)^{\frac{1}{2}} \text{ cm.} \tag{4.25}$$

4.45. The length λ_d, defined by equations (4.24) and (4.25), is called the *shielding* (or *screening*) *distance*.* It is a measure of the distance from an ion beyond which the atmosphere, with a net negative charge, screens off the Coulomb field of that ion from the field of another ion moving nearby. It is for this reason that the term "shielding distance" is used. The dependence of λ_d on the electron density, as derived from equation (4.25), is represented in Fig. 4.5 for the temperatures 1, 10, and 100 kev. It is evident that, for conditions of thermonuclear interest, e.g., an electron density of 10^{15} particles/cm³ and a temperature of 100 kev, the shielding length is 7.5×10^{-3} cm. This is large compared with the distance between the ions and electrons at densities of interest, so that the number of electrons (and ions) included in the spherical volume having a radius equal to the shielding length, i.e., the shielding volume, is considerable. In the case under consideration, for example, it is $\frac{4}{3}\pi \times (7.5 \times 10^{-3})^3 \times 10^{15} = 1.8 \times 10^9$ particles of each sign, making a total of 3.6×10^9. The values for other conditions are shown in Fig. 4.6 [7].

4.46. Since the number of charged particles in the shielding volume is quite large, a particle can interact with many others in traversing a distance equal to the shielding distance. For this reason, as will be shown below, the effect of so-called long-range interactions is more important than that of short-range encounters in producing large-angle scattering of any given charged particle as it passes through the plasma. In treating the long-range encounters, the combined effect of all the particles within the shielding volume is regarded as a collective or statistical interaction.

4.47. Although it has no direct connection with the problem of collisions in a plasma, it may be mentioned here that the shielding distance is roughly equal

---

* It is often referred to as the Debye shielding distance (or length) since its derivation is based on the Debye-Hückel treatment of electrolytes.

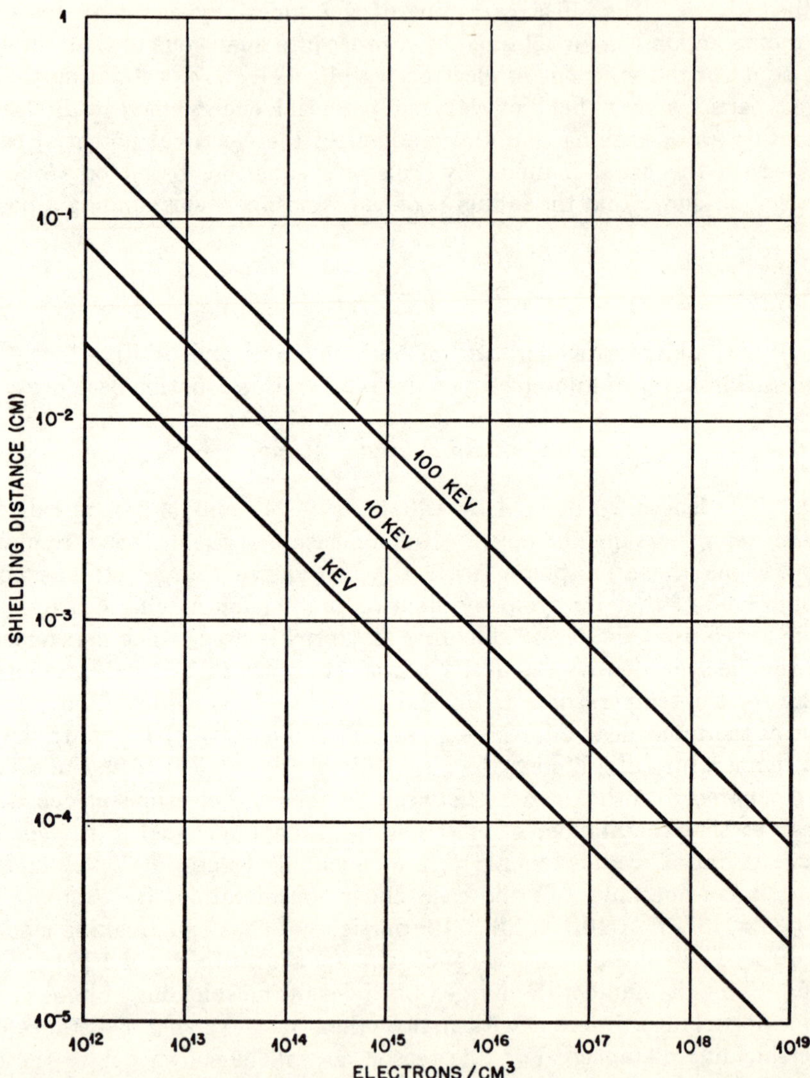

FIG. 4.5. Shielding distance at kinetic temperatures of 1, 10, and 100 kev.

to the thickness of the sheath which develops when a plasma is in contact with a solid surface (cf. §6.85). Since the electrons move faster than the ions, there will normally be a tendency for electrons to leave any given region of the plasma more rapidly than do the ions. At the floating potential of a solid surface in contact with a plasma, there is no net current flow to or from the surface, and so the rates at which positive and negative charges reach the

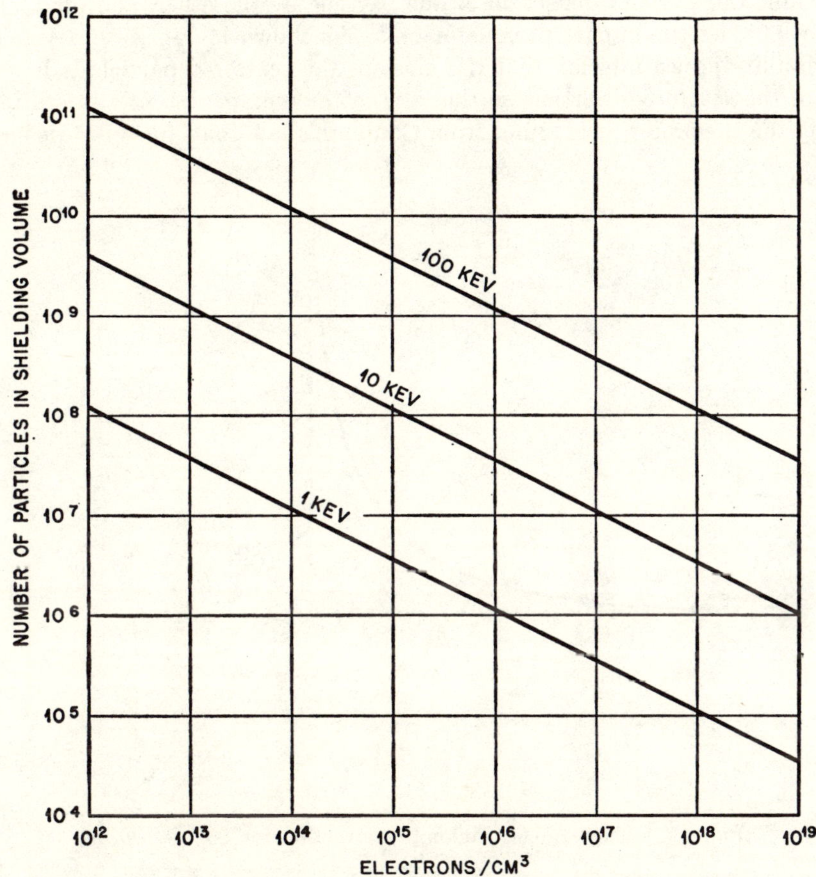

FIG. 4.6. Number of charged particles in shielding volume.

surface must be equal. This can be achieved only if there is a sheath of plasma near the solid surface where the number of electrons exceeds that of the ions. Within this sheath, which is somewhat analogous to the oppositely charged atmosphere surrounding a charged particle, electrical neutrality is not preserved.

SHORT-RANGE INTERACTIONS

4.48. An approximate value of the large-angle, single-collision cross section for short-range interaction (or close encounters) between charged particles may be obtained by a simple, classical treatment based on Coulomb's law. In the absence of any electrostatic forces, the distance of closest approach between two particles is called the *impact parameter*. The magnitude of this distance will determine the angle of deflection of one particle by the other, and for a deflection of 90°, let the impact parameter be b_0, as shown in Fig. 4.7. By making the simplifying assumption that the mass of the scattered particle is less than that of the scattering particle so that the latter remains essentially stationary during the encounter, it is found from Coulomb's law that, for a 90° deflection,

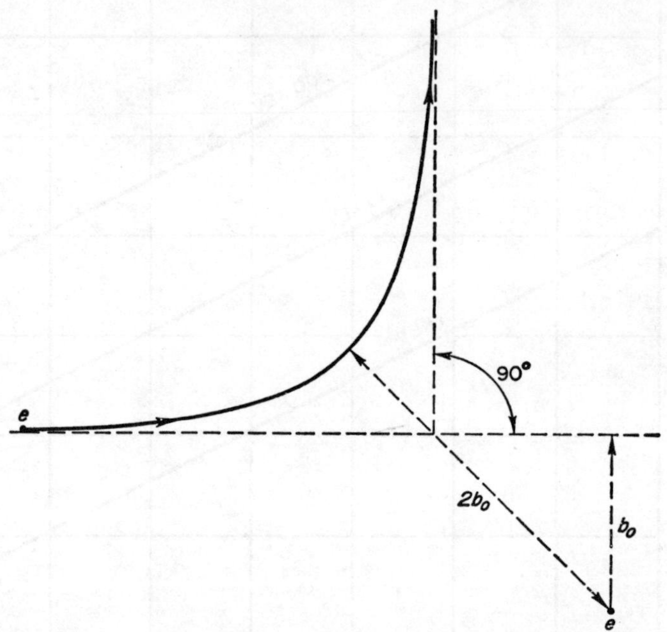

Fig. 4.7.   Short-range Coulomb interaction for 90° deflection.

the particles are a distance $2b_0$ apart at the point of closest approach. At this point the mutual potential Coulomb energy is equal to the center-of-mass (or relative) kinetic energy W of the interacting particles. In the case of a hydrogen isotope plasma, all the particles carry the unit charge e, and the mutual potential energy at the point of closest approach is $e^2/2b_0$, and so

$$W = \frac{e^2}{2b_0}$$

or

$$b_0 = \frac{e^2}{2W}. \tag{4.26}$$

4.49.   Such particles as approached each other with an impact parameter of b_0 or less will be scattered through large angles, i.e., 90° or more, in a single encounter. The cross section σ_c for close collisions of this type may be taken as roughly equal to the area of a disc with a radius b_0; hence,

$$\sigma_c \approx \pi b_0^2 = \frac{\pi e^4}{4W^2}, \tag{4.27}$$

so that σ_c is inversely proportional to the square of the relative kinetic energy. The cross section will be given in square centimeters if e is expressed in statcoulombs and W in ergs. If W is in kilo-electron volts, the equation for the classical cross section for close collisions becomes

$$\sigma_c \approx \frac{1.6 \times 10^{-20}}{W^2} \text{ cm}^2 = \frac{1.6 \times 10^4}{W^2} \text{ barns}. \tag{4.28}$$

4.50. An interesting application of equation (4.28) is that in which accelerated deuterons bombard a cold target, since this is the basis of some methods that have been proposed for obtaining nuclear fusion energy (§3.53 *et seq.*). The short-range Coulomb interactions between hot deuterons and either cold deuterons or electrons of the target is then important. The cross sections for these processes can be derived from equation (4.28).

4.51. In general, for a collision between two particles of masses m_1 and m_2, the relative kinetic energy W is given by (cf. §2.4)

$$W = \frac{m_1 m_2}{2(m_1 + m_2)} v^2, \tag{4.29}$$

where v is the relative velocity. If particle 2 is essentially at rest, e.g., in a cold target, v is equal to v_1, the velocity of the hot (or accelerated) particle. For a deuteron-deuteron interaction, equation (4.29) reduces to

$$W = \tfrac{1}{4} m_D v_D^2 = \tfrac{1}{2} W_D,$$

where W_D is the kinetic energy of the deuteron. Hence, equation (4.28) for the close-collision cross section in this case is

$$\sigma_c \approx \frac{6.4 \times 10^4}{W_D^2} \text{ barns}. \tag{4.30}$$

Thus, at an energy W_D of 100 kev, the value of σ_c is roughly 6.4 barns, i.e., about 200 times the total cross section for the D-D reactions at the same energy. For the bombardment of cold tritium by 100-kev deuterons, the close-collision cross section would be about 4.5 barns, which is not very different from that of the D-T fusion reaction.

4.52. For the collision of an accelerated ion with a cold electron in a target, $m_1 \gg m_2$, and equation (4.29) becomes

$$W \approx \tfrac{1}{2} m_e v_i^2$$

$$= \tfrac{1}{2} m_i v_i^2 \left(\frac{m_e}{m_i} \right) = W_i \frac{m_e}{m_i},$$

where m_i and m_e are the masses of the ion and electron, respectively, and W_i

is the energy of the accelerated ion. If this result is inserted into equation (4.28), it is seen that

$$\sigma_c \approx \frac{\pi e^4}{4W_i^2}\left(\frac{m_i}{m_e}\right)^2 = \frac{1.6 \times 10^4}{W_i^2}\left(\frac{m_i}{m_e}\right)^2 \text{ barns,} \tag{4.31}$$

with the ion energy W_i in kilo-electron volts. For deuterons of 100-kev energy, W_i is 100 and m_i/m_e is roughly 3660, so that the cross section for the short-range, large-angle ($> 90°$) scattering of cold electrons exceeds 10^7 barns. This is the value which was used in §3.54.

LONG-RANGE INTERACTIONS

4.53. The collision cross section derived above applies to a single Coulomb interaction that leads to scattering through an angle of 90° or more. As indicated in §4.39, however, consideration must also be given to the long-range, large-angle scattering (or distant collisions). These are the result of numerous small-angle Coulomb interactions, occurring at distances greater than the distance of closest approach defined by equation (4.26), that give a resultant large-angle scattering of about 90°. The effective cross section for this kind of scattering may be calculated in the following manner.

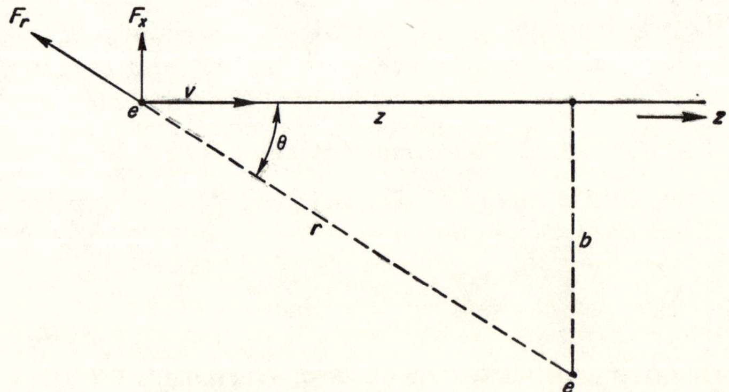

Fig. 4.8. Calculation of momentum transfer between charged particles.

4.54. Consider a (scattered) particle, with a charge e, moving in the z direction past another (scattering) particle having the same charge. Let r be the distance between the particles at any instant and b their impact parameter. The Coulomb force between the particles at the given instant is F_r and its component F_x at right angles to the z direction is equal to $F_r \sin \theta$, as may be seen from Fig. 4.8. The change of momentum Δp_x of the scattered particle in the x direction is then given by

$$\Delta p_x = \int_0^\infty F_x \, dt = \int_0^\infty F_r \sin \theta \, dt,$$

the integration being taken over all time. By Coulomb's law, $F_r = e^2/r^2$, and since $b/r = \sin \theta$, it follows that

$$\Delta p_x = \frac{e^2}{b^2} \int_0^\infty \sin^2 \theta \sin \theta \, dt. \qquad (4.32)$$

If v is the velocity of the scattered particle at the point under consideration, then $dz/dt = v$, so that $dt = dz/v$, and since $b/z = - \tan \theta$, it is readily found that

$$dt = \frac{b \operatorname{cosec}^2 \theta \, d\theta}{v}.$$

Upon making this substitution into equation (4.32) and changing the limits, the result is

$$\Delta p_x = \frac{e^2}{bv} \int_0^\pi \sin \theta \, d\theta = \frac{2e^2}{bv}. \qquad (4.33)$$

This gives the change in momentum in the x direction resulting from a single scattering; the change is the same in the y direction.*

4.55. In the course of its motion the scattered particle experiences many deflections, most of them being small. The total momentum change Δp_m in a particular direction, e.g., along the x axis, is then given by

$$\Delta p_x = (\Delta p_x)_1 + (\Delta p_x)_2 + \cdots + (\Delta p_x)_j + \cdots + (\Delta p_x)_N, \qquad (4.34)$$

where N is the total number of collisions made by the scattered particle with N scattering (or field) particles.† Since the successive small-angle collisions are assumed to be random in nature, it is impossible to predict the value of Δp_x. However, if many scattered particles are considered, each having the same initial velocity and direction of motion and making the same number of collisions, an average value, i.e., $\overline{\Delta p_x}$, can be determined. If the distribution of particle velocities in the plasma is isotropic, then $\overline{\Delta p_x}$ must vanish, from symmetry considerations, when all directions of motion are taken into account. However, $|\overline{\Delta p_x}|$ is not necessarily zero, and information concerning the change in momentum is best obtained by considering the mean square $\overline{(\Delta p_x)^2}$.

4.56. From equation (4.34), it is seen that

$$(\Delta p_x)^2 = (\Delta p_x)_1^2 + (\Delta p_x)_2^2 + \cdots + (\Delta p_x)_j^2 + \cdots + (\Delta p_x)_N^2$$
$$+ (\Delta p_x)_1(\Delta p_x)_2 + \cdots + (\Delta p_x)_j(\Delta p_x)_k + \cdots \qquad (4.35)$$

If $(\Delta p_x)_j$ in each collision is small, the average will be the same for all collisions; furthermore, taking into consideration the random nature of the deflections,

---

* It is of interest to note that the momentum change is equivalent to that resulting from the Coulomb force e^2/b^2, at the distance of the impact parameter b, operating over a path length of $2b$, i.e., for a time $2b/v$.

† The term "field particle" has been proposed because of its analogy with "field star," as used in astrophysics [15]. The scattered particle is sometimes called the "test particle."

the sum of the cross-product terms, such as $(\Delta p_x)_j(\Delta p_x)_k$, will vanish when averaged over all the particles. The same situation will apply, of course, to momentum changes along the y axis, and so equation (4.35) leads to the general result

$$\overline{(\Delta p)^2} = N\overline{(\Delta p)_j^2},$$

and upon differentiating it follows that

$$d\,\overline{(\Delta p)^2} = \overline{(\Delta p)_j^2}\,dN, \tag{4.36}$$

where $(\Delta p)_j^2$ may be taken as being equal to the square of Δp_x as given by equation (4.33).

4.57. If n is the number density of the scattering (or field) particles, the number dN of such particles contained in a cylindrical shell of length λ, radius b, and thickness db is

$$dN = 2\pi n\lambda b\,db.$$

Hence, equation (4.36) can be written as

$$d\,(\Delta p)^2 = \left(\frac{2e^2}{bv}\right)^2 2\pi n\lambda b\,db$$

$$= \frac{8\pi e^4}{v^2}\,n\lambda\,\frac{db}{b}. \tag{4.37}$$

If this is integrated over all values of the impact parameter, from a minimum of $b_{\min}$ to a maximum of $b_{\max}$, the result is

$$\overline{(\Delta p)^2} = \frac{8\pi e^4}{v^2}\,n\lambda \ln \Lambda, \tag{4.38}$$

where Λ is defined by

$$\Lambda \equiv \frac{b_{\max}}{b_{\min}}. \tag{4.39}$$

4.58. Provided the mass of the scattered particle is about equal to or less than that of the scattering particle, e.g., in the scattering of electrons by electrons or by ions and of ions by ions, it can be assumed that when $\overline{(\Delta p)^2}$ has increased to the point at which it has roughly the same magnitude as p^2, where p is the initial momentum of the scattered particle, then the particle has been scattered through a large angle, e.g., about 90°.* The net distance λ traveled by the particle is formally equivalent to the mean free path for this scattering process, but it is also called the *relaxation length*, i.e., the distance over which the cumulative effect of many small deflections in long-range encounters is to produce a scattering angle of about 90° (see §4.39).

---

*This cannot apply to the scattering of ions by electrons because the ion, being considerably heavier than the electron, can lose much of its energy in encounters with electrons without suffering appreciable deflection.

It is then possible to define a cross section σ_d for the large-angle, distant scattering by

$$\lambda = \frac{1}{n\sigma_d}. \tag{4.40}$$

4.59. If m is the mass of the scattered particle, then $p = mv$, so that, when $\overline{(\Delta p)^2}$ is approximately equal to $(mv)^2$, where the scattered particle velocity v is now an average value, the length λ may be replaced by $1/n\sigma_d$. Hence, from equation (4.38), after replacing $n\lambda$ by $1/\sigma_d$, in accordance with equation (4.40),

$$(mv)^2 \approx \frac{1}{\sigma_d} \cdot \frac{8\pi e^4}{v^2} \ln \Lambda$$

or

$$\sigma_d \approx \frac{8\pi e^4}{m^2 v^4} \ln \Lambda. \tag{4.41}$$

The kinetic energy W^* of the scattered particle is $\frac{1}{2}mv^2$, and so equation (4.41) reduces to

$$\sigma_d \approx \frac{2\pi e^4}{W^2} \ln \Lambda. \tag{4.42}$$

Upon comparison of this expression with equation (4.27), it is seen that, for equivalent energies, the cross section σ_c for close encounters leading to large-angle scattering is smaller than σ_d, the cross section for distant interactions, by a factor of $8 \ln \Lambda$.

4.60. In assigning values to $b_{\min}$ and $b_{\max}$, the ratio of which determines Λ, as defined by equation (4.39), it appears from the introductory discussion in §4.39 that $b_{\min}$ is equal to the impact parameter b_0 for a scattering of 90° in a single encounter, as defined by equation (4.26), and $b_{\max}$ is equal to the shielding distance λ_d, expressed by equation (4.24), beyond which the net effect of Coulomb forces becomes negligible. Hence, in accordance with equation (4.39),

$$\Lambda \approx \frac{\lambda_d}{b_0} = \left(\frac{kT}{4\pi n_e e^2}\right)^{\frac{1}{2}} \frac{2W}{e^2}.$$

If the energy distribution among the particles in the plasma is Maxwellian, W may be replaced by $\frac{3}{2} kT$, so that

$$\Lambda \approx \frac{3}{2e^3} \left(\frac{k^3 T^3}{\pi n_e}\right)^{\frac{1}{2}} \tag{4.43}$$

$$= 4.9 \times 10^{14} \frac{T^{\frac{3}{2}}}{n_e^{\frac{1}{2}}},$$

with T expressed in kilo-electron volts and n_e in electrons/cm^3.

---

* It should be noted that W *is* here the actual energy of the scattered particle, and not the relative energy as in equation (4.27).

4.61. Some values for ln Λ for plasma temperatures and electron densities in the range of thermonuclear interest, as calculated from equation (4.43), are given in Table 4.1.* It is seen that, under the probable conditions existing in a thermonuclear reactor, ln Λ would be in the vicinity of 20. Upon inserting this number into equation (4.41) and expressing W in kilo-electron volts, the result is

$$\sigma_d \approx \frac{2.6 \times 10^{-18}}{W^2} \text{ cm}^2 = \frac{2.6 \times 10^6}{W^2} \text{ barns.} \tag{4.44}$$

TABLE 4.1. APPROXIMATE VALUES OF ln Λ

T (kev)	Electron Density n_e (particles/cm³)		
	10^{12}	10^{14}	10^{16}
0.1	16.5	14.2	11.9
1	20.0	17.7	15.4
10	23.4	21.1	18.8
100	26.9	24.6	22.3

4.62. Although it is obvious from equations (4.28) and (4.44) that the cross section σ_d for large-angle scattering as a result of distant collisions must exceed σ_c for close collisions, the exact ratio depends upon the actual nature of the particles involved, since W in equation (4.28) is the relative energy of the interacting particles, whereas in equation (4.44) it is the energy of the scattered particles. A situation of interest, as a possible means of achieving nuclear fusion (§3.63), is that involving the head-on collision of two beams of accelerated deuterons. The relative energy is then twice the energy of the scattered deuterons, so that σ_d is about 640 times σ_c. Since at 100-kev energy, for example, σ_c is more than 200 times as large as the total D-D reaction cross section (§4.51), it is apparent that σ_d will exceed the latter by a considerable factor. It is obvious, therefore, that when two beams of accelerated deuterons collide, Coulomb scattering will be far more probable than combination.

4.63. The expressions derived above are not exact, because of various approximations, including the neglect of quantum mechanical factors. Nevertheless, the results may be expected to hold, approximately, for encounters between particles of equal mass or where the scattered particle is lighter than the scattering particles. Thus, in a plasma produced from hydrogen isotopes, σ_d as derived from equation (4.41) gives a rough estimate of the effects of distant ion-ion, electron-electron, and electron-ion encounters. For the scatter-

* Because of quantum mechanical effects, these values must be corrected for temperatures in excess of about 40 ev, especially for electron-ion collisions. The changes are, however, not significant for the present purpose. [16].

ing of ions by electrons (cf. §4.58, footnote), i.e., ion-electron encounters, however, a different approach must be used, as described below.

RELAXATION TIMES

4.64. The concept of collisional *relaxation times* cannot be very precisely defined;* it is nevertheless useful in the study of distant collisions for a known energy distribution. In general, the relaxation time is the time required for collisions to produce a large-angle scattering or a considerable change in the momentum (or energy) of a particle interacting with a background of other particles. Thus, it may be regarded as a measure of the rate of attempted approach of a nonequilibrium plasma distribution to an equilibrium state.

4.65. Three different cases are of special interest. The first is concerned with the time for a particular energetic ion or electron to be scattered in direction (or changed in energy) by encounters with field particles of the same kind. The second pertains to the effect of collisions of an electron with the ions as field particles, and the third applies to the reverse of this process, that is, the interaction of an ion with the background (or field) electrons. If λ is the relaxation length for a given interaction and v is the velocity of the scattered particle relative to the field particles, the average time elapsing before interaction occurs is λ/v or, since λ is equal to $1/n\sigma$,

$$t = \frac{1}{n\sigma v}, \tag{4.45}$$

where n is the number density of the field particles. Hence, the cross section σ_d for distant encounters derived above may be used to estimate relaxation times.

4.66. Although the individual field particles may have an arbitrary velocity distribution, the center of mass of a group of field particles is stationary. Hence, in determining the average value of the relaxation time, by means of the appropriate form of equation (4.45), v may be taken as the actual, rather than the relative, velocity of the scattered particle.

4.67. Using the subscripts e and i to indicate electrons and ions, respectively, it follows that t_{ii}, the mean collision time for an ion scattered by distant encounters with other ions, is thus

$$t_{ii} = \frac{1}{n_i \sigma_d v_i}, \tag{4.46}$$

where v_i is the average ion velocity. Similarly, for the scattering of electrons by electrons,

$$t_{ee} = \frac{1}{n_e \sigma_d v_e}, \tag{4.47}$$

---

* Relaxation times have long been used in astrophysics in the theoretical study of stellar encounters [2].

where v_e is the average velocity. Since the mass of a deuteron is about 3660 times that of an electron, the velocity of the latter is about $(3660)^{\frac{1}{2}}$, i.e., roughly 60 times that of the former, i.e., $v_e \approx 60v_i$, for equal particle energies. It follows, therefore, that in these circumstances $t_{ii} \approx 60t_{ee}$; hence, equilibrium distribution among the electrons is approached much more rapidly than for ions of the same energy.

4.68. By inserting the expression for σ_d, as given by equation (4.41), into equation (4.46), it is seen that, for singly charged ions,

$$t_{ii} \approx \frac{m_i^2 v_i^3}{8\pi e^4 n_i} \cdot \frac{1}{\ln \Lambda}, \tag{4.48}$$

or, since the ion energy W_i is equal to $\frac{1}{2} m_i v_i^2$, this result may be written as

$$t_{ii} \approx \frac{(2m_i)^{\frac{1}{2}} W_i^{\frac{3}{2}}}{4\pi e^4 n_i} \cdot \frac{1}{\ln \Lambda}. \tag{4.49}$$

A similar expression gives the electron-electron relaxation time; thus,

$$t_{ee} \approx \frac{(2m_e)^{\frac{1}{2}} W_e^{\frac{3}{2}}}{4\pi e^4 n_e} \cdot \frac{1}{\ln \Lambda}. \tag{4.50}$$

4.69. If the ions are deuterons and the conditions are such that $\ln \Lambda$ may be taken as approximately 20, equation (4.49), for deuteron-deuteron scattering, reduces to

$$t_{\mathrm{DD}} \approx 1.8 \times 10^{10} \frac{W_{\mathrm{D}}^{\frac{3}{2}}}{n_{\mathrm{D}}} \text{ sec,} \tag{4.51}$$

where W_{D} is the energy of the scattered deuteron in kilo-electron volts. The corresponding expression for the electron-electron relaxation time is

$$t_{ee} \approx 3.0 \times 10^8 \frac{W_e^{\frac{3}{2}}}{n_e} \text{ sec.} \tag{4.52}$$

The relaxation times thus vary directly as $W^{\frac{3}{2}}$ and inversely as n_e, so that they become longer at high energies and lower electron densities. It should be understood that the expressions given above provide estimates only of the relaxation times, since they may differ by factors of 2 or more from those derived by more exact methods [17].

4.70. From equation (4.51), it follows that, for a deuterium plasma in which the density is 10^{15} deuterons/cm³ and the kinetic energy W_{D} is 150 kev, the ion-ion relaxation time is about 0.018 sec. As seen in §2.38, the mean reaction time at a kinetic temperature of 100 kev, which corresponds to a mean particle energy of $\frac{3}{2} kT$, i.e., 150 kev, for the given deuteron density is about 40 sec. The latter is greater than the corresponding deuteron-deuteron relaxation time by a factor of over 2000, so that a deuteron must, on the average, suffer a considerable number of large-angle, long-range scattering collisions with other deuterons before it combines with another deuteron. This is, of course, another way of stating the conclusion reached earlier that the cross section for large-angle, distant D-D encounters greatly exceeds that for combination. A similar situation, although the discrepancy is not quite so large, arises in the D-T system.

4.71. Since recoil effects are negligible, the relaxation time for the scattering of an electron as a result of distant collisions with ions, i.e., t_{ei}, is given by

$$t_{ei} = \frac{1}{n_i \sigma_d v_e}, \tag{4.53}$$

so that, utilizing the appropriate form of equation (4.41) for σ_d, the result is

$$t_{ei} \approx \frac{(2m_e)^{1/2} W_e^{3/2}}{4\pi e^4 n_i} \cdot \frac{1}{\ln \Lambda}.$$

However, in a hydrogen isotope plasma n_e and n_i are equal, so that n_i in the denominator may be replaced by n_e. The result would then be the same as for electron-electron relaxation time t_{ee}. It follows that, to a good approximation, t_{ei} is equal to t_{ee} and thus can be represented by equation (4.50).

COLLISIONS OF IONS WITH ELECTRONS

4.72. In discussing the interaction of ions with the background (or field) electrons, two limiting situations may be considered: (a) the ion energy is much less than the mean electron energy and (b) the ion energy is greater than the mean electron energy. In the former case, the ion will gain energy by encounters with the more rapidly moving electrons at a rate determined essentially by the frequency of ion-electron collisions and the mean energy transfer per collision. On the other hand, in the latter situation the ion will lose energy to the electrons, although their mean velocity may be greater than that of the ion.

4.73. Although equation (4.42) for σ_d does not apply to the present case, the rate of energy transfer from electrons to the ion, when the energy of the latter is the smaller, may be estimated in a manner similar to that used to determine σ_d. Since the ions move very slowly compared to the average electron velocity, the problem is similar to that of the classical treatment of Brownian motion in which the suspended particles suffer impacts from the much lighter molecules of the medium. As each fast electron passes in the vicinity of the ion, the latter receives an increase in momentum which is numerically equal to the change in momentum of the electron, as given by equation (4.33). On the average, therefore,

$$\Delta p_i = \frac{2e^2}{bv_e} \tag{4.54}$$

is the momentum increase of the ion for each collision with an electron, where v_e is the mean electron velocity. In determining $\overline{(\Delta p_i)^2}$, use is made of equation (4.36), except that N is now taken as the number of field particles (electrons) colliding with the ion in a time Δt, instead of in a length λ.

4.74.   From kinetic theory, assuming a Maxwellian distribution, the number of electrons passing through unit area in unit time is approximately $\frac{1}{2}n_e v_e$; hence, the number dN of electrons traversing a ring of radius b and width db in a time Δt is $(2\pi b\, db)\,(\frac{1}{2}n_e v_e)\,\Delta t$.   Assuming, as before, that the ion-electron encounters are random, it follows, using equations (4.36) and (4.54), that

$$d\overline{(\Delta p_i)^2} = \frac{4\pi n_e e^4}{v_e}\,\Delta t\,\frac{db}{b}. \tag{4.55}$$

upon integration, it is found that

$$\overline{(\Delta p_i)^2} = \frac{4\pi n_e e^4}{v_e}\,\Delta t \ln \Lambda, \tag{4.56}$$

where Λ has the same significance as above.

4.75.   Assuming Maxwellian distribution among the electrons, $\frac{1}{2}m_e v_e^2 = \frac{3}{2}kT_e$; further, $\overline{(\Delta p_i)^2}$ may be set equal to $2m_i\overline{\Delta W_i}$, so that equation (4.56) becomes, after rearrangement and substitution of dW_i/dt for $\Delta W_i/\Delta t$,

$$\frac{dW_i}{dt} \approx \frac{2\pi n_e e^4}{(3m_e kT_e)^{1/2}} \cdot \frac{m_e}{m_i} \ln \Lambda. \tag{4.57}$$

This gives the rate of increase of energy of the ion as a result of collisions with electrons of higher energy, so that energy is transferred from the electrons to the ion.   It is seen that under these conditions the rate of gain of energy by the ion is independent of its actual energy and that it is inversely proportional to the square root of the electron temperature.

4.76.   The result just derived is of interest in connection with the possibility of heating cold ions by collision with hot electrons.   Since it is often easier to impart energy directly to the electrons of a plasma than to the ions, a possible method of increasing the temperature of a plasma would be to heat the electrons and then allow them to transfer energy to the ions by collision (§5.9).   Such a procedure is feasible at relatively low electron energies, but as the temperature increases, the rate of energy transfer decreases and the proposed method of heating the ions becomes increasingly less effective from the standpoint of time.

4.77.   If numerical values are inserted into equation (4.57) and m_i is taken as the mass of the deuteron, the result is

$$\frac{dW_\mathrm{D}}{dt} \approx 2.7 \times 10^{-14}\,\frac{n_e}{T_e^{1/2}} \ln \Lambda \quad \text{kev/sec.} \tag{4.58}$$

When the deuterium nuclei are being heated by the transfer of energy from hot electrons, the process becomes very slow at temperatures above 0.5 kev

(§5.12), and so a reasonable approximation is to take ln Λ as 15; then equation (4.58) becomes

$$\frac{dW_\mathrm{D}}{dt} \approx 4.0 \times 10^{-13} \frac{n_e}{T_e^{1/2}} \quad \text{kev/sec.}$$

Since W_D is roughly equal to $\frac{3}{2} T_\mathrm{D}$ for an approximately Maxwellian distribution, where T_D kev is the deuteron temperature, it follows that

$$\frac{dT_\mathrm{D}}{dt} \approx 2.7 \times 10^{-13} \frac{n_e}{T_e^{1/2}} \quad \text{kev/sec.} \tag{4.59}$$

4.78. In order to integrate this expression, a relationship is required between T_e and T_D. It appears that the rate of energy transfer from electrons to ions, in general, is a maximum when T_e is $3T_i$ [18]; in these circumstances equation (4.59) can be written as

$$\frac{dT_\mathrm{D}}{dt} \approx 1.6 \times 10^{-13} \frac{n_e}{T_\mathrm{D}^{1/2}} \quad \text{kev/sec.}$$

Upon integration, taking T_D as zero when t is zero, it is found that

$$t \approx 4 \times 10^{12} \frac{T_\mathrm{D}^{3/2}}{n_e} \quad \text{sec.}$$

An alternative treatment [19] leads to the equipartition time

$$t = 2.3 \times 10^{12} \frac{T_\mathrm{D}^{3/2}}{n_e} \quad \text{sec,} \tag{4.60}$$

and this will be used later (§5.15). The time required for the electrons to share their energy with the ions is seen to increase rapidly with the ion temperature.

TRANSFER OF ENERGY FROM IONS TO ELECTRONS

4.79. The situation in which the ion energy is large in comparison with the average electron energy is not susceptible of treatment by simple methods such as those given above. It happens that, although the mean electron velocity may be greater than the ion velocity because of the much smaller mass of the electron, the energy transfer is in fact dominated by the relatively small fraction of the electrons, in the Maxwellian distribution, having velocities less than that of the ion. The effect of these electron-ion collisions is to reduce the ion energy to a point where ion-ion interactions, between the high-energy ions and those whose energy has already been reduced by collisions with electrons, tend to establish equipartition.

4.80. An expression for dW_i/dt that covers both limiting cases of interaction of a hydrogen isotope ion with electrons [20, 21] is

$$\frac{dW_i}{dt} = \frac{4(2\pi)^{1/2} n_e e^4}{(m_e k T_e)^{1/2}} \cdot \frac{m_e}{m_i} \left(1 - \frac{W_i}{\frac{3}{2} k T_e} \right) \ln \Lambda. \tag{4.61}$$

When the ion energy is much less than the average electron energy, $W_i/\frac{3}{2}kT_e$ is negligible in comparison with unity, and equation (4.61) reduces to the same form as the approximate equation (4.57). The two expressions differ only by the factor $4\,(3/2\pi)^{1/2}$ or 2.76.

4.81. For the case in which the ion energy is considerably greater than the average electron energy, so that unity can be neglected in comparison with $W_i/\frac{3}{2}kT_e$, equation (4.61) becomes

$$\frac{dW_i}{dt} = -\frac{8(2\pi)^{1/2}n_e e^4}{3m_e^{1/2}(kT_e)^{3/2}} \cdot \frac{m_e}{m_i}\,W_i\ln\Lambda. \tag{4.62}$$

Upon inserting numerical values, with $\ln\Lambda$ equal to 20, as before, and expressing W_i and T_e in kilo-electron volts, this expression for a deuterium plasma is

$$\frac{1}{W_{\mathrm{D}}} \cdot \frac{dW_{\mathrm{D}}}{dt} = -0.9 \times 10^{-12}\frac{n_e}{T_e^{3/2}}, \tag{4.63}$$

and integration yields

$$W_{\mathrm{D}} = W_0 e^{-t/\tau}, \tag{4.64}$$

where τ is given by

$$\tau = 1.1 \times 10^{12}\frac{T_e^{3/2}}{n_e}. \tag{4.65}$$

This *time constant*, as it is often called, is the time required for the energy of the ion to be decreased by a factor of e; it is consequently sometimes referred to as the $1/e$-folding time.

4.82. It is evident from equations (4.64) and (4.65) that collisions with a low-energy electron distribution result in an exponential decrease with time of the energy of a fast-moving ion with a time constant proportional to $T_e^{3/2}$. Hence, the energy of the ion approaches asymptotically the value $\frac{3}{2}kT_e$, in agreement with classical equipartition considerations. At a density of 10^{15} electrons/cm³ and an electron temperature of 10 ev, which is typical of ordinary electrical discharges, τ is about 1 microsec; this may be compared with the mean lifetime of 40 sec for the D-D reactions at a temperature of 100 kev and a deuteron density of 10^{15} particles/cm³, and of a little over 1 sec for the D-T reaction under the same conditions. This result is the basis of the argument presented in §3.56 indicating that the release of fusion energy on a useful scale cannot be achieved by bombarding a moderately cold deuteron plasma by accelerated deuterons or tritons. However, by increasing the electron temperature in the plasma, possibly reducing the density, and increasing the energy of the tritons, τ might be increased until it was the same order of magnitude as the reaction lifetime.

4.83. An interesting consequence of the general equation (4.61) is that at high plasma temperatures the "equilibrium" electron temperature will lag somewhat behind the ion temperature. If no appreciable energy is being supplied to the plasma from thermonuclear reactions or from external sources, the electron temperature will be determined by a balance between the rate of energy loss as bremsstrahlung and the rate of transfer of energy from ions in collisions. By substituting $\frac{3}{2} kT_i$ for W_i on the right of equation (4.61), multiplying by the ion density and setting the result equal, but opposite in sign, to the equation for P_{br} in §2.64, it is found that, for a plasma of a hydrogen isotope of mass number A,

$$T_i - T_e \approx 4.4 \times 10^{-3} T_e^2 / A, \tag{4.66}$$

the temperatures being in kilo-electron volts. If T_i is 100 kev, the electron temperature T_e in a deuterium plasma would asymptotically approach 84 kev under the conditions specified. Equation (4.66) applies only if the plasma consists entirely of isotopes of hydrogen. If impurities of higher atomic number are present, the rate of energy loss as bremsstrahlung will be increased (§2.74) and so also will be the difference between ion and electron temperatures.

ELECTRICAL CONDUCTIVITY OF PLASMA

4.84. Since a plasma consists of a cloud of mobile positively and negatively charged particles, it will obviously possess electrical conductivity. The magnitude of this conductivity (or resistance) depends upon the frequency of collisions between ions and electrons and so it is appropriate to consider it here. In the absence of a magnetic field, the resistivity η may be defined by means of the conventional form of Ohm's law as

$$\eta = \frac{E}{j}, \tag{4.67}$$

where the electric field E and the current density j are in the same direction. An approximate expression for the resistivity of a fully ionized plasma may then be derived in the following manner [22, 23].

4.85. When an electric field is applied to the plasma, the electrons are accelerated in one direction and the ions in the opposite direction. As a result of Coulomb collisions, among which the long-range interactions are dominant, momentum is exchanged between the ions and electrons. But since the plasma as a whole is stationary, a steady state is rapidly attained in which the electrons have a constant momentum in one direction which is equal to that of the ions in the opposite direction.* The electrons and ions then move with constant average velocities, but since the mass of the electron is much less than that of an ion, the velocity of the electron will be proportionately greater.

4.86. In an applied field of E statvolts/cm, the force exerted on a particle carrying a charge of e statcoulombs is eE dynes. If t_d is the time between large-angle distant collisions, in which the electrons lose on the average a quantity of momentum equal to their mean momentum (cf. §4.58), the rate of

---

* The energy transferred by the electric field to the charged particles is dissipated as heat; i.e., the ordered motion produced by the field is degraded into random motion as a result of collisions. This is the basis of one method proposed for heating a plasma (§5.6).

change of momentum is equal to mv/t_d dynes, where m grams is the mass of the charged particle and v cm/sec is its velocity in the field direction. Consequently, in the steady state

$$eE = \frac{mv}{t_d},$$

so that

$$v = \frac{eEt_d}{m}. \tag{4.68}$$

This expression for the average velocity of the charged particle applies when the electric field does not change appreciably in the time t_d and over a distance equal to the collision mean free path.

4.87.　In a plasma consisting of hydrogen isotope ions and electrons,

$$j = e(n_i v_i - n_e v_e), \tag{4.69}$$

where j is expressed in statamp/cm², and the ion and electron velocities v_i and v_e are algebraic values, that is, they have opposite signs. Since, as seen above $|v_e| \gg |v_i|$, and the number densities n_i and n_e are equal in the hydrogen isotope plasma, the first term in the parentheses can be neglected. Hence, using equation (4.68) to express v_e, and allowing for the negative charge of the electron, equation (4.69) becomes

$$j = \frac{n_e e^2 t_d}{m_e} E,$$

and so, by equation (4.67),

$$\eta = \frac{m_e}{n_e e^2 t_d}. \tag{4.70}$$

In this expression, t_d may be taken as the relaxation time t_{ei} for large-angle, long-range scattering of an electron as the result of interaction with ions. As seen in §4.71, this is equal to t_{ee} in a hydrogen isotope plasma and so is given approximately by equation (4.50). Assuming a Maxwellian distribution of electron velocities, $\frac{3}{2}kT_e$ may be written for W_e, so that equation (4.70) becomes

$$\eta \approx \left(\frac{m_e}{2}\right)^{1/2} \frac{4\pi e^2}{(\frac{3}{2}kT_e)^{3/2}} \ln \Lambda.$$

4.88.　Upon inserting 9.11×10^{-28} g for m_e and 4.80×10^{-10} statcoulomb for e, and expressing the electron temperature in kilo-electron volts, it is found that

$$\eta \approx \frac{5.3 \times 10^{-19}}{T_e^{3/2}} \ln \Lambda \quad \text{statohm-cm}$$

$$\approx \frac{4.8 \times 10^{-7}}{T_e^{3/2}} \ln \Lambda \quad \text{ohm-cm}. \tag{4.71}$$

At electron densities and temperatures which might exist in a thermonuclear reactor, $\ln \Lambda$ is close to 20, and by adopting this value the results will certainly not be in error by more than a small factor. It follows that, for a highly ionized hydrogen isotope plasma,

$$\eta \approx \frac{9.6 \times 10^{-6}}{T_e^{3/2}} \text{ ohm-cm.} \tag{4.72}$$

A more exact treatment of the problem of electrical conductivity of a hydrogen-isotope plasma leads to a somewhat different numerical value in equation (4.71); in the absence of a magnetic field

$$\eta = \frac{1.65 \times 10^{-7}}{T_e^{3/2}} \ln \Lambda \quad \text{ohm-cm}$$

or, taking $\ln \Lambda$ to be 20,

$$\eta = \frac{3.3 \times 10^{-6}}{T_e^{3/2}} \quad \text{ohm-cm.} \tag{4.73}$$

In deriving this result, the assumption was made that the energy acquired by a charged particle from the electric field during the relaxation time is small compared with the mean kinetic energy $\frac{3}{2}kT$.

4.89. It is seen from equation (4.73) that the resistivity of a plasma decreases with increasing temperature. The physical reason is that in a plasma, contrary to the usual molecular behavior, the collision rate, as indicated by the electron-ion cross section for large-angle, distant scattering, is inversely proportional to T^2 (or to W^2), as is apparent from equation (4.42). The time between collisions thus increases with temperature, so that the resistance of the plasma is decreased.

4.90. The expressions given above apply only when the dimensions of the plasma are large compared with the relaxation length for electron-ion collisions in the plasma. This condition would not be satisfied at low densities and high temperatures in a thermonuclear reactor of reasonable size. Nevertheless, the qualitative conclusions that can be drawn from equation (4.73) remain valid in cases of interest. An important one is that, in the region of 10 or 100 kev, the resistivity of the plasma is very small. At 10 kev, for example, equation (4.73) implies that η would be less than 10^{-7} ohm-cm, so that it is only about one-tenth that of copper at ordinary temperature. In other words, at thermonuclear temperatures, plasmas are extremely good conductors of electricity. Some consequences of this fact will be apparent in later chapters.

4.91. The expressions for the resistivity of a plasma given above are based on the assumption that there is a Maxwellian distribution of electron velocities. In a magnetic field, the distribution normal to the field is, however, quite different from that parallel to it. In particular, the number of high-velocity electrons is decreased and so the average electron-ion scattering cross section is increased in the perpendicular direction. As a result, the rate of momentum transfer and, consequently, the resistivity are higher in the direction perpendicular to the field lines than they are in the parallel direction where the magnetic field has no effect. For a strong magnetic field the transverse resistivity has been calculated to be

$$\eta_\perp = \frac{6.6 \times 10^{-6}}{T_e^{3/2}} \quad \text{ohm-cm},$$

with T_e in kilo-electron volts. This is seen to be about twice the value given by equation (4.73), which applies to the direction parallel to the field lines or to the plasma in the absence of a magnetic field [24].

PLASMA OSCILLATIONS *

INTRODUCTION

4.92. The unusual diversity and complexity of the wave motions which a plasma is known to propagate result from the fact that two distinct gaseous species, namely, ions and electrons, are present, and that there are at least four varieties of driving and restoring forces, i.e., electric and magnetic forces, pressure gradients, and viscosity. Since, from the hydrodynamic point of view, a plasma retains the essential properties of a gas, it will transmit acoustic (and shock) waves like any other gas. In addition, there are wave motions which owe their origin to the fact that the gas particles are electrically charged. Three categories of such waves will be considered here: (a) electromagnetic, (b) electrostatic, and (c) hydromagnetic (or magnetohydrodynamic) waves.

4.93. This classification of the waves propagated through a plasma is somewhat artificial. It is based principally upon the tractability of the theoretical treatment. The waves which travel through stars and interstellar clouds, and even through many experimental plasmas in the laboratory, are certainly complicated combinations of waves belonging to two or more of the categories indicated above. Nevertheless, it is profitable to study the three simple (or basic) types, especially as, by carefully controlling the experimental conditions, it has been possible, to some extent, to generate them separately in the laboratory.

GENERAL WAVE EQUATION

4.94. In considering the transmission of oscillations in a plasma, the first step is to derive a general wave equation. For this purpose, use is made of three of Maxwell's equations which were given in §3.11 in a form particularly suitable to the treatment of a plasma; for convenience, these equations will be repeated here, as follows:

$$\nabla \cdot \mathbf{E} = 4\pi\sigma \tag{4.74}$$

---

*There are several treatments of plasma oscillations and wave phenomena in the literature [25-28]. The one given here follows that of L. Spitzer, Jr. [26].

$$\nabla \times \mathbf{E} = -\frac{1}{c} \cdot \frac{\partial \mathbf{B}}{\partial t} \tag{4.75}$$

and
$$\nabla \times \mathbf{B} = \frac{1}{c}\left(4\pi \mathbf{j} + \frac{\partial \mathbf{E}}{\partial t}\right), \tag{4.76}$$

where, as emphasized in §3.8 *et seq.*, $\mathbf{j}$ and σ refer to the *total* current and charge densities, respectively. As in preceding sections, these equations are expressed in Gaussian units.

4.95. Upon taking the curl of both sides of equation (4.75) and substituting equation (4.76), there is obtained

$$\nabla \times (\nabla \times \mathbf{E}) = \frac{1}{c^2}\left(4\pi \frac{\partial \mathbf{j}}{\partial t} + \frac{\partial^2 \mathbf{E}}{\partial t^2}\right),$$

and since

$$\nabla \times (\nabla \times \mathbf{E}) \equiv \nabla(\nabla \cdot \mathbf{E}) - \nabla^2 \mathbf{E},$$

it follows, after introducing equation (4.74), that

$$\nabla^2 \mathbf{E} - 4\pi \nabla \sigma - \frac{4\pi}{c^2} \cdot \frac{\partial \mathbf{j}}{\partial t} - \frac{1}{c^2} \cdot \frac{\partial^2 \mathbf{E}}{\partial t^2} = 0, \tag{4.77}$$

which is the required wave equation. It will be observed that, when the current density $\mathbf{j}$ and charge density σ are set equal to zero, equation (4.77) reduces to the ordinary electromagnetic wave equation for a vacuum. It may be surmised, therefore, that in a plasma of sufficiently low particle density an electromagnetic wave will suffer little perturbation.

4.96. The next problem is to express the third term in equation (4.77) in terms of the plasma particle densities and the electric field vector. If electron-ion collisions are neglected * and there is no steady magnetic field present, equation (4.1) for electrons and ions, respectively, in a hydrogen isotope plasma reduces to

$$m_e \frac{d\mathbf{v}_e}{dt} = -e\mathbf{E} \tag{4.78}$$

and

$$m_i \frac{d\mathbf{v}_i}{dt} = e\mathbf{E}. \tag{4.79}$$

Differentiation of equation (4.69), for the current density, with respect to time gives

$$\frac{\partial \mathbf{j}}{\partial t} = e\left(n_i \frac{\partial \mathbf{v}_i}{\partial t} - n_e \frac{\partial \mathbf{v}_e}{\partial t}\right)$$

---

* The absence of electron-ion collisions implies that the electrical conductivity of the plasma is infinite (cf. §4.84 *et seq.*).

and if equations (4.78) and (4.79) are substituted, the result is

$$\frac{\partial \mathbf{j}}{\partial t} = e^2 \left(\frac{n_i}{m_i} + \frac{n_e}{m_e} \right) \mathbf{E}$$

$$\approx \frac{n_e e^2}{m_e} \mathbf{E}, \tag{4.80}$$

since $n_e = n_i$ and $m_i \gg m_e$.

ELECTROMAGNETIC WAVES

4.97. If $\mathbf{E}$ and $\mathbf{j}$ are perpendicular to the propagation vector, there are no electrostatic restoring forces and typical *electromagnetic waves* are present. In this case, the j's at various distances along the propagation direction are sheets of current flowing parallel to each other; consequently, there is no accumulation of charge and σ is zero. The same result may be obtained more precisely by considering the equation for the conservation of charge,

$$\nabla \cdot \mathbf{j} + \frac{\partial \sigma}{\partial t} = 0, \tag{4.81}$$

which can be derived from Maxwell's equations. Now,

$$\nabla \cdot \mathbf{j} \equiv \frac{\partial j_x}{\partial x} + \frac{\partial j_y}{\partial y} + \frac{\partial j_z}{\partial z},$$

and since there is no variation in the current along its own direction in the case of a plane wave at present under consideration, it follows that $\nabla \cdot \mathbf{j}$ is zero and so also, therefore, is $\partial \sigma / \partial t$. Thus, the charge density remains constant in time and this constant value must be zero. It is in fact a general property of transverse waves that the particle densities remain unchanged; hence, gradients of charge density or particle pressure are not produced.

4.98. By setting the second term of equation (4.77) equal to zero and substituting equation (4.80) into the third term, the result is

$$\nabla^2 \mathbf{E} - \frac{4\pi n_e e^2}{m_e c^2} \mathbf{E} = \frac{1}{c^2} \cdot \frac{\partial^2 \mathbf{E}}{\partial t^2}, \tag{4.82}$$

and for an electromagnetic wave in which E_y is propagated along the x axis, this becomes

$$\frac{\partial^2 E_y}{\partial x^2} - \frac{\omega_p^2}{c^2} E_y = \frac{1}{c^2} \cdot \frac{\partial E_y^2}{\partial t^2}, \tag{4.83}$$

where ω_p is defined by

$$\omega_p \equiv \left(\frac{4\pi n_e e^2}{m_e} \right)^{1/2}. \tag{4.84}$$

The presence of the frequency term ω_p, called the *plasma frequency*, in equation (4.83) suggests that the plasma has dispersion properties, as will be shown below, and also characteristic vibrations which will be discussed in §4.104 *et seq.*

4.99. For a solution of the form

$$E_y = E_0 \exp\left[i(kx - \omega t)\right], \tag{4.85}$$

equation (4.83) reduces to

$$-k^2 - \frac{\omega_p^2}{c^2} = -\frac{\omega^2}{c^2}$$

or

$$k^2 = \frac{1}{c^2}\left(\omega^2 - \omega_p^2\right)$$

$$= \frac{\omega^2}{c^2}\left(1 - \frac{\omega_p^2}{\omega^2}\right). \tag{4.86}$$

For $\omega > \omega_p$, it is seen from equation (4.86) that k is real and so the wave will propagate through the plasma. On the other hand, for $\omega < \omega_p$, k is purely imaginary and the amplitude of the waves decreases by a factor of $1/e$ in a distance λ given by

$$\lambda = \frac{c}{(\omega_p^2 - \omega^2)^{\frac{1}{2}}}.$$

Thus, if electromagnetic radiations of frequency ω, where $\omega < \omega_p$, are incident on a plasma, they are almost completely reflected.

4.100. The phase velocity V of the waves propagated in a plasma is given by

$$V = \frac{\omega}{k} = \frac{c}{[1 - (\omega_p/\omega)^2]^{\frac{1}{2}}}. \tag{4.87}$$

The dielectric constant K of the plasma* is defined by

$$K = \frac{c^2}{V^2},$$

so that, from equation (4.87),

$$K = 1 - \left(\frac{\omega_p}{\omega}\right)^2. \tag{4.88}$$

PROPAGATION OF ELECTROMAGNETIC WAVES

4.101. Since ω_p is related to the electron density, as seen from equation (4.84), it is evident from equation (4.88) that the dielectric constant of the plasma for electromagnetic waves of frequency ω, where $\omega > \omega_p$, is also determined by the electron density. It is apparent, therefore, that the propagation of such waves may be used to measure electron densities in plasmas. The experimental method will be described in Chapter 6; for the present, however, it is of interest to determine what kind of electromagnetic waves could be used for the purpose.

4.102. The condition for the propagation of electromagnetic waves through a plasma can be expressed in a convenient form by replacing ω_p/ω by f_p/f,

---

* This is not a dielectric constant in the sense defined in §3.10; it is, rather, the square of the effective refractive index.

where f is the frequency of the waves in cycles/sec. Since f_p is equal to $\omega_p/2\pi$, it is found from equations (4.84) and (4.88), upon inserting numerical values for m_e $(9.1 \times 10^{-28}$ g) and e $(4.8 \times 10^{-10}$ esu), that

$$K = 1 - 8.1 \times 10^7 \frac{n_e}{f^2}.$$

The condition for propagation of the waves, i.e., for K to be positive, is thus

$$8.1 \times 10^7 \frac{n_e}{f^2} < 1$$

or

$$f > 9 \times 10^3 n_e^{1/2}.$$

For plasmas that might be used in the experimental study of controlled thermonuclear reactions, n_e would be about 10^{14} electrons/cm^3 and so only waves for which

$$f > 9 \times 10^{10} \text{ cycles/sec}$$

would be propagated. The frequency must then exceed 9×10^4 megacycles/sec, and the wave length must be less than about 0.3 cm; thus microwaves in the millimeter wave length region are required for the study of electron densities in thermonuclear research.

4.103. The fact that a plasma is not transparent to electromagnetic waves with frequencies below the characteristic plasma frequency can be explained in physical terms. If the radiation frequency is small, the period during which the electric field changes is long in comparison with that required for the plasma to readjust itself by the motion of the particles. Currents are then set up in the plasma so as to exclude the electric field, thereby reflecting the incident wave. On the other hand, if the time in which changes in the field occur is small, the inertia of the plasma will prevent it from making any appreciable response; radiation will then be propagated through the plasma without much attenuation.

ELECTROMAGNETIC WAVES IN PLASMA CONTAINING A MAGNETIC FIELD

4.104. If the plasma contains a magnetic field, the transmission of electromagnetic waves is somewhat more complicated than described above. A trivial special case is that for which the electric field vector **E** in the wave is parallel to the magnetic field direction; the magnetic field has then no effect at all, since the charged particles move along the field lines and the current density is unaffected. However, it is only for very limited situations that the electric vector in the wave can be everywhere parallel to the local magnetic field direction.

4.105. A relatively simple situation of some interest is that in which the electric vector is perpendicular to the steady magnetic field, so that the direction of propagation of the waves is parallel to the magnetic field. A dispersion relation governing the plasma response to polarized incident radiation will now be derived for this special case. The important effect of the magnetic field is to bend the paths of the charged particles, particularly of the electrons, and so give rise to current components perpendicular to the electric vector.

4.106. If the effect of collisions is negligible, the force equation (4.1) for electrons is applicable and may be written as

$$m_e \frac{d\mathbf{v}_e}{dt} = -e\left[\mathbf{E} + \frac{1}{c}(\mathbf{v}_e \times \mathbf{B})\right].$$

Suppose the magnetic field is parallel to the x direction, which is the direction in which the electromagnetic wave propagates; then $\mathbf{E}$, $\mathbf{v}_e$, and $\mathbf{j}$ will have only y and z components. Thus,

$$m_e \frac{dv_y}{dt} = -e(E_y + v_z B_x/c) \tag{4.89}$$

and

$$m_e \frac{dv_z}{dt} = -e(E_z - v_y B_x/c). \tag{4.90}$$

Further, since $\mathbf{j}_e = -n_e e \mathbf{v}_e$, it is readily found upon substituting for v in equation (4.89) that

$$\frac{dj_y}{dt} = \frac{n_e e^2}{m_e} E_y - \frac{eB_x}{m_e c} j_z$$

$$= \frac{\omega_p^2}{4\pi} E_y - \omega_{ge} j_z, \tag{4.91}$$

where ω_p is defined by equation (4.84), as before, and ω_{ge} is the gyromagnetic (or cyclotron) frequency of the electrons, as given by equations (4.3), i.e.,

$$\omega_{ge} = \frac{eB_x}{m_e c}.$$

The analogous expression derived from equation (4.90) is

$$\frac{dj_z}{dt} = \frac{\omega_p^2}{4\pi} E_z + \omega_{ge} j_y. \tag{4.92}$$

4.107. The effect of the magnetic field is to produce components of $\mathbf{j}$ and $\mathbf{E}$ perpendicular to their initial direction, assuming the wave to have been plane polarized initially. Consequently, the plane of polarization is rotated. The solution of equations (4.91) and (4.92) can then be simplified by considering two opposite circularly polarized waves. If $\mathbf{E}$ is circularly polarized with an angular frequency ω, then the same will be true for $\mathbf{j}$; it is thus possible to write for one direction of polarization

$$j_y = j \sin \omega t,$$

and

$$j_z = j \cos \omega t,$$

from which it follows that

$$\frac{dj_y}{dt} = \omega j_z \tag{4.93}$$

and

$$\frac{dj_z}{dt} = -\omega j_y. \tag{4.94}$$

For polarization in the opposite direction, the signs in equations (4.93) and (4.94) are interchanged. From equations (4.91) and (4.93) it is then found that

$$\frac{dj_y}{dt} = \frac{E_y}{4\pi} \cdot \frac{\omega_p^2}{1 - (\omega_{ge}/\omega)} \tag{4.95}$$

and similarly from equations (4.92) and (4.94),

$$\frac{dj_z}{dt} = \frac{E_z}{4\pi} \cdot \frac{\omega_p^2}{1 - (\omega_{ge}/\omega)}. \tag{4.96}$$

4.108.   Returning to equation (4.77)—with the second term eliminated since only transverse waves are being considered, i.e., σ is zero—and writing the electric field as $E_y + iE_z$ and the current as $j_y + ij_z$, the result is, for propagation in the x direction,

$$\frac{\partial^2 E}{\partial x^2} - \frac{4\pi}{c^2}\left(\frac{\partial j_y}{\partial t} + i\,\frac{\partial j_z}{\partial t}\right) = \frac{1}{c^2} \cdot \frac{\partial^2 E}{\partial t^2}.$$

Upon substituting equation (4.95) and (4.96) for the respective current derivatives, it follows that

$$\frac{\partial^2 E}{\partial x^2} - \frac{\omega_p^2}{c^2} \cdot \frac{E}{1 - (\omega_{ge}/\omega)} = \frac{1}{c^2} \cdot \frac{\partial^2 E}{\partial t^2}.$$

Assuming, as before, that

$$E = E_0 \exp\left[i(kx - \omega t)\right]$$

it is found that

$$k^2 = \frac{1}{c^2}\left[\omega^2 - \frac{\omega_p^2}{1 \pm (\omega_{ge}/\omega)}\right], \tag{4.97}$$

where the two signs in the denominator correspond to the two directions of rotation of the electric vector. The plus sign applies when $\mathbf{E}$ rotates in the direction opposite to the gyration of the electrons in the magnetic field; the wave is then referred to as the "ordinary" wave. When the electric vector rotates in the same direction as the electrons gyrate, the minus sign is applicable in equation (4.97) and this represents what is called the "extraordinary" wave. The corresponding phase velocities are given by

$$V^2 = \frac{c^2}{1 - \dfrac{\omega_p^2}{\omega^2} \cdot \dfrac{1}{1 \pm (\omega_{ge}/\omega)}}. \tag{4.98}$$

4.109. It can be seen from equation (4.97) or (4.98), upon comparison with equation (4.86) or (4.87) respectively, that the effect of the magnetic field is to introduce the factor $[1 \pm (\omega_{ge}/\omega)]^{-1}$. As a result, penetration of the plasma by waves of much lower frequency than ω_p is possible. When the magnetic field strength B_x is large, then ω_{ge} is high and electromagnetic waves of very low frequency can be transmitted. It is of interest to note that when B_x is small (or zero), so that ω_{ge} is negligible, equations (4.97) and (4.98) reduce, as they should, to equations (4.86) and (4.87), respectively, which apply in the absence of a magnetic field.

ELECTROSTATIC WAVES

4.110. Simple *electrostatic waves* in a plasma can arise from longitudinal oscillations associated with the displacements of charged particles of opposite sign, thus leading to a divergence in the current density. If there is a small momentary average displacement of charges of one sign with respect to those of the other sign in the plasma, an electrostatic (space-charge) field will be established in such a direction as to produce a restoring force which opposes the displacement. Inertia of the displaced charges then provides a mechanism for periodic oscillations about the point of zero average charge displacement. For simplicity of treatment, two types of electrostatic oscillations are distinguished, namely, those in which only the electrons oscillate while the ions remain stationary, called *electron oscillations*, and those in which both electrons and ions oscillate, referred to as *ion oscillations*. The latter are much lower in frequency and more difficult to detect than electron oscillations; they will not be considered further here.

4.111. To determine the frequency of the electron oscillations, the divergence is taken of both sides of equation (4.80) ; the result is

$$\frac{\partial}{\partial t} (\boldsymbol{\nabla} \cdot \mathbf{j}) = \frac{n_e e^2}{m_e} \boldsymbol{\nabla} \cdot \mathbf{E}. \tag{4.99}$$

Upon using equation (4.81) for $\boldsymbol{\nabla} \cdot \mathbf{j}$ and equation (4.74) for $\boldsymbol{\nabla} \cdot \mathbf{E}$, noting that σ is not zero in the present situation, it is found that

$$\frac{\partial^2 \sigma}{\partial t^2} + \frac{n_e e^2}{m_e} 4\pi\sigma = 0$$

or

$$\frac{\partial^2 \sigma}{\partial t^2} + \omega_p^2 \sigma = 0. \tag{4.100}$$

A solution of equation (4.100) of the form

$$\sigma = \sigma_0 \exp\left[i(kx - \omega t)\right]$$

then gives

$$\omega = \omega_p.$$

The frequency of the electrostatic (electron) oscillations is thus found to be equal to the so-called characteristic plasma frequency ω_p, defined by equation (4.84).

4.112.   It is apparent from equation (4.100) that the vibrations under consideration are such that the restoring forces are electrostatic in nature, resulting from charge separation (or imbalance) in the plasma.   It may be imagined that layers of net positive and negative charge occur periodically throughout the plasma.   However, the result obtained above implies that this configuration swings back and forth through charge neutrality at a frequency which does not depend on the spacing of the charge layers, i.e., on the wave length of the oscillations.

4.113.   It should be pointed out that in the foregoing treatment the effect of the gradient of the electron density (or pressure) has been ignored.   If this were taken into account, an additional term, involving $\nabla \cdot \nabla p_e$, which depends on the electron temperature, would be included on the right of equation (4.99).   The solution for the oscillation frequency would then contain a related temperature-dependent term.   At very low temperatures, however, the correction is negligible, and the frequency has only a simple value which is independent of the wave length.

HYDROMAGNETIC WAVES

4.114.   The frequencies of the wave motions discussed above are such that only electrons participate significantly.   There is, however, another type of situation in which ion motions play the predominant role when a plasma is immersed in a uniform magnetic field.   It will be postulated that the ion and electron gyromagnetic (or cyclotron) frequencies are much higher than any wave frequencies to be considered here.   In addition, the usual simplifying assumption will be made that there are no ion-electron collisions, so that the plasma has infinite conductivity.

4.115.   The mechanical effect of a magnetic field on a plasma is equivalent to a hydrostatic pressure transverse to the field lines and a tension along the lines.   The hydrostatic pressure can be balanced by the pressure of the plasma, thus leaving the lines of force effectively in a state of tension.   The system is then similar to a stretched string or elastic band and leads to the possibility of transverse waves that propagate along the lines of force in a plasma. Such waves, which arise from the interaction of the magnetic field with the inertial properties of the plasma fluid, are called *hydromagnetic (or magnetohydrodynamic) waves;* they are also commonly called *Alfvén waves* [3].

4.116.   In deriving the momentum balance (or force) equation (3.12), the tacit assumption was made that the plasma as a whole was stationary.   In considering wave formation, however, it is necessary to allow for the motion of the plasma; the force equation then takes the form

$$\rho \frac{\partial \mathbf{v}}{\partial t} + \nabla p = \frac{1}{c}\,(\mathbf{j} \times \mathbf{B}),$$

where $\mathbf{v}$ is the velocity vector of the plasma and ρ is the mass density, i.e., $n_e m_e + n_i m_i$. Suppose that the propagation vector of the hydromagnetic waves lies parallel to the steady magnetic field and that the displacement velocity $\mathbf{v}$ is perpendicular to it. As seen in §4.97, these "shear waves" are not accompanied by any pressure changes, so that ∇p is zero. Consequently, the force equation may be written as

$$\rho \frac{\partial \mathbf{v}}{\partial t} = \frac{1}{c} \, (\mathbf{j} \times \mathbf{B}).$$

4.117. If wave propagation is in the x direction with the electric field vector along the z axis, then

$$\rho \frac{\partial v_y}{\partial t} = \frac{1}{c} \, (j_z B_x), \tag{4.101}$$

the transverse component B_y of the magnetic field being neglected in comparison with B_x. Since the plasma is to be regarded as a perfect conductor, $\mathbf{E}$ must vanish in a frame moving with an element of plasma; or, in the laboratory frame of reference,

$$\mathbf{E} = \frac{1}{c} \, (\mathbf{v} \times \mathbf{B}),$$

so that, in the present case,

$$E_z = \frac{1}{c} \, (v_y B_x). \tag{4.102}$$

4.118. From equations (4.101) and (4.102) it is readily shown that

$$j_z = \frac{\rho c^2}{B_x^2} \cdot \frac{\partial E_z}{\partial t}$$

and therefore

$$\frac{\partial j_z}{\partial t} = \frac{\rho c^2}{B_x^2} \cdot \frac{\partial^2 E_z}{\partial t^2}. \tag{4.103}$$

Upon substituting equation (4.103) into the general wave equation (4.77), and remembering that no charge accumulations occur in the transverse waves, i.e., $\sigma = 0$, the result is

$$\frac{\partial^2 E_z}{\partial x^2} = \frac{1}{c^2} \left(1 + \frac{4\pi \rho c^2}{B_x^2} \right) \frac{\partial E_z^2}{\partial t^2}.$$

This is the wave equation for a medium of dielectric constant K^* given by

$$K = 1 + \frac{4\pi \rho c^2}{B_x^2}. \tag{4.104}$$

In plasmas of practical interest, this quantity is extremely large. For example,

---

*The quantity K defined here should not be regarded as an actual dielectric constant; it is merely equivalent to one in the manner in which it appears in the wave equation (cf. §11.50).

if the particle density of the plasma is 10^{15} per cm³, then the mass density ρ is roughly 10^{-9} g/cm³; if the magnetic field strength is 10^5 gauss, then K is seen to be of the order of 10^3.

4.119. The phase velocity of the hydromagnetic waves, often called the *Alfvén speed*, is

$$V = \frac{c}{K^{1/2}} = \frac{c}{\left(1 + \dfrac{4\pi\rho c^2}{B_x{}^2}\right)^{1/2}} \tag{4.105}$$

or, since the second term in the denominator is usually very large in comparison with unity,

$$V \approx \frac{B_x}{(4\pi\rho)^{1/2}}. \tag{4.106}$$

4.120. The velocity given by equation (4.106) can also be obtained by the simple analogy of treating the magnetic lines as loaded elastic strings, the loading consisting of the elements of plasma which are tightly attached to the strings (§4.115). In general, for a massive string

$$\frac{\partial^2 y}{\partial t^2} = T \frac{\partial^2 y}{\partial x^2}, \tag{4.107}$$

where T is the string tension. Magnetic lines exhibit a "tension" equal to $B^2/4\pi$ [29], and substitution of this value for T into equation (4.107) gives an equation for transverse waves which are propagated with a velocity equal to $B/(4\pi\rho)^{1/2}$.

4.121. A slight rearrangement of equation (4.105), after neglecting unity in the denominator, gives

$$\left(\frac{c}{V}\right)^2 = \frac{\rho c^2}{2(B^2/8\pi)},$$

where the expression on the right is the ratio of the rest mass energy density ρc^2 of the plasma to twice the magnetic energy density $B^2/8\pi$ (cf. §3.18). If, as in §4.118, ρ is taken to be 10^{-9} g/cm³ and B as 10^5 gauss, it is seen that c/V is about 30; the speed of the Alfvén waves along the field lines in plasmas of thermonuclear interest is therefore less than the speed of light.

4.122. If a longitudinal wave is transmitted in a direction perpendicular to the magnetic field, the ∇p term must be retained in the force equation (§4.116), and so compressive effects of both gas and field become important. The waves then have a mixture of acoustic and hydromagnetic properties. The velocity of such a wave in a hydrogen isotope plasma is given by

$$V^2 = \frac{c^2}{K} + \gamma \frac{p}{\rho}\left(1 - \frac{1}{K}\right), \tag{4.108}$$

where γ is the adiabatic constant equivalent to the ratio of the specific heats of an ordinary gas (§5.25). The first term on the right is the purely hydromagnetic contribution, whereas the coefficient of the second term is the square of the velocity of sound in the plasma. If the magnetic pressure $B^2/8\pi$ is large in comparison with the kinetic pressure, the velocity of the longitudinal hydromagnetic wave according to equation (4.108) becomes the same as the Alfvén speed, as given by equation (4.105) or (4.106). On the other hand, when the kinetic pressure of the plasma greatly exceeds the magnetic pressure, the velocity of the wave approaches the speed of sound.

REFERENCES FOR CHAPTER 4

1. S. Chapman and T. G. Cowling, *The Mathematical Theory of Non-Uniform Gases*, Cambridge University Press, 1939.
2. S. Chandrasekhar, *Principles of Stellar Dynamics*, University of Chicago Press, 1942.
3. H. Alfvén, *Cosmical Electrodynamics*, Clarendon Press, 1950.
4. L. Spitzer, *Physics of Fully Ionized Gases*, Interscience Publishers, Inc., 1956.
5. T. G. Cowling, *Magnetohydrodynamics*, Interscience Publishers, Inc., 1957.
6. R. Gunn, *Phys. Rev.*, **32**, 832 (1929).
7. R. F. Post, *Rev. Mod. Phys.* **28**, 338 (1956).
8. R. F. Post, *Ann. Rev. Nuclear Sci.*, **9**, 367 (1959).
9. A. Simon, *An Introduction to Thermonuclear Research*, Pergamon Press, Inc., 1959.
10. Reference 3, p. 21 *et seq.*
11. M. Creutz, USAEC Report TID-5447 (1956), p 7.
12. Reference 4, Chapter 5.
13. P. Debye and W. Hückel, *Physikal. Z.*, **24**, 183, 305 (1923).
14. Reference 4, p. 16.
15. Reference 2, p. 51.
16. Reference 4, p. 73.
17. Reference 4, p. 78.
18. T. H. Stix, USAEC Report TID-7536, p. 343 (1957).
19. Reference 4, p. 80.
20. M. H. Johnson, unpublished.
21. C. L. Longmire, Plasma Physics Lectures, unpublished.
22. Reference 4, p. 82.
23. Reference 9, p. 117.
24. Reference 4, p. 86.
25. Reference 3, Chapter IV.
26. Reference 4, Chapter 4.
27. D. Gabor, *Brit. J. Appl. Phys.*, **2**, 209 (1951).
28. G. Francis, *Ionization Phenomena in Gases*, Academic Press, Inc., 1959, Chapter 7.
29. Reference 5, p. 9.

Chapter 5

FORMATION AND HEATING OF PLASMA

FORMATION OF PLASMA: BREAKDOWN OF GAS

5.1. The procedures used for the formation of the plasma, i.e., the ionization of the deuterium (or mixed deuterium and tritium) gas at low pressure, and for heating the ions to temperatures in the useful thermonuclear range depend to a large extent on the method of confinement. As far as ionization and initial stages of heating are concerned, i.e., up to a few hundred electron volts, pinched-discharge and stellarator devices have much in common. But the heating in the later stages will probably have to be different. In thermonuclear reactors utilizing magnetic mirrors for confinement, another kind of approach appears to be necessary for producing the high-temperature ions required for the thermonuclear reactions.

5.2. For the initial ionization (or breakdown) of the gas in a stellarator or in a system in which the plasma is subsequently to be pinched, a high-frequency, e.g., radiofrequency, electrical discharge of moderate voltage is effective [1-3]. The frequency may be from 100 (or so) kilocycles to a few megacycles per second and the applied potential difference of the order of a hundred volts. The discharge can be induced in the gas which serves as the secondary of an air-core or ferrite-core transformer. The general principles will be apparent from Figs. 5.1 and 5.2; the former shows the type of arrangement which could be used in experimental work on linear pinches and the latter is that employed at one time in some stellarator studies. Alternatively, the output of the radiofrequency oscillator may be connected to two electrodes surrounding the tube, which may be linear or toroidal, containing the gas to be ionized (Fig. 5.3). The gas then acts as the dielectric of a capacitor and absorbs energy from the radiofrequency oscillations. When a metal torus is used, as in the later forms of the stellarator, the radiofrequency field is applied directly across a ceramic (insulating) break in the tube, as shown in Fig. 5.4.

5.3. No matter how applied, the energy of the radiofrequency field is taken up mainly by the electrons; the reason is that a particle carrying a charge e

116

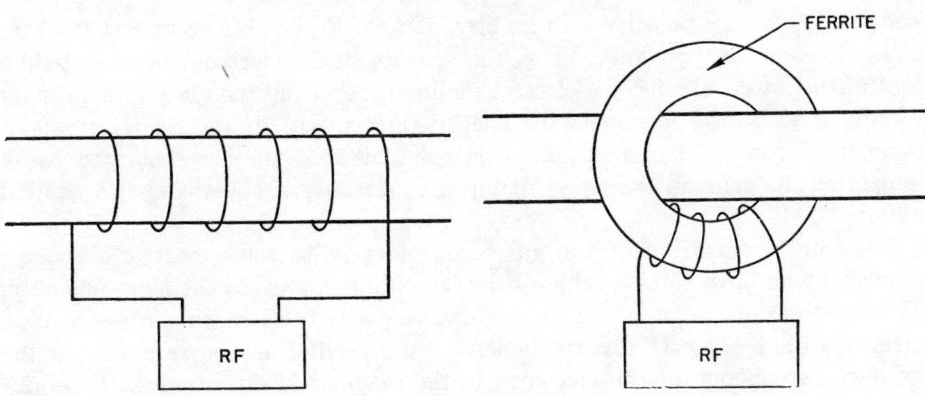

FIG. 5.1.   Radiofrequency breakdown with air-cored transformer.

FIG. 5.2.   Radiofrequency breakdown with ferrite-cored transformer.

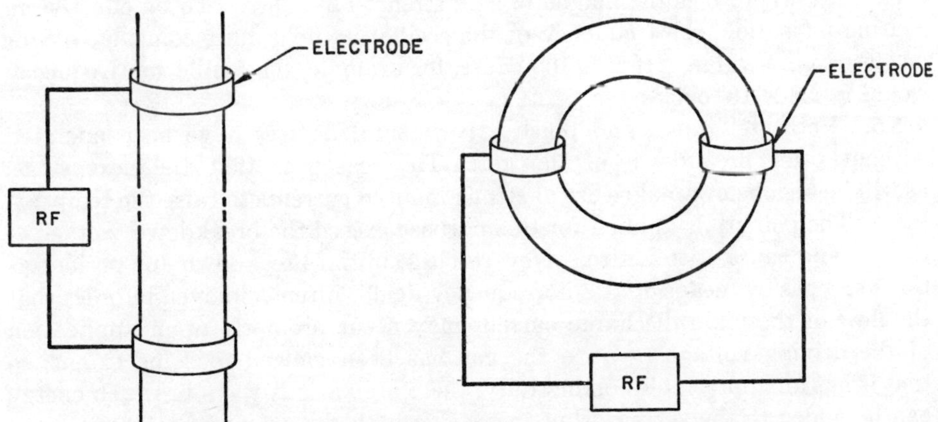

FIG. 5.3.   Radiofrequency breakdown with external electrodes.

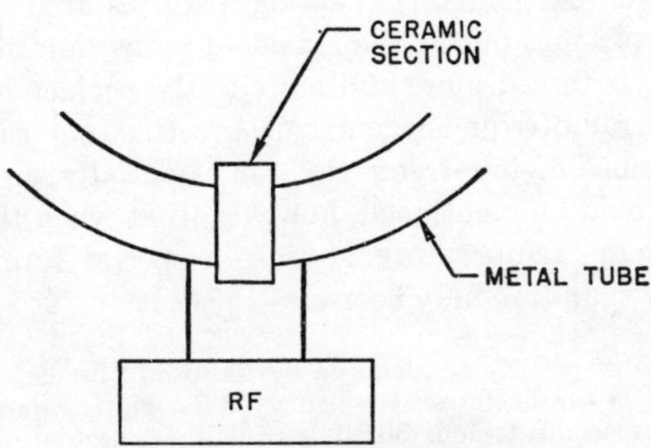

FIG. 5.4.   Radiofrequency breakdown with metal tube.

and moving with a velocity v in an electric field E acquires energy at the rate eEv.   Because of their lower mass, the velocity of the electrons is greater than that of the ions.   Hence, the electrons gain energy from the electric field more rapidly than do the ions.   As the energy (and speed) of the relatively small number of free electrons initially formed increases, they are able to cause ionization to occur, in the usual manner, as a result of collisions with neutral atoms.

5.4.   Since ionization of the gas is due largely to the energetic electrons, efficient breakdown can be achieved by supplying energy to the electrons only. This can be done by the use of microwave oscillations in the presence of a magnetic field [4].   If the frequency of the oscillations corresponds to the cyclotron frequency of the electrons in the magnetic field, as given by equation (4.3), then the electrons will readily acquire sufficient energy to ionize the gas.* Upon inserting the values of e (statcoulombs), m (grams), and c (cm/sec), the electron gyromagnetic frequency is found to be $2.8 \times 10^6\ B$ cycles/sec, where B is the magnetic field strength in gauss.   To be effective in causing ionization, the frequency of the oscillating field must coincide with or be close to this value.   If B is 100 gauss, for example, the oscillation frequency would be 2.8×10^8 cycles/sec.

5.5.   Both stellarator and pinch experimental devices have been operated without prior breakdown of the gas.   The reason is that the next stage, as will be seen below, makes use of strong induced currents to raise the temperature.   The potentials applied for this purpose exceed the breakdown voltage of the gas and cause it to ionize.   Nevertheless, initial breakdown (or preionization) of a gas by means of a radiofrequency field is often employed in order that the flow of the main discharge current may occur promptly upon application of the driving voltage.   Once the gas has been ionized to some extent, so that it has an appreciable conductivity, there are various ways in which energy can be added to the system, thus increasing both the extent of ionization and the temperature of the charged particles.   These results could be accomplished, in principle, by means of the radiofrequency oscillations, but the procedure is not usually effective. Satisfactory heating requires that large currents be passed through the plasma, but as the extent of ionization increases, the high-frequency discharge is forced more and more to the surface by the well-known "skin" effect. The radiofrequency currents are thus not efficient for heating the plasma and the use of low-frequency (or essentially direct) current is to be preferred.   It should be mentioned, however, that when the plasma is completely ionized and the temperature is high, a special form of heating with very high-frequency radiation may be useful (§5.61).

---

* The basic principles of particle acceleration by the use of oscillations having the same (resonance) frequency as the cyclotron frequency of the particles are described in §5.61. The treatment refers especially to ions but it is equally applicable to electrons.

OHMIC HEATING OF PLASMA

GENERAL PRINCIPLES

5.6.　An obvious method of heating and ionizing a gas after initial break-down is by *ohmic heating*.　This is the procedure commonly used to heat a resistance by the passage of current.　If j is the density of the current passing through a plasma having a resistivity η, the rate of heat dissipation per unit volume resulting from resistance loss is ηj^2.　It is because this is of the same form as the relationship which holds for conventional resistance heating, that the term ohmic heating is used.*　In order to heat a plasma, a pulse of current may be passed through a toroidal gas circuit by making it serve as the secondary of an iron-cored transformer (Fig. 5.5); the primary is energized either by

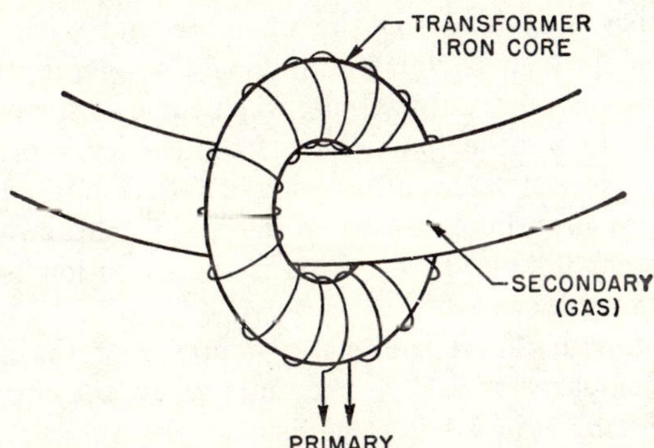

FIG. 5.5.　Ohmic heating of plasma in a torus.

the discharge from a capacitor bank or by means of a low-frequency generator.　If the plasma to be heated is in a magnetic field, as is often the case, the direction of the induced current must be parallel to the field lines in the plasma.　This requirement is usually satisfied automatically, since the directions of both the initial magnetic field and the circuit are longitudinal with respect to the tube containing the gas.

5.7.　In some experiments on the pinch effect in a toroidal geometry, the primary circuit is laid along the torus walls, either as a number of wires or in the form of conducting shells (§7.113).　The secondary (induced) current serves to heat and ionize the plasma and also to constrict it.　When working with linear (or straight) pinched discharges, the electrodes are inserted in the gas and the discharge is passed directly between them.　This technique is employed in experimental studies largely because of its simplicity; however, as indicated in §3.30, it might have deleterious consequences at very high temperatures.

---

*It is sometimes called *Joule heating*, because Joule's law is obeyed.

5.8.   The ohmic heating of a plasma can occur in a somewhat indirect form as a result of the intermixing or interdiffusion of magnetic fields.   Suppose that a plasma is confined by a magnetic field and that another field of different strength or direction is trapped within the plasma.   If the plasma has a finite (nonzero) resistance, there will be a tendency for the two fields to diffuse into one another.   As a result of the time-changing magnetic field, an electric field is generated in the plasma and the consequent flow of current through the finite resistance produces ohmic heating (cf. §7.39).

5.9.   It was stated above that the rate at which a charged particle gains energy from an electric field is proportional to the velocity of the particle. Since the electrons move much more rapidly than do the ions, because of their smaller mass, nearly all of the energy absorbed from the field will be taken up by the electrons.   As a result, the velocity of the electrons will increase further, so that they will tend to take up more and more energy from the electric field.   The electron is thus continuously accelerated until it makes a collision with an ion or a neutral particle, when some of the energy of the electron is transferred to the other particle.   If the energy gained by an electron from the electric field between collisions is small compared with its kinetic energy, the electrons may be expected to maintain a Maxwellian distribution corresponding to their temperature.   This is the condition for true ohmic (or Joule) heating to take place.

5.10.   On the other hand, if the energy increase of the electrons between collisions is large compared with their thermal energy, the electrons gain energy at a faster rate than it is lost by collisions.   In these circumstances, the electron velocities do not even approach a Maxwellian distribution and the electrons are said to "run away."   They then gain energy at an ever increasing rate until the gain is offset by losses, such as occur when the electrons strike the walls of the containing vessel [5-7].

5.11.   Although almost all the energy of the electric field is taken up directly by the electrons, the purpose of ohmic heating is, of course, to raise the temperature of the ions, e.g., deuterons and tritons.   The circumstances should consequently be such that transfer of energy occurs from the electrons to the ions.   Thus, electron runaway must be avoided.   An approximate indication of the relationship between the electric field strength and the temperature above which runaway occurs can be obtained by equating the energy gain of the electron between collisions to its kinetic energy.   If λ is the electron-ion collision mean free path (or relaxation length), the condition for runaway is then given by

$$\frac{eE\lambda}{c} \geqslant \frac{3}{2}kT, \tag{5.1}$$

where T is the electron temperature at which runaway occurs, and $\frac{3}{2}kT$ is the average kinetic energy of the electron, since a Maxwellian distribution applies up to this point. The relaxation length λ is equal to $1/n_i\sigma_d$, where σ_d is the collision cross section for distant electron-ion collisions, given by equation (4.42). It is seen that σ_d is inversely proportional to W^2, where W is essentially equal to the electron kinetic energy. Hence, it follows that σ_d is inversely proportional to T^2, and the inequality (5.1) yields the result

$$T \geqslant \text{constant} \times \frac{n_i}{E}.$$

The temperature at which electron runaway occurs is thus inversely proportional to the strength of the electric field used for the ohmic heating. A low field strength is, therefore, an advantage in preventing runaway.

5.12. In addition to the possibility of electron runaway, the ion temperatures that can usefully be obtained by ohmic heating are limited by the decrease in the rate of energy transfer in collisions between electrons and ions as the temperature increases. Thus, it was shown in Chapter 4 (see §4.76 *et seq.*) that the heating of cold ions by hot electrons is satisfactory only when the electron energy (or temperature) is relatively low. As a consequence of the decrease in the collision frequency, the resistivity of a highly ionized plasma decreases with increasing temperature, as is apparent from equation (4.73). Hence, the rate of energy absorption due to ohmic heating must decrease correspondingly. Because of the limited period over which experimental plasmas remain stable, ion temperatures of the order of 0.5 kev, i.e., about 5×10^6 °K, are probably the highest attainable by ohmic heating at the present time. If the lifetime of the plasma could be increased, higher temperatures would undoubtedly become possible.

5.13. In utilizing ohmic heating in a stellarator, the current strength is limited by the plasma instability referred to in §3.39, and possibly by other factors which will be considered more fully in Chapter 8. It may be mentioned here, however, that by suitable design, e.g., containment tube of large radius and strong confining field, the limiting current may be so high as not to represent a practical limitation on ohmic heating of the plasma in a stellarator system.

5.14. Although preliminary ionization and ohmic heating of the plasma are necessary in some experimental work and would be required in the start-up of certain thermonuclear reactors, it is possible that these stages would not be included in a continuously operating (or steady-state) system. In such a reactor, in which useful thermonuclear temperatures were being attained, injection of the cold gas into the interior, at a suitable rate, would probably lead to its immediate ionization and heating to a temperature sufficiently high to make the ohmic heating stage unnecessary.

RATE OF OHMIC HEATING

5.15. Since the energy acquired by the cold ions in a plasma subjected to ohmic heating is due almost entirely to the transfer from hot electrons, the rate at which the ion temperature increases is given by equation (4.59). The time t required to attain a deuteron temperature T_D kev can then be approximated by the electron-ion equipartition time, namely,

$$t = 2.3 \times 10^{12} \frac{T_D^{3/2}}{n_e} \text{ sec,} \tag{5.2}$$

where n_e is the density in electrons/cm³. From various experiments, it appears that the observed times for ohmic heating of deuterons by passage of a discharge through deuterium gas are less than those calculated from equation (5.2). Since the heating of the ions takes place more readily than estimated on the basis of collisional transfer of energy from electrons, it has been concluded that an additional mechanism of energy transfer is operative [7]. Some of the results are given in Table 5.1. The experimental data were obtained in studies made with the large pinched-discharge system called Zeta (§7.129).

TABLE 5.1.   CALCULATED AND OBSERVED TIMES FOR OHMIC HEATING OF DEUTERIUM IONS

T_D (°K)	T_D (kev)	n_e (per cm³)	t_{calc} (sec)	t_{obs} (sec)
3.3×10^6	0.28	8.6×10^{13}	4×10^{-3}	1×10^{-3}
2.5×10^6	0.21	1.1×10^{14}	2×10^{-3}	1×10^{-3}

5.16. It has long been known that the electrons in a conventional gas discharge approach a Maxwellian distribution much more rapidly than can be accounted for by collisions [8]. There is some evidence that oscillations in the plasma are to a great extent responsible for this effect, but the exact mechanism is not clear [9, 10]. It may be that the results mentioned above, namely, the more rapid increase in ion temperature than might have been expected from electron-ion collisions, may be due to a related phenomenon. In addition to plasma oscillations, it is possible that other cooperative effects, such as shocks and hydromagnetic waves (§4.92), may play a role.*

5.17. Referring back to the data in Table 5.1, it is seen that, with a particle density of 10^{14} electrons/cm³ in the plasma, about 10^{-3} sec is required to heat the ions to a temperature of about 300 ev, i.e., 0.3 kev. If the electron density were increased, the time would be reduced proportionately, since equation (5.2) shows it to be inversely proportional to the electron density. A precise knowledge of this heating time may prove to be important in the design and operation of thermonuclear reactors because of the possibility of the development of plasma instabilities.

---

* British scientists have used the expressions "wriggle heating" and "instability heating" to describe the general phenomenon.

HEATING BY MAGNETIC COMPRESSION

ADIABATIC COMPRESSION

5.18. Adiabatic heating results from the compression of a plasma at a relatively slow rate, as distinct from shock heating which may accompany a rapid compression (§5.35 *et seq.*). The compression is brought about by means of an increasing magnetic field, e.g., in a pinched discharge by the azimuthal self-magnetic field, or in a magnetic mirror system by increasing the central field strength (radial compression) or by bringing the reflecting fields closer together.

5.19. Consider, first, the case of radial compression of a plasma, i.e., a decrease in its radius, either in a toroidal discharge or in a mirror system, due to an increase in the strength of the confining (external) magnetic field [11]. If B is the value of the latter at any instant during the compression, then p_m, the pressure exerted by the external field, is equal to $B^2/8\pi$. Furthermore, if B_p is the strength of the field trapped within the plasma then, by equation (3.16),*

$$p_m = \frac{B^2}{8\pi} = nkT + \frac{B_p{}^2}{8\pi}, \tag{5.3}$$

where n is the particle density of the plasma. If N is the total number of plasma particles and V is the volume, equation (5.3) may be written as

$$p_m = \frac{B^2}{8\pi} = \frac{NkT}{V} + \frac{B_p{}^2}{8\pi}. \tag{5.4}$$

5.20. Assuming that there are no spatial gradients of temperature and density in the plasma, its internal energy is

$$W = \frac{f}{2} NkT + \frac{B_p{}^2}{8\pi} V, \tag{5.5}$$

where f is the number of translational degrees of freedom of the particles. The first term on the right of equation (5.5) is the particle kinetic energy and the second term is the magnetic energy of the trapped field. If the plasma density is relatively high and collisions among the particles are fairly frequent, that is, if the relaxation times in the given plasma are short in comparison with the time during which the magnetic field is increasing, then there will be an approach, at least, to the equipartition of energy among the particles and f in the equation (5.5) will be 3, and the kinetic energy will have the Maxwellian value of $\frac{3}{2} NkT$. On the other hand, for relatively low plasma densities, in which collisions among the particles are rare and the relaxation time is long compared with the rise time of the field, the particles will gain energy only in the two directions perpendicular to the field lines and f will then be 2.

---

* The symbols B and B_p are equivalent to B_0 and B, respectively, of §3.18.

5.21.   Allowing for the possibility that ions and electrons can escape from the plasma during radial compression, e.g., through the ends of a magnetic mirror system, the equation for energy conservation may be written as

$$\frac{dW}{dt} = \epsilon_i \frac{dN_i}{dt} + \epsilon_e \frac{dN_e}{dt} - p_m \frac{dV}{dt}, \tag{5.6}$$

where $-dN_i/dt$ and $-dN_e/dt$ represent the rates of loss of ions and electrons having mean energies ϵ_i and ϵ_e, respectively; the last term on the right is the work done by the external field on the plasma and the trapped field. Because of electrostatic (space-charge) effects the rates of loss of ions and electrons will be the same, and each is equal to $\frac{1}{2}(dN/dt)$. Hence, if ϵ is the average of ϵ_i and ϵ_e, equation (5.6) becomes

$$\frac{dW}{dt} = \epsilon \frac{dN}{dt} - \frac{NkT}{V} \cdot \frac{dV}{dt} - \frac{B_p{}^2}{8\pi} \cdot \frac{dV}{dt}, \tag{5.7}$$

after introducing the value of p_m from equation (5.4).

5.22.   If B_{p0} is the magnitude of the field trapped within the plasma of radius r_0 before compression and B_p and r are the corresponding values during the radial compression, then the constant flux condition derived in §4.24, which is applicable when the gyromagnetic frequency greatly exceeds the collision frequency, shows that

$$B_p = B_{p0} \left(\frac{r_0}{r}\right)^2.$$

Assuming that the length of the plasma remains constant during the adiabatic radial compression, then the volume is proportional to r^2, and hence

$$B_p = B_{p0} \frac{V_0}{V} = \frac{\text{constant}}{V}. \tag{5.8}$$

5.23.   Upon substituting this value for B_p in equation (5.5), differentiating with respect to time and comparing with equation (5.7), it is found that

$$\frac{1}{2}f\left(\frac{1}{T} \cdot \frac{dT}{dt}\right) + \left(\frac{1}{2}f - \frac{\epsilon}{kT}\right)\left(\frac{1}{N} \cdot \frac{dN}{dt}\right) + \frac{1}{V} \cdot \frac{dV}{dt} = 0 \tag{5.9}$$

and hence,

$$\frac{T}{T_0} = \left(\frac{V_0}{V}\right)^{2/f} \left(\frac{N_0}{N}\right)^{1-(2\epsilon/fkT)}. \tag{5.10}$$

This equation relates the temperature as a result of the radial compression to the volume of the plasma and the number of particles present. If there were no loss of particles during radial compression, e.g., in a toroidal tube, N_0/N would be unity, and equation (5.10) would reduce to

$$\frac{T}{T_0} = \left(\frac{V_0}{V}\right)^{2/f}, \tag{5.11}$$

where f is generally between 2 and 3, depending on the density of the plasma.

5.24. In the event that particles escape during compression, N_0/N will be greater than unity, and provided $1 - (2\epsilon/fkT)$ is positive, the increase of temperature, for a given compression ratio V_0/V, will be larger than if there were no particle loss. If the average energy ϵ of the escaping particles is the same as the mean particle energy in the plasma, i.e., $\tfrac{1}{2}fkT$, then the exponent of N_0/N in equation (5.10) is zero and the result is th: same as if there had been no loss of particles during compression. However, it will be seen later (§9.40) that the energy of the particles escaping through the ends in a mirror system, at least for low-pressure plasmas, is less than the average, i.e., $\epsilon <$ $\tfrac{1}{2}fkT$. The exponent of N_0/N is thus positive, and the temperature increase is greater than given by equation (5.11) for no particle loss.

5.25. If a quantity γ is defined by

$$\gamma \equiv \frac{f + 2}{f},\tag{5.12}$$

so that $2/f$ is equal to $\gamma - 1$, equation (5.11) can be written as

$$\frac{T}{T_0} = \left(\frac{V_0}{V}\right)^{\gamma-\cdot}\tag{5.13}$$

which is identical in form with the conventional expression for the reversible adiabatic compression (or expansion) of an ideal gas, with γ equal to the ratio of the specific heats. However, when the second factor on the right of equation (5.10) is not zero, owing to particle losses, then the magnetic radial compression does not obey the equation for a thermodynamically reversible adiabatic process.

5.26. In order to relate the temperature of the compressed plasma to the strength of the external magnetic field, it is simplest to consider the case in which the trapped field B_p is small, so that, by equation (5.4),

$$p_m = \frac{B^2}{8\pi} \approx \frac{NkT}{V}.$$

Upon differentiating with respect to time, dividing through by p_m, and using equation (5.9) to eliminate the dV/dt term, the result is

$$\frac{1}{p_m} \cdot \frac{dp_m}{dt} = \left(1 + \tfrac{1}{2}f\right)\frac{1}{T} \cdot \frac{dT}{dt} + \left(1 + \tfrac{1}{2}f - \frac{\epsilon}{kT}\right)\frac{1}{N} \cdot \frac{dN}{dt}.\tag{5.14}$$

Since p_m is equal to $B^2/8\pi$, the solution of this equation is found to be

$$\frac{T}{T_0} = \left(\frac{B}{B_0}\right)^{2(\gamma-1)/\gamma}\left(\frac{N_0}{N}\right)^{\alpha(\gamma-1)/\gamma}\tag{5.15}$$

where γ is defined by equation (5.12) and α by

$$\alpha \equiv 1 + \tfrac{1}{2}f - \frac{\epsilon}{kT}.$$

If there is no loss of particles during compression or if the particles which escape have the mean energy $\frac{1}{2}fkT$, then equation (5.15) becomes

$$\frac{T}{T_0} = \left(\frac{B}{B_0}\right)^{2(\gamma-1)/\gamma}. \tag{5.16}$$

However, as stated above, ϵ is usually less than $\frac{1}{2}fkT$, and thus α is positive; hence, so also is the exponent of N_0/N in equation (5.15). For a given relative increase B/B_0 in the strength of the confining magnetic field, the increase of temperature will then be greater when particle losses occur.

5.27. Since the foregoing results have been based on the postulate that the magnetic field inside the plasma is small, this is equivalent to assuming that β, the ratio of the kinetic pressure of the particles to the pressure of the external (confining) field, as defined in §3.20, is essentially unity. If, however, β is not exactly unity, it may vary during the compression; thus, from equations (5.8) and (5.10), together with the definition of β, it can be shown that

$$\beta = \frac{\beta_0}{\beta_0 + (1 - \beta_0)\left(\dfrac{T}{T_0}\right)^{\frac{1}{2}f-1}\left(\dfrac{N}{N_0}\right)^{2-\alpha}}.$$

If $f = 2$ and the particle losses during compression are negligible, then $\beta \approx \beta_0$, and there is essentially no change in β; on the other hand, for $f = 3$, the value of β decreases during compression even when there are no losses. Consequently, if β is even slightly less than unity at the beginning of the compression, it may become appreciably less by the time the compression is complete and then equation (5.13) will no longer be applicable.

5.28. In the general case, where p_m is given by equation (5.4), equation (5.13) takes the form

$$\frac{1}{p_m}\cdot\frac{dp_m}{dt} = \left[\beta + \frac{1}{2}f(2-\beta)\right]\frac{1}{T}\cdot\frac{dT}{dt} + \left[\beta + \left(\frac{1}{2}f - \frac{\epsilon}{kT}\right)(2-\beta)\right]\frac{1}{N}\cdot\frac{dN}{dt}, \tag{5.17}$$

which can be solved to give a result similar to equation (5.15) but with more complicated exponents. For the present purpose, however, it is sufficient to consider some of the qualitative consequences of β being less than unity. Whether particle loss causes an increase or decrease of temperature, as compared with the case of no loss, depends on whether the ratio of the coefficients of the two terms on the right side of equation (5.17) is positive or negative, respectively. Since the coefficient of the first term must always be positive, the condition that the ratio be positive is readily found to be

$$\frac{\beta}{2-\beta} > \frac{\epsilon}{kT} - \frac{1}{2}f.$$

Thus, for $f = 2$, the ratio is positive for all values of β, provided that ϵ is less than kT; however if ϵ is $\frac{3}{2}kT$, then the loss of particles during compression will cause an increase of temperature only if β exceeds 0.67. For $f = 3$, the ratio will be positive provided that ϵ is less than $\frac{3}{2}kT$.

5.29. The general effect of β being less than unity may be derived by considering the situation in which losses are small, so that the dN/dt term in equation (5.16) can be neglected. In this event, for $f = 2$,

$$\frac{1}{p_m} \cdot \frac{dp_m}{dt} = \frac{2}{T} \cdot \frac{dT}{dt},$$

so that $T/T_0 = B/B_0$, which is exactly the same as would be obtained from equation (5.16) for the case of $\beta = 1$. On the other hand, for $f = 3$, equation (5.17), with the neglect of losses, leads to $T/T_0 = (B/B_0)^{2/3}$, whereas equation (5.16) would yield $(B/B_0)^{4/5}$. Hence, in a relatively dense plasma, the heating effect of radial compression is smaller when β is less than unity.

5.30. *Axial compression* is, on the whole, a less useful method of heating a plasma than the radial compression described above. A simple case only will be considered in which the reflecting fields of a mirror system are brought closer together thereby decreasing the volume of the plasma confined between them. The escape of particles during compression will be ignored, and it will be postulated that the magnetic field strengths remain constant, so that the gain in energy of a particle is due only to its reflection from the relatively slowly moving mirror fields.

5.31. Let l be the distance between the mirrors, so that dl/dt is the relative rate at which they approach each other. Suppose, for simplicity, that one of the mirrors is stationary; then the other will be approaching it at the rate dl/dt. Each time a plasma particle strikes the moving mirror, the component $v_{\parallel}$ of its velocity in the direction parallel to the field lines will be increased by $2\ dl/dt$. The number of impacts per second made by a particle in the moving mirror is $v_{\parallel}/2l$, and so the rate of increase of $v_{\parallel}$ is given by

$$\frac{dv_{\parallel}}{dt} = -2\frac{dl}{dt} \cdot \frac{v_{\parallel}}{2l} = -\frac{v_{\parallel}}{l} \cdot \frac{dl}{dt},$$

the minus sign arising from the fact that $v_{\parallel}$ increases as l decreases. From this result, it follows immediately that $v_{\parallel}l$ is constant and hence

$$v_{\parallel}^2 l^2 = \text{constant.} \tag{5.18}$$

5.32. At relatively low plasma densities, when collisions among the particles are infrequent, the average of $v_{\parallel}^2$, which is proportional to $W_{\parallel}$, varies directly as T; hence, equation (5.18) may be written as

$$Tl^2 = \text{constant,}$$

or

$$\frac{T}{T_0} = \left(\frac{l_0}{l}\right)^2. \tag{5.19}$$

Furthermore, if no particles escape from the mirrors (cf. Chapter 9), l is proportional to the plasma volume V, so that

$$TV^2 = \text{constant}$$

or

$$\frac{T}{T_0} = \left(\frac{V_0}{V}\right)^2. \tag{5.20}$$

From the arguments presented in §5.25, this is seen to be equivalent to the equation for the adiabatic expansion or compression of a gas with $\gamma = 3$ or $f = 1$. This is to be expected in an axial compression with no collisions, since the particles have only one degree of freedom, namely, parallel to the field lines. For conditions under which collisions become significant, f would be 3 even for an axial compression, and then the relationship between T and V would be represented by

$$TV^{\frac{2}{3}} = \text{constant}.$$

5.33. In the absence of end losses, any type of adiabatic compression, axial or radial, can be represented by the general equation (5.13). Upon subtracting unity from both sides, the result is

$$\frac{T - T_0}{T_0} = \left(\frac{V_0}{V}\right)^{\gamma - 1} - 1,$$

and since $T - T_0$ is the increase of temperature ΔT and V_0/V is the compression C, it follows that

$$\Delta T = T_0(C^{\gamma - 1} - 1). \tag{5.21}$$

Hence, for a given compression, the increase of temperature is proportional to the temperature before compression, and thus ΔT becomes greater as the initial temperature is increased. Further, the particle collision time varies as $T^{3/2}$, as is shown in Chapter 4, so that the frequency of collisions decreases as the temperature is raised. This means that γ will tend to become greater than 5/3 at high temperatures. For a given initial temperature T_0 and compression C, it is apparent from equation (5.21) that the larger value of γ will mean a greater increase in temperature as a result of adiabatic heating. It may be noted, too, that since collisions occur more frequently among electrons than among ions, γ will tend to approach 5/3 for electrons more readily than for ions.

5.34. In any procedure based on the heating of a plasma by magnetic compression, it is necessary that the gas be highly ionized. Since the neutral atoms are not affected by the magnetic field, they will be left behind when the plasma is compressed. Such cold neutral atoms can then return to the plasma and undergo charge exchange with hot ions, leading to the formation of hot neutral atoms and cold ions. The hot neutral particles will not be confined by the magnetic field and so will escape from the plasma region and reach the walls of the tube.   Here, they will give up their energy and return as cold neutral atoms which can undergo further charge exchange with hot ions in the plasma. The presence of neutral particles can thus result in serious loss of energy to the heated plasma.

SHOCK HEATING

5.35.   Some attention has been paid to the possibility of attaining temperatures of thermonuclear interest by means of shock waves [12-14].   One area of immediate interest is that of the preliminary heating of a plasma, to be followed by another method which is best suited to the situation.   For example, shock heating may be useful in a magnetic mirror system where the conditions do not lend themselves readily to the employment of ohmic heating.   The temperature of the plasma preheated by a shock can then be further increased by magnetic adiabatic compression, as described in the preceding section.   Suggestions have also been made for the use of shock heating as the first stage prior to the passage of a pinch discharge.

5.36.   When the pressure in a localized region of a fluid is suddenly increased in some manner, e.g., as the result of an explosion, a disturbance of relatively large amplitude, i.e., a compression wave, is propagated through the medium. In this event, the adiabatic approximation applicable to ordinary, small-amplitude wave motion is no longer valid.   Initially, the leading edge of the wave travels with the velocity of sound in the medium.   But the velocity of sound increases with pressure, and so the following portions of the wave disturbance, moving through the region compressed by the leading portion, travel at a faster rate.   The center of the wave thus tends to catch up with the leading edge, so that the wave front becomes increasingly steeper.

5.37.   In the limit, for a large-amplitude disturbance, producing a rapid pressure increase of considerable magnitude, a virtually discontinuous wave front is formed.   The resulting disturbance (or pressure wave) is referred to as a shock (or shock wave).   The shock wave then travels, essentially without change of shape, although with steadily decreasing velocity, at a rate greater than the speed of sound in the medium.   The physical conditions, such as pressure, temperature, and density, change abruptly at the shock front, and the flow of the fluid is no longer adiabatic, since various irreversible phenomena are involved.   The point of immediate interest is that the temperature just behind the shock front is always greater than that which would result from adiabatic compression over the same pressure range.

5.38.   In addition to the shocks of hydrodynamic origin described above, resulting from the sudden release of a large amount of heat energy in a fluid, it is possible to produce hydromagnetic shocks in a plasma due to the propagation of hydromagnetic waves of finite amplitude.   That this is the case follows from the fact that the velocity of hydromagnetic waves increases with pressure. The general expression for the velocity of such waves of small amplitude may be written in the form of equation (4.108), i.e.,

$$v^2 = \frac{B^2}{4\pi\rho} + \gamma\,\frac{p}{\rho}, \tag{5.22}$$

where the first term is the square of the Alfvén speed and the second is that of the speed of sound. The latter increases in a rapid compression, and it remains to be shown that the former does so also. In a plasma of low density in a strong magnetic field, the particles behave as if they were tied to the field lines, as shown in §4.24; Br^2 is then constant, where r is the plasma radius. If the length is constant and there are no particle losses, r^2 is proportional to $1/\rho$, were ρ is the mass density of the plasma, and so B/ρ is constant. It follows from equation (5.22) that v^2 must increase with the field strength. Consequently, the velocity of the hydromagnetic wave in the plasma increases with the compression. As the wave moves through a region that has already been compressed by the leading portion, the following portion will travel at a faster rate and a shock will develop as described above.

5.39. Many properties of the medium behind a steady-state shock front, as well as the shock velocity, can be calculated by invoking the laws of conservation of mass, momentum, and energy and certain equations of state of the fluid carrying the shock. In hydrodynamics, these laws yield the Rankine-Hugoniot equations, and in the hydromagnetic case analogous relationships are obtained which give the pressure ratio across the shock front, the shock velocity, and the increase in the gyromagnetic energy, i.e., $W_\perp$, the energy perpendicular to the field lines, of the charged particles. The results show that $W_\perp$ is greater than is to be expected from adiabatic compression alone, so that the formation of the shock is accompanied by additional, nonadiabatic heating of the plasma. Although a detailed theoretical model of the hydromagnetic shock structure has not yet been worked out, it appears probable that, for shocks of moderate strength, the ions are heated more than the electrons [15]. This result is, of course, of considerable interest in connection with the realization of thermonuclear reactions.

5.40. Unfortunately, there exists in the literature of thermonuclear research some ambiguity concerning the significance of the term "shock heating." It has become the common practice to gather under this heading all heating processes which involve a disturbance moving faster than sound and producing an irreversible, nonadiabatic increase in the energy of the plasma. However, it appears desirable to make a distinction between "magnetic pistons," on the one hand, such as are operative in the dynamic (§7.12) and theta (§11.2) pinches, where a current-carrying sheath of plasma drives ahead of it all the mass it encounters, and, on the other hand, large-amplitude waves which, while leaving the plasma heated, do not act primarily to transport mass. The discussion given above and the treatment which follows apply to the latter situation, where the ordinary shock wave assumption of mass flow continuity across the front applies. Examples of the rapid, irreversible compressions by means of the magnetic piston effect will be given in Chapters 7 and 11.

5.41. In conventional hydromagnetic shocks, particle scattering is the mechanism whereby the dissipation of energy occurs. There are, in addition, what are known as "collisionless shocks" which are produced when the conditions are such, namely, low plasma density and high temperatures, that collisions between the particles are so infrequent as to be negligible. The dissipative mechanism is then the interaction of hydromagnetic waves of finite amplitude [16].

5.42. The shocks which have been used in controlled thermonuclear research have generally been partly hydrodynamic and partly hydromagnetic in nature. For example, a shocked plasma can be produced by the passage of a transient electrical discharge between two electrodes in a gas at low pressure [17, 18]. The very rapid ohmic heating causes a sudden local increase in the gas pressure which, together with a changing magnetic field, results in the development of a shock wave. The shocked gas is partially ionized and consequently it can be driven by a magnetic field.

5.43. A convenient apparatus for the study of electromagnetically driven shock waves is shown diagrammatically in Fig. 5.6 [19, 20]. It consists of

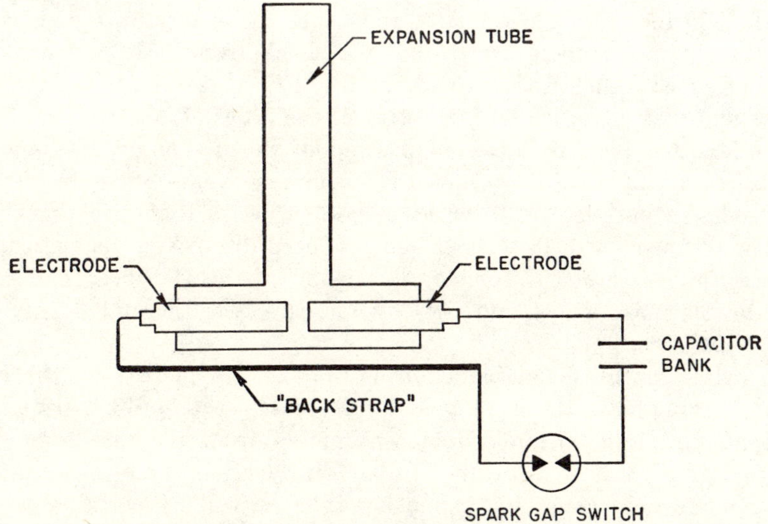

Fig. 5.6. Production of magnetically accelerated shocked plasma.

a T-shaped quartz tube about 2 or 3 cm in diameter, with a vertical leg (or expansion tube) of 30 to 90 cm or so in length. By means of electrodes inserted into the two arms of the T, a discharge voltage of 20 to 100 kilovolts or more, as required, is supplied from a capacitor bank through a spark-gap switch. The discharge currents may be as high as 100 to 300 kiloamp flowing for 0.3 to 0.5 microsec.

5.44. The passage of the electrical discharge produces a plasma which can be driven up the expansion tube by means of a magnetic field perpendicular to both the discharge and the (vertical) direction of the tube. The simplest way of applying such a field is depicted in Fig. 5.6; the return lead to one of the electrodes, sometimes called a "back strap," runs beneath and parallel to the discharge tube. The resulting Lorentz force then acts in such a manner as to drive the shocked and ionized gas up the expansion leg. Instead of (or in addition to) the magnetic field produced by the back strap, a field acting in the same direction may be applied by placing the shock tube in the middle of a solenoid with its axis perpendicular to both the discharge and expansion limbs of the T tube.

5.45. The velocity of the shock front as it moves up the expansion tube can be determined by making use of its luminosity, either by means of a rotating mirror (streak) camera (§6.129) or by locating a number of photo-multiplier tubes at various known distances from the discharge. Another experimental procedure utilizes the reflection of microwaves from the shock front to determine the velocity of the latter, and spectroscopic measurements can provide information concerning the temperatures and densities attained (see Chapter 6). There is definite evidence that the externally applied magnetic field increases the velocity, and hence the energy, in the shock wave and the maximum temperature behind the shock front. Using magnetic acceleration, shock velocities up to about 3×10^7 cm/sec have been observed in deuterium plasmas, with temperatures of the order of 100 ev.

5.46. Mention may be made in passing of the use of an H-shaped tube which is essentially two T tubes with a common expansion limb. Two sets of electrodes in parallel are inserted in both arms so that discharges can be passed simultaneously. Two shocks so produced are driven magnetically and collide in the center of the connecting portion of the H tube. The collision brings the plasmas to rest and thereby increases the thermal energy of the particles [21].

5.47. After the initial acceleration of the shocked plasma by the electromagnetic driving force, the shock velocity decreases because the thermal energy is dissipated rapidly within a few centimeters from the discharge, largely owing to loss of heat to the tube walls. However, if there is a time-rising axial magnetic field along the expansion tube, it will have the effect of insulating the plasma from the walls and will, in addition, accelerate the shock wave by compressing the plasma behind the shock front [21-23]. The tube is surrounded by a number of single-turn coils connected in parallel to a large capacitor bank; this bank is switched on when the shock wave enters the coil system. The resulting axial magnetic field, which cannot penetrate the highly ionized and highly conducting plasma, causes the plasma behind the shock front to move radially inward from the tube walls. This compression leads to an increase in the internal energy of the plasma and serves to maintain a high shock velocity and particle temperature. In addition to the energy due to compression, the plasma gains energy as the result of ohmic heating by the large induced currents flowing in its surface.

5.48. Another method for producing and accelerating hydromagnetic shocks makes use of what has been called a magnetic annular shock tube [24]. The driving force on the plasma is produced by the crossing of a radial current in the annulus with an azimuthal magnetic field. Since the force acts in the direction perpendicular to both the electric and magnetic fields, it serves to accelerate the pasma in the axial direction, i.e., along the length of the annular tube. A strong shock is observed to proceed through the gas ahead of the current layer.

5.49. A schematic drawing of the shock tube is shown in Fig. 5.7. The discharge from a capacitor bank is passed between the two cylindrical tungsten

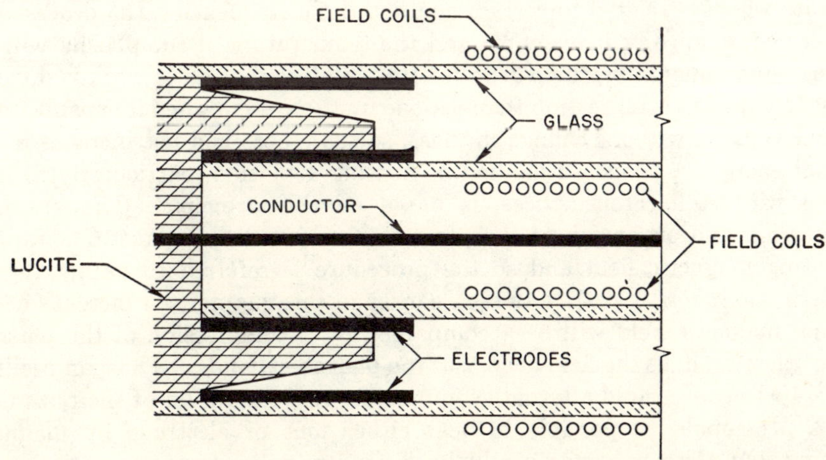

Fig. 5.7. Schematic representation of magnetically accelerated shock tube.

electrodes, the annular space between them containing the experimental gas. The latter is preionized by means of a subsidiary discharge about 1 microsec before the main discharge is triggered. Two types of "bias" field have been used to provide additional magnetic acceleration of the shocked plasma along the discharge tube. In one, a coil wound outside the tube produces an axial field, and in the other there is a combination of axial field from two coils, one outside and the other inside the annular region, and an azimuthal field obtained by passing a current through a conductor located at the axis of the system. With the latter arrangement, shock velocities of 4×10^7 cm/sec were obtained in hydrogen gas under such conditions that the gyromagnetic radius of the ions was small compared with the thickness of the annulus. The maximum temperature behind the shock front was estimated to be about 100 ev.

5.50. An important aspect of heating a plasma by the hydromagnetic shock produced by means of a time-rising magnetic field is the preionization and preheating of the gas. This is often necessary for the formation of a current sheath and a sharp boundary between the plasma and the magnetic field. The particular procedure employed varies according to the nature of the experiments being performed. In some cases a direct discharge between electrodes is used, as described above, whereas in others high-frequency electrodeless discharges serve the purpose. Use has also been made of a Philips Ionization Gauge (or Penning) type of discharge (§5.91) and of the first half-cycle of an oscillating magnetic field (see Chapter 11).

HEATING BY MAGNETIC PUMPING

5.51. If a plasma in a magnetic field is alternately compressed and expanded at a rate which is related to certain characteristic frequencies, the process will be thermodynamically irreversible and the temperature of the plasma will increase. Fundamentally, the reason for this result is that more work is done on the system during compression than is done by the system when it expands to its original volume, without change in mass, so that there is a net increase in the internal energy. A proposed method of heating a plasma, particularly in connection with stellarator devices, is based on this principle. The alternate compression and expansion of the plasma is achieved by means of a rapidly oscillating magnetic field and so the procedure is referred to as heating by *magnetic pumping* [25-28]. As seen earlier in this chapter, an increase in the external magnetic field will be accompanied by a compression of the plasma; on the other hand, as the field decreases the plasma expands. Thus, an oscillating field will cause rapid alternate compression and expansion of the plasma.

5.52. It should be possible to heat either ions or electrons by magnetic pumping, but the conditions would be different in the two cases. Since the direct heating of the ions is to be preferred, this particular aspect will be considered here. In the theoretical discussion of heating by magnetic pumping, three characteristic times, in addition to the gyromagnetic period of the ions, are important. They are (1) the average collision time t_c, i.e., the time for an ion to undergo a large-angle collision as a result of encounters with both ions and electrons, (2) the oscillation time t_o of the radiofrequency field, i.e., the reciprocal of the frequency, and (3) the transit time t_t, i.e., the time required for a typical ion to traverse the region over which the oscillating field is applied. If these times are of the same order of magnitude, the treatment becomes very complicated, and so three limiting cases have been considered.

5.53.   In the first case, it is postulated that the oscillation time t_o is of the same order of magnitude as the collision time t_c, and both are small in comparison with the transit time t_t; that is,

$$t_o \approx t_c \ll t_t.$$

All these characteristic times are large in comparison with the gyromagnetic period of the ions. The special case in which this is approximately equal to the oscillation time will be considered later, in connection with ion cyclotron resonance heating (§5.61).

5.54.   When the radiofrequency field is increased, there is an increase in the energy component of the ions in the direction perpendicular to the field. If an ion undergoes a collision before the field strength has returned to its initial value, there will be a transfer of energy from the perpendicular to the parallel degree of freedom. The situation is somewhat similar to that described in connection with adiabatic heating when the rise time of the magnetic field is of the same order as (or greater than) the collision time. When the field strength has completed a cycle there will have been a net transfer of energy from the alternating field to the plasma. This process is called *collisional heating.**   It has been shown theoretically that the rate of heating of the ions by this mechanism varies roughly as $T^{-1/2}$ so that it can be useful, as an alternative to ohmic heating, at relatively low plasma temperatures. At higher temperatures, collisional heating may be expected to be relatively inefficient from the standpoint of time.

5.55.   If the conditions are such, e.g., relatively low plasma density and high temperature, that the ion collision time is much larger than both the transit and oscillation times, which are of the same order, that is,

$$t_o \approx t_t \ll t_c,$$

then *transit time heating* can occur. In these circumstances, the ions suffer essentially no collisions in traversing the magnetic pumping region, but they gain energy due to the radial compression by the oscillating field. Calculations indicate that this is the situation of major interest for the attainment of high plasma temperatures, especially as the rate at which the ions absorb energy from the field increases in proportion to $T^{3/2}$. In order to obtain the maximum heating rate, the frequency should be optimized at each temperature.   In general, for plasma densities and temperatures of thermonuclear interest, the frequency of the oscillating field may be expected to be in the region of 50 to 100 kilocycles/sec. The oscillation time must always be somewhat

---

* This method for heating a plasma was proposed independently by A. Schlüter [29], who referred to it as heating by the "gyrorelaxation effect."   A similar suggestion was made by G. I. Budker in the U.S.S.R. [30].

less than the transit time; if it is greater, the ions will be cooled rather than heated [27].

5.56. A rough relationship between the energy increase and the amplitude of the oscillating magnetic field may be obtained from the following approximate considerations. Let the field strength B in the magnetic pumping region at any time t be represented by

$$B = B_0 + \Delta B \sin \omega t,$$

where ΔB is the maximum amplitude of the oscillating field of angular frequency ω, and B_0 is the steady strength of the magnetic field which confines the plasma. The field gradient, at any instant t, in the axial (or z) direction is then given very roughly by

$$\frac{dB}{dz} \approx \frac{\Delta B}{d} \sin \omega t,$$

where d is the length of the pumping region. Since, as has been postulated, the transit time of an ion is approximately equal to the oscillation time of the field, an ion moving in the axial direction will be subjected to a space-varying magnetic field, and the force F acting on the ion is given by

$$F = -\mu \frac{dB}{dz} \approx -\mu \frac{\Delta B}{d} \sin \omega t,$$

where μ is the magnetic moment of the ion (§9.18).

5.57. The force is proportional to the rate of change of velocity of the ion in traversing the magnetic pumping region, and so the transit time heating power P_{tt}, i.e., the rate of increase of energy, varies as F^2; hence,

$$P_{tt} \approx \text{constant} \times (\mu\,\Delta B)^2.$$

It will be shown in §9.6 that the magnetic moment is equal to $W_\perp /B$, and since $W_\perp$ does not change appreciably, it follows that

$$P_{tt} \approx \text{constant} \times \left(\frac{\Delta B}{B}\right)^2.$$

Thus, the transit time heating rate increases as the square of the ratio $\Delta B/B$. This ratio, which is represented by the symbol M, is called the *modulation* of the field; it is more precisely defined by

$$M = B_{rf}/B_{dc},$$

where B_{rf} is the peak value of the radiofrequency field and B_{dc} is the strength of the direct current (or confining) field. A more exact treatment than the one given above leads to the same conclusion, namely, that the transit time heating rate is proportional to the square of the modulation.

5.58. The third case to be considered is that in which the transit and oscillation times are approximately equal, as for transit time heating, but these times are now much larger than the collision time, i.e.,

$$t_t \approx t_o \gg t_c,$$

so that the ion undergoes many collisions during its passage through the radio-frequency field and in the course of a single oscillation. The short collision time or small effective mean free path of the ions may be due to relatively high densities, low temperatures, or cooperative plasma phenomena which are not yet well understood. In these circumstances, the oscillating field produces density variations which are equivalent to acoustic waves in the plasma. The resulting absorption of energy from the magnetic pumping field is consequently referred to as *acoustic heating*. The heating rate varies with the conditions in a manner similar to transit time heating, so that it is proportional to M^2 and to $T^{3/2}$. However, it is generally at low temperatures that the conditions for acoustic heating can exist, and then the $T^{3/2}$ factor provides little advantage.

5.59. It may be concluded from the foregoing discussion that, if magnetic pumping is used to heat the plasma in a system such as a stellarator, it would probably be best to choose the conditions for transit time heating of the ions, provided another procedure were used to preheat and ionize the gas. For the preheating stage, it is doubtful whether it is possible to improve on ohmic heating, in spite of the limitations mentioned in §5.11 *et seq.* The field modulation for magnetic pumping should be as large as possible; however, it must not be too large, otherwise the net strength of the magnetic field during the negative half cycles of the oscillating field would be insufficient to confine the plasma. It appears that a value of 0.5 for M is desirable, although it may be difficult to achieve at high-frequencies because the increased reactance of the radiofrequency coils makes the required voltages very large.

5.60. When the magnetic pumping heating concept was originally suggested, it was proposed to apply it at a somewhat widened portion (or bulge) of the tube containing the plasma. The purpose was to decrease the value of B_{dc}, the confining field strength, in this region and thereby make possible a larger value of M for a given radiofrequency field B_{rf}. It was realized later, however, that since the magnetic field would be larger at the ends of the bulge than in the center, the bulge would be in effect bounded by magnetic mirrors. It now appears that the effect of the bulge on magnetic pumping is quite complex and its influence on performance is not well understood.

ION CYCLOTRON RESONANCE HEATING

GENERAL CONSIDERATIONS

5.61. The proposal for increasing the energy of the ions in a plasma, known as *ion cyclotron resonance heating*, is based on a principle similar to that utilized to accelerate the charged particles in a cyclotron. If a plasma is confined by a static axial magnetic field of strength B, the ions will gyrate about the lines of force with an angular frequency given by equation (4.3) as

$$\omega_g = \frac{eB}{mc} \quad \text{radians/sec,}$$

so that

$$f_g = \frac{eB}{2\pi mc} \quad \text{cycles/sec.} \tag{5.23}$$

Since e is 4.80×10^{-10} statcoulomb, m is $1.67 \times 10^{-24}A$ g, where A is the mass number of the ion, and c is 3×10^{10} cm/sec, it follows that

$$f_g = 1.54 \times 10^3 \frac{B}{A} \quad \text{cycles/sec,} \tag{5.24}$$

where B is in gauss. For a deuteron, therefore,

$$f_g = 7.7 \times 10^2 B \quad \text{cycles/sec.} \tag{5.25}$$

Suppose a time-varying field of this frequency is superimposed on the static field B confining a deuterium plasma, by passage of a radiofrequency current through a coil which is concentric with that producing the axial field. Then in each half-cycle of its rotation about the field lines, the ions should acquire energy from the oscillating electric field. If B is 10,000 gauss, for example, the frequency of the field which is in resonance with deuterons in a plasma is found from equation (5.25) to be 7.7 megacycles/sec.

5.62. In comparing ion cyclotron resonance heating with heating by magnetic pumping, it should be observed, first, that the former requires an oscillating field having a definite frequency determined by the strength of the confining field. Although there is a preferred frequency (or frequency range) for magnetic pumping, dependent upon the existing conditions, the value is not critical. It should be noted, too, that in situations of present interest the ion cyclotron resonance frequency is much greater than both the transit frequency of the ions through the heating section and their collision frequency. Thus, the transit and collision times of the ions are long compared with the ion cyclotron (or gyromagnetic) period, and so the ions can undergo many gyrations in their passage through the oscillating field without suffering collisions. It would appear, therefore, that heating by ion cyclotron resonance should be very efficient, the energy of the ions increasing by a definite amount in each rotation.

5.63. A question arises, however, concerning the possibility that the oscillating field may be excluded from the plasma by internal electric fields produced by the gyration of the ions. Since the motions of all the separate ions are in phase, under the influence of the applied oscillating field, and their gyromagnetic radii increase with increasing energy, the ions have a net outward radial motion relative to the magnetic field lines. The electrons, on the other hand, do not gain energy directly from the oscillating field, because the electron cyclotron frequency is much higher than that of the applied field. Consequently, they remain tied to the lines of force, which they cannot cross except as a result of rare collisions. There will thus be a tendency for a space charge to be set up which produces a radial electric field. This radial field combined with the external azimuthal (oscillating) field results in the establishment of a circularly polarized electric field whose sense is opposite to the gyration of the ions. The induced electric field thus tends to decelerate the ions [27, 31].

5.64. The effect described above, whereby the plasma shields itself from the applied electric field which operates so as to increase the ion energies, increases with the density of the plasma. It has been estimated that at densities below about 10^{12} to 10^{13} particles/cm^3 the induced field effect is negligible and the simple, single-particle treatment of the motion of the ions under the influence of the oscillating field is applicable, to a good approximation. That is to say, each ion in the plasma may be regarded as acting independently of the other ions (and electrons), acquiring a certain amount of energy in each gyration about the magnetic field lines. At higher particle densities a cooperative phenomenon, leading to plasma oscillations, develops. As will be seen later, although the single-particle situation no longer holds, another efficient mechanism becomes available for heating the ions.

SINGLE-PARTICLE TREATMENT

5.65. If the plasma density is low, so that the single-particle treatment may be used, an approximate estimate of the amount of energy acquired by the ion per rotation, provided there are no charge exchange losses, can be made in the following manner [32]. The radiofrequency current produces a time-varying field and this induces an azimuthal electric field, which accelerates the charged particle. The strength of this field will vary with distance from the center of the tube holding the plasma, it being greatest at the outside near the current coil and falling to zero in the center. For simplicity, it will be supposed, however, that the field has a uniform (average) value in space. The time variation at all points can then be represented by

$$E = E_{\max} \sin \omega t,$$

where ω is the angular frequency of the applied field.

5.66. The rotational motion of the particle in the direction perpendicular to the axial magnetic field is shown in Fig. 5.8 and, when the ion is in resonance with the applied radiofrequency field, ωt is equal to θ at all times, so that

$$E = E_{\max} \sin \theta.$$

The energy gain by the ion per rotation is equal to the integral of $\mathbf{E} \cdot d\mathbf{s}$ around the circle, where

$$\mathbf{E} \cdot d\mathbf{s} = E \, ds \sin \theta$$
$$= E_{\max} \sin^2 \theta \, ds.$$

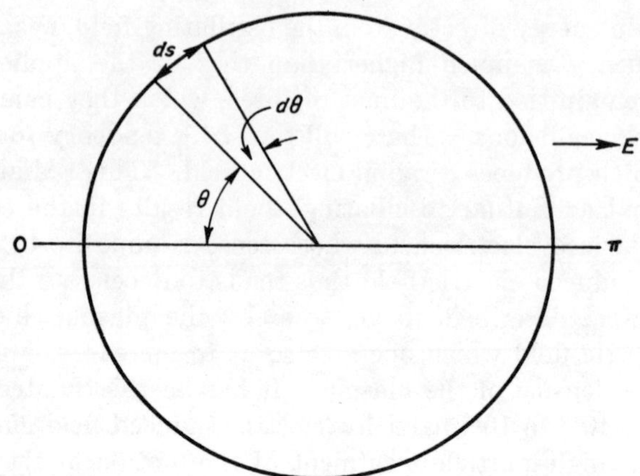

Fɪɢ. 5.8. Gyromagnetic (cyclotron) rotation of charged particle.

As is seen from Fig. 5.8 that $ds = r_g\, d\theta$, where r_g is the (gyromagnetic) radius of the ion path; hence,

$$\mathbf{E}\cdot d\mathbf{s} = E_{\max} r_g \sin^2\theta\, d\theta.$$

5.67. The energy gain per rotation $\Delta W_\perp$ in the direction perpendicular to the electric field, is thus given by

$$\Delta W_\perp = \oint \mathbf{E}\cdot d\mathbf{s}$$

$$= 2E_{\max} r_g \int_0^\pi \sin^2\theta\, d\theta$$

$$= \pi r_g E_{\max}.$$

For convenience this may be written as

$$\Delta W_\perp = a r_g, \tag{5.26}$$

where the constant a is defined by

$$a \equiv \pi E_{\max}.$$

Sinch r_g increases in every rotation of the ion, as shown below, the energy gain increases correspondingly, but it is always proportional to the maximum strength of the induced electric field.

5.68. The gyromagnetic radius of the ion is given by equation (4.5) as

$$r_g = \frac{mv_\perp c}{eB}$$

or, since the energy $W_\perp = \tfrac{1}{2}mv_\perp^2$,

$$r_g = \frac{(2m)^{\frac12}c}{eB}\, W_\perp^{\frac12}.$$

That is,

$$r_g = bW_\perp^{\frac12}, \tag{5.27}$$

where b is a constant defined by

$$b \equiv \frac{(2m)^{1/2}c}{eB}.$$

Upon differentiating equation (5.27) and replacing $dr_g/dW_\perp$ by $\Delta r/\Delta W_\perp$, omitting the subscript g for simplicity, it is seen that Δr, the increase in radius per turn in the radiofrequency field, is

$$\Delta r = \tfrac{1}{2}bW_\perp^{-1/2}\Delta W_\perp.$$

By combining this result with equations (5.26) and (5.27), it follows that

$$\Delta r = \tfrac{1}{2}ab^2, \tag{5.28}$$

so that, since a and b are constants, the radius of gyration increases by a constant amount in each turn. It should be noted that the frequency, given by equation (5.25), remains constant, since the energy (or speed) of the particle increases with the radius.

5.69. The number of turns n the ion must make in the radiofrequency field in order that it may acquire a certain amount of energy is given by

$$n = \frac{r - r_0}{\Delta r},$$

where r_0 is the initial radius of the ion path and r is the value when the ion has reached the required energy. If the corresponding perpendicular energies are W_0 and W, then, from equations (5.27) and (5.28),

$$n = \frac{2}{ab}\left(W^{1/2} - W_0^{1/2}\right). \tag{5.29}$$

The time t required for the particle to make the n turns is n/f, where f, the frequency of the oscillating field with which the gyrating ions are in resonance, is given by equation (5.25).

5.70. In view of the limited confinement time of a plasma, for various reasons, it is desirable that the heating period should be as short as possible. Since n is inversely related to ab, by equation (5.29), it follows that the heating time, for a given energy increase, is proportional to $1/abf$. Taking into consideration the definitions of a and b, and equation (5.25) for f, it is seen that, for given ions, e.g., deuterons, the time is independent of the confining field strength but is inversely proportional to $E_{\max}$. The latter varies directly as the strength of the oscillating field, and so a strong field is required to minimize the heating time. The limitation mentioned in §5.59, concerning the net field strength in the negative half-cycles, is, of course, applicable here also.

5.71. Ion cyclotron resonance heating at low plasma densities is particularly applicable to a magnetic mirror system where the density can be subsequently increased by axial or radial compression (§5.18 *et seq.*). Furthermore, by placing metal plates (or electrodes) at the ends of the tube containing the plasma, the electrons can travel along certain lines of force into the plates and return along other lines, thereby decreasing the space-charge effect. In experiments designed to test the heating of low-density hydrogen and deuterium plasmas confined in a magnetic mirror system, these plates also acted as electrodes for a Philips Ion Gauge type of discharge (§5.91) to produce, if necessary, preionization of the gas [37]. The resonance absorption of energy was studied by measuring the light output from the plasma glow. An oscillating field of constant frequency (about 3 megacycles/sec) was applied by means of a two-turn coil of copper strip, surrounding the discharge tube between the mirrors. The axial magnetic field strength was changed steadily and the corresponding light outputs were measured. It was found that when the field strength was approximately equal to that required for resonance, according to equation (5.25), there was a very marked increase in the light output, indicating that heating of the plasma had occurred at this point.

ION CYCLOTRON WAVES

5.72. Theoretical considerations indicate that, in moderately dense plasmas, there can occur natural oscillations at frequencies slightly lower than the ion (single-particle) cyclotron frequency in the given magnetic field [33]. These oscillations, called *ion cyclotron waves*, are the short wave length, low-density limit of transverse hydromagnetic waves (cf. §4.115) [34, 35]. If an external field of the proper frequency is applied to the plasma, the individual ions move in circles around the lines of force, just as at low densities. But the phases and amplitudes of the ion velocities now vary sinusoidally both in space and time, because the motions of the plasma particles are coupled electromagnetically to one another. The resulting oscillations (or waves) thus represent a cooperative organized motion of the plasma as a whole.

5.73. The ion flow is divergent, as in the single-particle picture, and this would be expected to produce a large space charge and accompanying radial

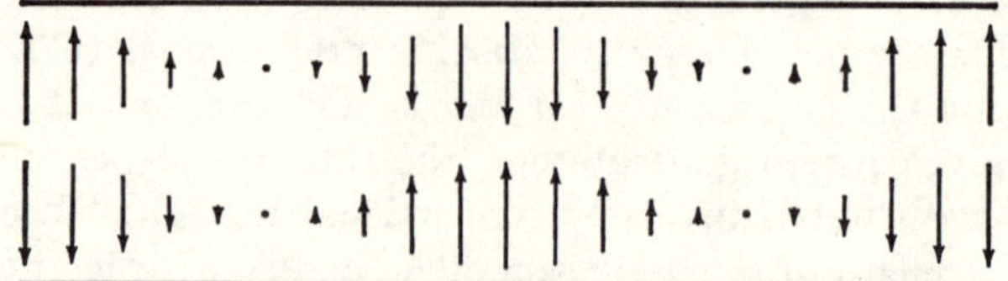

Fig. 5.9. Ion cyclotron wave motion.

electric fields. However, since, as a result of the wave motion, the ion flow pattern in the axial direction is periodic, as shown in Fig. 5.9, the electrons are able to flow along the lines of force and neutralize the space charge. Thus, if at some instant the ions at a given point have an inward motion, then at a distance of half a plasma wave length away they will be moving outward with the same amplitude.* The ion density will thus vary in the axial direction and so electrons can flow along the lines of force from a region where the ion density is low to one where it is high. There is experimental evidence, which will be mentioned below, that the flow of this neutralizing electron current produces ohmic heating of the plasma.

5.74. A much more important heating mode is expected to arise from the damping of the ion cyclotron waves. Normally when these waves are excited, they do not damp as they propagate, and no energy is transferred from the oscillating field to the plasma. However, the waves may be allowed to propagate into a region of slowly decreasing magnetic field. The decreasing field will cause the wave length to become shorter and shorter, and the frequency will approach the local ion cyclotron frequency. Any ions entering the plasma wave will then be subjected to an electric field oscillating at the resonant frequency, and so they will pick up energy from the field. As a result, the pasma wave amplitude will damp out; there is thus a decrease in the wave energy accompanied by an increase in the ion cyclotron particle energy. This energy is perpendicular to the field lines, but it soon becomes random in direction, i.e., it is thermalized. A number of other factors also facilitate the thermalization of the ion energy. It should be emphasized that the heating by cyclotron damping, as it is called, is thus not direct but involves the intermediate action of the ion cyclotron waves.

5.75. The damping (or thermalization) of ion cyclotron waves in a region of lower magnetic field has been compared to the transfer of the energy in water waves to turbulent motion as they roll onto a beach. As the waves move toward the shore, the wave lengths become shorter and shorter because

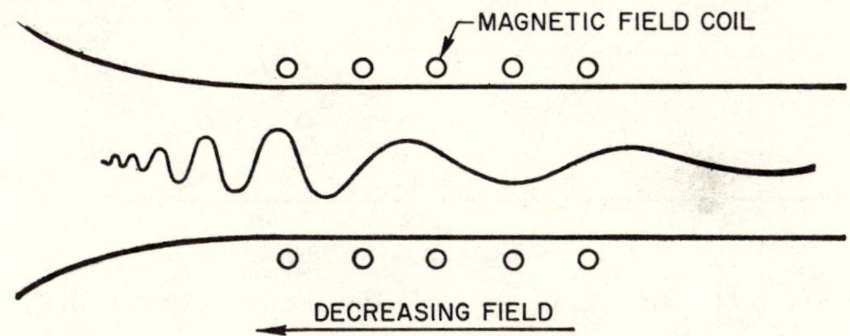

FIG. 5.10.   Thermalization of ion cyclotron wave on "magnetic beach."

the water is getting shallower. Finally, in the shallow water the waves are unable to propagate, and so they form breakers and the wave energy is transformed into heat [36, 37]. Because of the analogy, the situation which should lead to damping and thermalization of ion cyclotron waves has been referred to as a "magnetic beach" (Fig. 5.10).

*A method for exciting these plasma waves, by using coils of alternate phase, is described in §8.108.

5.76. Heating by means of ion cyclotron waves is suitable for application to plasmas in stellarator and magnetic mirror systems. Experiments on the B-65 stellarator (§8.108 *et seq.*) indicate that both single-particle direct heating and ion cyclotron wave heating can be produced. The method used was to apply an electric field of fixed frequency (11.5 megacycles/sec) and then to increase the strength of the confining magnetic field through the region

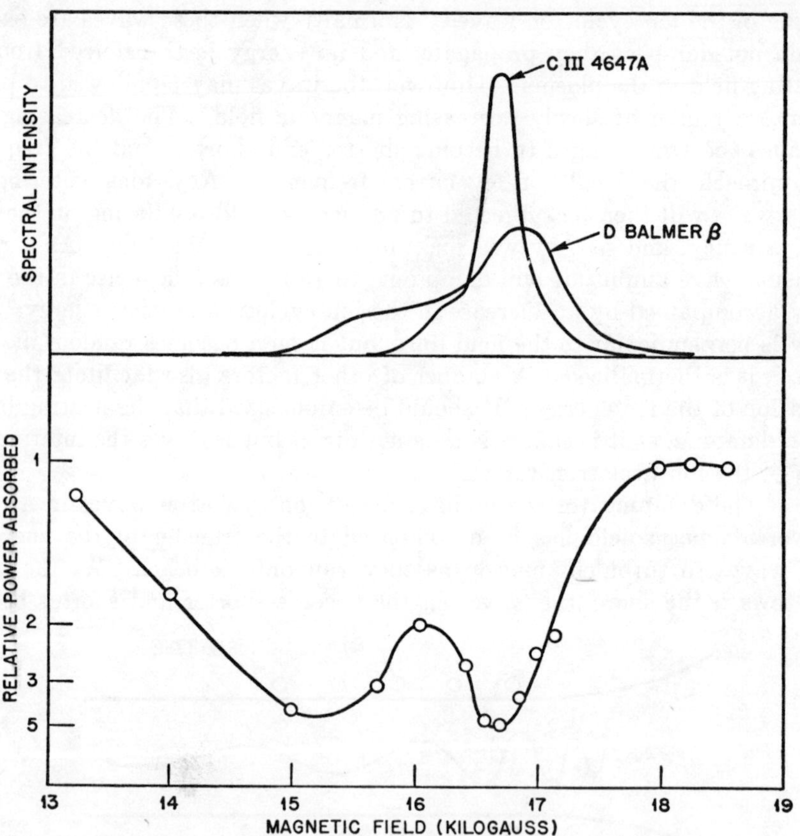

Fig. 5.11. Results of ion cyclotron heating experiments.

where ion cyclotron resonance was to be expected. From the observed input impedance corresponding to each magnetic field value, the plasma loading was expressed as the ratio of the radiofrequency power absorbed by the plasma to that which would be lost in an induction coil having a Q of 300. The results obtained with a deuterium plasma and a radiofrequency power of 200 kw are shown in Fig. 5.11 [37].

5.77. It is seen that the plasma presents particularly heavy loads at magnetic field strengths of 15 and 16.5 kilogauss. At the lower field value, which corresponds to the single-particle ion cyclotron frequency, neutrons were produced, presumably due to the acceleration of deuterons in the low-density region outside the main plasma. The loading observed at 16.5 kilogauss, representing a lower oscillation frequency, is attributed to the formation of ion cyclotron waves and their absorption of energy from the exciting radio-frequency field. At this point strong Balmer lines of the deuterium spectrum are observed as well as the C III lines of carbon. The Doppler broadening of the C III line at 4647A indicates an ion temperature of approximately 50 ev, and the electron temperature is probably not greatly different.* The heating is attributed to the electron neutralizing current mentioned in §5.73, since the experimental conditions were not such as to lead to damping of the ion cyclotron waves on a "magnetic beach."

PLASMA SOURCES FOR MAGNETIC MIRROR SYSTEMS

INTRODUCTION

5.78. In certain magnetic mirror systems, it is not convenient to form the plasma within the containment chamber. In these circumstances, the ions and electrons are produced, and sometimes accelerated, outside; they are then injected between the mirrors where they are trapped in a suitable manner (§9.49 *et seq.*). Two general approaches to the situation have been considered: (1) the injection of a plasma of relatively low temperature, i.e., a few electron volts, which is subsequently heated by adiabatic compression in the confining magnetic fields, and (2) the introduction of high-energy atomic or molecular ions or neutral atoms, with enough electrons, in the former cases, to permit space charge neutralization. The atomic ions are trapped by changing the magnetic field, the molecular ions by dissociation, and the neutral atoms by ionization. If the particles are injected with sufficiently high energies, i.e., 100 kev or more, it is possible that a thermonuclear reaction system might result after Maxwellian distribution is attained.

5.79. One of the problems related to injection of a plasma into a magnetic mirror system arises from the fact that the direction of motion of the charged particles should be as far as possible perpendicular to the magnetic field lines. The reason is that if the energy component of a particle in the direction parallel to the field exceeds a certain fraction of its total energy, the particle will pass through the mirrors and not be reflected (§9.16). It would appear desirable, therefore, that energetic charged particles should be injected transversely to the field direction. For the simple injection of ions into a steady magnetic field, however, trapping will not be accomplished; the ions, after following a curved path in the field, must necessarily leave it again. Some of the ways which have been proposed for overcoming this aspect of the problem of plasma

---

* Owing to a typographical error, the temperature was given in the original reference as about 5×10^6 °K, i.e., nearly 500 ev.

injection into a mirror geometry will be considered in Chapter 9. The discussion below will refer primarily to the design of the plasma sources.

OCCLUDED-GAS, HIGH-ENERGY ION SOURCES [38-40]

5.80. Where a pulsed source is required, it is advantageous to have the gas, e.g., deuterium, which is to be ionized, adsorbed (or occluded) by a suitable material, such as titanium. An occluded source of this kind has the advantage, over that using the deuterium in its normal gaseous form, in the respect that no neutral particles are introduced in the periods between pulses. The forms of occluded-gas source to be described here have the merit of compactness and of a high ratio of atomic to molecular ions in their output. The operation involves the passage of a discharge across a series of "sandwiches" consisting of discs (or washers) of titanium, containing deuterium, separated by insulators. The titanium washers of suitable dimensions are first heated in a good vacuum to remove adsorbed gases. They are then allowed to cool in an atmosphere of deuterium; as a result, approximately 300 cm³ of gas are taken up per gram of metal. The washers are then assembled to form a stack, each pair being separated by a layer of mica or ceramic of slightly larger outer diameter.

5.81. In one type of assembly, which is about an inch long, the titanium washers have a diameter of 0.2 inch with a central hole of less than 0.1 inch diameter. The stack of washers constitutes one electrode (anode); the other

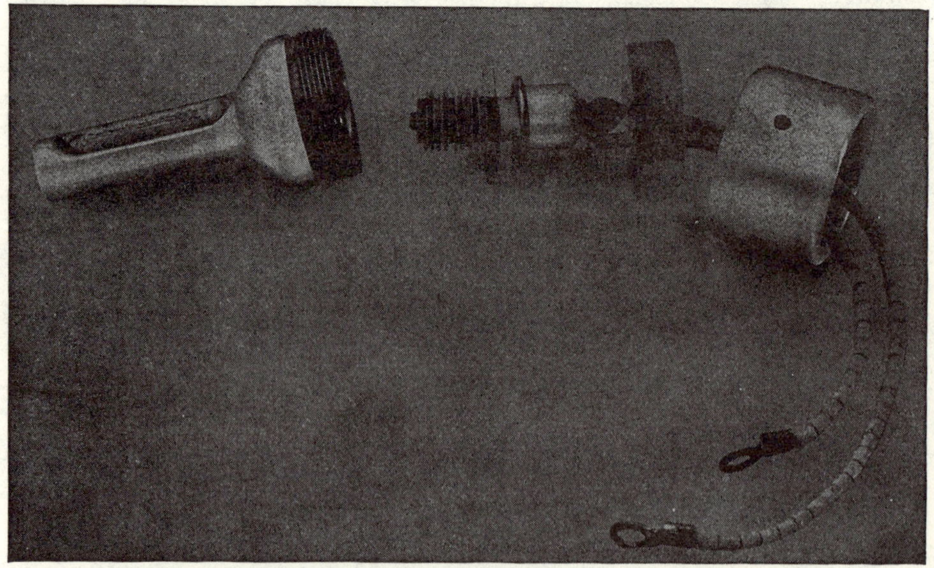

Fig. 5.12. Occluded-gas ion source. (Courtesy of Lawrence Radiation Laboratory, Livermore)

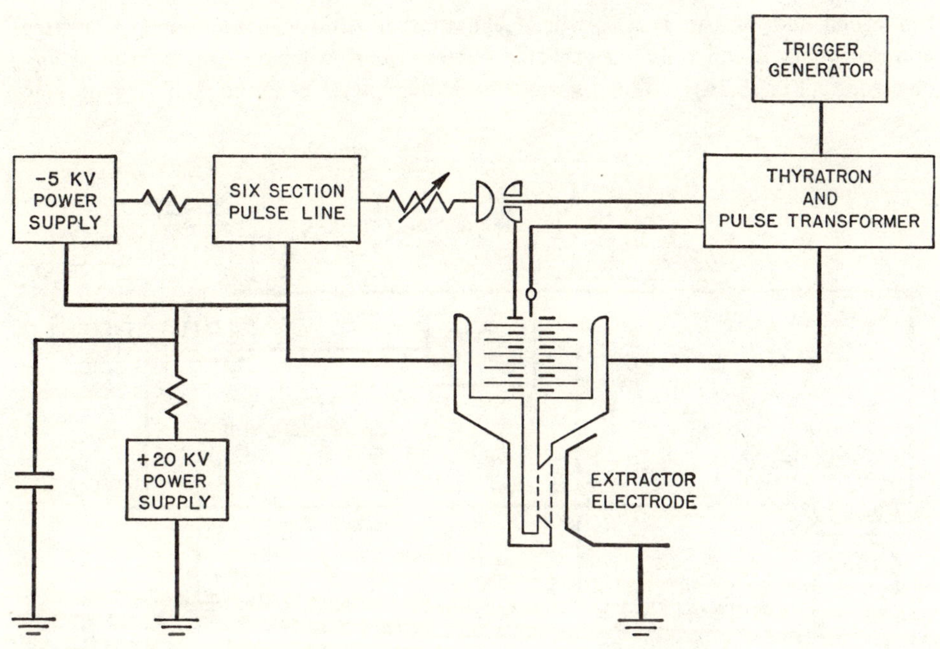

Fig. 5.13. Electrical circuit for ion source.

electrode is a funnel-shaped cup, with an axial slit aperture through which the plasma is extracted (Fig. 5.12) [38]. A block diagram of the electrical circuit is shown in Fig. 5.13. The pulse from the transformer triggers an arc between the electrodes and at the same time fires a series gap which in turn triggers the main ball gap, thus allowing the six-section pulse line to discharge across the titanium washer assembly. The plasma, apparently led by the electrons, extends down the tube into the region of extraction. In order to extract the plasma, the source and its associated equipment are made positive, whereas the extractor electrode is connected to ground, thus allowing ions to stream out radially through the slit in the center electrode mentioned above. The plasma contains a certain proportion of titanium ions, which are not desirable, in addition to deuterons (D^+) and deuterium-molecule (D_2^+) ions. Ion sources of this type can be operated in the absence of a magnetic field, but it seems that the use of a field of a few thousand gauss, in a direction parallel to that of the arc, is advantageous. The ratio of deuterons to deuterium-molecule ions in the plasma is high and the proportion of titanium ions is low.

5.82. A more powerful source of the occluded gas type consists of a 2-inch-long stack of deuterated washers, of $\frac{3}{4}$-inch outer diameter with $\frac{1}{2}$-inch diameter holes, separated by insulators. At one end, where the hole in the washer is only $\frac{1}{4}$ inch in diameter, an insulated trigger electrode is inserted;

the other end of the stack, where the plasma emerges, is open. Electrical connection is made only to the two extreme end washers and to the trigger electrode (Fig. 5.14). The passage of a discharge between the trigger elec-

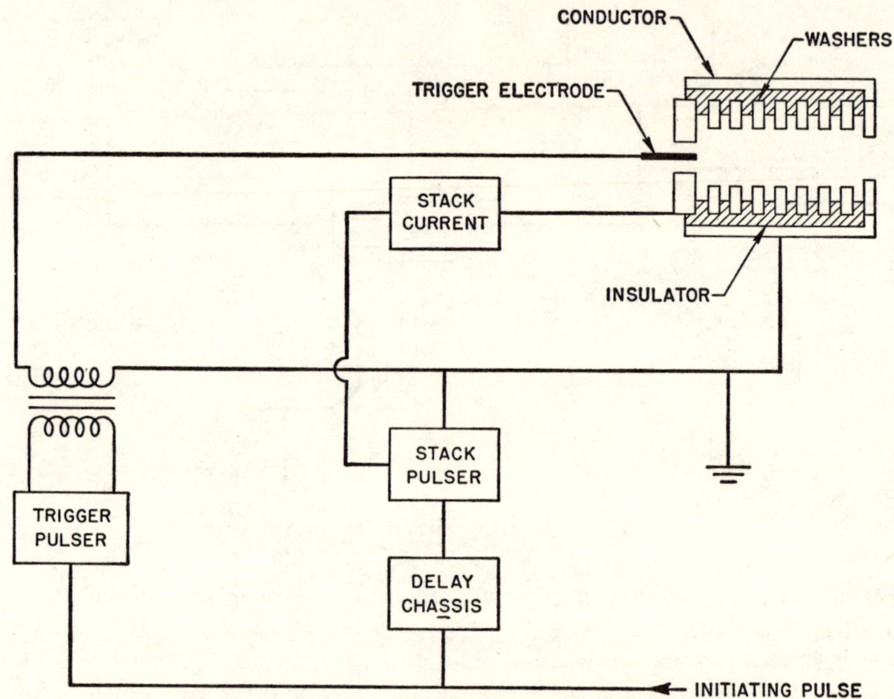

Fig. 5.14.   Electrical circuit for more powerful ion source.

trode and the washer at the left end generates enough plasma within the hole to insure passage of the main discharge, between the two end electrodes, through the center of the stack rather than outside [40].

OCCLUDED-GAS RING SOURCE

5.83.   To study the effects of injecting a ring-shaped plasma at the median plane of a magnetic mirror system, the occluded-gas source shown in Fig. 5.15 has been used [41]. In one form it consists of 100 titanium wire rings of $\frac{1}{2}$-inch diameter, loaded with deuterium, mounted approximately $\frac{1}{2}$ inch apart in a Lucite ring of 17-inch inside diameter. The rings are connected electrically in such a manner as to divide them into two equal sections, with a common initial electrode and trigger, in the first gap of each section, and a common final electrode (Fig. 5.16). With this method of connection, only one source pulse line and one trigger source are needed. Between each pair of titanium rings there is connected in series a small resistor of stainless steel wire.

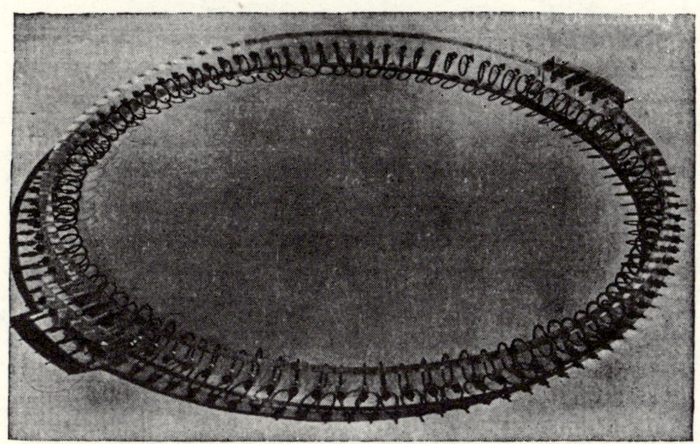

FIG. 5.15. Occluded-gas ring source. (Courtesy of Lawrence Radiation
Laboratory, Livermore)

5.84. The discharge is initiated by means of a trigger pulse which causes
the discharge of an auxiliary capacitor, shown in Fig. 5.16. This insures
sufficient gas evolution for reliable operation of both halves of the source, thus
completing the ring discharge.

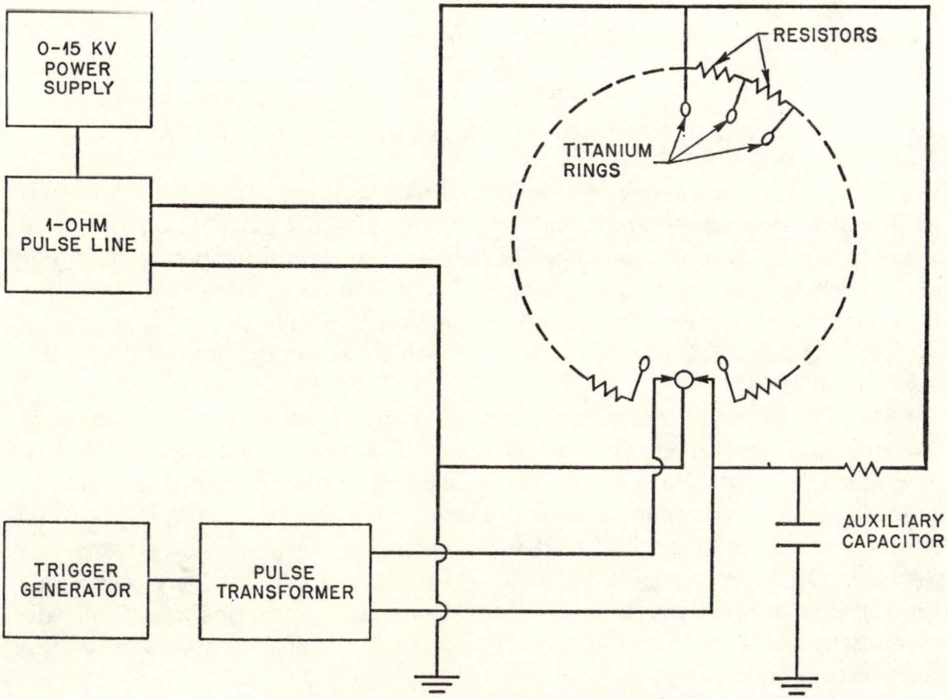

FIG. 5.16. Electrical circuit for ring source.

OCCLUDED-GAS, LOW-ENERGY ION SOURCES: PLASMOIDS

5.85. In another kind of pulsed occluded-gas source, which produces a plasma with particles of relatively low energy, i.e., a few electron volts, a high-current discharge is passed between the tips of deuterium-loaded titanium wires. The wires, about 0.04 inch in diameter, are first heated by radio-frequency current in a vacuum and allowed to cool. The process of heating and cooling is then repeated in an atmosphere of deuterium gas. Two wires, which serve as the electrodes for the source, are then mounted in a lava or alumina ceramic disc, about $\frac{3}{8}$ inch in diameter and $\frac{1}{8}$ inch thick, with the ends of the wires 0.005 inch apart and flush with the face of the disc (Fig. 5.17).

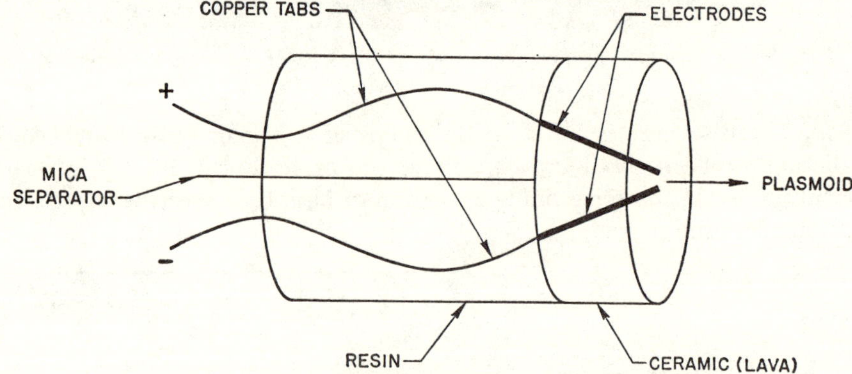

Fig. 5.17. Button source for production of plasmoids.

Copper strips, to serve as connections and also to prevent motion of the titanium wires, are soldered to the latter, separated by a thin sheet of mica, and then "potted" in a resin cement. The "button" so formed is about $\frac{1}{2}$ inch long and $\frac{3}{8}$ inch in diameter. The passage of a discharge of 1000 to 10,000 amp between the deuterium-loaded titanium electrodes results in the formation of a plasma pulse of 0.1 to 0.5 microsec duration, containing from 10^{15} to 10^{18} ions per pulse [42].

5.86. The plasma, apparently in the shape of a torus, to which the name *plasmoid* has been given, is expelled from the source in a forward direction at speeds up to 2×10^7 cm/sec. It contains both deuterons and titanium ions. A study of the light emitted indicates that the mean energy of the thermal motion of the ions is much less than the energy of motion of the plasmoid as a whole. In other words, the kinetic temperature of the ions is considerably smaller than what corresponds to the velocity of propagation of the plasma system; the temperature of the deuterium ions is probably not more than a few electron volts.

5.87. The plasmoids have the remarkable property of being able to travel

across a magnetic field, although with some change of shape. The reason
is that the temperature of the electrons is relatively low and the resistance is
consequently high. As a result, the magnetic field is able to penetrate into
the plasma. In this field, the electrons gyrate in one direction and the ions
in the opposite direction, and the tendency for the charges to separate leads
to the formation of a space-charge electric field (Hall effect). The combined
action of the electric and magnetic fields, which are at right angles, then forces
the plasma to move in the direction perpendicular to both, i.e., transversely
to the magnetic field [43-47].

5.88. In the plasma source described above, the fact that the ends of the
electrodes are in one plane and close together causes the discharge to be curved,
and this leads to the characteristic shape of the emitted plasma. In order to
avoid this curvature, a modified vacuum spark plasma source has been designed (Fig. 5.18). The energy source is a 0.05-microfarad capacitor, con-

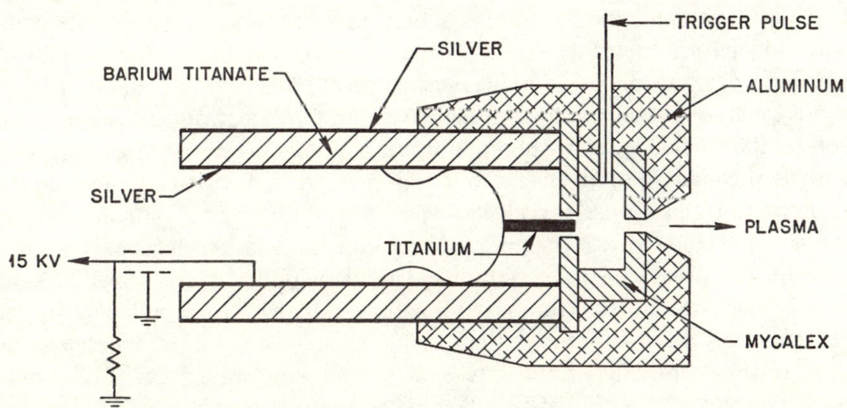

Fig. 5.18. Vacuum spark plasma source.

sisting of a cylinder of barium titanate, silvered inside and outside, capable
of being charged to 15 kilovolts. The outer silver layer makes electrical
contact with an aluminum housing which also acts as one terminal of the discharge. The other terminal is a $\frac{1}{16}$-inch thick titanium rod, loaded with
deuterium, and this is connected to the inner silver layer of the condenser.
The discharge is triggered by a third electrode and takes place within a Mycalex chamber. The plasma formed in the discharge is expelled in a forward
direction, with high velocity, through a hole in the Mycalex chamber and the
surrounding aluminum housing [48].

PLASMA GUNS

5.89. A plasma gun is a device for producing and accelerating into a
vacuum bursts of plasma at speeds in excess of 10^7 cm/sec. Two types of

such guns are described below; both make use of hydromagnetic pistons to accelerate the gas. The plasma bursts formed resemble plasmoids in some respects, but they have the advantage of containing few, if any, extraneous ions. In one form of plasma gun a quantity of gas, e.g., about 1 cm³ of deuterium at atmospheric pressure, is admitted to the system by means of a special quick-acting mechanical valve in a time period of the order of 100 microsec. This gives a pressure of roughly 1 mm of mercury at the input end while the remainder of the system remains temporarily under high vacuum. As the deuterium proceeds down a cylindrical tube, the gas is ionized, compressed, and heated hydromagnetically by means of a discharge passing through a single-turn coil surrounding the tube. In this way, a short cylinder of plasma is formed which is then driven by a magnetic field somewhat in the manner described in §5.47. In this way, plasma bursts of high velocity can be obtained [49].*

5.90.   Another, simpler, type of plasma gun is based on the same principle as that used in the magnetic annular shock tube described in §5.49. It consists of two concentric metal tubes, about 15 to 30 cm long, which act as the electrodes. By means of a quick-acting mechanical valve about 1 cm³ of gas is admitted to the annular space between the electrodes. After a delay of 100 to 400 microsec after the opening of the valve, thus giving the gas time to spread through the gun barrel but not beyond, a low-inductance capacitor bank is discharged across the electrodes. A radial layer of current flows between the electrodes near the rear of the gun, and is accelerated toward the muzzle by the pressure of the azimuthal field behind it. As a result, a burst of plasma containing more than 10^{19} ions is produced with a velocity in excess of 10^7 cm/sec. It is estimated that about 40 per cent of the electrical input energy at the terminals of the gun appears as kinetic energy of the plasma burst. The velocity attained is sufficient to permit penetration of a magnetic field up to 10,000 gauss [50].

PLASMA FORMATION IN P.I.G. DISCHARGE

5.91.   A method for continuously producing a plasma with particles of moderate energy makes use of the Philips Ionization Gauge (P.I.G.) or Penning type of discharge [51, 52]. Basically, the system consists of two cathodes, with a tubular anode between them, in a magnetic field parallel to the axis of the anode tube (Fig. 5.19). The cathodes are grounded and a potential of about 1 kilovolt is applied to the anode. Deuterium gas is introduced through perforations in the cathode at the left, which is made of tantalum; the cathode at the right is a hollow cylinder through which the plasma emerges. Some of the ions formed in the discharge are accelerated back to the left cathode, where they produce secondary electrons. These secondaries are pre-

---

*The delay line described in the original report on this plasma gun probably does **not** play the part attributed to it.

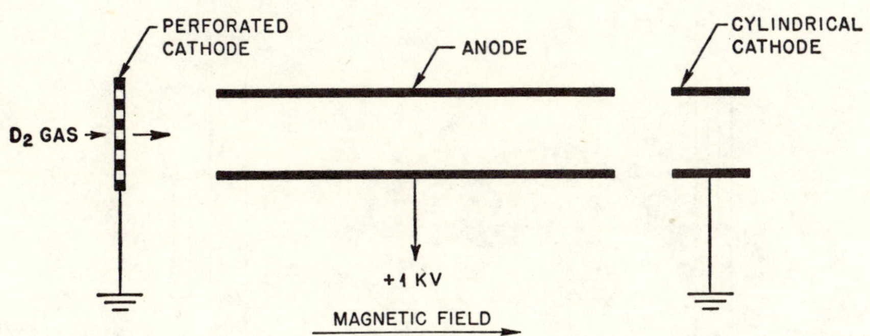

FIG. 5.19. Philips Ion Gauge (or Penning) source.

vented by the magnetic field from reaching the anode immediately and they oscillate back and forth between the two cathodes, producing more ions.

5.92. The plasma from a P.I.G. discharge contains a high proportion of neutral particles. A method proposed for reducing their number is to allow the plasma to flow down a long cylinder to which is applied an axial magnetic field. This field confines the plasma to the cylinder, but the neutral particles diffuse freely to the walls, where they are pumped away [53].

PRESSURE-GRADIENT ARC SOURCE

5.93. A novel type of vacuum arc, in which there is a pronounced pressure gradient along the arc column,* has special characteristics that make it of possible interest as a continuous ion source for a magnetic mirror system. The arc is struck along the field axis, but the ions are trapped by virtue of the fact that most of their energy is perpendicular to the field. Such an arc constitutes what is called a "virtual" source, since there are no solid electrodes at the point of ion emission, which is within the confining field (Fig. 5.20). Hence, the ions are not lost when they return to the region of their origin, as a result of deflection by the magnetic field.

5.94. The discharge passes between the tungsten cathode at the left, which is biased negatively and is heated by ion bombardment, and the floating plate electrode at the right, which acts as an electron reflector. The arc is collimated in the magnetic field (2000 to 8000 gauss) by passage through a defining hole in a grounded plate. The pressure gradient is achieved by feeding deuterium gas directly into the hole which collimates the arc at such a rate that the pressure at the hole is 10^{-3} mm of mercury (or more) whereas in the vacuum chamber in which the discharge is propagated it is less than 10^{-5} mm.

5.95. When the proper pressure gradient is attained, the appearance of the discharge changes suddenly. The arc remains bright near the defining

* This has been called a Mode II arc, to distinguish it from the Mode I arc in which there is no pressure gradient [54].

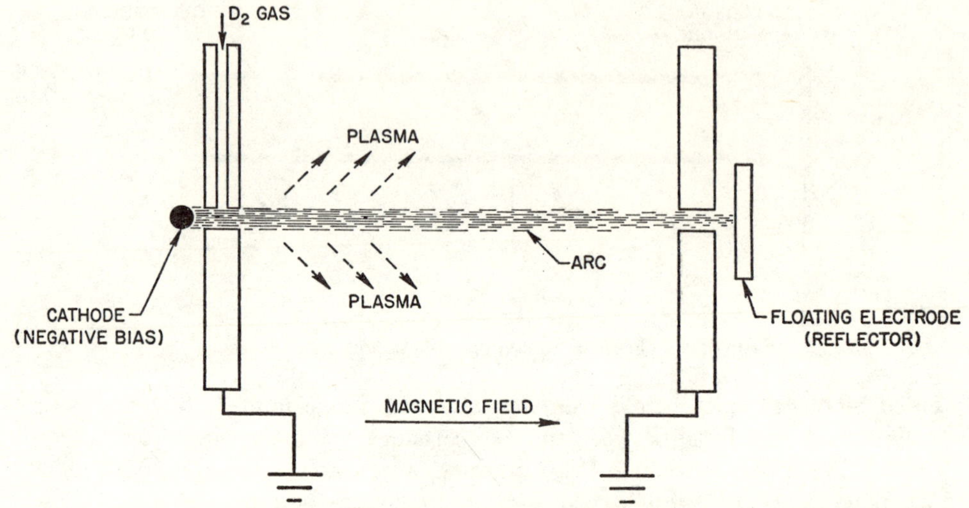

FIG. 5.20. Pressure-gradient arc source.

hole where the gas is admitted, but the central column becomes diffuse and nearly as dim as the surrounding plasma, which is now somewhat brighter than before. Time-varying electric fields apparently exist in and around the central arc column which accelerate some ions to at least twice the bias voltage. The arc currents observed are in the range of 5 to 10 amp [54].

HIGH-CURRENT, HIGH-VOLTAGE SOURCES

5.96. An ion source capable of producing beam currents of several amperes of deuterons with about 10 kev energy has been designed for use in connection with magnetic mirror systems [55]. The deuterium gas is passed through a direct current (or pulsed) arc wherein it is dissociated and ionized to yield deuterons. The arc is kept collimated by a magnetic field parallel to the arc axis. The deuterons are accelerated and extracted radially by the combined action of a system of electrodes between which exists a potential difference of some kilovolts. A special feature of the design is that the plasma is extracted at a floating (or insulated) electrode located between the cathode and a grounded extractor (or accelerator) electrode.

5.97. A typical ion source of this kind, with its associated power supply, is represented schematically in Fig. 5.21. The arc is struck between a tungsten filament (cathode), heated by 1000 amp or more of direct current, acting as a source of electrons, and a stainless steel, floating electrode (anode). The arc voltage is applied between the filament and a defining electrode, so called because a channel in this electrode serves to define the arc. In operation, the floating electrode charges to a potential negative with respect to the defining electrode, and so primary electrons are reflected back into the arc and are

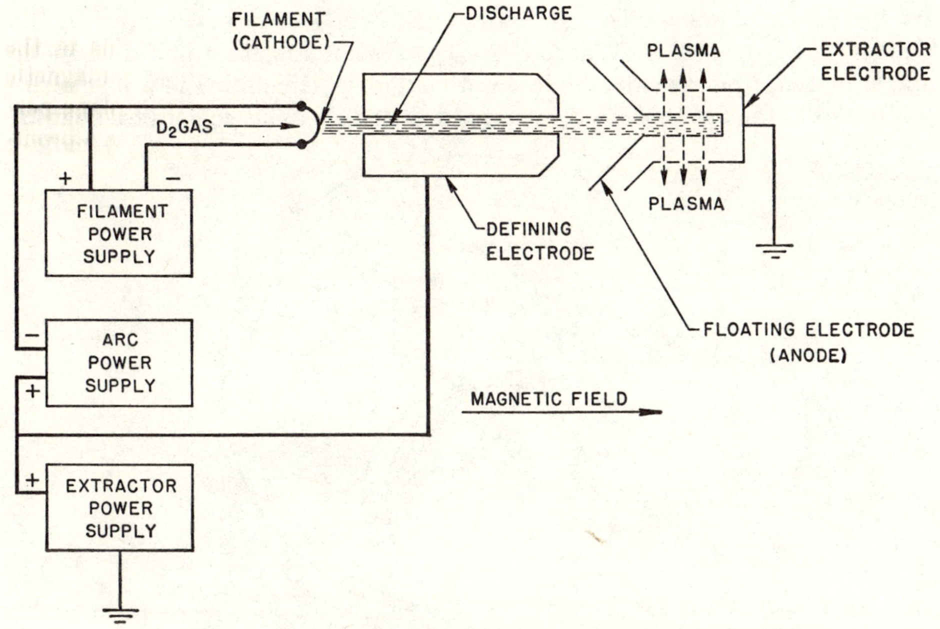

Fig. 5.21.   Electrical circuit for high-current arc source.

thus conserved.   The accelerating voltage is applied between the defining electrode, which is at a potential of about 10 kev, and the grounded extractor electrode.

5.98.   A stream of deuterium gas at low pressure is passed into the arc from the filament end.   The molecules are dissociated and ionized in the arc and are then accelerated to approximately 10 kev energy.   The high-energy ions, together with some electrons, are extracted radially through a number of longitudinal slits in the floating and extractor electrodes.   There is a voltage drop of a few hundred volts from the arc to the floating electrode and the resulting space charge sets a limit on the ion current that can be drawn from the system [56].

DISSOCIATION OF DEUTERIUM MOLECULAR IONS

5.99.   An important step in the production of high-energy plasmas has resulted from the development of high-current, vacuum arcs and the discovery that such discharges are very effective for the dissociation of deuterium-molecule ions (D_2^+) to form deuterons (D^+) [57, 58].   In a magnetic field, the D_2^+ ions will describe circles, with their planes perpendicular to the field, having gyromagnetic radii given by equation (4.4).   However, the deuterons resulting from dissociation have only half the mass of the D_2^+ ions and carry half the energy.   Consequently, m/e for the deuterons will be half that for the deuterium-molecule ions, and so also will be the path radius in the same magnetic field (Fig. 5.22). If the $D_2{}^+$ beam is injected into a magnetic mirror system from outside, the beam of deuterons, circulating in a plane perpendicular to the field lines, can be trapped (§9.89).* This constitutes a prom-

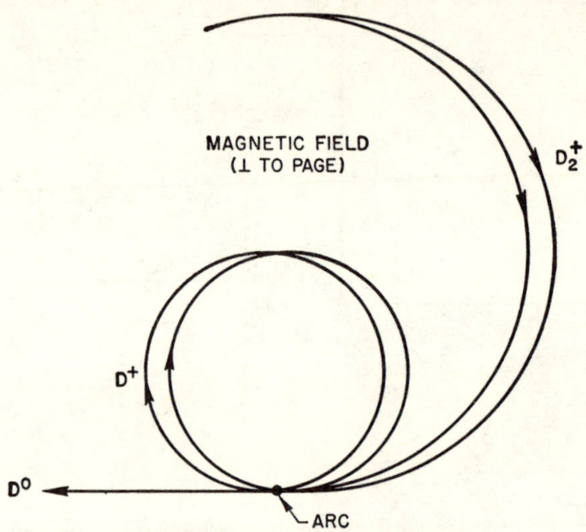

Fig. 5.22. Dissociation of deuterium-molecule ions in magnetic field.

ising method for injection of high-energy plasma into a mirror device. Furthermore, the trapped, high-energy deuterons ionize, and thus facilitate the removal of, neutral particles from the plasma.

5.100. The production of large currents of high-energy deuterons by the dissociation of deuterium-molecule ions falls into three distinct phases: (a) formation of strong $D_2{}^+$ ion beams, (b) acceleration of the $D_2{}^+$ ions, and (c) dissociation of the $D_2{}^+$ ions. Since the second phase involves fairly conventional techniques, it will be treated briefly here before taking up the other aspects of the problem. It appears that acceleration of the $D_2{}^+$ ions to about 600 kev is desirable, so that the deuterons finally produced have an energy of 300 kev before entering the magnetic mirror system. Although accelerators for such energies are not difficult to construct, there are two special requirements for the present purpose that should be mentioned. First, the device must be capable of accelerating currents of the order of tenths of amperes of deuterium-molecule ions; and, second, the extreme voltage variation must not be greater than 2 per cent, since the path of the accelerated particles in the trapping magnetic field (cf. Fig. 5.22) can be permitted to fluctuate by a distance which is no greater than the diameter of the arc in which the $D_2{}^+$ ions are dissociated.

*The result will be the same irrespective of whether the $D_2{}^+$ ion dissociates into $D^+ + D^0$ or into $2D^+ + e$, except that there are no neutral atoms in the latter case.

5.101.   A common form of molecular-ion source makes use of a radiofrequency discharge to ionize the deuterium gas, but it is considered doubtful that the ion currents produced in this manner can exceed a few milliamperes. Several alternative procedures have been suggested.   One proposal is based on the high efficiency for ionization of the trapped electrons in a P.I.G. type of discharge (§5.91) [59].   A schematic representation of a possible arrangement is shown in Fig. 5.23.   By making the distance between the cathode at the left

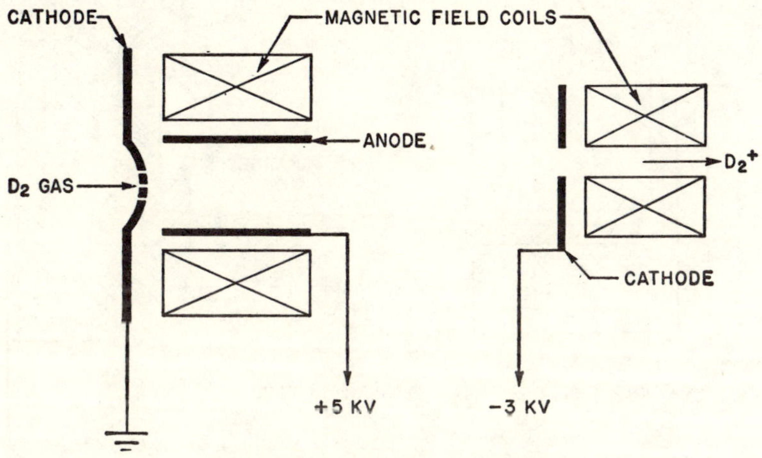

Fig. 5.23.   Molecular-ion source.

and the anode as short as possible and that between the anode and the cathode at the right, where ion acceleration occurs, as long as possible, it is believed that ion losses can be minimized.

5.102.   Other sources of molecular ions which are being studied make use of various types of arcs to ionize the deuterium.   Both the so-called Mode II arc, described in §5.93, and the arc between tungsten electrodes, of the type referred to below (§5.108) for the dissociation of $D_2{}^+$ ions, seem to offer prospects for success.   In the latter case, operation at low deuterium gas pressure appears to favor the production of molecular over atomic ions.*

5.103.   After the $D_2{}^+$ ions are accelerated, they must be dissociated into deuterons.   For this purpose, the direct current arcs, operating at very low gas pressures (10^{-4} to 10^{-6} mm mercury), in a magnetic field of a few thousand gauss, have been shown to be very effective.   A schematic drawing of an arc of this type is given in Fig. 5.24.   The electrodes between which the discharge is passed are made of carbon, since the most successful arcs, as long as 6 feet and carrying currents up to 5000 amp, have been obtained with such electrodes.

---

* The Von Ardenne source is also attracting attention as a possible source of molecular deuterium ions [60].

The voltages required are in the range of 50 to a few hundred volts, depending on the length of the arc [58].

5.104. Two general procedures, both of which are quite conventional, have been used for striking the arc. In one method, direct current and a radiofrequency voltage (about 250 volts) are applied between the electrodes, and gas is slowly passed into the inlet in the cathode (see Fig. 5.24). A radiofrequency

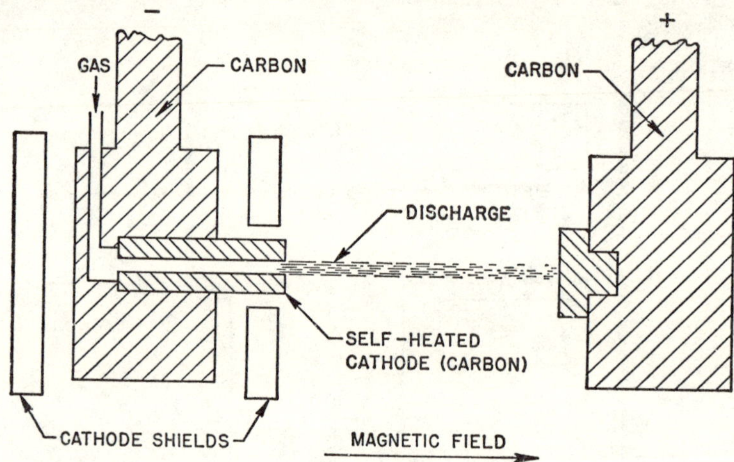

FIG. 5.24. Dissociation of deuterium-molecule ions in carbon arc.

arc is immediately formed and is soon followed by the direct current arc; when this arc forms, both the radiofrequency voltage and the gas flow are shut off. The direct current arc subsequently maintains itself by carbon vaporized from the electrodes. The second method that has been used to strike the arc is to bring the anode and cathode together until they just touch and then to separate them slowly.

5.105. When first struck, the basic characteristics of the arcs are similar to those of large direct-current, gas arcs in a magnetic field. Under these conditions, a secondary plasma is observed around the arc. As the pressure drops, however, the nature of the high-vacuum discharge changes. It is then no longer a true arc, in the respect that it ceases to have a stable homogeneous structure. The discharge apparently breaks up into a number of separate filaments which are often observed to undergo a continual slow rotary motion in the region of their termination at the electrodes. The life of each filament is limited and new ones are constantly forming. Small particles of hot carbon are visible in the discharge and they appear to be torn from the cathode and ejected at high speed in wide angles toward the anode. These carbon particles adsorb gas as they cool and so act as a pumping system (cf. §9.161). In addition, carbon ions reach the walls of the containing vessel and assist in occluding neutral gas.

5.106. Passage of deuterium-molecule ions through the discharge results in effective dissociation, especially at low energies. Approximately 90 per cent dissociation has been reported with arc currents of 700 amp for 20-kev deuterium-molecule ions, but the efficiency is decreased to 10 to 20 per cent for 600-kev ions. Apparently, the D_2^+ ions break up into $D^+ + D^0$ as well as into $2D^+ + e$, and it may be possible, by a proper choice of conditions, to dissociate the molecular ions preferentially in one way or the other.

5.107. Although the carbon arc described above is a very valuable device, it has the great disadvantage of introducing carbon atoms into the plasma, thus increasing the possible bremsstrahlung and other energy losses from a thermonuclear reaction system. Early experiments, in which attempts were made to dissociate D_2^+ ions by means of arcs in deuterium gas at low pressure, were not successful; but a modification in the shape of the anode has given promising results.

5.108. By making this electrode in the form of a long tube, instead of flat, as in Fig. 5.24, the deuterium gas must pass through a considerable length of arc before it enters the vacuum chamber and so the probability of ionization is increased. Both anode and cathode are made of tungsten, and are mounted in water-cooled copper blocks, as shown in Fig. 5.25. Intense local heating of

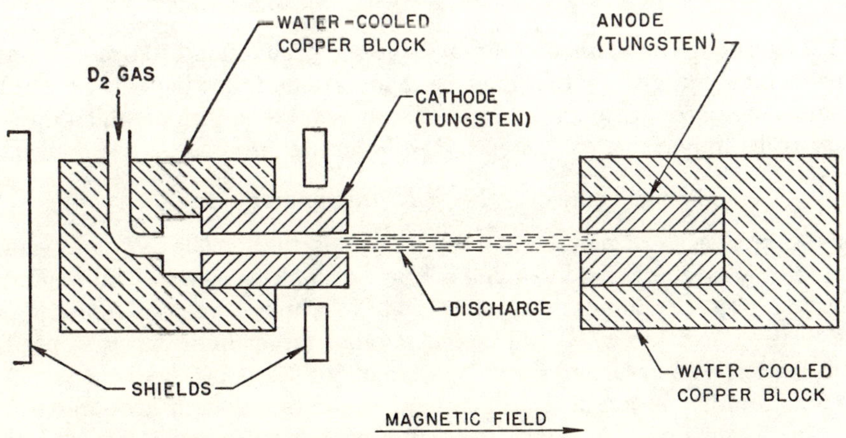

Fig. 5.25. Dissociation of deuterium-molecule ions in deuterium arc.

the anode occurs so that effective cooling is necessary. The gas which feeds the arc is supplied through either of the electrodes [57, 61].

5.109. In preliminary observations with the deuterium arc, a 10 per cent dissociation of 20-kev deuterium-molecule ions was achieved with an arc current of 150 amp and a potential of 200 volts applied across a 4-inch gap. The operating pressure was 5×10^{-5} mm of mercury, although it is found that a considerably higher pressure must be used to strike the arc. With improvements in the design, a somewhat better dissociation efficiency (25 per cent) has been realized, but it is possible that tungsten impurities may have been present.

HIGH-ENERGY BEAMS OF NEUTRAL ATOMS

5.110. If a plasma contained in a magnetic mirror system is already hot, e.g., as a result of thermonuclear reaction, injection of neutral deuterium atoms would appear to have certain advantages. They could be injected into the interior of the plasma across the magnetic field lines without difficulty and, provided ionization occurred rapidly, the resulting deuterons would be trapped by the field. The introduction of cold neutral particles would lead to charge-exchange losses, for which the cross sections are very high at low energies. It would therefore be desirable that, if neutral deuterium atoms are to be injected, they should have high energies.

5.111. It is necessary, however, that the energy of the deuterium atoms should not be too high. As just indicated, trapping by the magnetic field is dependent upon the rapid loss of an electron from the atom to form an ion. The cross section for such ionization decreases with increasing energy (§12.66), and so the higher the energy of the deuterium atom, the greater is the probability that it will escape to the wall and give up its energy there. The return of the cooled atom into the plasma would then result in energy loss by charge exchange, as explained earlier.

5.112. Among the methods proposed for the production of strong beams of atoms of high energy, mention may be made of the following. If accelerated deuterons are passed through a target of cold gas, the nature of which does not appear to be important, charge exchange occurs; thus,

$$D^+ \text{ (high energy)} + X^0 \rightarrow D^0 \text{ (high energy)} + X^+,$$

where X represents a gas-target atom (or molecule). The X^+ ions can be diverted in a magnetic field, leaving a beam of high-energy neutral deuterium atoms [62, 63]. Preliminary measurements in which a beam of 100-kev deuterons was passed through a cold mercury vapor target indicated a 15 per cent neutralization as a result of charge exchange [64].

5.113. Another proposal is to bombard a gas target with accelerated deuterium-molecule ions. In a collision between a high-energy D_2^+ ion and a neutral atom or molecule, the former would be mainly dissociated into D^+ and D^0, the energy being shared equally between these two particles. Although there are at least two significant competing processes, namely, loss of D_2^+ by dissociation into $2D^+ + e$ and loss of D^0 as a result of ionization, it appears that there could be appreciable formation of high-energy neutral atoms [58, 65, 66]. Instead of using a gas target to dissociate the energetic D_2^+ ions, it has been suggested that collisions with residual neutral particles or with ions may serve this purpose if the deuterium-molecule ions are injected directly into a plasma [67].

5.114. If high-energy deuterium-molecule ions are to be used as the source of neutral atoms, dissociation in the high-current, vacuum arc, described above (§5.103), may be of interest. In this case, however, the deuterons produced would be deflected by a magnetic field whereas the high-energy deuterium atoms, carrying half the energy of the original D_2^+ ions, would enter the reacting system. By the proper choice of conditions, the undesirable dissociation into $2D^+ + e$ might perhaps be minimized.

REFERENCES FOR CHAPTER 5

1. J. M. Meek and J. D. Craggs, *Electrical Breakdown of Gases*, Clarendon Press, 1953, Chapter IX.
2. S. C. Brown, *Handbuch der Physik*, Springer-Verlag, 1956, Vol. 22, p. 531.
3. G. Francis, *Ionization Phenomena in Gases*, Academic Press, Inc., 1959, Chapter 4.
4. B. Lax, W. C. Allis, and S. C. Brown, *J. Appl. Phys.*, **21**, 197 (1950).
5. A. Gibson, UKAEA Report AERE GP/R-2371 (1957) Chapter 27; *Third Int. Cong. on Ionization Phenomena in Gases* (1957), p. 365.
6. H. Dreicer, *Phys. Rev.*, **115**, 238 (1959); **117**, 329, 343 (1960).
7. L. Spitzer, *Nature*, **181**, 221 (1958).
8. I. Langmuir, *Phys. Rev.*, **26**, 585 (1925); *Z. Physik*, **46**, 271 (1928).
9. D. Gabor, E. Ash, and D. Dracott, *Nature*, **176**, 916 (1955).
10. D. Reagan, *Phys. Fluids*, **3**, 33 (1960).
11. A. C. Kolb, *Int. Symp. on Magneto-Fluid Dynamics* (1960).
12. R. K. M. Landshoff, Ed., *Magnetohydrodynamics*, Stanford University Press, 1957, pp. 75-103.
13. R. K. M. Landshoff, Ed., *The Plasma in a Magnetic Field*, Stanford University Press, 1958, pp. 87-116.
14. A. C. Kolb, *Fourth Int. Cong. on Ionization Phenomena in Gases*, p. 1021 (1959).
15. C. S. Gardner, H. Goertzel, H. Grad, C. S. Morawetz, M. H. Rose, and H. Rubin, *Proc. Second U.N. Conf. on Peaceful Uses of Atomic Energy*, **31**, 230 (1958).
16. A. Kantrowitz, R. M. Patrick, and H. E. Petschek, AVCO Research Report 63 (1959).
17. R. G. Fowler et al., *Phys. Rev.*, **82**, 879 (1951); **87**, 966 (1952); **88**, 137 (1952).
18. V. Josephson, *J. Appl. Phys.*, **29**, 30 (1958).
19. A. C. Kolb, reference 12, p. 76.
20. A. C. Kolb, *Phys. Rev.*, **107**, 345 (1957).
21. A. C. Kolb, *Phys. Rev.*, **112**, 291 (1958).
22. A. C. Kolb, *Phys. Rev.*, **107**, 1197 (1958).
23. A. C. Kolb, *Proc. Second U.N. Conf. on Peaceful Uses of Atomic Energy*, **32**, 337 (1958).
24. R. M. Patrick, *Phys. Fluids*, **2**, 589 (1959).
25. L. Spitzer and L. Witten, USAEC Report NYO-999 (1953).
26. J. M. Berger and W. A. Newcomb, USAEC Report NYO-6046 (1954).
27. J. M. Berger, W. A. Newcomb, J. M. Dawson, E. A. Frieman, R. M. Kulsrud, and A. Lenard, *Phys. Fluids*, **1**, 301 (1958); also in *Proc. Second U.N. Conf. on Peaceful Uses of Atomic Energy*, **31**, 112 (1958).
28. L. Spitzer, *Phys. Fluids*, **1**, 253 (1958); also in *Proc. Second U.N. Conf. on Peaceful Uses of Atomic Energy*, **32**, 188 (1958).
29. A. Schlüter, *Z. Naturforsch.*, **12a**, 822 (1957).

30. G. I. Budker, *Plasma Physics and The Problem of Controlled Thermonuclear Reactions*, Pergamon Press, Inc., Vol. I, 1960.
31. E. S. Chambers, A. A. Garren, D. O. Kippenham, W. A. S. Lamb, and R. J. Riddell, USAEC Report UCRL-5286 (1958).
32. E. S. Chambers, W. A. S. Lamb, and D. O. Kippenham, USAEC Report UCRL-5015 (1957).
33. T. H. Stix, *Phys. Rev.*, **106**, 1146 (1957).
34. H. Alfvén, *Arkiv Mat. Astron. Fysik*, **29B**, No. 2 (1942).
35. E. Åstrom, *Arkiv Fysik*, **2**, 443 (1951).
36. T. H. Stix, *Phys. Fluids*, **1**, 308 (1958); also in *Proc. Second U.N. Conf. on Peaceful Uses of Atomic Energy*, **31**, 125 (1958).
37. T. H. Stix and R. W. Palladino, *Phys. Fluids*, **1**, 446 (1958); also in *Proc. Second U.N. Conf. on Peaceful Uses of Atomic Energy*, **31**, 282 (1958).
38. K. W. Ehlers, USAEC Report TID-7520 (1956), p. 426.
39. K. W. Ehlers, J. D. Gow, L. Ruby, and J. M. Wilcox, *Rev. Sci. Instr.*, **29**, 614 (1958).
40. F. H. Coensgen, W. F. Cummins, and A. E. Sherman, *Phys. Fluids*, **2**, 350 (1958).
41. D. F. Martin and M. M. Hill, USAEC Report TID-7520 (1956), p. 155.
42. V. G. McIntosh and W. H. Bostick, USAEC Report UCRL-4688 (1956).
43. W. H. Bostick, *Phys. Rev.*, **106**, 292 (1956); **106**, 104 (1957).
44. W. H. Bostick and O. Twite, *Nature*, **179**, 214 (1957).
45. E. Harris, R. Theus, and W. H. Bostick, *Phys. Rev.*, **105**, 46 (1957).
46. W. H. Bostick, *Proc. Second U.N. Conf. on Peaceful Uses of Atomic Energy*, **32**, 427 (1958).
47. K. D. Sinelnikov et al., *Proc. Second U.N. Conf. on Peaceful Uses of Atomic Energy*, **31**, 292 (1958).
48. D. Finkelstein, USAEC Report TID-7520 (1956), p. 589.
49. J. Marshall, *Proc. Second U.N. Conf. on Peaceful Uses of Atomic Energy*, **31**, 341 (1958).
50. J. Marshall, *Phys. Fluids*, **3**, 134 (1960).
51. F. M. Penning, *Physica*, **4**, 71 (1937).
52. A. Guthrie and R. K. Wakerling, Eds., *The Characteristics of Electrical Discharges in Magnetic Fields*, NNES Div. 1, Vol. 5, McGraw-Hill Book Co., Inc., 1949, p. 346.
53. C. M. Aplin, A. L. Gardner, L. S. Hall, and H. H. Vandenmark, USAEC Report TID-7558 (1958), p. 463.
54. R. V. Neidigh and C. H. Weaver, *Proc. Second U.N. Conf. on Peaceful Uses of Atomic Energy*, **31**, 315 (1958).
55. C. W. Blue, USAEC Report TID-7503 (1956), p. 326.
56. J. A. Fasolo, USAEC Report UCRL-5285 (1958).
57. J. S. Luce, *Proc. Second U.N. Conf. on Peaceful Uses of Atomic Energy*, **31**, 305 (1958).
58. C. F. Barnett, *Proc. Second U.N. Conf. on Peaceful Uses of Atomic Energy*, **32**, 398 (1958).
59. C. F. Barnett, P. M. Stier, and G. E. Evans, *Rev. Sci. Instr.*, **24**, 394 (1953).
60. M. von Ardenne, *Tabellen der Elektronenphysik, Ionenphysik, und Übermikroskopie*, Deutscher Verlag der Wissenschaften, 1956; cf. USAEC Report ORNL-2926 (1960), p. 60.
61. R. P. Bell and J. S. Luce, USAEC Report ORNL-2457 (1957).
62. E. J. Lauer, USAEC Reports TID-7503 (1955), p. 339; UCRL-4554 (1955).
63. G. Gibson, W. A. S. Lamb, and E. J. Lauer, *Proc. Second U.N. Conf. on Peaceful Uses of Atomic Energy*, **32**, 275 (1958); *Phys. Rev.*, **114**, 937 (1959).

64. R. E. Hester, W. A. S. Lamb, and E. J. Lauer, USAEC Reports TID-7558 (1958), p. 460; UCRL-5088 (1958).
65. J. R. Hiskes, USAEC Report UCRL-3188 (1954).
66. G. Gibson and E. J. Lauer, USAEC Report UCRL-4646 (1956).
67. I. V. Kurchatov, *J. Nuclear Energy*, **8**, 168 (1958).
68. I. N. Golovin, *Proc. Inst. Elec. Eng. (London)*, **106A,** Suppl. No. 2, 95 (1959).

Chapter 6

PLASMA DIAGNOSTIC TECHNIQUES*

ELECTRICAL CHARACTERISTICS OF THE DISCHARGE

CURRENT AND VOLTAGE MEASUREMENTS

6.1. The most obvious measurements to be made on a gas discharge are those which determine its properties as an electrical circuit element. These measurements are indeed important, since such parameters as resistance and inductance are related, respectively, to the plasma temperature and to the distribution of currents within the discharge. Consideration will first be given to the determination of current and voltage, and then it will be seen what other properties of the discharge can be inferred from the results.

6.2. In practical discharge systems, current measurements are made either with resistive shunts or with current transformers. In a typical form of the latter, the secondary is a toroidally wound coil through which the main current is threaded (Fig. 6.1).† Provided the current path does not approach the coil too closely, the magnetic field which couples the toroidal coil is independent of the manner in which the current threads the coil. For a current of I amp and a secondary coil of major radius r meters, the magnetic field strength B, in mks units, is obtained from the Biot-Savart law (equation 3.23) as

$$B = 2 \times 10^{-7} \frac{I}{r}.$$

Apart from sign the voltage V_c at the coil terminals is given by Faraday's law of induction as

$$V_c = \frac{d\phi}{dt}, \tag{6.1}$$

where ϕ is the magnetic flux through the coil. Hence, if the coil has n turns of cross-sectional area A, whose minor radii are small compared to the major radius, equation (6.1) becomes

---

* For reviews, see references 1–4 at the end of the chapter.
† In the Russian literature this is referred to as a Rogovsky coil (or belt).

164

$$V_c = nA \frac{dB}{dt}$$

$$= 2 \times 10^{-7} \frac{nA}{r} \cdot \frac{dI}{dt}. \tag{6.2}$$

By connecting the terminals of the coil directly to an oscilloscope, which indicates V_c, the value of dI/dt in the plasma, and its variation during the course of the discharge, can be determined.

6.3.   If an RC integrating circuit is inserted between the coil and the oscilloscope, the output voltage V_o is given by

$$V_o = \frac{1}{RC} \int V_c \, dt$$

$$= 2 \times 10^{-7} \frac{nAI}{rRC}. \tag{6.3}$$

In this expression R and C are the resistance and capacitance, respectively, of the integrator circuit, RC being larger by at least an order of magnitude than the longest time interval of observation.   Thus, according to equation (6.3), the oscilloscope reading provides a measure of the discharge current.

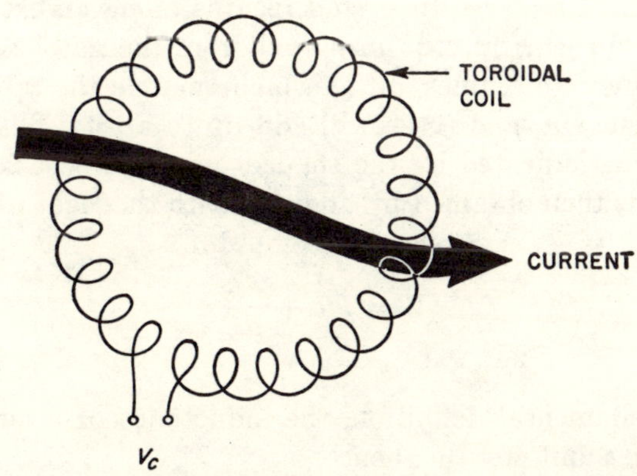

FIG. 6.1.   Transformer for measuring discharge current.

6.4.   For linear discharges, with electrodes within the gas, voltage measurements are usually made with conventional resistive dividers.   These must be arranged, however, so as to minimize inductive pickup, and self-inductances of the measurement circuitry must not adversely affect the necessary frequency response.   If the gas is contained in an endless (toroidal) tube, the discharge is induced in the plasma by making it the secondary of a transformer.   To determine the voltage across the plasma, one or more turns of a conductor are placed around and close to the torus, so that they are both coupled by exactly the same time-changing magnetic field produced by a discharge through the primary winding. The voltage per turn developed across the conductor surrounding the torus, which can be measured directly, is then equal to the voltage induced in the plasma (cf. §7.113 *et seq.*).

6.5. In order to see what information can be derived, in principle, from measurement of voltage and current at the terminals of a discharge tube, consider a simple linear discharge, as shown in Fig. 6.2. The results will, however,

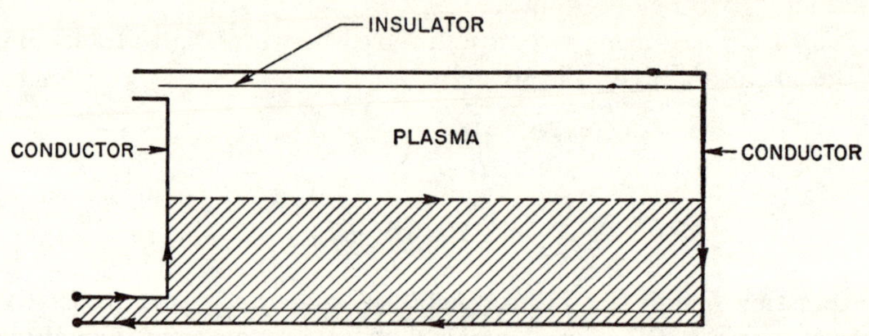

Fig. 6.2. Simple linear discharge system.

apply to discharges of other types. The current passes from the electrode at the left, through the discharge tube, and returns along an external cylindrical conductor which will generally be insulated from the gas. Some distribution of transverse magnetic flux lines, all of which encircle the tube axis, will then be set up by the current, and these will add up to a total flux ϕ which passes through the contour indicated by the shaded area in Fig. 6.2. The total circuit voltage drop is then obtained by adding IR to the right of equation (6.1); thus,

$$V = IR + \frac{d\phi}{dt} \tag{6.4}$$

in mks units.

6.6. By its fundamental definition, the inductance of a circuit is the total flux threading it per unit current; hence,

$$\phi = LI$$

and so equation (6.4) can be written as

$$V = IR + \frac{d(LI)}{dt}$$

or, after rearrangement,

$$V = I\left(R + \frac{dL}{dt}\right) + L\frac{dI}{dt}. \tag{6.5}$$

6.7. In conventional electric circuits, inductances are usually constant, so that dL/dt is zero. In plasma systems this is generally not the case, and much of what might be regarded as novel electrical behavior is the result.   It is common, for example, for rapidly moving plasmas to generate an $I(dL/dt)$ term that is several times larger than the voltage applied to the apparatus from an external supply.   Some ways in which current and voltage measurements can be used to provide information concerning the characteristics of a plasma are described below.

6.8.   The inductance, for example, can be determined from equation (6.5) when the current I is zero or very small.   The first term on the right of equation (6.5) is then negligible, so that

$$L = \frac{V}{dI/dt} \quad \text{for} \quad I \to 0.$$

Alternatively, under such conditions that the inductance is essentially constant, differentiation of equation (6.5) with respect to time gives

$$\frac{dV}{dt} = R\frac{dI}{dt} + L\frac{d^2I}{dt^2}.$$

At or near the current maximum dI/dt is zero, so that

$$L = \frac{dV/dt}{d^2I/dt^2} \quad \text{for} \quad I \to I_{\text{max}}.$$

The values of dV/dt and d^2I/dt^2 can be derived from the observed variations of V and dI/dt with time in the vicinity of I_{max}.

RESISTANCE AND ELECTRON TEMPERATURE

6.9.   The resistance of a plasma is related to the electron temperature, but, in view of equation (6.5), the problem arises concerning how R may be evaluated from observations of the current and terminal voltage.   Even if measurements of V, I, and dI/dt are all made, equation (6.5) shows that it would still not be possible to obtain anything more than the sum of the resistance and dL/dt.   The ideal solution to the problem is to determine $d\phi/dt$ by magnetic probe measurements, as described in §6.107, and then to make use of equation (6.4).   In the case of steady or low-frequency discharges, the problem becomes simpler.   In these circumstances both dI/dt and dL/dt are small and equation (6.5) reduces to $V \approx IR$; the resistance is then approximately equal to V/I.   It is this latter procedure which is the basis of the method for determining electron temperatures described below.   The use of magnetic probe measurements will be referred to later.

6.10.   The resistivity of a completely ionized gas, either in the absence of a magnetic field or parallel to the field direction, was given in §4.89 as

$$\eta = \frac{1.65 \times 10^{-7}}{T_e^{3/2}}\ln \Lambda \quad \text{ohm-cm,}$$

where T_e is the electron temperature in kilo-electron volts and Λ is a dimensionless ratio, some typical values of which were quoted in Table 4.1. From this relationship it should be possible to calculate the electron temperature from the measured resistivity [5]. Actually, Λ depends on both the electron density and the temperature; but since it is not a very sensitive function of the latter, it is generally satisfactory to set $\ln \Lambda$ equal to 10, as a first step, and to obtain better values, if required, by successive approximations. The electron density is calculated from the initial pressure of the gas and the assumption of complete dissociation and ionization. The resistivity of the plasma during the actual discharge can be derived from measurements of the terminal voltage and the plasma current, provided the latter varies relatively slowly with time.

6.11. An interesting application of this principle is the determination of changes in temperature of a stellarator plasma after the cessation of the ohmic heating pulse. Such data provide information concerning the period of confinement of the hot plasma. In a torus (or other endless tube) the resistance can be determined by a scheme depicted in the form of a block diagram in Fig. 6.3 [6]. Since the measurements are to be made after the primary heating

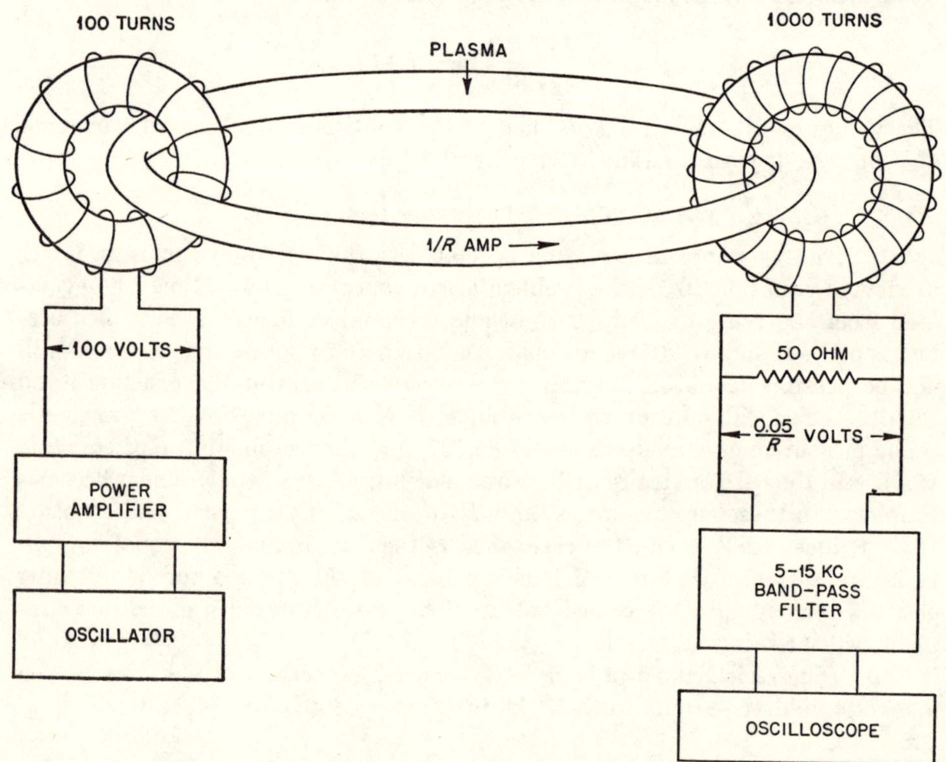

FIG. 6.3. Determination of electrical resistance of plasma.

discharge has ceased, the circuits in Fig. 6.3 are left open until an appropriate time has elapsed and then they are closed by relays actuated by a delay circuit.

6.12.   A signal from a 10-kilocycle/sec oscillator is amplified and fed into the 100-turn primary of a transformer, shown at the left of Fig. 6.3, of which the plasma is the secondary circuit.   If the primary signal has a peak-to-peak voltage of 100 volts, the voltage induced in the plasma will be 1 volt.   Consequently, the plasma current will then be $1/R$ amp, where R ohms is the plasma resistance.   The current induced as a result in the 1000-turn transformer, which is unity-coupled to the plasma circuit, at the right of Fig. 6.2, is thus $1/R$ milliamp, and a peak-to-peak voltage of $50/R$ millivolt, i.e., $0.050/R$ volt, appears across the 50-ohm resistance.   The signal is then fed through a 5- to 15-kilocycle/sec band-pass filter to eliminate extraneous signals, and is observed on an oscilloscope.   The voltage recorded is equal to $0.050/R$ and hence the resistance R of the plasma can be determined.   From the length and cross-sectional area, the resistivity can now be evaluated.

6.13.   The frequency of 10 kilocycles/sec for the current was chosen as being the lowest practical value for reasonable time resolution; the times of interest are of the order of 1 millisecond.   Higher frequencies can give better resolution, but would limit too severely the range of resistance that could be measured.   At high temperatures, the resistance is very low and the time constant L/R becomes so long that the current would be inductively limited.

TEMPERATURE IN A PINCHED DISCHARGE

6.14.   In a pinched discharge, an estimate of the plasma temperature can be made either from the value of the maximum discharge current or from the applied voltage, utilizing theoretical relationships.   According to equation (3.25), provided no magnetic field is trapped within the plasma, the pinch current I amp is related to the mean plasma temperature T by $I^2 = 200\ NkT$, where N is the total number of particles, i.e., ions and electrons, per centimeter length of the discharge and kT is expressed in ergs.   Since N may be calculated from the initial pressure of the gas in the discharge tube, assuming complete dissociation and ionization, it should be possible, in the ideal case of a stable discharge, to determine the temperature from a measurement of the maximum current in the plasma when it is pinched.   Unfortunately, a completely stable pinch, in which the magnetic force on the plasma just balances the kinetic pressure of the particles, has not been realized and so this method for estimating the temperature gives only approximate values.

6.15.   Another approximate method of calculating the temperature is based on the applied voltage in the case of a pinched toroidal plasma stabilized by an axial magnetic field (§7.26).   The plasma is constricted by discharging a condenser bank through a primary circuit closely coupled with the torus containing the plasma.   By making a number of assumptions concerning the characteristics of the discharge and neglecting all heat losses, it has been shown that the temperature, in kilo-electron volts, reached at the peak current is given by

$$T \approx 2.5 \, \frac{J^{\frac{3}{5}}}{(pEd^4)^{\frac{2}{5}}} \quad \text{kev},$$

where J is the energy, expressed in joules per centimeter of the discharge tube length, p is the initial gas pressure in microns of mercury, E is the initial applied electric field in volts per centimeter, and d is the internal diameter of the discharge tube in centimeters [7]. Temperatures calculated from this expression have been found to be in general agreement with those derived from the Doppler broadening of spectral lines (§6.59), which gives the ion, rather than the electron, temperature. The agreement is probably fortuitous, however, since both methods involve assumptions which are certainly not met in an actual discharge.

MICROWAVE TECHNIQUES

ELECTRON DENSITY

6.16. A number of experimental procedures based on the use of microwaves have been developed for the study of plasmas in which the ionized gas is contained within a resonant cavity [8, 9]. These methods are, however, not altogether suited to thermonuclear research because it is often difficult, if not impossible, to adapt the apparatus to the requirements of a resonant cavity. Consequently, other microwave techniques have been proposed which depend upon the transmission of the oscillations through a plasma. They are simple in principle, although a quantitative interpretation of the results is not always possible [10-15].

6.17. Among the advantages of microwave methods is the fact that the probing signals can be of such low power that they cause practically no disturbance of the conditions in the plasma. Furthermore, quantities which can be measured in terms of oscillation frequencies or of frequency variations can be determined with considerable accuracy. One of the most useful applications of microwaves is the evaluation of electron (number) densities in plasmas and of their variation with time. However, with the oscillators at present available, which have a minimum wave length of a few millimeters, an upper limit of about 10^{14} electrons/cm³ is set upon the densities which can be studied. Plasmas of lower electron density can, of course, be readily investigated.

6.18. It was stated in §4.100 that, in the absence of a magnetic field, the dielectric constant of a plasma for oscillations of frequency ω is given by

$$K = 1 - \left(\frac{\omega_p}{\omega}\right)^2,$$

and the characteristic angular frequency ω_p of the plasma is defined, in cgs-esu, by equation (4.84) as

$$\omega_p \equiv \left(\frac{4\pi n_e e}{m_e}\right)^{1/2} = 5.6 \times 10^4 n_e^{1/2} \quad \text{radians/sec} \qquad (6.6)$$

or, in terms of cycles/sec,

$$f_p = \frac{\omega_p}{2\pi} = 9.0 \times 10^3 n_e^{1/2} \quad \text{cycles/sec,}$$

where n_e is expressed in electrons/cm³.   Provided $\omega > \omega_p$, so that K is positive, the oscillations are transmitted through the plasma, but if $\omega < \omega_p$, essentially complete reflection occurs.   The so-called "cut-off" condition, between transmission and reflection, is that for which $\omega = \omega_p$ or $f = f_p$; consequently, it is related to the electron density by

$$f(\text{cut-off}) = 9.0 \times 10^3 n_e^{1/2} \quad \text{cycles/sec.} \qquad (6.7)$$

6.19.   The determination of the cut-off frequency of a given plasma for electromagnetic waves of short wave length, i.e., microwaves, thus offers, in principle, a possible method for obtaining the electron density. Because of the large frequency range which the oscillator would have to cover, this procedure is not very practical.   Nevertheless, a crude form has been used to a considerable extent to provide a rough indication of the instantaneous electron density in a plasma.   If it is found that microwaves of frequency f are largely reflected by the plasma, then that frequency is less than the cut-off value for the existing electron density; hence, $f < f_p$ so that, from equation (6.7),

$$n_e > 1.2 \times 10^{-8} f^2.$$

6.20.   A different approach to the use of microwaves in plasma studies involves the determination of the phase change accompanying transmission of the oscillations through the plasma with only slight attenuation, i.e., when $\omega > \omega_p$.   In the propagation of a plane wave through a dielectric medium, the electric field strength can be expressed as

$$\mathbf{E}_z = \mathbf{E}_0 \exp{(\gamma x - i\omega t)},$$

where γ is the *propagation constant* of the medium.   It has been shown [16] that γ is related to the frequency of the oscillations and the electron density of the medium by

$$\gamma^2 = -\frac{\omega^2 K'}{c^2}, \qquad (6.8)$$

K' being the complex dielectric constant of the medium (plasma), defined by

$$K' = 1 - \left(\frac{\omega_p}{\omega}\right)^2 \left[\frac{1}{1 + (\nu/\omega)^2} + \frac{i(\nu/\omega)}{1 + (\nu/\omega)^2}\right], \qquad (6.9)$$

where ν is the collision frequency of the electrons.   It is assumed here that ν is independent of the velocity of the electrons.*   If $(\nu/\omega)^2 \ll 1$, as is generally the

---

* This cannot be true in a fully ionized plasma where Coulomb collisions are important.

case in plasmas employed in thermonuclear research, equation (6.9) reduces to

$$K' = 1 - \left(\frac{\omega_p}{\omega}\right)^2 - i\left(\frac{\omega_p}{\omega}\right)^2 \frac{\nu}{\omega}. \tag{6.10}$$

6.21. The propagation constant is, in general, a complex number which can be written as

$$\gamma = \alpha + i\beta, \tag{6.11}$$

where α is the *attenuation coefficient* of the waves as they are transmitted through the plasma and β is the *phase coefficient* which determines the phase shift in radians per unit length.   Upon substituting equations (6.10) and (6.11) into (6.8) and comparing coefficients, it is found that

$$\beta = \frac{\omega}{c}\left[1 - \left(\frac{\omega_p}{\omega}\right)^2\right]^{1/2} \tag{6.12}$$

and

$$\alpha = \frac{\omega}{c} \cdot \frac{(\omega_p/\omega)^2(\nu/\omega)}{[1 - (\omega_p/\omega)^2]^{1/2}}. \tag{6.13}$$

Consequently, if the phase shift can be determined when the plane wave, of known frequency, passes through a certain length of plasma, it should be possible to evaluate ω_p from equation (6.12).   The electron density can then be obtained from equation (6.6).   With ω_p known, measurement of the wave attenuation makes it possible to calculate ν, the electron collision frequency.

6.22.   If the electron density were uniform throughout the plasma, the total phase shift would be obtained upon multiplying the coefficient β by the transmission length d.   However, the quantity determined experimentally is the difference between the shift due to the plasma and that for the same path length in the absence of plasma; thus,

$$\Delta\phi = -(\beta - \beta_0)d,$$

where β_0 is given by equation (6.12) as ω/c.   In situations of interest in the study of thermonuclear reactions, when the plasma is confined by a magnetic field, the electron density is rarely uniform, and the experimental phase shift is then represented by

$$\Delta\phi = \frac{\omega d}{c} - \int_0^d \beta\, dx.$$

Utilizing equations (6.6) and (6.12) to express β, this becomes

$$\Delta\phi = \frac{\omega}{c}\left\{d - \int_0^d \left[1 - \frac{4\pi e^2}{m_e\omega^2} n_e(x)\right]^{1/2} dx\right\}, \tag{6.14}$$

where $n_e(x)$ represents the electron density distribution along the transmission path [9, 13].

6.23.   The integral in equation (6.14) has been evaluated in some simple cases of postulated electron density distributions [13, 17].   A reasonably good approximation, for not too high electron densities, results from expressing the integrand of equation (6.14) as a binomial series and neglecting all terms beyond the second; thus,

$$\Delta\phi = \frac{\omega}{c}\left\{d - \int_0^d\left[1 - \frac{2\pi e^2}{m_e\omega^2}n_e(x)\right]dx\right\}$$

$$= \frac{\omega}{c}\left[\frac{2\pi e^2}{m_e\omega^2}\int_0^d n_e(x)\,dx\right]$$

$$= \frac{\omega}{c}\left(\frac{2\pi e^2}{m_e\omega^2}N_e\right), \tag{6.15}$$

where N_e is the total number of electrons per square centimeter of cross-sectional area of the plasma, in the whole path traversed by the microwaves. The observed phase shift, over a fixed path length, is thus directly proportional to the average electron density.

6.24.   For the experimental determination of the phase shift, a microwave interferometer system is used; a simplified schematic representation of the arrangement is given in Fig. 6.4 [10-15, 18].   The microwaves are transmitted

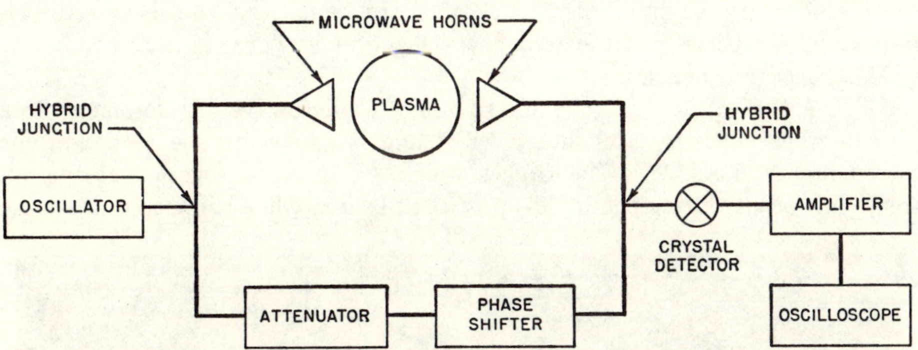

FIG. 6.4.   Schematic representation of microwave interferometer.

through the plasma by means of suitable horns, as shown.   For transmission through a medium in which the electron density is n_e, the frequency of the microwaves must be such that

$$f > 9.0 \times 10^3 n_e^{1/2} \quad \text{cycles/sec}$$

or, in terms of wave lengths,

$$\lambda = \frac{c}{f} < 3.3 \times 10^6 n_e^{-1/2} \quad \text{cm.}$$

Since microwaves of about 4-mm wave length are about the shortest at present available, it is seen that techniques dependent upon microwave transmission are restricted to electron densities less than 10^{14} electrons/cm^3, as stated above.

6.25. Upon leaving the oscillator (Fig. 6.4), the microwaves are split into two paths; in one, which may be referred to as the transmission path, are the transmitting and receiving horns, and in the other, called the nulling or reference path, there are an attenuator and a phase shifter. The latter are adjusted so that the reference signal interferes destructively with the transmitted signal yielding a null output in the absence of plasma. When a plasma of sufficient density is present in the transmission path, a phase shift occurs and this unbalances the circuit; the resulting output signal is recorded on the oscilloscope. The procedure just described is generally applied to a plasma in which the density is changing. The output then appears as a series of waves or interference fringes, each of which represents a phase shift of 2π radians. An illustration of this behavior and a discussion of its interpretation are given below.

6.26. In making measurements with the microwave interferometer of a plasma in a magnetic field, the microwaves should be polarized so that the electric vector is exactly parallel to the field. If the electric vector is not strictly parallel to the magnetic field, the dielectric constant is a much more complicated function than given in §6.18 and contains the gyromagnetic frequency of the electrons. Hence the expressions for the coefficients α and β are considerably altered from the values given by equations (6.13) and (6.12), respectively, especially if the frequency of the microwaves is near one of the gyromagnetic frequencies.

6.27. A typical example of the response of a microwave interferometer to a plasma whose density is changing with time is shown by the oscillographic record in Fig. 6.5 [11]. The fringes which are crowded together at the extreme left occur during the build-up of the plasma, while the electron density

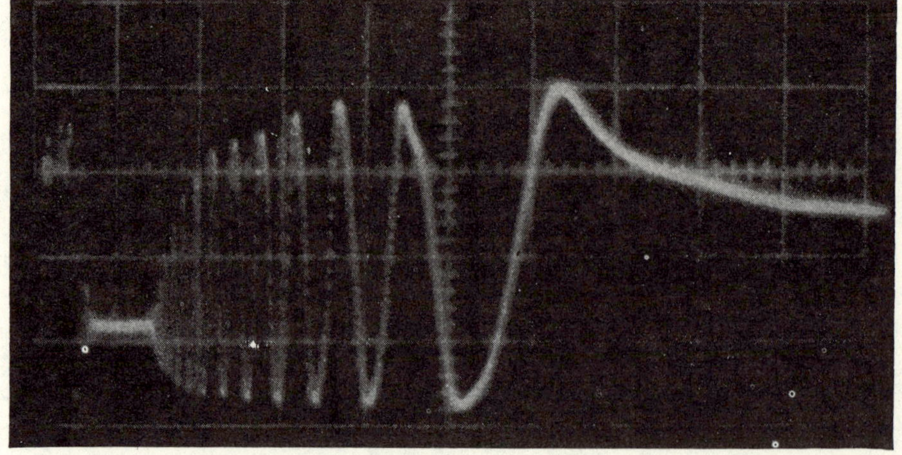

FIG. 6.5. Microwave interferometer trace. (Courtesy of Lawrence Radiation Laboratory, Livermore)

is increasing. There follows a short region where the microwaves are reflected because the electron density then exceeds the cut-off value n_c, which is
given by equation (6.7) as $1.2 \times 10^{-8}f^2$ electrons/cm³. During this period
the steady signal observed is that passing through the reference path of the
interferometer. Subsequently, as the plasma decays, the density falls to the
cut-off value and below, and the oscillatory signal again appears and consists
of some 12 fringes while the electron density falls essentially to zero. In the
experiments during which Fig. 6.5 was obtained, the frequency of the micro-
waves was 3.5×10^{10} cycles/sec, and so the cut-off density was about 1.5×10^{13}
electrons/cm³.

6.28. In order to explain more fully the significance of the results described
above, an idealized situation shown in Fig. 6.6 will be considered. The upper

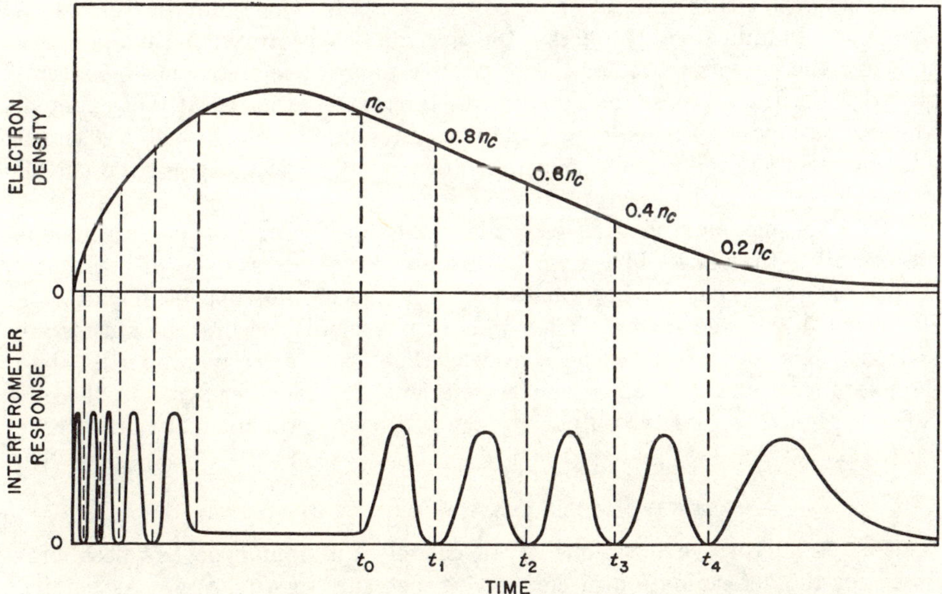

Fig. 6.6. Interpretation of microwave interferometer response.

portion represents the variation with time of the electron density of the plasma
and the lower portion indicates the corresponding interferometer response. It
is seen that there are five fringes while the plasma is building up to the cut-off
density, and the same number while it is decaying. The crowding of the fringes
before cut-off, as compared with those after, shows that the plasma was building up at a faster rate than it decayed. Because of the larger spacing of the
fringes during decay, the interpretation given here will be based on these
fringes, although the general conclusions apply equally to those observed during the build-up of the plasma.

6.29. At the time t_0 the density is just the cut-off value, which is known from the microwave frequency, and during the decay to essentially zero density the total phase shift corresponds to five fringes, i.e., to $5 \times 2\pi$. If, as implied by equation (6.15), the phase shift is taken to be proportional to the electron density, each fringe represents a change of $0.2\ n_c$ in the density. Thus, at t_1 the electron density is $0.8\ n_c$, at t_2 it is $0.6\ n_c$, and so on. Hence, from the locations of the successive minima (or null points) on the time scale of the oscilloscope record it is possible to plot a curve, like the upper part of Fig. 6.6, showing the rate of decay of the electron density in the plasma. The same procedure can, of course, be applied to the build-up phase of the plasma. It will be realized, however, that no information can be obtained concerning the plasma density during the period when it exceeds the cut-off value.

6.30. The microwave circuit of Fig. 6.4 has been modified in various ways which make it either simpler or more complicated. One simplification is its adaptation to indicate only whether the electron density in the plasma is above or below the cut-off value for the particular microwave frequency being employed. For this purpose the signal in the reference path is greatly attenuated. The oscilloscope then records only the transmission, if any, through the transmission path. If there is no transmission then the density exceeds the cut-off value.

6.31. Another modification is to adapt the circuit so that the oscilloscope pattern gives the phase shift directly. In one technique a series of horizontal stripes are produced, for zero phase shift, on the oscilloscope by means of a frequency modulated raster. Then when a phase shift occurs, the stripes are deflected in a vertical direction, the deflection being directly proportional to the phase shift and, hence, to the electron density. In other systems, the circuit is designed so that the output voltage is directly proportional to the phase shift. [11, 15].

MICROWAVE NOISE AND ELECTRON TEMPERATURE

6.32. A microwave technique of an entirely different type has been proposed for the determination of electron temperature in a plasma. As pointed out in §2.57, a heated plasma is far from behaving as a black-body radiator in the frequency range where most of the radiation would be emitted. Nevertheless, even for plasmas of limited dimensions, there is a narrow frequency range, in the vicinity of the plasma frequency, in which the rate of energy emission may be expected to correspond to the black-body value at the existing temperature [12, 15, 19-24].

6.33. A simple relationship between the power of the black-body radiation, emitted as microwave "noise," and the plasma temperature can be obtained in the following manner. Since the frequency f of the radiation is of the order of 10^{10} cycles/sec and Planck's constant h is 6.62×10^{-27} erg-sec, the magnitude of the energy quantum hf is roughly $6.6 \times 10^{-27} \times 10^{10} \approx 10^{-16}$ erg.

The Boltzmann constant k is 1.38×10^{-16} erg/°K, and so for plasma temperatures of interest, e.g., 10^5 °K or more, the microwave radiation energy quantum is much less than kT. In these circumstances, the rate of emission of energy by a black body can be expressed by the Rayleigh-Jeans equation. For the present purpose this may be written in the form

$$dP = \frac{8\pi kT}{\lambda^2} \, df, \tag{6.16}$$

where dP is the radiation power, per unit area, in the frequency range df from f to $f + df$, and λ is the wave length in free space corresponding to the frequency f.

6.34. A receiving microwave antenna can be characterized by an effective area A_{eff} which, in the case of horns or parabolic reflectors whose diameters are many wave lengths, is very close to the geometrical area. If such an antenna is placed near enough to the plasma so that the latter more than covers this area, equation (6.16) may be multiplied by A_{eff} to obtain the received noise power in the frequency interval df. However, the wave guide following the antenna transmits only one plane of polarization of the microwave radiation, and so the power of the transmitted noise is

$$\frac{1}{2} \times \frac{8\pi kT}{\lambda^2} A_{\text{eff}} \, df = CT \, df,$$

where C is constant. For a finite frequency band width Δf, the noise power emitted from a plasma at a temperature T is then

$$P = CT \int_f^{f+\Delta f} df = CT \, \Delta f. \tag{6.17}$$

Thus the microwave power received in a given band width is directly proportional to the electron temperature, assuming that there is a Maxwell distribution of electron energies [25] and that the plasma is behaving effectively as a black body in this radiation frequency range.*

6.35. In order to avoid the necessity for determining the actual power of the microwave noise or the frequency band width over which the noise is received by a low-noise receiver, a relative method has been suggested. In this procedure the noise from the plasma is compared with that received over the same band width by a standard noise source of known electron temperature. Since the plasma temperatures of interest will generally be much higher than the standard noise source, the noise will be proportionately larger, in accordance with equation (6.17). By means of a calibrated microwave attenuator, the plasma noise is then reduced until it exactly matches that of the standard.

---

*The same result has been derived for a plasma by postulating that the noise can be ascribed to the electron current fluctuations due to collisions of electrons with atoms and ions. The temperature is then that of the electrons [26].

The degree of attenuation required is thus equal to the ratio of the electron temperature of the experimental plasma to that of the standard source.

6.36.   A block diagram of the apparatus, which consists of a typical heterodyne circuit, is shown in Fig. 6.7, with the horn antenna for receiving noise from

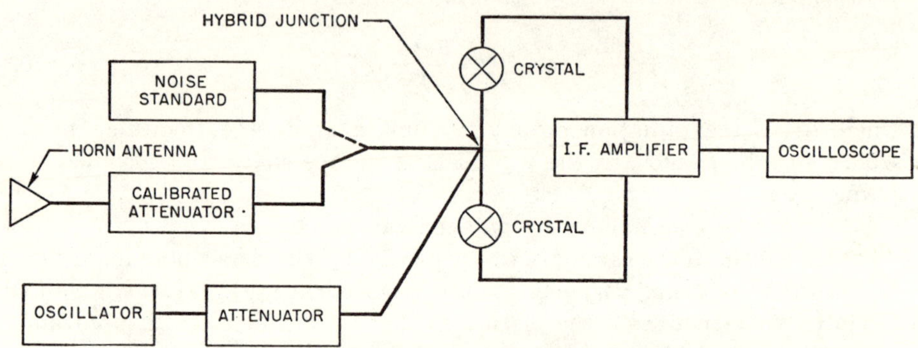

Fig. 6.7.   Heterodyne circuit for microwave noise measurement.

the plasma at the extreme left.   The horn angle and its direction should be chosen so as to avoid, as far as possible, the interfering effects of reflection of the microwave radiation within the tube containing the plasma.   The klystron oscillator, with a suitably attenuated output, supplies local oscillator power to the balanced mixer circuit consisting of two crystals and the hybrid junction. The intermediate frequency (I.F.) amplifier responds in the frequency region of about 30 megacycles/sec, the band width being a few megacycles/sec to provide the necessary transient response.

6.37.   For calibration purposes, the horn antenna is replaced by the standard noise source, e.g., a high-pressure discharge crossing a standard wave guide [27, 28].   The I.F. amplifier gain is then adjusted to give a suitable signal on the oscilloscope.   With the plasma source and horn antenna, the noise received is reduced by the calibrated attenuator until the same oscilloscope signal is obtained.   The plasma temperature is then derived from the calibrated attenuator reading in the manner described above.

6.38.   When the plasma temperature is changing rapidly during the course of an observation, use is made of the good dynamic range of the superheterodyne system.   In such cases the ratio of the plasma noise to that of the standard source may be estimated from the shape of the oscilloscope trace, after making a single adjustment of the calibrated attenuator.

6.39.   Although the microwave noise method for determining temperatures is simple in principle, it is subject to a number of theoretical and experimental uncertainties.   Basically, there is some doubt as to whether the plasma is really radiating like a black body in the microwave region.   Furthermore, the presence of magnetic fields, used for confinement of the plasma, may introduce complicating effects.   For example, plasma oscillations and similar cooperative phenomena can set up stray electric fields and in the presence of a magnetic field will emit high-frequency radiations.   Nonthermal noise of this kind may be much more intense than the thermal noise which is to be measured.

6.40.   As far as the measurements are concerned, consideration must be given to possible disturbances in the noise due to reflection.   In addition, the extent of the plasma is not many wave lengths across and its effective area may change during heating and compression.   Thus, focusing of the antenna onto the noise source may be poor and may vary in the course of the observations.   In spite of the uncertainties, microwave noise measurements must be regarded as a plasma diagnostic technique, especially as reliable methods for temperature determination are somewhat rare.   It should be emphasized that the procedure has validity only because the noise frequencies are near the cut-off frequency of the plasma, and so the latter radiates effectively like a black body (§6.32).

SPECTROSCOPIC MEASUREMENTS [29, 30]

STARK EFFECT BROADENING AND ION DENSITIES

6.41.   It is well known that in an electric field the spectral lines of an atom are often resolved into a number of components; this is called the Stark effect. In a plasma an emitting atom is in an inhomogeneous, although satistically isotropic, electric field due to the neighboring ions and electrons.   For the purpose of the present somewhat approximate treatment the effect of the electrons will be neglected, but this may not be altogether justifiable, as will be seen in §6.57. The electronic transition in an atom, which is accompanied by the emission of a specific spectral line, occurs so rapidly that the change in the electric field due to the surrounding ions during the transition is small.   It is a good approximation, therefore, to consider the ions to be stationary.   A spectral "line," representing the summation of energy quanta emitted by many atoms in the inhomogeneous field due to ions distributed at random is thus broadened so that it takes the form of a band, whereas in a uniform electric field the Stark components appear as discrete lines.

6.42.   For a given energy level, the energy change ΔW resulting from the interaction between a hydrogen atom and a uniform, constant electric field of strength E is represented by

$$\Delta W = aE + bE^2 + \cdots, \tag{6.18}$$

where a is called the first-order Stark coefficient, b is the second-order coefficient, and so on.   If ΔW is expressed in ergs and E in statvolts/cm, then it is found from the theory of the Stark effect that

$$a = \frac{3h^2}{8\pi^2 m_e e}\, nn_F,$$

where h is the Planck constant in erg-sec, and m_e and e are the mass in grams and the charge in statcoulombs, respectively, of the electron; n is the principal quantum number of the energy level and n_F, which is also an integer, is sometimes called the electric (field) quantum number. It is related to the electric dipole moment of the atom and the field direction. For an electronic transition between the principal quantum levels n and n', with n_F and n_F' as the corresponding values of the electric quantum number, the shift $\Delta\lambda$ in the wave length of λ cm, in a field of E volts/cm, due to the first-order Stark effect, is represented by

$$\Delta\lambda \; (\text{cm}) = 6.43 \times 10^{-5}\lambda^2(nn_F - n'n_F')E. \tag{6.19}$$

6.43. As implied above, the energy change and the associated splitting of the spectral lines in the Stark effect depend on the value of the electric dipole moment of the emitting species. If the latter does not have a permanent moment, the electric field induces a moment which is proportional to E, and hence the energy change caused by the field varies as E^2. This is the origin of the second-order Stark effect. In hydrogen and hydrogenlike atoms (or ions), i.e., a nucleus and a single electron, electric dipoles, resulting from the noncoincidence of the centers of the positive (nuclear) and negative (electronic) charges, are produced by extremely small fields. As a result, such atoms behave as if they have a permanent dipole moment which is independent of the strength of the electric field. The Stark splitting is then proportional to the applied field, giving the first-order (or linear) Stark effect. In the spectra of hydrogen (and hydrogenlike) atoms, the first-order Stark splitting is very pronounced. The broadening of the first-order lines is thus of special interest in thermonuclear research because the plasmas are mainly hydrogenic in nature.

6.44. Assuming the effect of the electrons can be neglected, the local (or microscopic) electric field strength in the vicinity of a hydrogen isotope atom may be expected to be related to the ion density [31-38]. It is apparent, therefore, that there should be a connection between the ion density and the broadening of the Stark lines. This is the underlying principle of the method, to be described below, for determining ion densities from measurements of the first-order Stark effect in plasmas. It may be mentioned that the second-order effect in hydrogenlike atoms does not become appreciable except for relatively strong fields, i.e., for high ion densities. However, plasmas with fairly high ion densities are necessary for the study of Stark broadening—otherwise it is too small to be measured—and so there is a possibility that the results may be influenced by the second-order effect.

6.45. The expressions given above for the Stark splitting apply to a uniform and constant electric field. In a plasma, however, the fields arise from moving charged particles and they are not constant either in space or in time. One way of treating this situation is based on a statistical theory, in which it is postulated that the ions are randomly distributed in space and remain fixed, as stated in §6.41. The field distribution is then expressed in terms of an average or "normalized" field and an appropriate probability function. The following simplified approach will provide an indication of the treatment employed [39].

6.46. Suppose that there are N charged particles (ions) within a sphere of radius R; the probability of finding all these particles within a radius r is equal to $(r^3/R^3)^N$. On the other hand, the probability of finding none of the particles within a sphere of radius r is $[1 - (r/R)^3]^N$. Hence, the probability that at least one of the N particles shall fall within this sphere is given by

$$p = 1 - \left[1 - \left(\frac{r}{R}\right)^3\right]^N \approx 1 - \exp\left[-\left(\frac{r}{R}\right)^3 N\right], \qquad (6.20)$$

since N is large. If n_i is the number of ions per cm³, then N is equal to $\frac{4}{3}\pi R^3 n_i$, so that equation (6.20) becomes

$$p \approx 1 - \exp\left(-\tfrac{4}{3}\pi r^3 n_i\right).$$

The probability that at least one ion shall be found in the shell from r to $r + dr$ is obtained upon differentiation as

$$dp = e^{-x}\, dx, \qquad (6.21)$$

where x is defined by

$$x \equiv \left(\frac{r}{r_0}\right)^3, \qquad (6.22)$$

and the mean distance of separation r_0 is determined by

$$\tfrac{4}{3}\pi r_0{}^3 n_i = 1.$$

The "average" or "normalized" electric field strength corresponding to this mean distance in a hydrogen isotope plasma, where all the particles carry a single charge, is then

$$E_0 = \frac{e}{r_0{}^2} = 2.60\, e n_i{}^{2/3} \qquad (6.23)$$

in cgs-esu. A more detailed treatment, in which allowance is made for the simultaneous interaction of several particles, gives a numerical factor of 2.61 in equation (6.23) and upon conversion into practical units this becomes

$$E_0 = 3.74 \times 10^{-7} n_i{}^{2/3} \quad \text{volts/cm},$$

with n_i expressed as ions per cm³.

6.47. The electric field E at a point r may be stated in terms of the average field by means of a ratio β, i.e.,

$$\beta \equiv \frac{E}{E_0},$$

and from Coulomb's law,

$$\beta = \frac{E}{E_0} = \left(\frac{r_0}{r}\right)^2,$$

so that x, defined by equation (6.22), becomes

$$x = \beta^{-\frac{3}{2}}.$$

For a given first-order Stark component, equation (6.19) shows that $\Delta\lambda$ is proportional to E, and so the shift in the wave length is related to that ($\Delta\lambda_0$) which would be produced by the average field by $\Delta\lambda/\Delta\lambda_0 = \beta$. It follows, therefore, by integration of equation (6.21), that the probability of a displacement $\Delta\lambda$ of a specific first-order Stark component is given by

$$H(\beta) = \tfrac{3}{2}\beta^{-\frac{5}{2}} \exp\,(-\beta^{-\frac{3}{2}}),$$

where $H(\beta)$ is known as the Holtsmark function [40]. The actual Holtsmark function is more complicated than that given here; however, the results are in general agreement, especially in indicating that for large values of $\Delta\lambda$, that is, when β is large, the function $H(\beta)$ varies as the $-5/2$ power of β and, hence, also of $\Delta\lambda$.

6.48. For a given, e.g., the kth, component of the Stark-split lines, the shift $\Delta\lambda_k$ in the wave length in an electric field E may be written, according to equation (6.19), as

$$\Delta\lambda_k = C_k E, \tag{6.24}$$

where C_k depends on the initial and final values of the quantum numbers and the wave length λ of the line in the absence of an electric field. Hence, since β is equal to E/E_0 it follows that

$$\beta = \frac{\Delta\lambda_k}{C_k E_0}. \tag{6.25}$$

6.49. If I_k is a function representing the intensity distribution of the broadened kth component of the first-order Stark spectrum and $I_k{}^0$ is the intensity of the unbroadened line, the value of which is known from theoretical considerations, then it has been shown that

$$I_k = \frac{I_k{}^0}{C_k} H(\beta)$$

or, utilizing equation (6.25) for β,

$$I_k = \frac{I_k{}^0}{C_k} H\left(\frac{\Delta\lambda_k}{C_k E_0}\right). \tag{6.26}$$

The total intensity distribution of the broadened Stark line in the inhomogeneous field of a plasma is then obtained by summing the I_k values, as derived from equation (6.26), for all the components of a particular line. In practice it is found that beyond the first three or four components the contribution to the intensity can be neglected.

6.50. The procedure for determining the ion density in a deuterium plasma, for example, is then as follows [32]. First, the Stark pattern for deuterium in a uniform field and the intensities of the component lines are derived from the standard wave mechanical theory of atomic spectra. Taking a number of arbitrary values of $\Delta\lambda_k/E_0$, the corresponding I_k is calculated for each of these values, utilizing equation (6.26) and the appropriate magnitude of I_k^0 and of the Holtsmark function [41]. This is done for each of the component lines and the results are plotted as in Fig. 6.8 and then added. The full

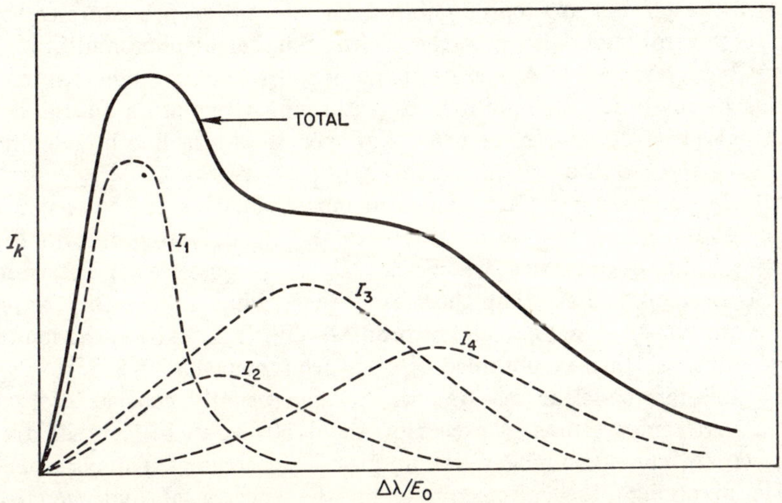

Fig. 6.8. Calculation of intensity of Stark spectrum.

line in the figure, which represents the sum of the I_k values, shows the expected profile of the broadened Stark lines as a function of $\Delta\lambda_k/E_0$.

6.51. Because of the assumptions involved in the foregoing treatment, various corrections should be made in the curve for the total intensity in Fig. 6.8. The postulate of random distribution of the ions is objectionable, in particular, since it leads to too many closely spaced pairs of ions. In fact, this distribution can be expected to hold only at infinite temperature. Since the Holtsmark function varies as the $-5/2$ power of $\Delta\lambda$ for large $\Delta\lambda$, the total intensity curve falls off in this manner at the right of Fig. 6.8. However, when the correction which conforms to finite temperature is applied, the curve falls off faster than $(\Delta\lambda)^{-5/2}$.

6.52. The next step in the procedure is to evaluate E_0 from the experimental observations of the Stark spectrum of the plasma under consideration. The image of the plasma is first formed on the slit of a spectroscope. The plateholder is then replaced by a carriage holding a photomultiplier tube behind a narrow slit which can be moved by means of a micrometer screw so that it is exposed to different wave lengths across the line profile. From the output of the tube it is possible to draw a graph showing the relative intensity profile of the broadened Stark spectrum as a function of the wave length. For the present purpose, the results should be expressed in terms of the Stark shift, i.e., $\Delta\lambda$, the difference between any particular wave length and that of the original spectral line in the absence of an electric field.

6.53.   The experimental technique described above requires that the plasma discharge be repeated for each wave length position of the slit. The reliability of the results thus depends upon the uncertain premise that the conditions in the plasma are reproducible. An alternative procedure is to use a system of several slits, with a corresponding number of photomultiplier tubes, each slit being exposed to a narrow range of the spectrum. Several exposures can thus be made for a single discharge through the plasma [42]. In addition, streak spectrographs have been employed to obtain line broadening data with a time resolution of 0.1 microsec [30].

6.54.   By comparing the experimental curve, showing the intensity of the Stark spectrum as a function of $\Delta\lambda_k$, with the theoretical profile in Fig. 6.8, which is a plot of intensity versus $\Delta\lambda_k/E_0$, it is possible to determine E_0. It is assumed, of course, that there is a reasonably satisfactory agreement between experimental and calculated profiles. With E_0 known, the ion density in the plasma can then be obtained directly from equation (6.23).

6.55.   In thermonuclear research the plasma usually consists of hydrogen isotopes. At temperatures in excess of about 50 ev, these isotopes are completely ionized and so no longer give an atomic spectrum. Thus, spectroscopic methods involving observations made on the spectra of hydrogen isotopes cannot be used at temperatures greater than 50 ev.

6.56.   The situation may be overcome to some extent by introducing a small amount of helium into the system. The hydrogenlike spectrum of singly ionized helium, i.e., He^+, can then be used to study the broadening of the Stark lines. A difficulty arises, however, in interpreting the ion density, since the value obtained from equation (6.23) is really the sum of the $n_i Z_i^{3/2}$ terms, where n_i is the particle density and Z_i is the charge, for all the ions present. The fact that the exact amounts of He^+ and He^{++} ions in the plasma are not known is thus a complication. However, the resulting error will probably not be serious, in comparison with others, if the proportion of helium in the gas is relatively small.

6.57.   The approximate nature of the Stark effect method for determining ion densities in a plasma is evident from the fact that the experimental and calculated profiles generally do not fit over the whole width of the broadened spectrum. It appears that the estimated ion densities may be too high by a factor of from 2 to 5. There are several causes for this discrepancy [42-47].

In the first place, some of the assumptions made, e.g., that concerning the random distribution of the ions, are not justified; in addition, certain factors, such as the influence of the electrons on the field strength and the second- (and higher-) order Stark effect, have been neglected [43-47].

6.58. If, instead of requiring absolute ion densities, relative values are adequate, then the foregoing procedure may give reliable results in spite of errors in the individual data. A case in point is the determination of the degree of compression occurring in a pinched discharge. This has been obtained from a comparison of a Balmer line profile observed at the time of the maximum pinch with that appearing in the afterglow when the plasma has expanded to fill the tube [48].

DOPPLER BROADENING AND ION TEMPERATURES

6.59. Since the atoms present in a gas (or plasma) are moving with different velocities, they emit light of slightly different wave lengths for a given electronic transition, as a result of the Doppler effect. Consequently, each line of the spectrum is somewhat broadened, so that it extends over a range of wave lengths. The nature of the Doppler broadening will evidently depend upon the distribution of velocities among the emitting particles, and if this is Maxwellian in nature it will be directly related to the temperature. It is seen, therefore, that a study of the Doppler broadening of the spectral lines from a plasma might provide information concerning the temperature [1, 4, 35, 36, 49-53].

6.60. Before showing how the broadening is related to the temperature, consideration must be given to the possible difficulties arising from simultaneous Stark broadening of the lines and how they may be overcome. In the first place, the shift in wave length due to the Doppler effect is most evident close to the position of the unperturbed spectral line, i.e., at the center (or core) of the Stark system, whereas the consequences of the Stark broadening are chiefly apparent in the wings. It has been suggested, therefore, that the experimental profile, e.g., of a Balmer line, which should be capable of representation by a Voigt function, [54], might be analyzed so as to separate the Doppler and Stark effects. In this event, ion density and temperature might be obtainable from the same set of observations.

6.61. Another approach to the problem is to utilize the fact that the Stark effect depends upon the ion density whereas the Doppler broadening does not. In many plasmas, therefore, the densities may be so low that the Stark broadening of the lines is insignificant. A third solution is to choose for study a spectral line for which the first-order Stark effect, at least, is negligible or absent. In hydrogen isotope plasmas this unfortunately requires the introduction of an impurity; thus, for temperature determinations based on Doppler broadening use has been made of the spectral lines of oxygen IV and V, nitrogen IV and V, and argon II.

6.62. If a particle at rest emits a spectral line of wave length λ, then the corresponding wave length λ' from the particle moving with a velocity v in the line of sight is given by

$$\lambda = \lambda' \left(1 + \frac{v}{c}\right).$$

Hence, the Doppler shift $\Delta\lambda = \lambda - \lambda'$, corresponding to a particular velocity, is

$$\Delta\lambda = \frac{v}{c}\lambda' \approx \frac{v}{c}\lambda,$$

so that

$$v \approx c\,\frac{\Delta\lambda}{\lambda}. \tag{6.27}$$

6.63. Assuming a Maxwellian distribution of velocities among the emitting species, the spectral intensity is given by

$$I(\lambda) = \text{constant} \times \exp\left(-mv^2/2kT\right),$$

where m is the mass of the emitting atom (or ion) and T is the kinetic temperature. Upon introducing equation (6.27), this becomes

$$I(\lambda) \approx \text{constant} \times \exp\left[-mc^2(\Delta\lambda/\lambda)^2/2kT\right].$$

The half-intensity width $\Delta\lambda_{\frac{1}{2}}$ of the line, i.e., the width at which the intensity is half the maximum, expressed in the same units as λ, is then found to be

$$\Delta\lambda_{\frac{1}{2}} \approx 1.67\,\frac{\lambda}{c}\left(\frac{2kT}{m}\right)^{\frac{1}{2}},$$

with m in grams, k in ergs/°K, and T in °K. If T is in kev, then

$$\Delta\lambda_{\frac{1}{2}} \approx 2.44 \times 10^{-3}\lambda\left(\frac{T}{M}\right)^{\frac{1}{2}}, \tag{6.28}$$

where M is now the conventional atomic weight of the emitting species [55].

6.64. Hence, from a study of the Doppler broadening of impurity lines, it should be possible to determine, by means of equation (6.28), the temperature of the emitting atoms (or ions). Since a Maxwellian distribution of velocities has been postulated, it is reasonable to suppose that the result obtained will also apply to the temperature of hydrogen isotope ions in the plasma, since in many cases equipartition can be assumed to have taken place.

6.65. An indication of the magnitude of the Doppler broadening may be obtained by considering a species with a mass number of 20, so that M is approximately 20. Suppose the wave length of the line being studied is in the vicinity of 4000 A, and let the temperature be 0.1 kev. The breadth of the spectral line is then found from equation (6.28) to be less than 1 A. It is evident that measurements of this kind require the use of a spectrograph of high dispersion. One technique which has been devised in order to obtain the necessary resolving power is to employ a Fabry-Pérot interferometer in conjunction with a prism spectrograph [56]. The apparent Doppler broadening, as observed, must always be corrected for the normal instrumental broadening which is appreciable even when the dispersion is high.

6.66. As is the case with much plasma research, the exact significance of the results obtained from Doppler effect measurements may be somewhat in doubt [57]. The distribution of velocities among the emitting particles may not be Maxwellian in a pulsed discharge of short duration, and the kinetic temperature of these particles may not be the same as that of the ions which constitute the bulk of the plasma. Furthermore, mass motion and turbulence of the plasma and the drift of the ions in a magnetic field will be sources of error. It is of interest to note in this connection that the rate of mass motion of a plasma, e.g., rotation or motion in the direction of the electric field, has been measured from observations on the Doppler broadening of the spectral lines of He II, C II, O V, etc. (cf. §§11.47, 11.56) [58, 59].

COMPARISON OF LINE INTENSITIES AND PLASMA TEMPERATURES

6.67. A method for obtaining an approximate indication of the temperature in a plasma depends upon the fact that the relative intensities of the spectral lines from two different atomic (or ionic) states of a given emitter will generally be a function of the temperature. A case of interest that lends itself particularly to theoretical treatment arises when one of the lines originates from an atom, e.g., He, and the other from a singly ionized form of the same species, e.g., He$^+$. The helium lines at 5875 A and 4686 A, respectively, satisfy this requirement.

6.68. Suppose that the electronic transition responsible for the He line is represented by $l \to k$, and that for the He$^+$ line by $j \to i$ (see Fig. 6.9). Let the

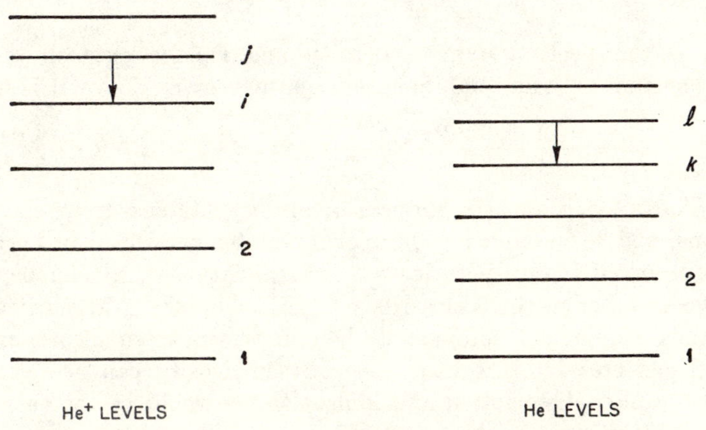

FIG. 6.9. Electronic transitions in He$^+$ and He.

corresponding frequencies be ν_{lk} and ν_{ji}, respectively. Provided self-absorption of the lines is negligible, the intensities may then be written as

$$I_{ji} = \nu_{ji}A_{ji}n_j$$

and

$$I_{lk} = \nu_{lk}A_{lk}n_l,$$

where A_{ji} and A_{lk} are the respective transition probabilities, which can be derived from theoretical considerations, and n_j and n_l are the densities of the emitting particles in the indicated electronic levels. If the Boltzmann equation is assumed to be applicable, the values of n_j and n_l may be expressed in terms of the number densities n^+ and n^0, respectively, of the ground states of He^+ and He (neutral); thus

$$n_j = \frac{g_j}{g^+} n^+ \exp\left[-(W_j - W^+)/kT\right]$$

and

$$n_l = \frac{g_l}{g^0} n^0 \exp\left[-(W_l - W^0)/kT\right],$$

where the g's are the statistical weights and the W's are the energies of the indicated states; the absence of a specific subscript implies the ground state of the particular emitting species.

6.69. If thermodynamic equilibrium may be assumed between helium atoms, on the one hand, and helium ions and electrons, on the other hand, i.e.,

$$He \rightleftharpoons He^+ + e,$$

and the energy distribution is Maxwellian in character, the ratio of the particle densities in the respective ground states may be expressed by means of the Saha equation

$$\frac{n^+n_e}{n^0} = \frac{2g^+}{g^0} \cdot \frac{(2\pi m_e kT)^{3/2}}{h^3} \exp\left[-(W^+ - W^0)/kT\right],$$

where n_e is the electron number density and m_e the electron mass; h is Planck's constant. Upon combining the equations given above, it is found that

$$\frac{I_{ji}}{I_{lk}} = \frac{\nu_{ji}A_{ji}g_j}{\nu_{lk}A_{lk}g_l} \cdot \frac{2(2\pi m_e kT)^{3/2}}{n_e h^3} \exp\left[-(W_j - W_l)/kT\right]. \qquad (6.29)$$

6.70. The ν, A, g, and W terms can all be obtained from spectroscopic tabulations, and so, provided n_e is known, the temperature can be evaluated from the observed intensity ratios. A determination of n_e can be made by microwave or other methods described in this chapter. Alternatively, if the gas is mainly a hydrogen isotope, with helium present in small proportion, and the initial gas pressure is known, the electron density can be estimated by assuming complete dissociation and ionization, as would be the case for temperatures of 50 ev, i.e., about 5×10^5 °K or more [60, 61].

6.71.   Since Maxwellian equilibrium has been assumed in the derivation of equation (6.29), the calculated temperatures should presumably be those of both ions and electrons.   However, it is not certain that such equilibrium will exist in a plasma heated by a pulsed discharge, especially at low plasma densities and high temperatures.   Furthermore, the conditions may not be favorable to the establishment of the Maxwell-Boltzmann distribution, so that, here again, the precise significance of the results is by no means certain.

6.72.   Instead of calculating the relative intensities of the spectral lines as a function of temperature, use may be made of experimental data on cross sections for the excitation of certain singlet and triplet lines by bombardment with electrons of known energy [62].   The measured cross sections are for specific electron energies, and these must be integrated over a postulated distribution of energies such as might exist in a plasma.   In the absence of more definite information, a Maxwellian distribution is assumed.   From the ratio of the cross sections, the ratio of the intensities of the lines can then be plotted in terms of the electron kinetic temperature.   The necessary excitation data are available for the helium triplet at 4713 A and the singlet at 4921 A, and from these the curve in Fig. 6.10 has been calculated [3].   Hence, if

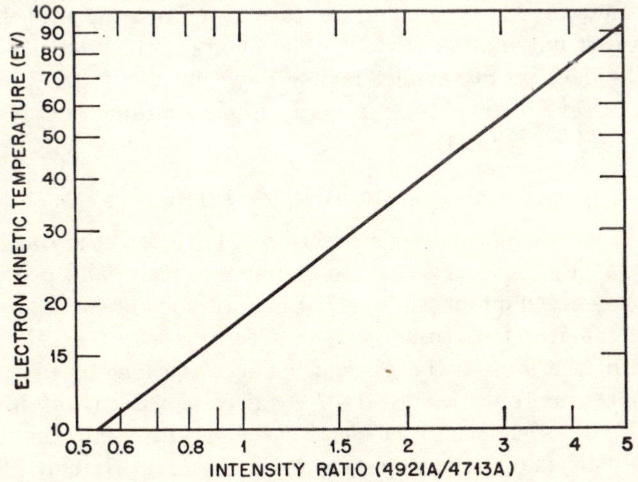

FIG. 6.10.   Relative intensities of helium lines.

the intensity ratio of the two lines, as emitted from a heated plasma containing some helium, is determined experimentally, the electron temperature can be estimated [63-65].   It has been found that, from the observational standpoint, the singlet line at 4921 A is not ideal; the line at 5047 A would be preferable, but the electron excitation cross sections have not been determined.

6.73.   There are several sources of error in the methods for deriving electron temperatures from a comparison of the intensities of spectral lines.   Brief mention may be made here of some of the more obvious ones. In the first place, in a heated plasma the distribution of electron energies, especially in the high-energy region responsible for most of the excitation, is probably not Maxwellian. Second, there is no certainty that, in the determination of the excitation cross sections, direct electron bombardment was the sole cause of excitation. Finally, in the plasma the spectra are probably excited by still other mechanisms, so that estimates of temperature, based even upon reliable electron excitation data, would be in error.

SPECTROSCOPIC OBSERVATIONS OF CONTAMINANTS

6.74. A somewhat qualitative application of spectroscopic observations on a heated plasma is to determine the nature and time of arrival of contaminants in the discharge. If the material of the tube holding the plasma contains silica (or silicates), e.g., as in glass, quartz, or porcelain, then contact of the hot plasma with the wall will generally result in the appearance of various lines of silicon and usually also of oxygen. The time at which a line appears, as observed with a monochromator in conjunction with a photomultiplier tube and an oscilloscope, can thus provide information concerning the contaminant-free time of the plasma. This can then be correlated with other phenomena, e.g., the appearance of X-rays and an increase in the resistance of the discharge due to the decrease in temperature. Measurements of the intensity of impurity radiation in the vacuum ultraviolet region, down to about 100 A or less, have proved useful in several respects, especially in determining energy losses from pinched discharges (cf. §7.124).

CONTINUUM RADIATION MEASUREMENTS

6.75. The spectroscopic methods described above for studying the properties of a plasma have involved the use of line spectra. The procedure to be considered next is based on measurements of continuous bremsstrahlung spectra accompanying the free-free transitions of electrons when accelerated in the Coulomb field of the ions in the plasma. The wave lengths of these spectra extend over a considerable range, from a fairly sharp cut-off in the X-ray region, which is very sensitive to the electron temperature, into the visible range and presumably beyond, as is apparent from Fig. 2.8. It can be seen from equations (2.25) and (2.29), respectively, that both the total power radiated as bremsstrahlung and the emission at a given wave length depend on the electron temperature and density. Hence, bremsstrahlung measurements provide, in principle, a means for determining these two important properties of a plasma. The basic assumption, as in other spectroscopic techniques, is that the distribution of energies is Maxwellian in nature. It is also assumed, with complete justification, that the experimental plasma is optically thin to the bremsstrahlung radiation [1, 30, 66, 67].

6.76. In order to determine the electron temperature a treatment must be used which is independent of the plasma density. This is achieved by operating in the short wave length (X-ray) region, where the results are most sensitive to temperature. From observations on the transmission intensities of the plasma radiation through absorbers of various materials and thicknesses, curves of relative intensity versus absorber thickness are plotted. For the experimental intensity measurements both scintillation counters and ionization chambers containing argon have been used.

6.77. From the known absorption coefficient μ_λ of the given absorber material as a function of wave length, the total relative intensity $I(t)$ transmitted through an absorber of thickness t can be calculated by using the relationship.

$$I(t) = \int_0^\infty \left(\frac{dP_\lambda}{d\lambda}\right) e^{-\mu_\lambda t}\, d\lambda, \tag{6.30}$$

where $dP_\lambda/d\lambda$ is given by equation (2.29); the integration is performed numerically for a series of temperatures and absorber thicknesses using any arbitrary plasma particle density. If a semilogarithmic plot is then made of $I(t)$ against the absorber thickness, the *shape* of the curve is a function of the temperature only. The electron temperature of the plasma is then obtained by finding the best fit between the experimental curve and those computed in the manner described. An example of the results obtained in a determination of this kind with carbon (polyethylene) as absorber is shown in Fig. 6.11; the broken line represents the experimental curve located where it appears to fit best into the series of curves calculated for electron temperatures ranging from 100 to 500 ev. The electron temperature in the experimental plasma is evidently about 240 ev [67].

6.78. An absorption method, which does not depend on curve fitting, makes use of the selective transmission of X-rays that occurs at wave lengths above the K, L, etc., edges for various absorbers. It is then possible to choose two absorbers such that the ratio of the bremsstrahlung transmissions, at equal areal densities, changes rapidly with temperature in the region of interest. By means of equation (6.30), the ratio can be evaluated for a number of temperatures, the results being independent of the plasma density. The point at which the experimentally determined ratio for the two given absorbers falls on the calculated curve of transmission ratio versus temperature then gives the required electron temperature. The data obtained for the same conditions to which Fig. 6.11 applies are depicted in Fig. 6.12; the curves are calculated and the experimental points for four pairs of absorbers are indicated by the circles. The mean value of the electron temperature is seen to be in the neighborhood of 230 ev [67].

6.79. It should be mentioned that the results given above were based on classical considerations with the Gaunt factor g being taken as unity in equations (2.28) and (2.29). Actually, for free-free transitions, g is somewhat larger than unity in the short wave length region of the spectrum in which the measurements were made. In the present situation, it has been estimated that introduction of the Gaunt factor would increase the estimated temperature by some 13 per cent, i.e., to above 260 ev, so that the correction is within the rather large experimental errors of the intensity measurements.

6.80. With the electron temperature known, it should be possible to derive the electron (or ion density) from an experimental determination of the absolute total intensity of the bremsstrahlung, by making use of equation (2.29). This measurement is, however, very sensitive to the effects of impurities.

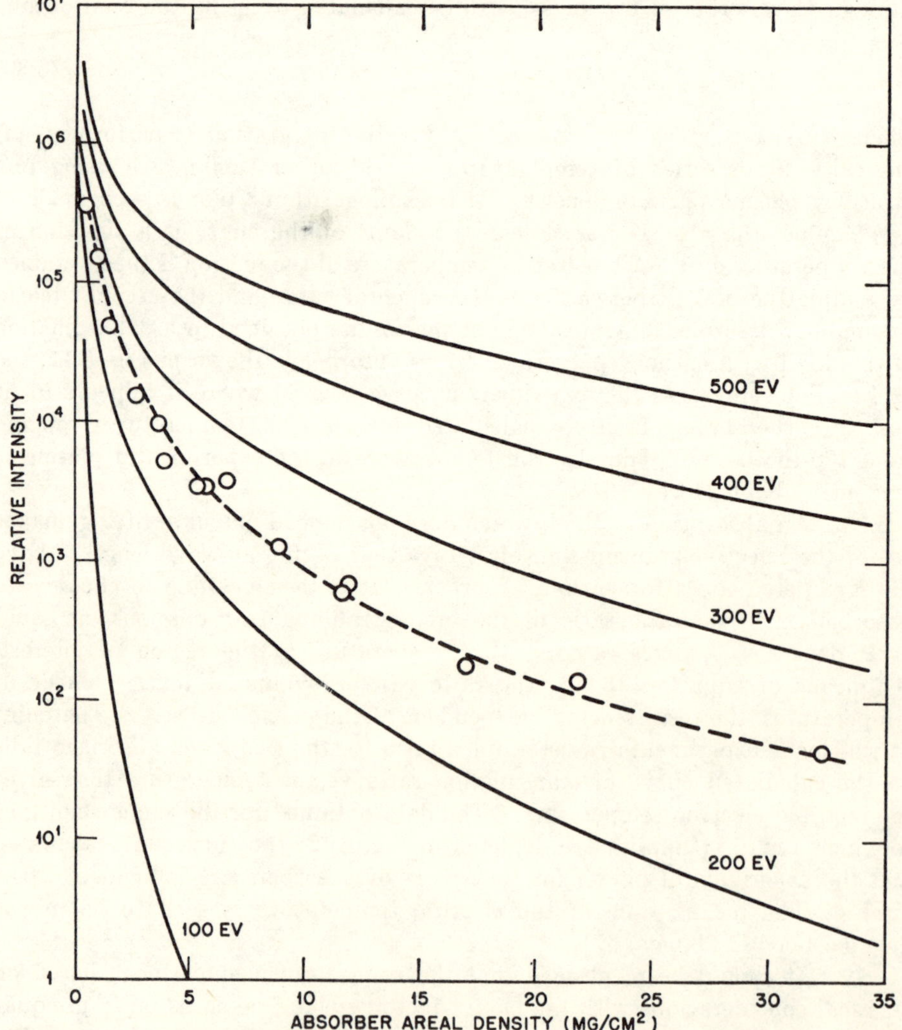

Fig. 6.11. Electron temperature from relative bremsstrahlung transmission intensities.

Not only does the presence of impurities greatly enhance the free-free bremsstrahlung intensity (§2.75), but there is also emission of free-bound (or recombination) radiation, resulting from electron capture, and also of bound-bound (or line) radiation. Fortunately, these radiations do not appreciably

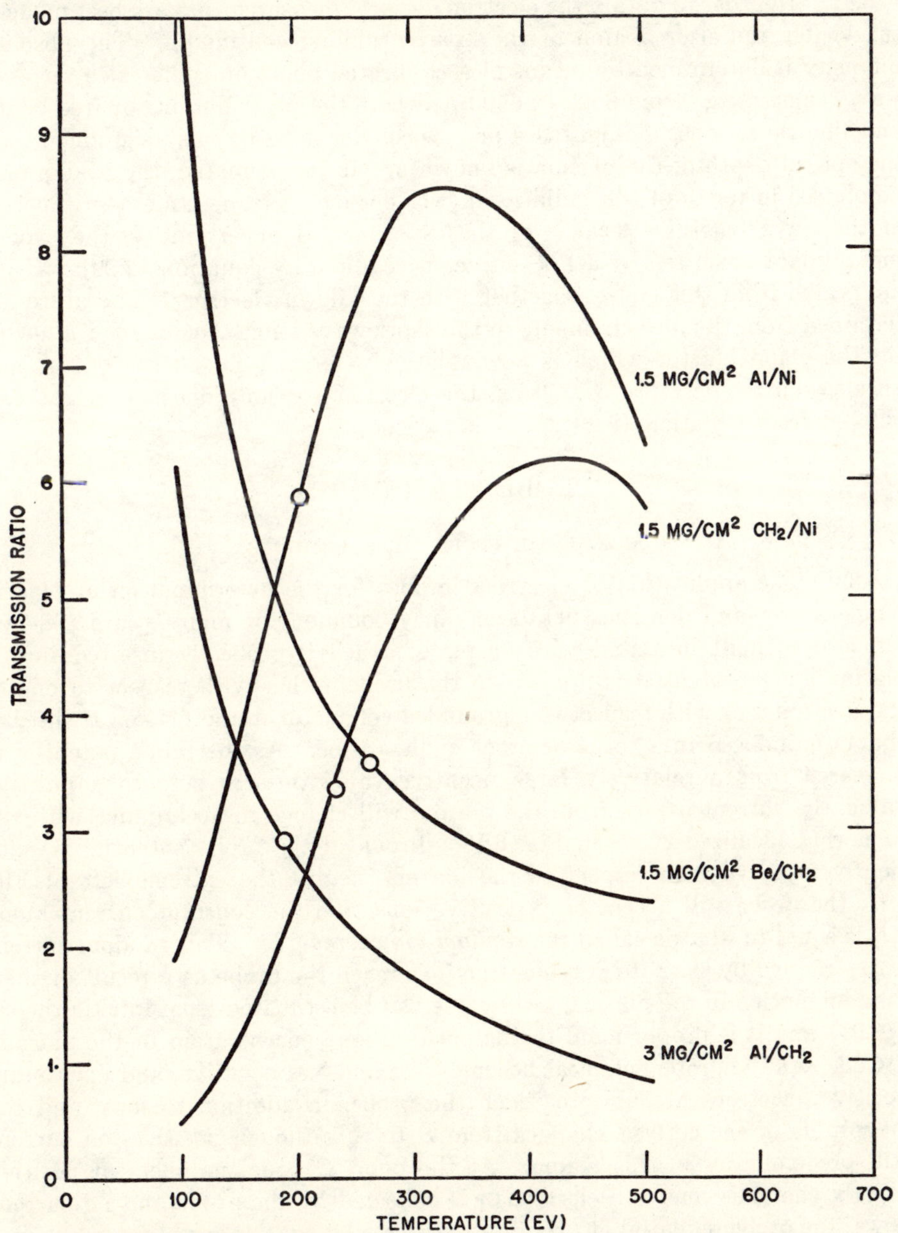

Fig. 6.12. Electron temperature from bremsstrahlung transmission ratios.

affect the estimation of the temperature, provided the observations are made on radiations of sufficiently short wave length. The line radiation is then not important and the free-bound radiation intensity has the same distribution with wave length as does that of the free-free bremsstrahlung radiation.

6.81. In order to obtain the electron density, measurements are best made in the visible radiation region of the bremsstrahlung continuum. The absolute intensity is determined by means of a calibrated photomultiplier at a series of wave lengths, e.g., from 3500 to 6500 A, where the contributions of free-bound and line radiations of impurities are expected not to be very significant. If the optical depth of the plasma is known or can be estimated, the results may be plotted in terms of the radiation power per cm³ per angstrom as a function of the wave length. Because of the experimental uncertainties, the experimental data are fitted by a $1/\lambda^2$ curve, as required by equation (2.29). Using any value from this curve, together with the known electron temperature determined from the measurements in the short wave length region, and allowing for the Gaunt factor which is now relatively large, e.g., 3.0 at 5000 A and an electron temperature of 250 ev, the electron (or ion) density can be calculated from equation (2.29).

PROBE TECHNIQUES

ELECTRICAL PROBE MEASUREMENTS

6.82. The application of electrical probes for the determination of plasma temperatures and densities has been fairly common for many years [68-71]. Suppose a small insulated wire or plate, called a probe, is inserted into a plasma and a potential is applied to the probe, either with respect to one of the electrodes or with respect to a grounded conductor in the plasma if the discharge is induced in a torus or other endless tube. As the probe potential is increased from a relatively large negative value through zero to a positive value, the current drawn from the plasma will change in accordance with the somewhat idealized curve in Fig. 6.13. Provided the probe potential is sufficiently negative with respect to the plasma, i.e., at the extreme left of Fig. 6.13, the probe will accept only positive ions, and the constant current along AB is equal to what is called the *random ion current, i_{ri}*. This random current is determined by the rate at which the ions reach the probe as a result of their random motion in the plasma. Its value can be estimated from kinetic theory (§6.91) and it is proportional to the positive-ion concentration in the plasma.

6.83. As the probe potential becomes less and less negative, and eventually positive, electrons are able to reach the probe, in addition to ions, and the magnitude of the current changes from B to C, although positive-ion current still predominates in this region. At the point C equal numbers of positive (singly charged) ions and electrons are collected by the probe and no current flows; the probe potential at C is referred to as the *floating potential*.

6.84. When the probe potential is made more positive, beyond C, the number of electrons collected exceeds the number of ions and the current direction is reversed. The current now increases rapidly with increasing applied potential as more and more electrons are accepted by the probe. Ultimately, at a point such as D, the electrons reach the probe at their maximum rate, determined by the electron density in the plasma, and the current becomes constant, as shown by DE, even though the probe voltage is increased. At E the current may increase rapidly along EF; the high positive potential then accelerates the electrons to such an extent that they can cause secondary ionization of any neutral particles that may still be present.

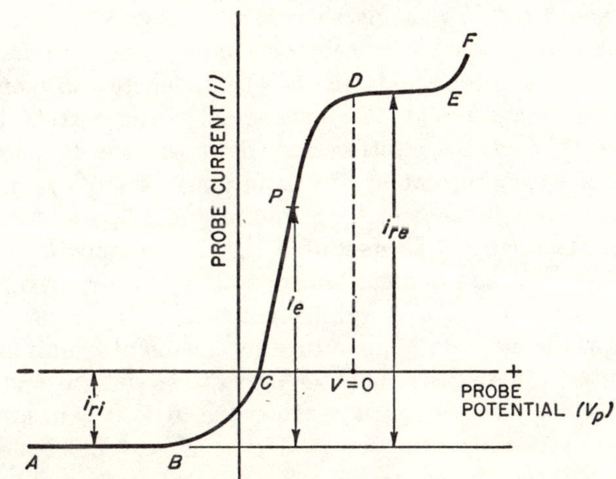

Fig. 6.13. Idealized curve of probe potential and current.

6.85. Consider the region CD in Fig. 6.13; the number of electrons reaching the probe exceeds the number of positive ions, so that the probe may be regarded as being surrounded by a sheath in which electrons predominate. Only electrons with sufficient energy to overcome the repulsion of the sheath will then be able to penetrate it and reach the probe. If V is the potential difference across the sheath, i.e., between the bulk of the plasma and the probe surface, and e is the electronic charge, the required energy is eV. Assuming the Boltzmann distribution law to apply, the probability that an electron will have this amount of energy is $\exp\ (-eV/kT_e)$. Consequently, if n_e is the electron density in the bulk of the plasma and n_p is the density at the probe surface, then

$$n_p = n_e \exp\ (-eV/kT_e), \qquad (6.31)$$

where T_e is the electron temperature in the plasma.

6.86. The electron current i_e at any stage is proportional to the rate at which electrons reach the probe surface, that is, to the electron density at the surface; hence, from equation (6.31),

$$i_e = i_{re} \exp\left(-eV/kT_e\right), \qquad (6.32)$$

where i_{re} is the *random electron current*, exactly analogous to the random ion current described in §6.82.* The random electron current is equal to the electron current at the point D in Fig. 6.13, when the potential across the sheath has been decreased essentially to zero by the external potential applied to the probe. In these circumstancs all the electrons reaching the sheath are able to penetrate to the probe and contribute to the electron current.

6.87. It should be observed in determining i_e and i_{re} from experimental data, such as from those used in plotting the curve in Fig. 6.13, that ions as well as electrons reach the probe surface in the region from B to D. This is the case because electrons predominate in the sheath up to the point D, so that ions are attracted to the probe. The measured probe current is thus equal to the difference between the electron current i_e and the random ion current i_{ri}. The actual electron current at any point, for example P in Fig. 6.13, is then obtained from the distance of P above the line AB, as shown. The same applies to the random current i_{re} measured between D and E.

6.88. Since i_e and i_{re} can be determined experimentally, from a plot of the probe current against the applied probe potential (Fig. 6.13), it should be possible to obtain the electron temperature by means of equation (6.32), provided the potential difference V between the probe surface and the bulk of the plasma is known. As seen earlier, the value of V is zero at the point D, where the applied probe potential is V_d. Hence, at any point along the curve

$$V = V_d - V_p,$$

where V_p is the applied probe potential, which is plotted as abscissa in Fig. 6.13. In principle, both i_{re} and V_d can be obtained from the experimental curve, but the point D is usually difficult to identify with any degree of accuracy, and so an alternative procedure must be used to determine the temperature.

6.89. Upon replacing V in equation (6.32) by $V_d - V_p$ and taking logarithms of both sides, it is seen that

$$\ln i_e = \ln i_{re} - \frac{eV_d}{kT_e} + \frac{eV_p}{kT_e}$$

$$= \text{constant} + \frac{eV_p}{kT_e}, \qquad (6.33)$$

since the first two terms on the right are constant in a given set of measurements. It follows that a plot of the experimental values of $\ln i_e$ against V_p

---

* The random ion current is, of course, less than the random electron current because the ions are heavier, and hence have smaller velocities, than the electrons at a given plasma temperature.

should be a straight line with a slope e/kT_e. Hence, the electron temperature in the plasma can be readily derived. Incidentally, any departure of the plot from linearity implies that a Boltzmann distribution does not exist among the electrons. The significance of the results obtained from the measurements is then uncertain.

6.90. Although it is difficult to determine the exact location of the point D at which the probe-plasma potential difference is zero, a reasonably good value can be obtained for i_{re} and this can be related to the electron density of the plasma. From purely kinetic considerations, it can be shown that, for a system of particles in which the velocity distribution is Maxwellian, the particles will strike an area of 1 cm^2 at the rate of $n(kT/2\pi m)^{1/2}$ per sec, where m is the particle mass in grams, kT is in ergs, and n is the density in particles/cm^3. The random electron current to a probe of area A cm^2 is thus

$$i_{re} = A e n_e \left(\frac{kT_e}{2\pi m_e}\right)^{\frac{1}{2}}, \tag{6.34}$$

where i_{re} is expressed in statamperes if e is taken as 4.80×10^{-10} statcoulomb and k as 1.38×10^{-16} erg/°K; in order to convert the result into amperes, it is divided by 3×10^9. The electron temperature can be obtained by means of equation (6.33) and so the electron density can be calculated from equation (6.34) since all the other quantities are known.

6.91. The random ion current i_{ri} can be determined fairly accurately from the experimental data, since the region AB of the curve in Fig. 6.13 is generally flat. This may be related to the ion density, temperature, and mass by an expression analogous to equation (6.34). But since the ion temperature is not known and cannot be derived from probe measurements, the result is not of direct use. It has been shown, however, that

$$i_{ri} \approx 0.40 A e n_i \left(\frac{2kT_e}{m_i}\right)^{\frac{1}{2}}, \tag{6.35}$$

so that the ion density n_i can be calculated provided T_e is known [69]. Although equation (6.35) appears to be merely an alternative to equation (6.34), since in a hydrogen isotope plasma n_e and n_i are identical, there are certain circumstances, as will be seen shortly, in which equation (6.35) is to be preferred.

6.92. In spite of certain limitations to the use of probes, some of which will be mentioned below, it seems that the treatment described above is reasonably satisfactory in the absence of magnetic fields. In thermonuclear research, however, the plasma is confined by such a field, and consideration must be given to its possible effect on the probe measurements. An important requirement in the application of probes is that the dimensions of the probe should be small in comparison with the mean free path of the collected particles. For this reason, wires or small plates are generally used as probes and the necessary condition is thereby satisfied.

6.93. In the presence of a magnetic field, however, the situation may be completely changed. The principal effect of a strong field is to restrict the motion of the particles to within a gyromagnetic radius, given by equation (4.4), of the lines of force. For electrons in a field of a few thousand gauss and a temperature of about 10^6 °K, this distance may be of the order of 10^{-3} cm. Except for collisions with other particles in the plasma which disturb its motion along the field lines, the effective free path of an electron is thus only 10^{-3} cm, and this is considerably less than the dimensions of a practical probe.

6.94. When the probe is much larger than the mean free path, the plasma surrounding the probe is depleted of electrons at a rate greater than that at which they can be replaced by random motion, i.e., by diffusion. Consequently, in a magnetic field, electron collection is markedly decreased because of the small effective mean free path. The electron current then depends upon a number of factors, including electron temperature, probe radius, ion energy and density, and magnetic field strength, as well as on the electron density. In this event, equation (6.34) for the random electron current is completely erroneous. Nevertheless, in spite of the fact that there is no theoretical basis for the result, it appears from experiments that the slope of the portion CD of the curve of probe current versus applied probe potential (Fig. 6.13) is determined by the electron temperature in the same way as in the absence of a magnetic field. Thus, the electron temperature can be obtained from the slope of the linear plot of $\ln i_e$ against V_p, as stated in §6.89.

6.95. Because of the greater mass of an ion, its gyromagnetic radius is larger than that of an electron in the same field at a given temperature. For magnetic fields and temperatures of the order postulated above, the gyration radius of a deuteron, for example, about the lines of force will approach 0.1 cm, and this is more nearly like a practical size for a probe. Hence, it is to be expected that equation (6.35) for the random ion current will hold fairly well. This equation can consequently be used to determine the ion (and electron) density in a plasma from probe measurements, since the value of T_e is known.

6.96. The simple theory of electrical probes is based on the tacit assumption that, although the observed probe currents are dependent upon the acceptance of ions and electrons by the probe, this does not affect the plasma in any way. Unless the probe is extremely small and draws negligible current which could not be measured with any degree of accuracy, it is evident that the conditions in the region of the probe may be quite different from those elsewhere in the plasma. The temperature and density values obtained under these conditions may thus be misleading. A possible method for overcoming this situation is by the use of a floating double-probe system, to be described below.

6.97. Apart from the quantitative (or semiquantitative) applications of probes, they may be used to provide other data, e.g., concerning the motion of the plasma in a discharge tube. Comparison of the currents at probes inserted to different depths in the plasma or oriented in different directions may also supply information on possible inhomogeneities in density and temperature.

6.98. In the double-probe technique to be described here, no current is drawn from the plasma [72-75]. The observed current is supplied by a battery and the plasma essentially provides the conducting medium containing positive ions and electrons. The situation is somewhat similar to electrolysis; the plasma is equivalent to the electrolyte and the probes behave as inert electrodes.

6.99. A schematic representation of the double-probe circuit is given in Fig. 6.14. By means of a battery B, a potentiometer P, and a reversing

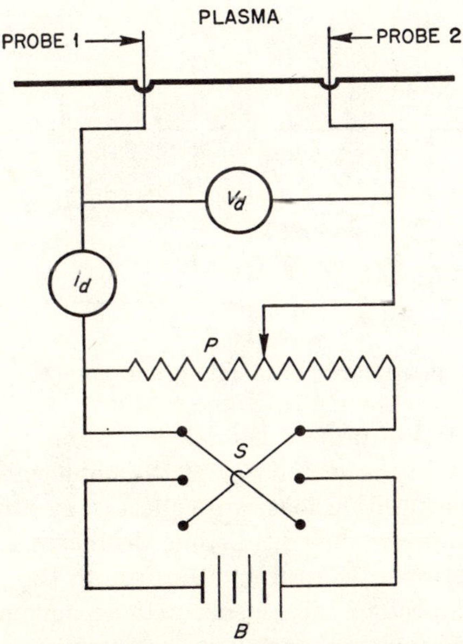

FIG. 6.14. Electrical circuit for double-probe measurements.

switch S, a voltage, variable in both magnitude and direction, can be applied between probes 1 and 2 inserted into the plasma. The total current i_d flowing between the probes is observed on a suitable meter, as is also the difference of potential V_d between the probes.

6.100. An idealized plot of the measured current as a function of the potential difference, as the latter changes from a fairly large value in one direction to a similar value in the opposite direction, is shown in Fig. 6.15. For simplicity, it is assumed for the present that the two probes are identical, so that the two portions of the curve have the same shape. This is not a necessary requirement, as will be apparent in due course.

6.101. Suppose the voltage applied from the battery is such as to make probe 2 highly negative with respect to probe 1. Probe 2 will then collect positive ions at the maximum rate, so that the observed current is equal to i_{ri_2}, the random ion current to probe 2. Probe 1 will then collect an equivalent number of electrons, but no ions. However, since the ions move more slowly than do the electrons, it is the maximum ion current, rather than the maximum possible electron current, which determines the maximum (absolute) value of the current flowing through the plasma between the probes.

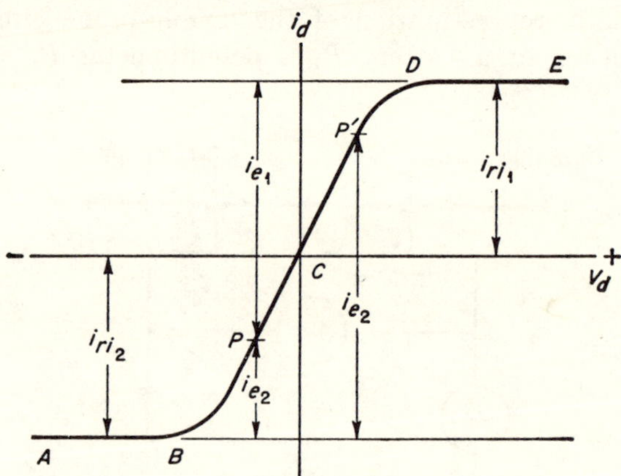

FIG. 6.15. Idealized curve of double-probe potential and current.

6.102. As the potential of probe 2 is made less negative, the current remains constant along AB, but in the vicinity of B the potentials are such that some electrons, in addition to positive ions, are collected at probe 2. The absolute magnitude of the net plasma current i_d thus decreases from B to C. At C, the external voltage between the probes is zero; both the probes then have the same potential with respect to the plasma and no current flows. As the direction of the applied voltage is changed, so that probe 1 now becomes negative with respect to probe 2, the situation is exactly reversed, and the curve CDE is obtained for the current i_d, which now flows in the opposite direction, as a function of the voltage V_d applied between the probes. Thus, DE gives i_{ri_1}, the random ion current to probe 1, and this will be the same as i_{ri_2} if the probes are identical.

6.103. The electron current i_e to a given probe can be determined from the fact that the absolute value of the total current i_d, i.e., regardless of direction, must be equal to the difference between the absolute values of the ion and electron currents to that probe. Thus, for probe 2, for example,

$$\pm i_d = i_{ri_2} - i_{e_2}$$

or

$$i_{e_2} = i_{ri_2} \mp i_d,$$

$$(6.36)$$

the negative sign applying when i_d is in the same direction as i_{ri2} and the positive sign when the directions of current flow are opposite. Examples of these two situations are illustrated by the points P and P', respectively, in Fig. 6.15.

6.104. Since there is no net removal of ions or electrons from the plasma, it follows that the total positive ion current to the probes must always equal the total electron current; thus,

$$i_{ri_1} + i_{ri_2} = \sum i_{ri} = i_{e_1} + i_{e_2},\tag{6.37}$$

as will be apparent from a consideration of Fig. 6.15. From the treatment in §6.85, based on the postulated Maxwell-Boltzmann distribution, expressions analogous to equation (6.32) can be written for i_{e_1} and i_{e_2}; they are

$$i_{e_1} = i_{re_1} \exp\left(-\,e\,V_1/kT_e\right)\tag{6.38}$$

and

$$i_{e_2} = i_{re_2} \exp\left(-\,e\,V_2/kT_e\right),\tag{6.39}$$

where V_1 and V_2 are the potentials across the sheaths surrounding the probes 1 and 2, respectively (Fig. 6.16). Allowing for any unknown potential (or

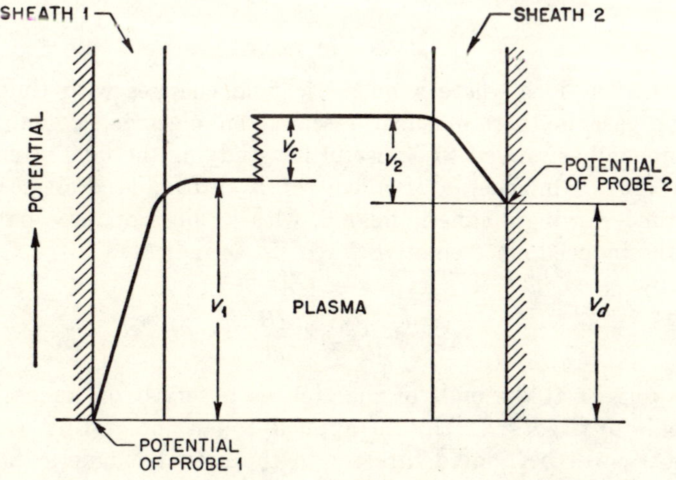

FIG. 6.16. Potentials in double-probe system.

potentials) V_c in the plasma, due to contact phenomena or other causes, it is seen that the applied potential difference V_d, which is measured, is related to the other potentials by

$$V_d = V_1 + V_c - V_2$$

or

$$V_1 - V_2 = V_d - V_c.\tag{6.40}$$

6.105. By combining equations (6.37), (6.38), (6.39), and (6.40), it is found that

$$\ln \frac{i_{e_1}}{i_{e_2}} = \ln \left(\frac{\sum i_{ri}}{i_{e_2}} - 1 \right) = \ln \frac{i_{re_1}}{i_{re_2}} + \frac{eV_2}{kT_e} - \frac{eV_d}{kT_e}$$

$$= \text{constant} - \frac{eV_d}{kT_e}.$$

Hence, a plot of $\ln\left[(\Sigma i_{ri}/i_{e_2}) - 1\right]$, which can be obtained from the experimental data, against V_d should yield a straight line and from its slope the kinetic electron temperature can be derived. It will be noted that the preceding treatment has been made quite general and is indpendent of whether the two probes are identical or not.

6.106. The double-probe method provides no indication concerning the magnitude of the random electron currents and so it is not possible to utilize equation (6.34), even if it were applicable, to obtain the electron density. However, since the random ion currents to the probes are known, use may be made of equation (6.35), together with the known electron temperature, to derive the ion density. In a plasma, this may be taken as equal to the electron density.

MAGNETIC PROBES

6.107. In situations where a magnetic field changes with time, as is frequently the case in thermonuclear research on plasmas, a simple magnetic probe technique has proved to be useful for studying the field strength and its variations [76]. In general, if a coil of n turns and cross-sectional area A is in a time-varying magnetic field B, with its lines of force parallel to the coil axis, the induced electromotive force V_i, as indicated in §6.2, is given in mks units by

$$V_i = nA \frac{dB}{dt}.$$

Thus, the voltage at the ends of the coil, as observed on an oscilloscope, is proportional to dB/dt. If this is applied to an integrating (RC) circuit, the output V_o will be related directly to the actual magnetic field strength at any instant; thus,

$$V_o = \frac{nAB}{RC},$$

where R and C are the resistance and capacitance, respectively, of the integrating circuit. If V_o is expressed in volts, R in ohms, C in farads, and A in square centimeters, then B is given in gauss by

$$B = 10^8 \times \frac{V_o RC}{nA}.$$

6.108. Magnetic probe measurements have been applied in investigations of the pinched discharge for studying the time and space variations of both axial and azimuthal magnetic fields associated with the plasma. In order to cause a minimum disturbance of the conditions in the plasma, a probe (or pickup) coil of very small diameter, but with a moderately large number of turns, so as to give an appreciable voltage, is used. A typical coil consists of 20 to 50 turns of 0.003-inch copper wire wound on a 1-mm diameter form. This is electrostatically shielded, enclosed in a quartz tube, and placed at a known location in the cylinder or torus within which the plasma is to be formed.

6.109. If the coil is oriented with its axis parallel to the tube axis, it will measure the time variation of the axial (or longitudinal) field, as indicated at Fig. 6.17A. But if the coil axis is perpendicular to the tube, as in Fig. 6.17B,

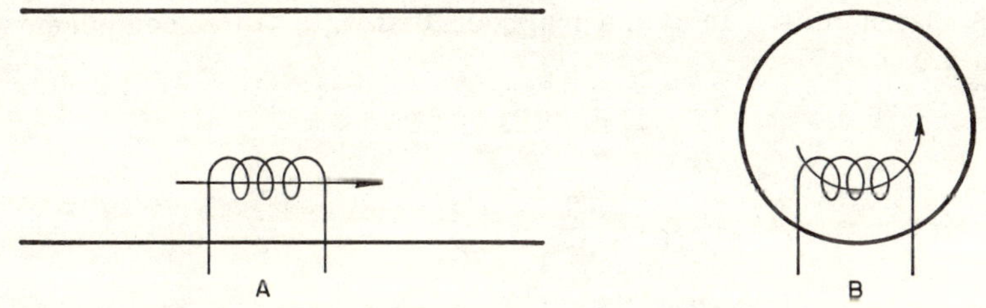

FIG. 6.17. Pickup coils for measurement of axial (A) and azimuthal (B) magnetic fields.

which shows a section through the tube, the measurements will provide information about the azimuthal self-magnetic field accompanying a pinched plasma. By changing the location of the probe, the spatial variation of the fields can be investigated.

6.110. In another type of measurement of the axial field in a pinched discharge, the pickup coil (or loop) is wound around the discharge tube. If the tube is surrounded by a conductor (cf. Fig. 6.2), the loop must lie between this and the insulating wall, e.g., quartz, porcelain, etc., of the discharge tube, as shown in Fig. 6.18. When the discharge is pinched, the axial magnetic

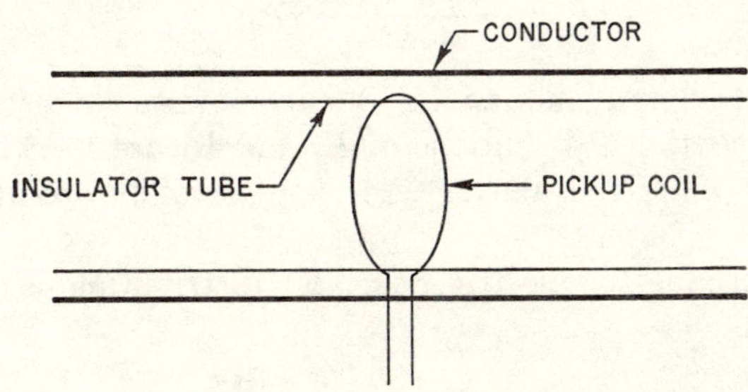

FIG. 6.18. Pickup coil for study of axial magnetic field.

lines within the plasma are trapped, but the field previously contained in the space between the external conductor and the insulating walls now expands to fill the entire annular region around the pinch. The result is an increase in the total flux through the coil and a consequent increase in the output voltage signal. The integral of this signal gives a measure of the pinch radius, but its greatest usefulness has been in the detection of helical pinch instabilities (§7.87).

6.111. From a knowledge of the space and time distributions of the magnetic fields associated with the discharge, the corresponding distributions of current density and electric field can be calculated. For slowly varying electric fields, one of Maxwell's equations, as seen in §3.17, is

$$\nabla \times \mathbf{B} = \frac{4\pi}{c}\, \mathbf{j},$$

where $\mathbf{B}$ and $\mathbf{j}$ are the magnetic field and current density vectors, respectively, in Gaussian-cgs units. In cylindrical coordinates, the three components of this equation are

$$\frac{4\pi}{c}\, j_r = \frac{1}{r} \cdot \frac{\partial B_z}{\partial \theta} - \frac{\partial B_\theta}{\partial z}$$

$$\frac{4\pi}{c}\, j_\theta = \frac{\partial B_r}{\partial z} - \frac{\partial B_z}{\partial r}$$

and

$$\frac{4\pi}{c}\, j_z = \frac{1}{r} \cdot \frac{\partial}{\partial r}\, (rB_\theta) - \frac{1}{r} \cdot \frac{\partial B_r}{\partial \theta}.$$

6.112. To illustrate the use of these equations, it will be supposed, as an example, that a straight cylindrical discharge, such as a linear pinch, is under consideration. It is then reasonable to assume that the magnetic fields are independent of θ and z, and depend only on r; since the $\partial B/\partial \theta$ and $\partial B/\partial z$ terms are zero, it follows that

$$j_r = 0$$

$$j_\theta = -\frac{c}{4\pi} \cdot \frac{\partial B_z}{\partial r},$$

and

$$j_z = \frac{c}{4\pi r} \cdot \frac{\partial}{\partial r}\, (rB_\theta).$$

Hence, the azimuthal (j_θ) and axial (j_z) current density distributions can be calculated if the spatial distributions of B_z and B_θ are known from measurements of the type described above. An example of the results obtained will be given in §7.103.

6.113. To determine the electric field ($\mathbf{E}$) distribution, use is made of the Maxwell equation

$$\nabla \times \mathbf{E} = -\frac{1}{c} \cdot \frac{\partial \mathbf{B}}{\partial t},$$

which, in cylindrical coordinates, has the components

$$-\frac{1}{c} \cdot \frac{\partial B_r}{\partial t} = \frac{1}{r} \cdot \frac{\partial E_z}{\partial \theta} - \frac{\partial E_\theta}{\partial z} \tag{6.41}$$

$$-\frac{1}{c} \cdot \frac{\partial B_\theta}{\partial t} = \frac{\partial E_r}{\partial z} - \frac{\partial E_z}{\partial r} \tag{6.42}$$

and

$$-\frac{1}{c} \cdot \frac{\partial B_z}{\partial t} = \frac{1}{r} \cdot \frac{\partial}{\partial r} (rE_\theta) - \frac{1}{r} \cdot \frac{\partial E_r}{\partial \theta}. \tag{6.43}$$

6.114. In a cylindrically symmetrical system, it may be accepted that all quantities are θ-independent; in addition, it has been assumed above that the magnetic fields are z-independent, which is in agreement with experimental observations. But this simplification cannot necessarily be made for the electric field distribution. In a linear discharge, for example, there is a possibility that the electric fields may have a z dependence, since the voltage is applied at some localized point in a coaxial circuit, as can be seen in Fig. 6.2. A special situation arises, however, as far as the E_r component is concerned. In a simple shorted metallic coaxial line, E_r increases from zero to its maximum value with increasing distance from the shorted end of the line. A linear discharge in a plasma is similar in many respects to a coaxial circuit, except that an insulating cylinder lies between the discharge and the outer metallic wall (cf. Fig. 6.2). As a result, charges accumulate on this wall with a density that is just sufficient to cancel E_r completely. Hence, in a linear discharge E_r may be taken as zero and only E_θ and E_z remain.

6.115. If all the E components are θ-independent and in addition E_r is taken to be zero, equations (6.41), (6.42), and (6.43) reduce to

$$-\frac{1}{c} \cdot \frac{\partial B_\theta}{\partial t} = -\frac{\partial E_z}{\partial r} \tag{6.44}$$

and

$$-\frac{1}{c} \cdot \frac{\partial B_z}{\partial t} = \frac{1}{r} \cdot \frac{\partial}{\partial r} (rE_\theta). \tag{6.45}$$

Integration of equation (6.44) gives the difference $E_z(r) - E_z(0)$, that is, the difference between the electric fields at a radial distance r and at the axis of the discharge tube; thus,

$$E_z(r) - E_z(0) = \frac{1}{c} \int_0^r \frac{\partial B_\theta(r')}{\partial t} \, dr'. \tag{6.46}$$

This expression could be used to determine the distribution of E_z in the r direction provided $E_z(0)$, which is the result of the IR drop in the discharge, were known. To derive $E_z(0)$ use is made of the fact that E_z at the tube wall, i.e., $E_z(r_0)$, multiplied by the distance between the electrodes is equal to the voltage at the tube terminals. Hence, the measurement of this voltage makes it possible to obtain $E_z(r_0)$. From magnetic probe observations, from the wall to the discharge axis, $\partial B_\theta(r')/\partial t$ can be determined and hence $E_z(r_0) - E_z(0)$ can be evaluated from equation (6.46). Since $E_z(r_0)$ is known, $E_z(0)$ can thus be obtained. There is a tacit assumption involved here that $\partial B_\theta/\partial t$ is independent of z, but this seems to be justified by other measurements.

6.16. The distribution of E_θ is given by integrating equation (6.45); there is no integration constant in this case since E_θ must vanish when $r = 0$. The result is, therefore,

$$E_\theta(r) = - \frac{1}{rc} \int_0^r \frac{\partial B_z(r')}{\partial t} r'\, dr', \cdot$$

from which E_θ at any radial distance r can be determined.

6.117. Since the E and j values can be obtained from magnetic probe measurements, a procedure is available, in principle, for deriving the resistivity of the plasma. However, as implied in §4.91, it is only the component of the electric field parallel to the magnetic field lines that arises from resistivity alone. It is necessary, therefore, to resolve $\mathbf{E}$ and $\mathbf{j}$ into components parallel and perpendicular, respectively, to the *total* magnetic vector. This resolution, while somewhat laborious, is straightforward, and the plasma resistivity η, as a function of r, is given by $E_{\parallel}/j_{\parallel}$. With the result so obtained, the approximate plasma (electron) temperature can be calculated using equation (4.73).

6.118. Another application of magnetic probe measurements is to determine the kinetic pressure and, from this, the average particle temperature. For this purpose, use is made of equation (3.14), namely,

$$\frac{1}{4\pi} (\nabla \times \mathbf{B}) \times \mathbf{B} = \nabla p. \tag{6.47}$$

In cylindrical coordinates, the components of $\nabla \times \mathbf{B}$ are

$$(\nabla \times \mathbf{B})_r = \frac{1}{r} \cdot \frac{\partial B_z}{\partial \theta} - \frac{\partial B_\theta}{\partial z} = 0$$

$$(\nabla \times \mathbf{B})_\theta = \frac{\partial B_r}{\partial z} - \frac{\partial B_z}{\partial r} = - \frac{\partial B_z}{\partial r},$$

and

$$(\nabla \times \mathbf{B})_z = \frac{1}{r} \cdot \frac{\partial}{\partial r} (rB_\theta) - \frac{1}{r} \cdot \frac{\partial B_r}{\partial \theta} = \frac{1}{r} \cdot \frac{\partial}{\partial r} (rB_\theta),$$

since, as before, the magnetic field is taken to be both θ- and z-independent. Utilizing these results, it is found that

$$(\nabla \times \mathbf{B}) \times \mathbf{B} = \hat{\mathbf{r}} \left[\frac{B_\theta}{r} \cdot \frac{\partial}{\partial r} (rB_\theta) + B_z \frac{\partial B_z}{\partial r} \right],$$

where $\hat{\mathbf{r}}$ is the unit radial vector. It follows, therefore, from equation (6.47) that

$$\frac{\partial p}{\partial r} = - \frac{1}{4\pi}\left[\frac{B_\theta}{r}\cdot\frac{\partial}{\partial r}\,(rB_\theta) + B_z\,\frac{\partial B_z}{\partial r}\right].$$

Upon expanding $\partial(rB_\theta)/\partial r$ and rearranging, the result

$$\frac{\partial}{\partial r}\left[p + \frac{1}{8\pi}\,(B_\theta{}^2 + B_z{}^2)\right] + \frac{1}{4\pi}\cdot\frac{B_\theta{}^2}{r} = 0$$

is obtained, so that integration yields

$$\left[p + \frac{B_\theta{}^2 + B_z{}^2}{8\pi}\right]_{r_1}^{r_2} + \frac{1}{4\pi}\int_{r_1}^{r_2}\frac{B_\theta{}^2}{r}\,dr = 0. \tag{6.48}$$

6.119. The reasonable assumption is usually made that the pressure at the tube wall, radius r_0, is zero; in addition, the integration is replaced by summation, so that equation (6.48) becomes

$$p(r) = \frac{1}{8\pi}\left[B_z{}^2(r_0) + B_\theta{}^2(r_0) - B_z{}^2(r) - B_\theta{}^2(r) + 2\sum_{r'=r}^{r'=r_0}\frac{B_\theta{}^2(r')\,\Delta r'}{r'}\right], \tag{6.49}$$

where $\Delta r'$ represents the radial increment between successive probe measurements of B_θ. It is seen, therefore, that the pressure in the plasma in a linear discharge can be determined as a function of radial distance (cf. §7.103).

6.120. With a knowledge of the pressure distribution it becomes possible to obtain an approximate indication of the corresponding kinetic temperature distribution. For this purpose use is made of the equilibrium relationship

$$p(r) = n(r)kT(r),$$

where n is the total particle density, i.e., electrons, ions, and neutral particles.

6.121. In the foregoing treatment the effect of the plasma inertial forces, which are negligible at low densities, has been ignored; strictly speaking (cf. §4.116), the term $\rho\,\partial^2 r/\partial t^2$, where ρ is the mass density of the plasma, should be added to the right side of equation (6.47). If p^* is the pressure calculated, as above, by neglecting the inertial forces, then the true pressure p is given by

$$p(r) = p^*(r) + \int_{r_1}^{r_2}\rho\,\frac{\partial^2 r'}{\partial t^2}\,dr'. \tag{6.50}$$

For a plasma moving with high acceleration, as in the early stages of a pinched discharge, the second term on the right of equation (6.50) can assume significant proportions.* However, from measurements of the radius of the current sheath as a function of time it is possible to determine the acceleration and so apply the necessary correction [77].

6.122. For a toroidal system, the integration of the equations relating the current density, electric field, and pressure to the space and time variations of the magnetic field is much more complex than for the cylindrical geometry con-

---

*Fine-grained local turbulence in the plasma could affect the magnetic probe measurements and hence also appear as a spurious pressure.

sidered above. By making certain approximations, it has been found that, for magnetic probe measurements made along a minor diameter of the torus,

$$j_\theta = \frac{c}{4\pi} \cdot \frac{\partial B_z}{\partial R} + \frac{cB_z}{R}$$

and

$$j_z = \frac{c}{4\pi} \cdot \frac{\partial B_\theta}{\partial R} + \frac{2Rc}{R^2 - A^2} B_\theta,$$

where R is the distance from the major axis to the point under consideration and the quantity $(R^2 - A^2)/2R$ is the minor radius of the toroidal coordinate surface through that point [78]. The equation for the pressure distribution is

$$\left[p + \frac{B_\theta^2 + B_z^2}{8\pi} \right]_{R_1}^{R_2} + \frac{1}{4\pi} \int_{R_1}^{R_2} \left\{ \frac{2R}{R^2 - A^2} B_\theta + \frac{B_z^2}{2R} \right\} dR = 0.$$

6.123. An entirely different kind of application of magnetic probes which has been suggested is their use in the determination of the mass density of a plasma. Local high-frequency disturbances can be generated in a plasma, e.g., by means of a single-turn coil inside a pinched discharge column, and the speed of propagation of the resulting hydromagnetic waves can be measured by suitably located magnetic probes. Since the magnitude of the field is measured directly by the probes, a value of the Alfvén wave velocity along the field lines yields the mass density by means of equation (4.106) [79].

ELECTRON BEAM PROBES

6.124. Since material probes must inevitably disturb the plasma to some extent, the suggestion has been made, although it does not yet appear to have been successfully applied, that the current distribution in a pinched discharge could be studied by utilizing a stream of electrons as a probe [80, 81]. If a narrow beam of electrons of high energy is injected into the discharge, the beam will be deflected by the strong self-magnetic field of the pinched plasma. The deflection can be observed on a fluorescent screen located in the wall of the discharge tube, and from the magnitude of this deflection the pinch current can be calculated. Somewhat complicated expressions have been derived relating the deflections z and θ, in cylindrical coordinates, to the current, but these must be solved numerically by machine techniques. By means of a variable external magnetic field, the position of the electron beam probe can be changed. The beam deflection, and hence the current strength, at various distances from the pinch axis can thus be observed.

6.125. In order that the beam may not be turned around by the magnetic field of the plasma, it appears that electrons of several hundred kev energy are required for even modest plasma currents. The accelerating voltage can be obtained by charging a capacitor bank and discharging through a pulse transformer. Electrons are produced in a standard gun and are accelerated through a series of collimating apertures in a drift tube to produce a narrow beam. The gun-accelerator system must then be attached to the discharge tube so that the electrons enter through a screen. This screen must permit passage of the electrons, but must not disturb the pinch system under investigation.

HIGH-SPEED PHOTOGRAPHY [82, 83]

INTRODUCTION

6.126. Photographic techniques provide information only on the shape of the discharge and its brightness. Nevertheless, such techniques have been used extensively, e.g., for the examination of the changes occurring in pinched discharges due to their instability. Since these changes take place with great rapidity, high-speed photography must be used if the results are to be significant. Thus, the exposures should be of the order of a microsecond or less, and successive exposures should follow at very short intervals. The fact that discharges often emit a comparatively small amount of light adds to the difficulty of obtaining good photographs.

6.127. The luminosity of the discharge in hydrogen (or its isotopes) decreases with increasing temperature and is relatively weak at temperatures above 50 ev or so, when the gas is completely ionized. For this reason much of the photographic work has been done with argon and heavier gases. Discharges in these gases are still luminous at high temperatures. The results obtained by photographic methods with such gases must be regarded as qualitative only, as far as their applicability to thermonuclear problems is concerned.

6.128. In addition to the use of high-speed motion picture cameras, capable of taking pictures at the rate of several million frames per second, three main photographic techniques have been applied to the study of plasmas. They are (1) streak (or smear) cameras, (2) image converter tube cameras, and (3) Kerr cells. Electro-optical systems utilizing Kerr cells as shutters have been employed with ordinary cameras to secure accurately timed exposures of 0.1 microsec or less. The Kerr cell, however, has the drawback of retaining a small transmission, e.g., 10^{-5} or so, even after it has been turned off. If highly luminous events occur before or after the exposure time, spurious images may be recorded on the photographic emulsion.

STREAK CAMERAS

6.129. A streak (or smear) camera, also called a rotating mirror camera, is designed to provide a continuous exposure for a short time interval, resolution being achieved by a rapidly rotating mirror to change the position of the image. A typical schematic arrangement of a streak camera is shown in Fig. 6.19. An image of the discharge is focused on a narrow slit and then onto a circular arc of photographic film by way of a rotating plane mirror at its center of curva-

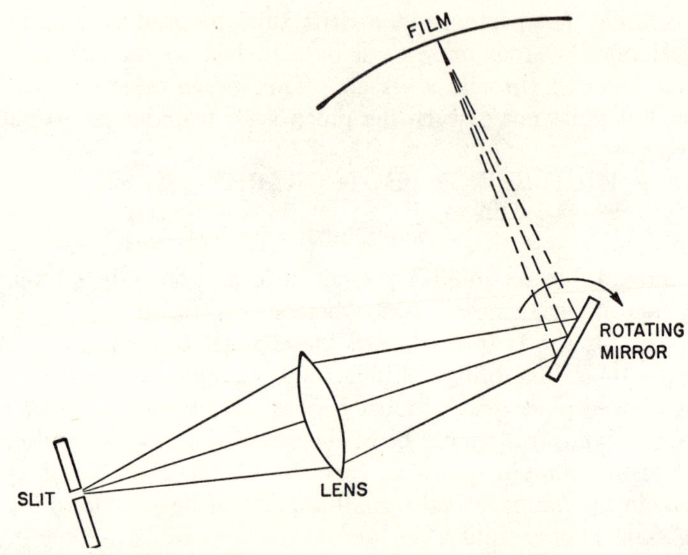

FIG. 6.19.   Schematic arrangement of streak (rotating mirror) camera.

ture.   The slit is oriented parallel to the mirror axis.   As the mirror rotates, at speeds often greater than 1000 revolutions per second, the light from the discharge is reflected to continuously changing, adjacent locations on the film, so that the image obtained upon development is a continuous streak (or smear). The changing appearance of the streak may, for example, indicate the changing diameter of the discharge or, more correctly, of the luminosity of the discharge. As a general rule, the writing (or streaking) speed of a rotating mirror camera is several millimeters per microsecond.   Photographs having good time resolution are usually obtained without further magnification.

6.130.   An example of a streak photograph from a pinched discharge is shown in Fig. 6.20.   (The horizontal bars have no significance; they arise from the fact that the observations were made through a set of apertures rather than through a continuous slit in the discharge tube.)   Reading from left to right, the photograph shows that the plasma became sufficiently hot to emit light within less than a microsecond; then rapid contraction occurred to form a pinched discharge with a diameter about one-third that of the original gas. The first contraction was followed by a slight expansion and a further contraction; subsequently there was a smaller expansion and still another contraction. At this stage, just over 3 microsec after initiation of the discharge, the onset of instabilities caused the discharge to expand and fill the tube.

IMAGE CONVERTER TUBE CAMERAS

6.131.   An image converter tube is an electronic device for amplifying the light falling upon a photoelectric emitter and displaying the result, as a repro-

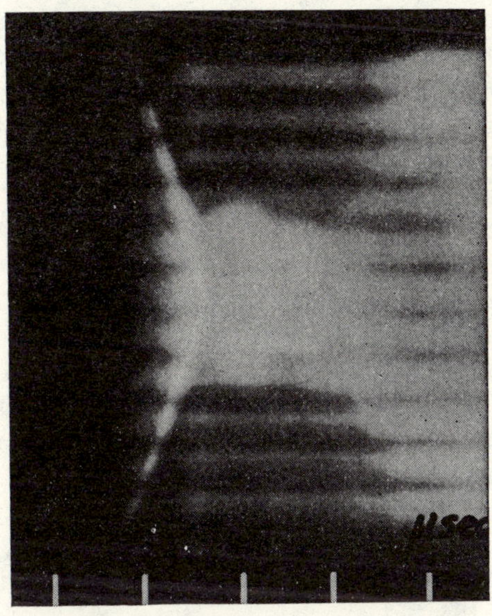

Fig. 6.20. Streak-camera photograph of pinched discharge. (Courtesy of Los Alamos Scientific Laboratory)

duction of the original, on a fluorescent screen. Although the original exposure can be extremely short, down to a small fraction of a microsecond, the persistence of the image on the screen makes it possible to photograph it by conventional methods. By the use of suitable delay lines and electronic timing, the

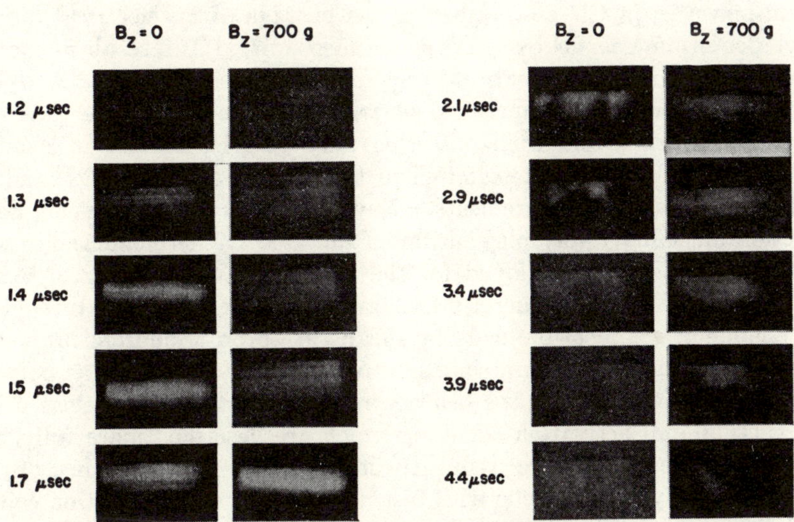

Fig. 6.21. Image converter photographs of pinched discharge. (Courtesy of Los Alamos Scientific Laboratory)

exposures can be timed to take place at precisely known intervals after initiation of a discharge and for the required short exposure periods, usually a microsecond or less. An image converter tube camera employing several tubes, each with its own delay line, can be arranged to obtain successive and separate pictures of a given discharge.

6.132. An interesting sequence of pictures of a pinched discharge obtained with an image converter tube camera is given in Fig. 6.21; the delay times are indicated in each case. The right sets of pictures were produced in the presence of an applied axial (or longitudinal) magnetic field of 700 gauss; there was no axial field for the left sets. The formation of the pinched discharge is clearly seen in the early stages, but by the end of 2.1 microsec the break-up, in the absence of an axial magnetic field, is well on the way. The onset of the so-called "sausage" type of instability is apparent at this time. (It should be noted that the pictures are not symmetrical, as they should be, because they were slightly clipped at the upper edges by the 1.5 cm × 3 cm slot through which the pictures were taken.) With an applied axial field, marked instability does not occur until about 4 microsec, thus proving the partial stabilizing effect of the magnetic field [84, 85].

NUCLEAR RADIATION MEASUREMENTS

NEUTRON DETECTION AND COUNTING

6.133. Since most reactions of thermonuclear interest are accompanied by the production of high-energy neutrons (Chapter 2), it would appear that the detection of neutrons might indicate the occurrence of such reactions. It will be seen, however, in Chapter 7 that in the passage of an electrical discharge through deuterium gas, neutrons are sometimes emitted as a result of the interaction of deuterons which have been accelerated by high potentials due to certain fortuitous conditions in a pinched discharge. Such neutrons are thus not of thermonuclear origin and their detection may be misleading. It is evident, therefore, that care must be exercised in the interpretation of the results obtained from neutron measurements. Nevertheless, observations on neutron production must inevitably play an important role in thermonuclear research.

6.134. The methods employed for the detection and counting of neutrons and for determining their energies are fairly conventional in nature [1, 86]. They include direct measurements by means of boron trifluoride proportional counters in conjunction with a solid hydrogenous material, such as polyethylene or paraffin, to slow down the fast neutrons (about 2.4 Mev) released in the D-D reaction, and activation counters, which are described more fully below. In addition, observations are frequently made of the recoil protons, produced by collisions of hydrogen nuclei with fast neutrons, using scintillation counters, nuclear track emulsions, and cloud chambers containing hydrogen or methane gas under pressure. The types of measurements fall broadly into four categories, namely, time of emission, total number (or yield) of neutrons, energy distribution, and spatial distribution.

6.135. Information concerning the time and duration of the period of neutron emission is often of interest, since it may be correlated with other characteristics of the deuterium plasma, e.g., applied voltage, discharge current, magnetic field strength, etc. Since the neutron pulse is often of short duration, the indicating system must have a rapid response. For this purpose, the best arrangement is a scintillation counter, which may contain either a liquid, crystalline, or plastic organic fluor, in combination with an oscilloscope. The counter is usually surrounded by lead to absorb X-rays which would also affect the scintillator. An example of the records obtained during the first few microseconds of a linear pinch discharge are shown in Fig. 6.22 [87]. It is seen that

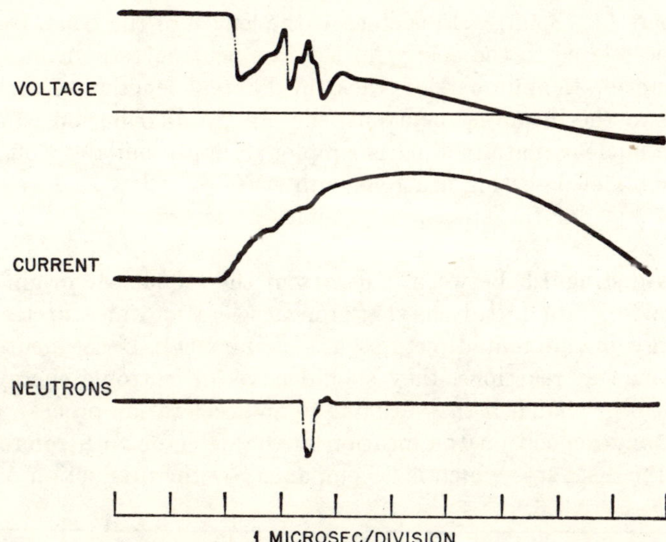

FIG. 6.22. Emission of neutrons in a pinched discharge.

neutron emission coincides with breaks in both the current and voltage curves. The latter curves were obtained by the methods described at the beginning of this chapter (§6.1 *et seq.*).

6.136. Provided the rate of neutron emission is not too high, the total neutron yield can be determined with a boron trifluoride proportional counter. This type of instrument has the merit of providing good discrimination against X-rays. For moderate rates of neutron emission, a scintillation counter connected to a fast scaler is satisfactory. If the total number of neutrons is sufficiently large, it is often convenient to use an activation counter to determine the total yield. The device consists of a standard instrument, such as a proportional counter or Geiger tube for counting beta particles, surrounded by a foil of an element, e.g., silver or indium, which becomes radioactive as a result of the capture of neutrons. By suitable calibration of the counter, observation of the activity of the foil can be used to determine the total neutron emission.

6.137. The advantage of this activation procedure is that the counting can be done after the discharge which produces the neutrons has occurred, thereby eliminating the possible effect of spurious counts due to extraneous causes. A somewhat similar situation arises when a scintillation counter surrounded by cadmium is used to detect the delayed gamma rays accompanying the capture of slow neutrons by cadmium and hydrogen. The delay, which is much less than in the activation method, is due to the time required to slow down the fast neutrons to the low-energy range where they are readily captured.

6.138. Determination of the distribution of energies among the neutrons produced has been made with both nuclear emulsion plates [1, 88-91] and cloud chambers [1, 78, 92]. In each case, the length of the track from a recoil proton can be related to the energy of the fast neutron responsible for the recoil. By counting the numbers of tracks of different lengths, at various angles with respect to the incoming neutrons, the energy distribution can be determined. The nuclear plate method is simpler to apply but the cloud chamber, which is capable of detecting neutrons with energies as low as 2 kev compared with about 7 kev for the emulsion, provides better resolution because of the longer tracks.

6.139. To distinguish between neutrons of thermonuclear origin and those which may arise from fortuitous deuteron acceleration, measurements of the neutron energy in different directions are of interest. If the neutrons result from thermonuclear reactions, they should have an isotropic energy distribution. On the other hand, if they are due to an acceleration process, the energy distribution may depend on the location of the detector with reference to the direction of the discharge which is accompanied by the production of neutrons.

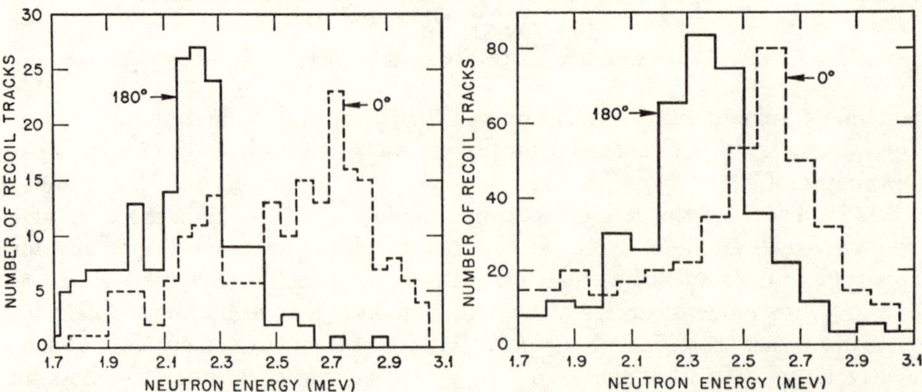

FIG. 6.23. Nuclear plate measurement of neutron energy distribution. FIG. 6.24. Cloud chamber measurement of neutron energy distribution.

Both the nuclear emulsion plate technique and the cloud chamber method have been used to try to distinguish in this manner between thermonuclear and non-thermonuclear neutrons.*

6.140.   The results of some observations made by those two procedures are depicted in Figs. 6.23 and 6.24; the former was obtained from nuclear plate measurements accompanying a linear pinched discharge [89] and the latter from cloud chamber photographs during passage of a discharge through a torus [78].   The numbers of recoil tracks of different lengths, representing different neutron energies, are plotted as a function of the energy.   In each figure the full lines give the energy distribution in a direction opposite to that of the applied electric field whereas the dotted lines refer to the same direction as the field.   The shift in the energy distribution, which is similar in the two cases, is quite marked and shows that accelerated deuterons were probably responsible for a large fraction of the neutrons.   If the neutrons had been of thermonuclear origin and the plasma stationary, the two sets of observations would have fallen on a single curve with an energy distribution around a mean value of 2.45 Mev, which is the D-D reaction energy carried by the neutrons (§2.25).

6.141.   Use can also be made of the spatial distribution of the neutrons to determine the region of the discharge from which the particles originate [1, 78]. Such measurements are generally made with a scintillation counter and a collimator which is directed at various regions of the discharge.   An example of the results obtained in this manner with a toroidal pinch discharge is given in Fig. 6.25.   These particular data were obtained as the result of a vertical scanning of the minor diameter of the torus; essentially the same values were obtained in a horizontal scan, indicating that the source region is probably symmetrically located about the minor axis.   The observed distribution is consistent with the neutrons being produced mainly either in a hollow shell of plasma of 3 cm outer and 2 cm inner diameter or in a central cylinder of 3 cm diameter.   The neutron data alone cannot distinguish between these two alternatives [78].

6.142.   If it could be established that the neutrons emitted from a deuterium plasma are definitely thermonuclear in origin, an estimate could be made of the ion temperature.   For this purpose it is necessary to know the total yield of 2.45-Mev neutrons, the duration of the emission, and the volume of the emitting region in order to determine the rate of neutron production per cm^3 per sec.   From the initial pressure of the gas and the radius of the emitting plasma, as compared with that of the containing tube, the number density of the deuterons can be calculated, assuming ionization to be complete when neutron emission occurs.   By means of equation (2.11) it is now possible to evaluate $\overline{\sigma v}$ for the neutron branch of the D-D reactions and the corresponding kinetic temperature can then be estimated from the D-D curve in Fig. 2.4.   It should

---

* Strictly speaking, it is not possible, without further information, to distinguish between acceleration of individual deuterons and the mass motion of the plasma as a whole.

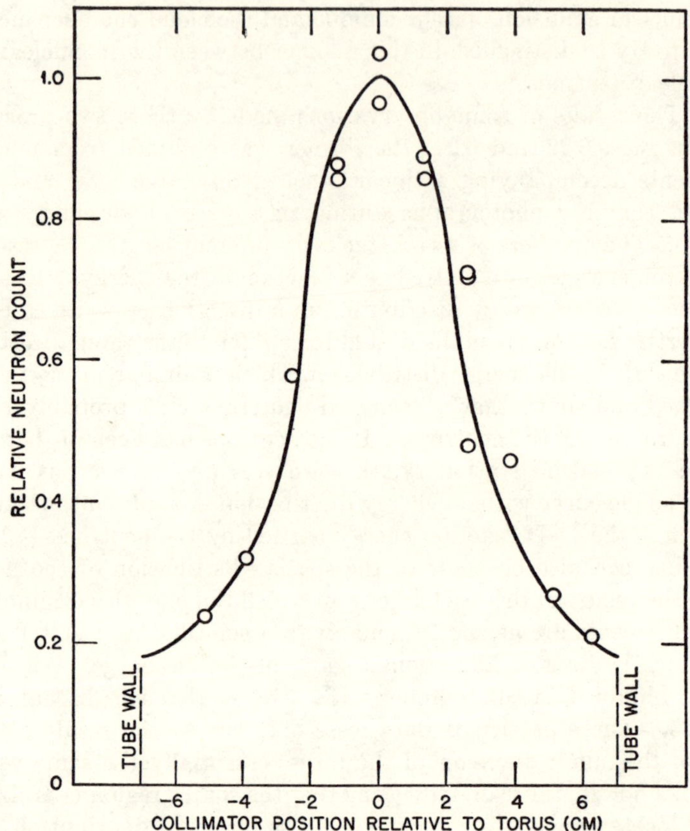

Fig. 6.25.   Spatial distribution of neutrons from a toroidal pinch.

be recalled that this curve gives the total $\overline{\sigma v}$ for the two D-D reactions, and the value for the neutron branch is almost exactly half of the total at temperatures of thermonuclear interest.

PROTON MEASUREMENTS

6.143.   Instead of studying the neutrons from the D-D reaction, it is equally possible to observe the protons from the other branch [90, 93, 94].   Nuclear emulsion plates have been used to determine the energy distribution, and the results obtained with pinched discharges are quite similar to those for neutrons, as described above.   The numbers of tracks of lengths corresponding to various proton energies, in directions at angles of 135° (full lines) and 45° (broken lines) to the positive current through the gas, are shown as a function of energy in Fig. 6.26 [94].

6.144.   A more complete study of proton energy distribution has been made of the particles emitted from a deuterium plasma heated by means of rapid

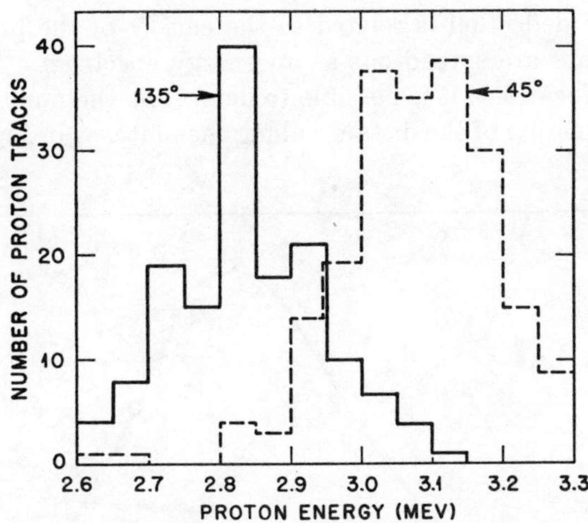

FIG. 6.26. Nuclear plate measurement of proton energy distribution.

hydromagnetic compression in the device known as Scylla (§11.17). A diagrammatic representation of the experimental arrangement is depicted in Fig. 6.27 [95]. A narrow beam of protons emerging from a small hole at the object position of a magnetic spectrograph is deflected by the magnetic field, so that the protons are dispersed along a nuclear emulsion located on the focal surface.

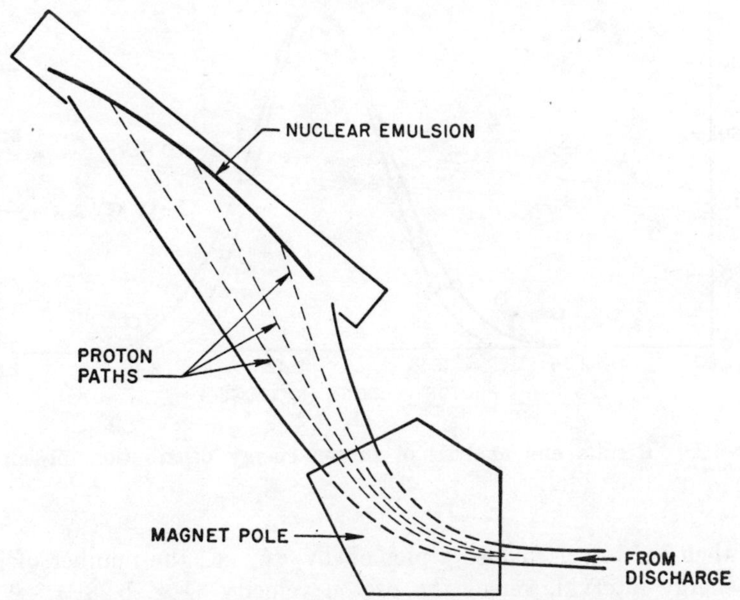

FIG. 6.27. Energy distribution measurements of protons from a compressed plasma.

Since the angle of deflection is related to the energy of the proton, the tracks on the nuclear plate are spread out as an energy spectrum. By counting the tracks at various locations, it is possible to determine the number of tracks per unit length as a function of the distance along the plate, as in Fig. 6.28A. These

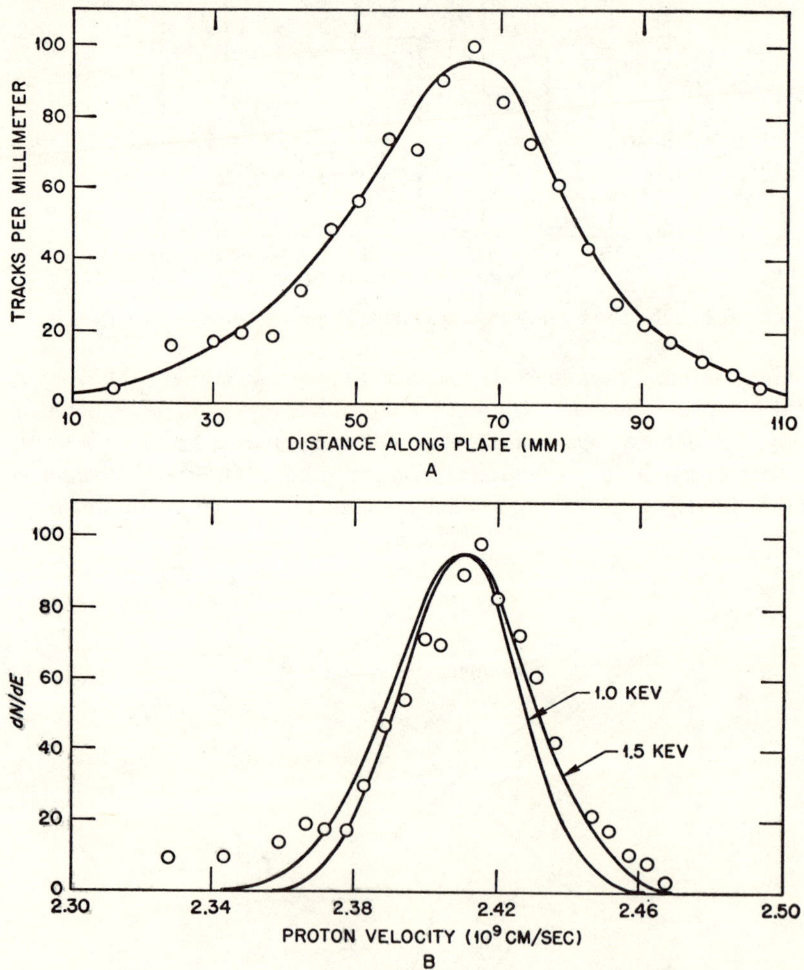

Fig. 6.28. Results and analysis of proton energy distribution measurements.

data can then be converted into a plot of dN/dE, i.e., the number of protons per unit energy interval, versus the proton velocity (Fig. 6.28B). The two curves indicate what would be expected for thermonuclear reactions, i.e., with a Maxwellian distribution of deuteron energies, at kinetic temperatures of 1.0

and 1.5 kev, respectively.   The results are seen to be consistent with a thermonuclear D-D reaction in the plasma at a temperature of about 1.3 kev [95].

X-RAY MEASUREMENTS

6.145.   The application of thin-target bremsstrahlung radiation, in the X-ray region, to the determination of plasma temperatures was described in §6.76 *et seq*.   In addition, observations on X-rays have been used to provide general information on the behavior of the discharge.   In particular, runaway electrons (§5.10) striking the walls of the containing tube will produce X-rays, as thick-target bremsstrahlung, of high energy.   For the detection of such radiation, frequent use has been made of scintillators, e.g., sodium iodide activated with thallium, and of photographic film techniques [1, 4, 91].   With the former, the energy of single X-ray photons can be determined directly, and even with the latter method a rough indication of the energy can be obtained from experiments with absorbers, e.g., aluminum, copper, and lead.   The maximum energy of the X-rays may be taken as being roughly equal to that of the electrons producing them.

6.146.   If it is required to identify the region of the discharge where the X-rays originate, the detector may be combined with a collimator consisting of an absorption barrier of a dense metal, such as lead, with an aperture through which the radiation can pass.   Small pinhole cameras, about 1 cm^2 in cross section, have been inserted into the walls of a metal torus in order to determine the source of X-rays.   The plate is protected from other radiations by a thin sheet of copper foil.

6.147.   In order to study the energy distribution of the X-rays, the most satisfactory procedure developed so far is based on the use of a cloud chamber

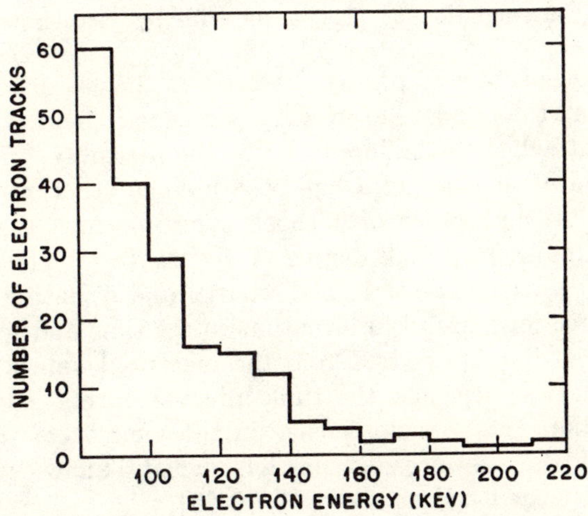

FIG. 6.29.   Cloud chamber study of X-rays by Compton recoil.

containing hydrogen gas at low pressure.   The advantage of the cloud chamber technique in this connection lies in the fact that it is particularly useful for
studying radiation bursts of short duration such as often accompany the passage
of large currents through a gas.   The cloud tracks produced in the chamber by
Compton recoil electrons arising from interaction of the X-ray photons with
the gas in the chamber are photographed and measured.   From the known
range-energy data, the electron energy distribution can be determined.   The
combined results obtained from a number of photographs made during the
passage of a 40-kilovolt discharge through hydrogen gas are presented in
Fig. 6.29 [91].   It is seen that many of the electrons have energies of more than
80 kev and even as high as 200 kev; calculations based on the theory of the
Compton effect show that the X-rays producing the electrons must have had
energies up to 320 kev.   Since the discharge voltage was only 40 kilovolts, these
X-rays presumably originated from runaway electrons.

OTHER DIAGNOSTIC TECHNIQUES

DIRECT PRESSURE MEASUREMENTS

6.148.   Direct measurement of the pressure in a pinched plasma have been
made by means of a ceramic piezoelectric pickup, the output of which is recorded by means of an oscilloscope (Fig. 6.30).   The instrument can be cali-

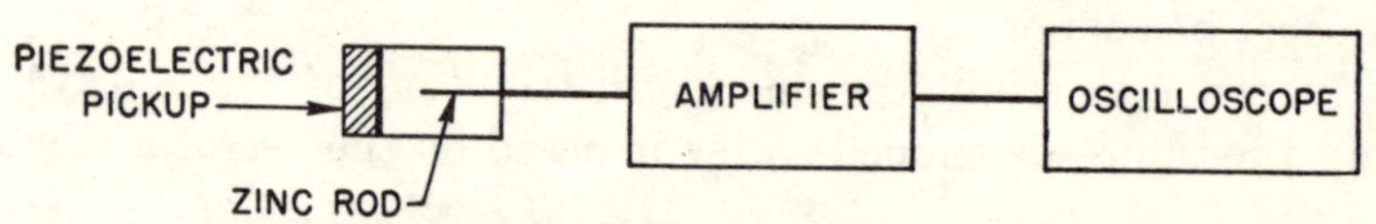

Fig. 6.30.   Determination of plasma pressure by piezoelectric method.

brated either by means of a diaphragm detector in the gas discharge or, outside
the discharge, from the pressure exerted by a magnetic field of known strength
on a metal disc attached to the input end of the assembly.   The best method
of calibration is probably one that has been used in other studies with piezoelectric probes; this makes use of a shock tube in which the sharp pressure
discontinuity is known to a high degree of accuracy.   By locating the piezoelectric pickup at various points in a discharge and orienting it in either the
radial or axial directions, useful information on the time and space distribution
of the pressure can be obtained.   From the measured values of the pressure
pulse, the particle velocity, and the time interval during which the pressure
acted on the pickup, the particle density in the vicinity of the latter can be
calculated.   With the pressure and particle density known, the temperature
can be evaluated in the usual manner [91, 96-98].

MASS SPECTRA STUDIES

6.149. A number of mass spectrometer methods have been described for determining the nature and relative abundances of the ions present in a plasma. In each case a narrow beam of particles is allowed to enter a special chamber where it is subjected to the action of electric and magnetic fields. In one procedure, the fields are applied simultaneously and typical Thomson parabolas are obtained on a photographic plate. From the location of each parabola the ratio of mass to charge, i.e., m/e, of the ions producing it can be determined, and from its extent the range of the particle energies can be calculated [91].

6.150. In another technique, the charged particles are first accelerated by means of an electric field and then bent into semicircular paths by a magnetic field. Two methods have been used for collecting the ions and measuring the ion currents corresponding to different mass-to-charge ratios. If the collector electrode is fixed then, for given values of the electric and magnetic fields, only ions having a particular value of the mass-to-charge ratio are collected. By keeping the magnetic field strength constant and changing the electric field steadily, particles of different m/e values are continuously brought up to the collector electrode and the corresponding ion currents recorded [99]. In this manner a complete mass analysis can be carried out very rapidly. An alternative procedure is to maintain both electric and magnetic fields constant during the course of a set of observations and to move the collector electrode. The measurements can then be repeated, if desired, with an electric field of different strength. From the ion currents it is possible to determine the nature and relative abundances of the ions entering the spectrometer chamber [100].

REFERENCES FOR CHAPTER 6

1. G. N. Harding, A. N. Dellis, A. Gibson, B. Jones, D. J. Lees, R. W. P. McWhirter, S. A. Ramsden, and S. Ward, *Proc. Second U.N. Conf. on Peaceful Uses of Atomic Energy,* **32,** 365 (1956).
2. C. B. Wharton, J. C. Howard, and O. Heinz, *Proc. Second U.N. Conf. on Peaceful Uses of Atomic Energy,* **32,** 388 (1958).
3. M. A. Heald, *Trans. Inst. Radio Eng.,* **NS6,** No. 3, 33 (1959).
4. R. M. Payne and S. Kaufman, *Proc. Inst. Elec. Eng. (London),* **106A,** Suppl. No. 2, 36 (1959).
5. S. M. Goldman, USAEC Report WASH-289 (1955), p. 272.
6. T. Coor and J. W. Brault, USAEC Report TID-7503 (1956), p. 380.
7. B. S. Liley, quoted by T. E. Allibone et al., *Proc. Second U.N. Conf. on Peaceful Uses of Atomic Energy,* **32,** 169 (1958); see p. 178.
8. S. C. Brown, *Trans. Inst. Radio Eng.,* **PGTT7,** 69 (1959).
9. K. B. Persson, *Phys. Rev.,* **106,** 191 (1957).
10. R. F. Post, T. F. Prosser, and C. B. Wharton, USAEC Report UCRL-4477 (1955).
11. C. B. Wharton, USAEC Report UCRL-4836 (1957).
12. C. B. Wharton, USAEC Report UCRL-5129 (1958).

13. C. B. Wharton and D. M. Slager, USAEC Reports UCRL-5244 (1958), UCRL-5400 (1958); *Trans. Inst. Radio Eng.*, **NS6**, No. 3, 20 (1959).
14. M. A. Heald, *Inst. Rad. Eng., Natl. Conf. Record*, **9**, 14 (1958).
15. M. A. Heald, USAEC Report MATT-17 (1959).
16. H. Margenau, *Phys. Rev.*, **69**, 508 (1946).
17. R. Motley and M. A. Heald, USAEC Report MATT-2 (1959).
18. R. F. Whitmer, *Phys. Rev.*, **104**, 572 (1956).
19. K. S. Knol, *Philips Research Reports*, **6**, 288 (1951).
20. M. A. Easley and W. W. Mumford, *J. Appl. Phys.*, **22**, p. 846 (1951).
21. C. B. Wharton and A. L. Gardner, USAEC Report TID-7520 (1956), p. 480.
22. A. N. Dellis, UKAEA Report AERE GP/R-2265 (1957); *Third Int. Cong. on Ionization Phenomena in Gases* (1957), p. 228.
23. A. N. Dellis and A. P. Willmore, UKAEA Report AERE GP/R-2324 (1957).
24. J. E. Drummond, in H. Fischer and L. P. Mansur, Eds., *Conference on Extremely High Temperatures*, John Wiley & Sons, Inc., 1958, p. 97.
25. L. W. Davis and E. Coucher, *Australian J. Phys.*, **8**, 108 (1955).
26. P. Parzen and L. Goldstein, *Phys. Rev.*, **82**, 724 (1951).
27. W. W. Mumford, *Bell System Tech. J.*, **28**, 608 (1949).
28. T. J. Bridges, *Proc. Inst. Radio Eng.*, **42**, 818 (1954).
29. H. Wulff, *Nuclear Instr.*, **4**, 352 (1959).
30. S. Y. Lukyanov and V. I. Sinitsyn, *Proc. Second U.N. Conf. on Peaceful Uses of Atomic Energy*, **32**, 358 (1958).
31. J. D. Craggs et al., *Proc. Phys. Soc. (London)*, **A59**, 755 (1947); **A64**, 574 (1951).
32. H. Dreicer, USAEC Report LA-1631 (1955).
33. E. J. Stovall, USAEC Report WASH-184 (1955), p. 62.
34. C. Breton et al., *Third Int. Conf. on Ionization Phenomena in Gases* (1957), p. 163.
35. A. M. Andrianov et al., *Plasma Physics and the Problem of Controlled Thermonuclear Reactions*, Pergamon Press, Inc., Vol. I, 1960.
36. V. I. Kogan, *Plasma Physics and the Problem of Controlled Thermonuclear Reactions*, Pergamon Press, Inc., Vol. IV, 1960.
37. A. Lochte-Holtgreven, *Repts. Progr. in Phys.*, **21**, 312 (1958).
38. P. J. Dickerman, in H. Fischer and L. C. Mansur, Eds., *Conference on Extremely High Temperatures*, John Wiley & Sons, Inc., 1958, p. 77.
39. L. H. Aller, *Astrophysics*, The Ronald Press Co., 1953, Vol. 1, p. 308.
40. J. Holtsmark, *Ann. Physik*, **58**, 577 (1919); *Physikal. Z.*, **20**, 162 (1919); **25**, 73 (1924).
41. S. Chandrasekhar, *Rev. Mod. Phys.*, **15**, 1 (1943); see p. 73.
42. F. R. Scott, A. T. Brousseau, E. M. Little, and A. E. Schofield, USAEC Report TID-7520 (1950).
43. H. Griem, *Z. Physik*, **137**, 280 (1954).
44. H. Griem, in H. Fischer and L. C. Mansur, Eds., *Conference on Extremely High Temperatures*, John Wiley & Sons, Inc., 1958, p. 93.
45. A. C. Kolb and H. R. Griem, *Phys. Rev.*, **111**, 514 (1958).
46. H. R. Griem and A. C. Kolb, *Fourth Intl. Conf. on Ionization Phenomena in Gases*, p. 808 (1959).
47. H. R. Griem, A. C. Kolb, and K. Y. Shen, *Phys. Rev.*, **116**, 4 (1959).
48. E. J. Stovall, USAEC Report WASH-184 (1955), p. 62.
49. A. H. Turnbull, W. R. Hindmarsh, and T. K. Allen, UKAEA Report AERE GP/R-1793 (1955).
50. A. H. Turnbull and B. B. Jones, UKAEA Report AERE GP/R-2035 (1956).
51. S. P. Cunningham, USAEC Report TID-7503 (1956), p. 19.

52. K. Weiner, USAEC Report NYO-7885 (1957).

53. C. R. Burnett, D. J. Grove, R. W. Palladino, T. H. Stix, and K. E. Wakefield, *Phys. Fluids*, **1**, 438 (1958); also in *Proc. Second U.N. Conf. on Peaceful Uses of Atomic Energy*, **32**, 225 (1958).

54. H. C. Van de Hulst and J. J. M. Reesinck, *Astrophys. J.*, **106**, 121 (1947).

55. L. H. Aller, *Astrophysics*, The Ronald Press Co., 1953, Vol. I, p. 58.

56. K. W. Meissner, *J. Opt. Soc. Am.*, **31**, 405 (1941).

57. J. R. McNally, USAEC Report TID-7503 (1956), p. 385.

58. T. P. Hughes and S. Kaufman, *Nature*, **183**, 7 (1959).

59. J. Hirschberg and E. Hinnov, Division of Plasma Physics APS, Monterey Conference, D 14 (1959).

60. J. E. Allen and P. Reynolds, USAEC Report TID-7536 (1957), p. 543.

61. E. A. McLean, C. E. Faneuff, H. R. Griem, and A. C. Kolb, *Phys. Fluids*, **3**, (1960).

62. J. H. Lees, *Proc. Roy. Soc.*, **A137**, 173 (1932).

63. S. P. Cunningham, USAEC Reports WASH-289 (1955), p. 279; TID-7503 (1956), p. 19.

64. J. M. Berger, USAEC Report NYO-6371 (1955).

65. J. C. Howard, USAEC Report UCRL-5265 (1958).

66. H. R. Griem, A. C. Kolb, and W. R. Faust, *Phys. Rev. Letters*, **2**, 281 (1959).

67. F. C. Jahoda, E. M. Little, W. E. Quinn, C. A. Sawyer, and T. F. Stratton, *Phys. Rev.*, **119** (1960).

68. I. Langmuir and H. Mott-Smith, *Gen. Elec. Rev.*, **27**, 449, 538, 616, 762, 810 (1924).

69. D. Bohm, E. H. S. Burhop, and H. S. W. Massey, in A. Guthrie and R. K. Wakerling, Eds., *The Characteristics of Electrical Discharges in Magnetic Fields*, NNES, Div. 1, Vol. 5, McGraw-Hill Book Co., Inc., 1949, p. 13.

70. J. E. Allen, R. L. F. Boyd, and P. Reynolds, *Proc. Phys. Soc. (London)*, **B70**, 297 (1956).

71. I. B. Bernstein and I. N. Rabinowitz, *Phys. Fluids*, **2**, 112 (1959).

72. E. O. Johnson and L. Malter, *Phys. Rev.*, **80**, 58 (1950).

73. S. Kojima et al., *J. Phys. Soc. Japan*, **4**, 349 (1949); **8**, 55 (1953).

74. K. Yamamoto and T. Okeeda, *J. Phys. Soc. Japan*, **11**, 57 (1956).

75. V. I. Pistonovich, *Plasma Physics and the Problem of Controlled Thermonuclear Reactions*, Pergamon Press, Inc., Vol. IV, 1960.

76. R. H. Lovberg, *Ann. Physics*, **8**, 311 (1959).

77. L. C. Burkhardt and R. H. Lovberg, *Proc. Second U.N. Conf. on Peaceful Uses of Atomic Energy*, **32**, 29 (1958).

78. J. P. Conner, D. C. Hagerman, J. L. Honsaker, H. J. Karr, J. P. Mize, J. E. Osher, J. A. Phillips, and E. J. Stovall, *Proc. Second U.N. Conf. on Peaceful Use of Atomic Energy*, **32**, 297 (1958).

79. H. P. Furth, unpublished.

80. R. E. Dunaway and J. A. Phillips, USAEC Report LAMS-1682 (1954).

81. R. F. Whitmer and H. J. Karr, USAEC Report TID-7503 (1956), p. 427.

82. *Proc. Third Int. Cong. on High-Speed Photography*, Butterworth's Scientific Publications, London, 1957; J. D. McGee and W. L. Wilcock, Eds., *Photo-Electronic Image Devices*, Advances in Electronic Physics, Vol. XII, Academic Press, Inc., 1959.

83. A. Folkierski, *Nucl. Instr.*, **4**, 346 (1959).

84. R. H. Lovberg, L. C. Burkhardt, G. A. Sawyer, and T. F. Stratton, USAEC Report TID-7520 (1956), p. 308.

85. L. C. Burkhardt, R. H. Lovberg, G. A. Sawyer, and T. F. Stratton, *J. Appl. Phys.*, **29**, 964 (1958).

86. L. H. T. Rietjens and C. M. Braams, *Nucl. Instr.*, **4**, 363 (1959).

87. R. Dunaway and J. A. Phillips, *J. Appl. Phys.*, **29**, 1137 (1958).

88. J. W. Mather and A. H. Williams, *Proc. Second U.N. Conf. on Peaceful Uses of Atomic Energy*, **32**, 26 (1958).

89. O. A. Anderson, W. R. Baker, S. A. Colgate, H. D. Furth, J. Ise, R. V. Pyle, and R. E. Wright, *Phys. Rev.*, **110**, 1375 (1958).

90. T. E. Allibone, D. R. Chick, G. P. Thomson, and A. A. Ware, *Proc. Second U.N. Conf. on Peaceful Uses of Atomic Energy*, **32**, 169 (1958).

91. A. M. Andrianov et al., *Proc. Second U.N. Conf. on Peaceful Uses of Atomic Energy*, **31**, 348 (1958).

92. B. Rose, A. E. Taylor, and E. Wood, *Nature*, **181**, 1630 (1958).

93. W. M. Jones, A. C. L. Barnard, S. E. Hunt, and D. R. Chick, *Nature*, **182**, 216 (1958).

94. A. A. Ward, *Proc. Inst. Elec. Eng.*, **106A**, Suppl. No. 2, 30 (1959).

95. D. P. Nagle, W. E. Quinn, W. B. Riesenfeld, and W. Leland, *Phys. Rev. Letters*, **3**, 318 (1959).

96. V. S. Komelkov and V. J. Sinitsyn, *Plasma Physics and the Problem of Controlled Thermonuclear Reactions*, Pergamon Press, Inc., Vol. I, 1960.

97. N. V. Filippov, *Plasma Physics and the Problem of Controlled Thermonuclear Reactions*, Pergamon Press, Inc., Vol. III, 1959, p. 280.

98. M. O. Stern, unpublished.

99. L. Smith, *Seventh Meeting on Mass Spectroscopy*, Los Angeles, 1959, quoted in reference 3.

Chapter 7

THE PINCH EFFECT

THEORETICAL TREATMENT

THE QUASI-EQUILIBRIUM PINCH

7.1. The theory of the constricted gas current was first developed by W. H. Bennett in 1934 [1]. A few years later, a different approach was presented independently by L. Tonks [2], who used the term "pinch effect" in the sense in which it is employed at the present time.* Subsequently, a number of other theoretical analyses have been published of the constriction of a current filament due to the action of the azimuthal magnetic field generated by the current itself [1-6]. A somewhat simplified treatment will be given here that will serve to indicate the basic conclusions [7].

7.2. Although the pinched discharge is an unstable system, as was indicated in Chapter 3 and as will be further elaborated below, it may be treated as a succession of quasi-equilibrium states. In any one of such states, it may be supposed that there is a balance, at any point in the plasma, between the electromagnetic forces and those due to gradients of the kinetic pressure; hence, equations (3.9) and (3.10), which are applicable, can be written in Gaussian-cgs units as

$$-\nabla p_i + n_i e \left[\mathbf{E} + \frac{1}{c} \left(\mathbf{v}_i \times \mathbf{B} \right) \right] = 0 \qquad (7.1)$$

for the ions and

$$-\nabla p_e - n_e e \left[\mathbf{E} + \frac{1}{c} \left(\mathbf{v}_e \times \mathbf{B} \right) \right] = 0 \qquad (7.2)$$

for the electrons in a hydrogen isotope plasma. Since the discharges to be considered are cylindrically symmetric, with no dependence on the axial (z) coordinate, it is only the radial force terms which are of interest. Furthermore, it is only the azimuthal (B_θ) magnetic field, arising from the longitudinal (or

---

*The constriction of a liquid metal conductor by the passage of a large current was observed many years earlier by C. Hering, who referred to it as the "pinch phenomenon." The effect was further studied both experimentally and theoretically by C. L. Northrup [3].

225

axial) current elements, which is present. Consequently, equations (7:1) and (7.2) become

$$\frac{-dp_i}{dr} + n_i e \left(E_r + \frac{v_i}{c} B_\theta \right) = 0 \tag{7.3}$$

and

$$\frac{-dp_e}{dr} - n_e e \left(E_r + \frac{v_e}{c} B_\theta \right) = 0, \tag{7.4}$$

where the velocities v_i and v_e are in the z-direction.

7.3.   It will now be postulated that the plasma is electrically neutral, so that $n_e = n_i$; the field E_r is then produced by an otherwise negligible difference between n_e and n_i.   Further, T_i and T_e are assumed to be independent of position, so that the plasma is isothermal.   If there is a Maxwellian distribution of velocities, then for the particles of each kind $p = nkT$, and addition of equations (7.3) and (7.4) leads to

$$-k(T_i + T_e)\frac{dn}{dr} + \frac{ne}{c}(v_i - v_e)B_\theta = 0, \tag{7.5}$$

where n is the number density of either ions or electrons.   The ion and electron axial velocities, v_i and v_e, are maintained, in general, by an applied electric field.   But since the plasma is electrically neutral, this field cannot cause any net momentum of the plasma as a whole.   Consequently, as seen in §4.85, v_e is much larger than v_i, so that the latter can be neglected in comparison with the former; hence, equation (7.5) reduces to

$$k(T_i + T_e)\frac{dn}{dr} + \frac{nev_e}{c}B_\theta = 0.$$

Upon multiplying through by r/n and rearranging, it is found that

$$\frac{r}{n} \cdot \frac{dn}{dr} + \frac{ev_e r B_\theta}{ck(T_e + T_i)} = 0.$$

Assuming the drift velocity of the electrons to be independent of position, differentiation with respect to r then gives

$$\frac{d}{dr}\left(\frac{r}{n} \cdot \frac{dn}{dr}\right) + \frac{ev_e}{ck(T_e + T_i)} \cdot \frac{d}{dr}(rB_\theta) = 0. \tag{7.6}$$

7.4.   By rearrangement of the equation in §6.112, it follows that

$$\frac{d}{dr}(rB_\theta) = \frac{4\pi}{c}rj_z, \tag{7.7}$$

where j_z is the current density in the axial direction.   If the drift velocity of the ions is negligible, as indicated above, then

$$j_z = env_e. \tag{7.8}$$

Upon combining equations (7.6), (7.7), and (7.8), the result is

$$\frac{d}{dr}\left(\frac{r}{n}\cdot\frac{dn}{dr}\right) + \frac{4\pi e^2 v_e^2}{c^2 k(T_e + T_i)}\, nr = 0. \tag{7.9}$$

7.5. Since the discharge will presumably be symmetrical about the central axis, where $r = 0$, it follows that, if $n(r)$ is a smoothly varying function of (r),

$$\frac{dn}{dr} = 0 \quad \text{when} \quad r = 0.$$

With this boundary condition, and setting n equal to n_0 at the axis, the solution of equation (7.9) is

$$n = \frac{n_0}{(1 + n_0 b r^2)^2}, \tag{7.10}$$

where b is defined by

$$b \equiv \frac{\pi e^2 v_e^2}{2c^2 k(T_e + T_i)}. \tag{7.11}$$

The variation of the particle density n with distance r from the axis of the discharge, as represented by equation (7.10), is referred to as the *Bennett distribution*. It is seen that, in any quasi-equilibrium state, the particle density is a maximum at the axis, where $r = 0$, and falls off rapidly with increasing dis-

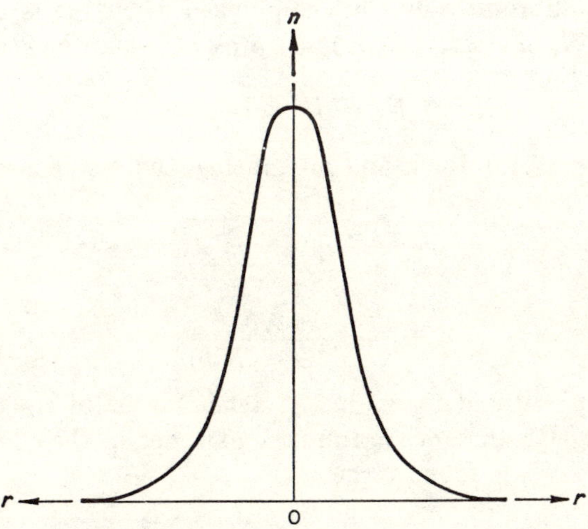

FIG. 7.1. Bennett distribution of particle density in constricted current.

tance (Fig. 7.1). Thus, the plasma is concentrated (or constricted) toward the discharge axis. According to equation (7.10), the particle density falls to zero only when r is infinite. However, in a finite system, provided b is relatively large, the rate of decrease of particle density with distance from the axis is so great that the plasma may be regarded as being essentially confined within a channel of finite radius.

7.6.   If p_i and p_e, and consequently T_i and T_e, are approximately equal, then the term involving v_i in equation (7.3) can be neglected in comparison with the others; hence,

$$\frac{-dp_i}{dr} + n_i e E_r \approx 0$$

or

$$-kT_i \frac{dn}{dr} + neE_r \approx 0.$$

Upon integration, it is found that

$$V(r) - V_0 = \frac{kT_i}{e} \ln \frac{n(r)}{n_0},$$

where, as before, the zero subscripts refer to the values on the axis of the discharge.   Hence, a potential difference exists across the radial width of the pinch which corresponds to particle energies of the order of the mean thermal energy.

THE BENNETT EQUATION

7.7.   If the equilibrium radius of the pinched discharge is taken as r_p, then the total number N_e of electrons (or ions) present in unit length is given by

$$N_e = \int_0^{r_p} 2\pi r n dr.$$

Utilizing equation (7.10) for n and integrating, the result is

$$N_e = \frac{n_0 \pi r_p^2}{1 + n_0 b r_p^2},$$

so that

$$n_0 = \frac{N_e}{r_p^2(\pi - N_e b)}.$$

If this value is substituted for n_0 in the denominator of equation (7.10), the particle density n_p at a distance r_p from the axis, i.e., at the effective exterior of the pinched discharge, is found to be

$$n_p = n_0 \left(1 + \frac{N_e b}{\pi - N_e b}\right)^{-2}.$$

Since this is essentially zero, it follows that

$$\pi - N_e b \approx 0$$

or

$$N_e b \approx \pi. \tag{7.12}$$

7.8. Upon combining this result with equation (7.11), which defines b, it is found that

$$\left(\frac{N_e e v_e}{c}\right)^2 = 2N_e k(T_e + T_i). \tag{7.13}$$

The total current I in the discharge is given by

$$I = N_e e v_e$$

if, as before, the contribution of the ions is neglected. Hence, equation (7.13) becomes

$$I^2 = 2N_e k(T_e + T_i)c^2, \tag{7.14}$$

which is known as the *Bennett relation*. This expression may be put in a slightly different form by replacing $T_e + T_i$ by $2T$, where T is the average temperature of the particles in the plasma, and $2N_e$ by N, where N is the total number of particles per centimeter length of the pinched discharge; hence,

$$I^2 = 2NkTc^2, \tag{7.15}$$

with the current in statamperes and kT in ergs. If I is multiplied by $c/10$ to convert it to ordinary amperes, equation (7.15) becomes

$$I^2 = 200NkT, \tag{7.16}$$

which is identical with equation (3.25). Alternatively, if the temperature is expressed in kilo-electron volts, the Bennett relation may be written as

$$I \text{ (amp)} = 5.7 \times 10^{-4} N^{1/2} T^{1/2}. \tag{7.17}$$

7.9. It would appear from equation (7.16) or (7.17) that the attainment of a sufficiently high temperature in a pinched discharge depends on the ability to pass a current of the required magnitude. Theoretical arguments, based on somewhat uncertain postulates, but which, nevertheless, lead to results that must be at least qualitatively correct, indicate that there is a limiting value to such a current. In very simple terms, this result is largely determined by the fact that, as the temperature of the plasma increases, the energy loss as bremsstrahlung increases (§2.64). Since the temperature in the quasi-equilibrium pinch varies as the square of the current, it is evident that the rate of energy loss will increase rapidly with the pinched discharge current. The loss will consequently limit the temperature that can be attained and also the current that can be passed [7a].

7.10. The limiting value of the pinch current has been calculated to be about 1.7×10^6 amp. Hence, if N is taken as 10^{18} particles/cm^2, which is as low as would appear reasonable (§3.29), the maximum temperature which can exist in a pinched discharge is found to be about **9 kev**. If this conclusion is correct, it would imply that the temperatures required for the operation of a useful thermonuclear reactor utilizing deuterium as the feed gas cannot be realized with the pinch effect alone.

7.11. It should be understood that the results given here can apply only if there is no magnetic field inside the pinched plasma. For the purpose of stabilizing the discharge, an axial (or longitudinal) field is often included within the plasma (§7.26). When this is the case, the discharge current and average plasma temperature are no longer related by equations (7.16) and (7.17). Since the force due to the external self-magnetic field of the discharge must now balance the force due to the included field as well as that due to the kinetic pressure of the plasma particles, the current required to pinch the discharge is larger than would be the case in the absence of the axial field (cf. §7.37). The temperature corresponding to the limiting pinch current would be correspondingly lower than calculated above.

DYNAMICS OF A PINCHED PLASMA

7.12. The foregoing discussion of the quasi-equilibrium pinch applies only to a discharge with an essentially constant current or, more exactly, a discharge whose current does not vary appreciably in times comparable to the traversal time for a sound wave or the time required for a magnetic field to diffuse through the plasma. Although this condition holds in some experimental studies of the pinch effect (cf. §7.48 *et seq.*), a considerable amount of work has been done with "fast discharges," in which the constricted plasma is established in a time that is short in comparison with the sound transit and field diffusion times. It is the behavior of discharges of the latter type which will now be considered [8-13].

7.13. For the purpose of the present treatment of the dynamic behavior of a pinch produced by a rapidly increasing current, the simplifying assumption will first be made that the gas has been preionized and is a perfect conductor.* It will also be assumed that the discharge tube is an insulator, so that the initial current flow is in the plasma. In these circumstances, when the tube is connected to a voltage source, e.g., a capacitor bank, current begins to flow in a thin layer of the plasma adjacent to the walls of the cylindrical containing vessel. The behavior is essentially the same as that exhibited by a metallic conductor when a high-frequency voltage is applied to it.

7.14. The flow of current, parallel to the tube axis, in the plasma sheath sets up an azimuthal magnetic field B_θ just outside the current layer (Fig. 7.2). Since there is no field on the inside, the sheath experiences an inward pressure $B_\theta^2/8\pi$ and so it begins to contract. As it moves inward, the sheath behaves like a "magnetic piston" and sweeps up all the charged particles it encounters. The rate of momentum change of the plasma, balanced against the external

---

* The treatment given here is a modification of what has been called the "infinite conductivity" theory of the pinch. It has also been called the "M" (for "motor") theory because of a certain resemblance to the theory of the electric motor.

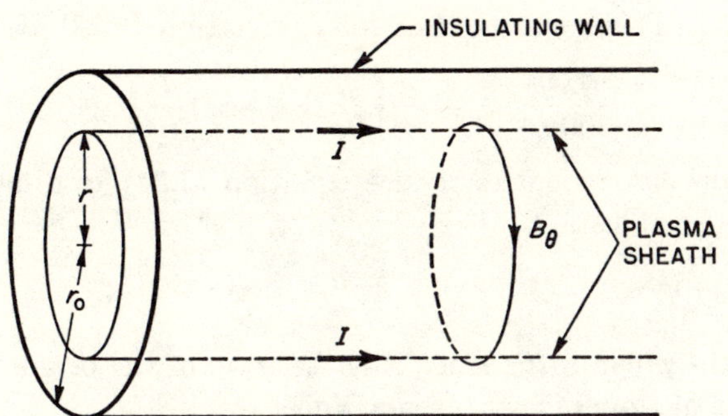

FIG. 7.2. Model of perfectly conducting pinched plasma.

magnetic pressure, then gives the inward velocity of the sheath as a function of time. This is the basis of what is known as the *snowplow model* of the constricted discharge [8, 9].

7.15. The magnetic field strength in gauss just outside the sheath of radius r cm, carrying a current I amp, is given by equation (3.23) as

$$B_\theta = \frac{I}{5r} \tag{7.18}$$

and so the inward pressure in dynes/cm² is

$$p = \frac{B_\theta^2}{8\pi} = \frac{I^2}{200\pi r^2}. \tag{7.19}$$

The next step is to express the pressure in terms of the density and velocity of the plasma sheath. According to the snowplow model, momentum balance at the surface requires that

$$2\pi r p = -\frac{d}{dt}\left(M\frac{dr}{dt}\right) \tag{7.20}$$

where M is the mass per unit length of the particles swept up by the sheath as it moves inward; the right side of equation (7.20) represents the time rate of change of momentum of the sheath, i.e., the force exerted on unit length of the plasma by the inward pressure of the magnetic field. If ρ is the initial mass density, in grams/cm³, of the gas in the discharge tube, then

$$M = \pi(r_0^2 - r^2)\rho, \tag{7.21}$$

where r_0 is the tube radius. Hence, it follows from equations (7.19), (7.20), and (7.21) that

$$\frac{I^2}{200\pi r^2} = -\frac{d}{dt}\left[\frac{\rho}{2r}(r_0^2 - r^2)\frac{dr}{dt}\right]. \tag{7.22}$$

7.16. Suppose that the discharge current increases linearly with time, as is approximately the case when a capacitor bank is switched into the circuit, i.e.,

$$I = I_0 \sin \omega t \approx I_0 \omega t, \tag{7.23}$$

when ωt is small.   Then, writing $I_0\omega t$ for I, equation (7.22) becomes

$$\frac{I_0^2\omega^2 \ell^2}{200\pi r^2} = -\frac{d}{dt}\left[\frac{\rho}{2r}\,(r_0{}^2 - r^2)\,\frac{dr}{dt}\right]. \qquad (7.24)$$

7.17.   It is now convenient to reduce equation (7.24) to dimensionless form by defining the variables

$$\kappa \equiv \frac{r_0}{r},$$

which is called the *pinch ratio,* since it is the ratio of the original radius of the gas to the radius of the pinched plasma, and

$$\tau \equiv t\left[\frac{I_0{}^2\omega^2}{100\pi\rho r_0{}^4}\right]^{\frac14},$$

so that equation (7.24) becomes

$$\frac{d}{d\tau}\left[(1 - \kappa^2)\,\frac{d\kappa}{d\tau}\right] = \frac{-\tau^2}{\kappa},$$

which is plotted in Fig. 7.3 [12].   It is seen from the curve that the first pinch occurs when τ is approximately 1.4, i.e., when

$$t_p = 1.4\left[\frac{100\pi\rho r_0{}^4}{I_0{}^2\omega^2}\right]^{\frac14}. \qquad (7.25)$$

The average sheath velocity is then

$$\bar{v} \equiv \frac{r_0}{t_p} = \frac{1}{1.4}\left[\frac{I_0{}^2\omega^2}{100\pi\rho}\right]^{\frac14}. \qquad (7.26)$$

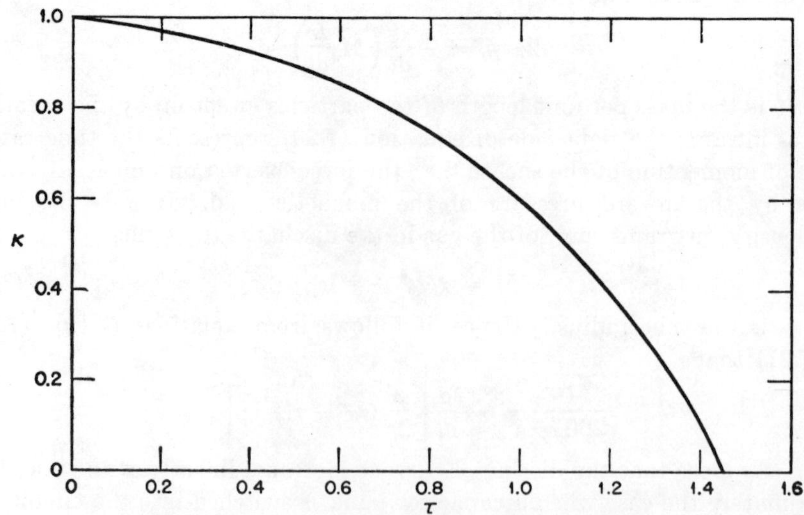

FIG. 7.3.   Time dependence of pinch ratio.

Alternative forms of equations (7.25) and (7.26) are sometimes used in which $I_0\omega$ is replaced by dI/dt, in accordance with equation (7.23).

7.18. The foregoing results are applicable to common practical situations in which the connecting circuitry from the capacitor bank to the discharge tube has an inductance larger than that of the plasma, thus insuring that the current will be nearly sinusoidal throughout the pinching process [12]. The opposite extreme situation is also encountered, in which the external inductance is very small and the source capacitance is so large that the voltage applied to the discharge tube is essentially constant. In this latter situation [8, 9], the average inward velocity of the sheath is given in Gaussian-cgs units by

$$\bar{v} \approx \left(\frac{c^2 E^2}{4\pi\rho}\right)^{\frac{1}{4}}$$

or, if E is the constant field strength in volts/cm,

$$\bar{v} \approx 10^4 \left(\frac{E^2}{4\pi\rho}\right)^{\frac{1}{4}}. \tag{7.27}$$

Both equations (7.26) and (7.27) are useful, although the former has the advantage of corresponding more closely in its approximations to the situations most frequently encountered in pinched discharge work.

7.19. The *dynamic pinch* (or *shock pinch*) as described above has received a great deal of attention experimentally since it offers, in principle, a way to obtain high average particle energy in a plasma. When the pinch reaches its smallest radius, it has imparted to all the particles in the plasma a radial velocity somewhere near its own maximum inward velocity, i.e., in excess of $\bar{v}$ as calculated above. If the pinched discharge column can be kept contracted long enough to permit randomization, i.e., thermalization, of the ion velocities to take place, a high temperature will result.

7.20. An estimate of the conditions that might be required in a thermonuclear reactor based on this concept may be obtained from equation (7.26) which, upon rearrangement, gives

$$I_0\omega = \frac{dI}{dt} \approx 20\bar{v}^2(\pi\rho)^{\frac{1}{2}}.$$

In a deuterium plasma at a kinetic temperature of 10 kev, the value of $\bar{v}$ is approximately 10^8 cm/sec; hence, for a particle density of 10^{15} per cm^3, which corresponds to a mass density of 1.7×10^{-9} g/cm^3, it follows that

$$I_0\omega = \frac{dI}{dt} = 1.4 \times 10^{13} \text{ amp/sec.} \tag{7.28}$$

Since $I_0\omega$ is approximately equal to V/L, where V is the capacitor bank voltage and L is the nearly constant circuit inductance (cf. §6.7), it is seen that, with a reasonable lower limit of 10^{-7} henry for L, the extremely high voltage of

1.4×10^6 volts would be necessary to satisfy equation (7.28). Some reduction in voltage might result from operating with plasmas of lower density; however, the temperature of 10 kev used in the calculation is certainly too low for a self-sustaining thermonuclear reaction system using deuterium only. The practical problems of attaining thermonuclear temperatures by means of an extremely fast pinch are thus formidable.

PINCH INSTABILITIES AND STABILIZATION

INSTABILITY OF THE PINCHED DISCHARGE

7.21. One of the major problems associated with the constricted discharge— and, in fact, with any system of charged particles confined by a magnetic field —is that of instability. There is probably no stable condition of an ordinary pinched plasma and it can exist, at best, only in a state of unstable equilibrium. In this state, the discharge is in the form of a straight-walled (or uniformly curved) cylinder, but any perturbation, e.g., due to a slight asymmetry in the discharge current or to any other transient phenomenon, will tend to increase rather than to decrease. The extra force on the pinched plasma resulting from a distortion of the uniform cylinder has the effect of amplifying that distortion. In other words, there is a tendency for minor instabilities to grow into large ones, thus leading to complete break-up of the pinched discharge.

7.22. A fuller theoretical treatment of plasma instabilities and stabilization will be given in Chapter 13, but some aspects of these subjects, of special interest in connection with pinched discharges, will be considered here. There are two main types of instability which lead to the destruction of a pinched plasma. The first, often called the *sausage type* of instability, results from an azimuthally symmetrical localized constriction (and expansion) of the plasma

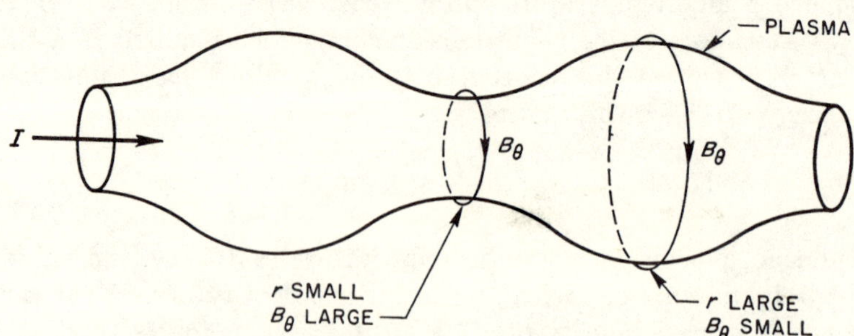

FIG. 7.4. Sausage-type ($m = 0$) instability of plasma column.

column, as shown in Fig. 7.4. As seen earlier, the strength B_θ of the azimuthal self-magnetic field is proportional to I/r, where r is the radial distance. Since the value of the current I is the same in all parts of the discharge, it is evident that B_θ will be larger than normal in the constricted regions, where r is smaller, and the reverse will be the case at the bulges where the radius is larger. Consequently, the constrictions tend to become more and more constricted, whereas the bulges will grow larger and larger until the pinched discharge is completely disrupted.

7.23. The *kink type* of instability originates with the formation of a bend or kink in the plasma column while the latter retains its uniform, circular cross section (Fig. 7.5). It is seen that the lines of force of the azimuthal self-

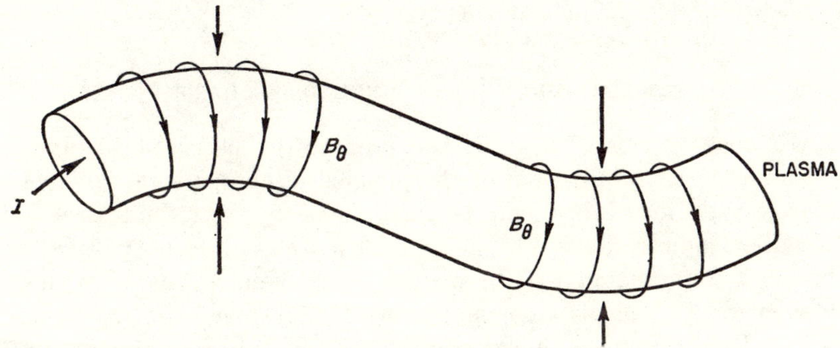

Fig. 7.5. Kink-type ($m = 1$) instability of plasma column.

magnetic field, due to the current in the plasma, are brought closer together on the inside, but they are farther apart on the outside of a bend. As a result, the magnetic pressure is greater on the inside, as indicated by the length of the arrows in Fig. 7.5, and there is a net force which acts in such a direction as to increase the bend. Thus, once a slight kink develops, it will grow in size until the pinched discharge strikes the walls and is cooled. The plasma then becomes diffuse and fills the containing tube.

7.24. It it doubtful whether the kink instability takes the simple form in which the discharge channel is merely bent while remaining in the plane containing the axis. Magnetic probe and photographic studies show that, in the common type of kink instability, the pinch is deformed into a corkscrew or helical form, although it largely retains its circular cross section. Experimental results indicate that the helix twists in the same direction as the self-magnetic field of the discharge. Such behavior is in harmony with the fact that less work is required to produce a deformation which follows the magnetic lines than in the opposite direction. Because of the motion of the plasma, the term "wriggling" has been largely used in the British literature to describe the phenomenon which has been frequently observed with pinched discharges.

7.25. The rate of growth of instabilities will generally be determined by the velocity of hydromagnetic waves in the plasma. If equation (4.108) is used to express this velocity, it is seen that the value approaches the Alfvén speed in the extreme of low kinetic pressure of the plasma and the acoustic (sound) speed when the magnetic pressure is relatively unimportant within the plasma. Suppose that in a nonstabilized pinch the perturbations grow with the speed of sound at the existing conditions. At a temperature of 1 kev, for example, sound speed in a deuterium plasma is about 4×10^7 cm/sec; consequently, in a discharge tube of 50 cm radius, the pinched plasma would be expected to reach the walls in less than a microsecond. This result is sufficient to indicate that, unless the constricted discharge can be stabilized in some manner, its lifetime would be much too short to be of practical value for the release of thermonuclear energy.

STABILIZATION OF THE PINCHED DISCHARGE

7.26. Extensive theoretical work and laboratory observations have shown that at least partial stabilization of the pinched discharge is possible [14-16]. The sausage-type instability can be suppressed by including an axial (or longitudinal) magnetic field, i.e., a field with lines of force parallel to the discharge axis, represented by B_z, within the plasma. This is achieved by passing a current through a solenoidal winding around the discharge tube (Fig. 7.6). By applying the axial magnetic field before the discharge is

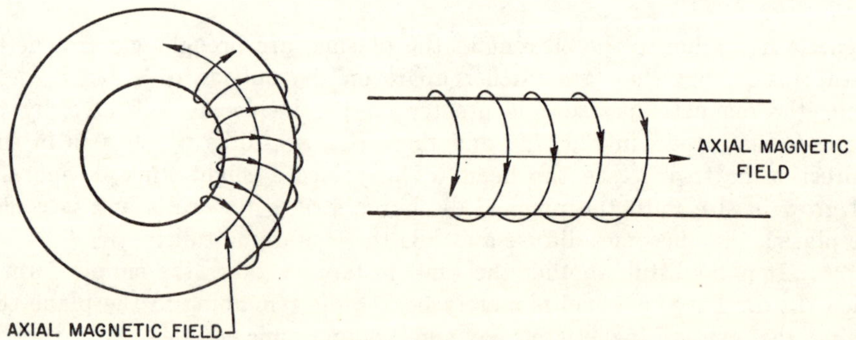

FIG. 7.6. Stabilization of pinched discharge by axial magnetic field.

passed, this field is largely trapped within the plasma as it becomes pinched. The formation of a sausage-type constriction then requires that work be done in further compression of the included magnetic field. Consequently, the presence of the axial field tends to inhibit the development of sausage-type instabilities.

7.27. The axial magnetic field also helps to prevent the occurrence of kink instabilities, since the twisting of the plasma into a helix is accompanied by stretching of the magnetic lines and this involves the performance of work. The kink instabilities are further counteracted by means of a conducting shell surrounding the discharge tube. As the current column moves away from the axis of the discharge tube toward the wall, the azimuthal magnetic field lines are compressed between the discharge and the wall, assuming that the unstable motion occurs over a time much shorter than is needed for the field lines to penetrate the conductor. This compression requires work to be done and so results in a restoring force. An alternative point of view is that the pinch is repelled by an opposite image current induced in the conducting wall. Kinks of wave length that are short compared with the tube radius will not be suppressed in this manner, since the resulting undulations in the field will have smoothed out at the wall.

7.28. Theoretical treatments of the pinch stabilization problem have of necessity involved a number of simplifying assumptions concerning the nature of the discharge. One of these, which has led to the results to be described below, is based on the postulate that the plasma has zero resistivity and consequently carries the longitudinal current in an infinitesimally thin boundary layer (cf. §7.13). It has been shown, on this basis, that completely stable conditions are then possible for what is sometimes called the "B_z pinch," i.e., a pinched discharge in a tube surrounded by conducting walls with an axial (B_z) field included within the plasma.

7.29. When the discharge is passed, the initial current is carried in a sheath of negligible thickness adjacent to the wall of the tube, which may either be of insulating material or of segmented (and insulated) metal construction. As the sheath begins to contract, it will compress inside it the axial flux which originally filled the tube. None of the field lines will leak out of this perfectly conducting cylinder, but some may be left behind if there is an annular space between the solenoid producing the B_z field (or the conducting shell) and the initial plasma radius. For simplicity, however, it will be assumed here that there is no such space and that there is no axial field external to the constricted discharge.

7.30. The general equation of motion of the plasma sheath is obtained by adding two terms to that derived in §7.15, based on the ideal snowplow model; thus,

$$\frac{B_\theta^2}{8\pi} = \frac{B_z^2}{8\pi} + nkT + \frac{d}{dt}\left[\frac{\rho_0}{2r}\left(r_0^2 - r^2\right)\frac{dr}{dt}\right], \tag{7.29}$$

where the first term on the right is the magnetic pressure exerted by the B_z field trapped by the plasma and the second represents the outward kinetic pressure of particles within the sheath cylinder. The use of nkT in this connection supposes that the plasma has a significant temperature; its value will depend on the discharge current strength, the pinch radius, losses, etc. The particle density n will be determined by the behavior of the gas particles upon encountering the sheath as it moves inward, i.e., whether they are all retained, as implied by the last term in equation (7.29), or whether some bounce off when they strike the sheath.

7.31. The value of B_θ in terms of the discharge current is given by equation (7.18). Upon introducing r_0, the initial radius of the plasma, this becomes

$$B_\theta = \frac{I}{5r} = \frac{I}{5r_0}\left(\frac{r_0}{r}\right)$$

$$= \frac{I}{5r_0}\kappa, \tag{7.30}$$

where r_0/r is the pinch ratio κ, defined in §7.17. Furthermore, if B_{z0} is the initial strength of the axial stabilizing field then, according to equation (4.14),

$$B_z = B_{z0}\left(\frac{r_0}{r}\right)^2 = B_{z0}\kappa^2, \tag{7.31}$$

provided, as postulated above, that the axial field is completely retained within the plasma during compression. Upon combining equations (7.29), (7.30), and (7.31) the result is

$$\frac{I^2\kappa^2}{25r_0^2} = B_{z0}^2\kappa^4 + 8\pi nkT - \frac{d}{dt}\left[\frac{4\pi\rho_0}{r}(r_0^2 - r^2)\frac{dr}{dt}\right]. \tag{7.32}$$

7.32. An interesting consequence of the inclusion of the first two terms on the right of equation (7.32) is that it is now possible to obtain equilibrium solutions of the equation of motion, as well as strictly dynamic ones. It is the former which are of immediate interest in connection with the investigation of the properties of the steady-state stabilized pinch. Thus, by setting dr/dt equal to zero, the inertial term in equation (7.32) disappears and the three remaining terms give the value of the pinch radius (or pinch ratio) as determined by a balance of three static pressures; thus,

$$\frac{I^2\kappa^2}{25r_0^2} = B_{z0}^2\kappa^4 + 8\pi nkT. \tag{7.33}$$

If the internal particle pressure is small in comparison with the magnetic pressure of the trapped B_z field, as is generally the case, the second term on the right may be neglected in comparison with the first, so that

$$\kappa \approx \frac{I}{5r_0B_{z0}}, \tag{7.34}$$

a result widely used in experimental work to establish a B_z-stabilized pinch of any desired pinch ratio. It should be noted, incidentally, that the neglect of the particle pressure term in equation (7.33) means that, in the steady-state B_z-pinch, the strength of the trapped B_z field is equal to that of the B_θ field just outside the pinched discharge column.

7.33. In the normal mode analysis of pinch stability (see Chapter 13), it is usual to describe small-amplitude perturbations of the plasma in terms of the Fourier components. Any displacement ξ may then be represented by

$$\xi = \xi(r) \exp\left[i(m\theta + kz + \omega t)\right], \tag{7.35}$$

where m is zero or an integer, representing the azimuthal periodicity of the particular mode of deformation, k is the longitudinal wave number, and ω determines the growth rate of the perturbation. If ω^2 is positive, the perturbation is periodic in time and is a simple wavelike disturbance; but if ω^2 is negative, then one of the associated modes grows exponentially in time and the system is unstable. The latter situation holds in a simple pinched discharge, but the inclusion of the axial field and the presence of conducting walls leads to conditions under which ω^2 is positive and the pinch should be stable.

7.34. For perturbations of the plasma which are symmetrical about the axis, i.e., the sausage type of instability, the displacement is independent of the azimuthal angular coordinate θ; hence, for this instability, m in equation (7.35) must be zero. For this reason, it is often referred to as the $m = 0$ instability. For the simplest kink (or spiral) instability, $m = 1$; the pinch is deformed into a corkscrew or helical shape but retains a circular cross section. Although other perturbations, with $m = 2$, etc., are theoretically possible it is the $m = 0$ and $m = 1$ modes which are of major interest in experimental plasma instability studies.

7.35. The results of a theoretical treatment [14] of the B_z-pinch, for dif-

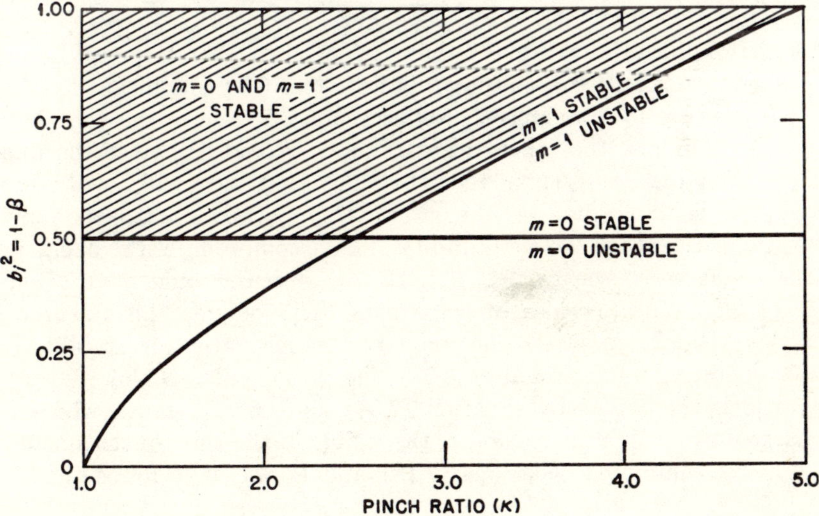

FIG. 7.7. Stable and unstable regions for $m = 0$ and $m = 1$ perturbations.

ferent m values, are presented in Fig. 7.7 in terms of the pinch ratio κ and a parameter b_i defined by

$$b_i \equiv \frac{B_z}{B_\theta}. \tag{7.36}$$

The data in the figure are based on the postulate that the current sheath is infinitely thin and that the B_z field is completely trapped by the plasma, so that there is no axial field outside it. The horizontal and sloping lines are divisions between domains of stability and instability against $m = 0$ and $m = 1$ perturbations, respectively. For stability against both of these modes, κ and b_i must be chosen so that the conditions of operation lie in the cross-hatched area of the figure. Since, for the steady state, the last term in equation (7.29) is zero, it is seen that, if there is no B_z field exterior to the plasma,

$$B_\theta^2 = B_z^2 + 8\pi nkT,$$

and so, utilizing equation (7.36),

$$b_i^2 \equiv \frac{B_z^2}{B_\theta^2} = 1 - \frac{8\pi nkT}{B_\theta^2}.$$

However, by definition, e.g., equation (3.17), the ratio β of the kinetic pressure (or energy density) of the plasma to that of the confining field is, in the present case, given by

$$\beta = \frac{nkT}{B_\theta^2/8\pi} = \frac{8\pi nkT}{B_\theta^2}. \tag{7.37}$$

so that

$$b_i^2 = 1 - \beta.$$

7.36. It follows from Fig. 7.7 that stable pinch operation can be expected only when b_i^2 exceeds 0.5, so that β is less than 0.5; this means that at least half the pressure in the pinched discharge must be due to the contained axial field. Furthermore, the pinch ratio κ cannot exceed 5.0, since the maximum possible value of b_i^2 is 1.0; but then the plasma pressure will be zero—in other words, there is no plasma. At the other extreme, when $\beta = 0.5$, the pinch ratio can be no greater than 2.6 for stability. As a compromise, the operating conditions might be chosen so that $\beta = 0.2$, for which κ could have a maximum value of 3.8. The presence of some axial field outside the pinch, which is almost inevitable, decreases the maximum pinch ratio for stability and so a safer operating value would be 3.0. The initial value B_{z0} of the stabilizing axial field is then given by equation (7.34) as 0.06 I/r_0 gauss, where I is the discharge current in amperes and r_0 the radius of the tube in centimeters.

7.37. By combining equations (7.18) and (7.37), it is found that

$$I^2 = \frac{200NkT}{\beta},$$

where N, as before, is the total number of particles per unit length of the discharge. Comparison of this result with equation (7.16) shows that since β must be appreciably less than unity when an axial magnetic field is used to stabilize the pinch, the discharge current, for given values of N and T, is greater than for a nonstabilized pinch. For example, if β is 0.2, as suggested above, the current must be more than doubled.

7.38. The effect of leaving an appreciable axial flux outside the discharge when it contracts is to decrease the range of both b_i and κ over which stable operation may be expected. This subject is considered more fully in Chapter 13, and numerical values for the stable operating range as a function of the external and internal axial fields are given in Fig. 13.2.

INTERMIXING OF MAGNETIC FIELDS

7.39. So far, it has been assumed that the plasma possesses zero resistivity, and that, as a consequence, the currents j_z and j_θ, which account for the discontinuities in the B_θ and B_z fields, respectively, at the plasma sheath, do indeed flow in a layer of zero thickness. In any actual plasma, however, this condition is not satisfied; for any temperature short of infinity, the plasma will have a finite resistivity and the B_θ and B_z fields, even though they are discretely separated initially, will diffuse into each other. There are two important consequences of this interdiffusion (or intermixing) of the fields.

7.40. First, the new distribution of fields, which is now continuous across the current layer, is unstable. But even if the thickened current shell were not intrinsically unstable, the leakage of the axial field into the region between the pinched discharge and the outer conducting shell would narrow the permissible stable range for b_i and κ, as seen above, and would finally produce instability.

7.41. The second result of the field intermixing process is beneficial. It can be shown that the B_z flux, for example, while being conserved within the discharge tube as a whole, has a lower total energy after some has leaked out of the pinched plasma, assuming the radius of the latter to remain constant. The difference in energy appears as a resistance loss in the current sheath and serves to heat the plasma. This is, of course, ordinary ohmic heating; however, since the values of the field energy are known before and after diffusion has occurred, it is possible to determine the extent of heating which should result in the absence of losses. In particular, the estimate does not depend in any way on a knowledge of the time variation of the resistivity, although this must be taken into account in any calculation of the heating time.

7.42. The contribution made by the azimuthal (B_θ) field to the intermixing heating will depend upon the experimental conditions. If the voltage input connections to the discharge tube are "crowbarred," i.e., short-circuited (§7.118), during the heating phase, then all of the B_θ field energy before diffusion will be available for heating. In the limit, all the B_θ will disappear, with the flux lines vanishing into the axis, as the axial current j_z dies out. On the other hand, if the tube is supplied with constant current, an unlimited amount of energy may be added to the plasma by the continual replenishment of the B_θ field.

7.43. An estimate will now be made of the time over which the initial, separated B_θ and B_z field configuration may be expected to last by calculating the time constant for leakage of the axial field flux out of a cylindrical shell. For this purpose, it will be assumed that once the lines of force have escaped, they will disappear. This is not to be expected for a stabilized pinch, in which case the escaping magnetic lines may be expected to accumulate in the annular region outside the current sheath. The resulting error is minor, however, and the conclusions reached below are adequate for the present purpose.

7.44. Consider a cylindrical plasma sheath of unit length and radius r; end effects will be ignored, so that the following treatment involves no dependence on the z-coordinate. In mks units, the field energy W is given by

$$W = \frac{B^2}{2\mu_0}\,\pi r^2$$

so that

$$-\frac{dW}{dt} = -\frac{\pi r^2}{2\mu_0}\cdot\frac{\partial B^2}{\partial t}. \tag{7.38}$$

This rate of energy loss from the field may be set equal in magnitude to the resistive power P in the sheath, which is

$$P = I^2 R = I^2 \eta\,\frac{2\pi r}{\delta}, \tag{7.39}$$

where δ is the thickness of the sheath and η is its resistivity; hence, from equations (7.38) and (7.39),

$$I^2 = -\left(\frac{r\delta}{4\eta\mu_0}\right)\frac{\partial B^2}{\partial t}. \tag{7.40}$$

7.45. Since the plasma sheath is equivalent to a single-turn solenoid, the current and magnetic field are related by $I = B/\mu_0$, so that equation (7.40) becomes

$$B^2 = -\left(\frac{\mu_0 r\delta}{4\eta}\right)\frac{\partial B^2}{\partial t}.$$

Upon integration, this gives

$$B^2 = B_0^2 e^{-t/\tau_c}$$

where B_0 is the initial value of the field and the time constant τ_c is defined by

$$\tau_c \equiv \frac{\mu_0 r\delta}{4\eta}.$$

This is the time constant (or $1/e$-folding time) for B^2, i.e., for the energy; the field time constant, i.e., for B, is then $2\tau_c$; hence,

$$\tau = \frac{\mu_0 r\delta}{2\eta}.$$

In this equation the various quantities are in mks units; if r and δ are expressed in centimeters and η in ohm-cm, it becomes

$$\tau = 2 \times 10^{-9} \frac{r\delta}{\eta} \quad \text{sec.} \tag{7.41}$$

7.46. In a fully contracted pinch, r and δ are not very different, and so, as a rough approximation for qualitative purposes, r and δ may be taken to be equal; consequently the $1/e$-folding time for the decay of the magnetic field is given by

$$\tau \approx 2 \times 10^{-9} \frac{r^2}{\eta}.$$

It is evident, therefore, that the escape of the confined magnetic field from the plasma can be delayed if η is small and r is large. In a fully stabilized pinch r is related to the radius of the discharge tube r_0, because κ, i.e., r_0/r must be about 3 (§7.36). Hence, it will take longer for the axial field to diffuse out of the B_z-stabilized pinched discharge in a wide tube than in a narrow one.

7.47. Since the resistivity η of the plasma decreases with increasing temperature (cf. §4.88), it is evident that the rate of intermixing of the fields is greatest at low temperatures. For this reason, the attainment of a stable pinched discharge requires rapid initial heating and contraction of the plasma, so that the stage in which the axial magnetic field tends to escape most readily is made as short as possible. When the electron temperature has reached 100 ev, the value of η has decreased to about 10^{-5} ohm-cm, and if r is about 10 cm, the $1/e$-folding time of the confined field is about 0.02 sec. This is quite long in comparison with the expected lifetime of the pinched discharge.

LINEAR PINCHED DISCHARGE

GENERAL EXPERIMENTAL PROCEDURES

7.48. A considerable amount of work on the pinch phenomenon has been done with plasmas in linear discharge tubes, the electrodes being inserted directly into the plasma at the ends of the tube [10, 17-50]. Significant advantages of this technique are its simplicity and flexibility, as compared with that involving a toroidal discharge tube, permitting observations to be made under a wide variety of conditions. In addition, the analytical treatment of experimental data, e.g., space and time variations of the magnetic fields (cf. §6.107), is much simpler in a straight tube than in a torus.

7.49. It is realized that, with the linear discharge, impurities are probably introduced from the electrodes. But when a discharge tube of moderate length is used, the behavior of the plasma at some distance from the ends, in pulsed discharges of short duration, appears to be unaffected by the presence of the electrodes. Consequently, the results obtained, which do not differ essentially from those secured by inducing an electrical discharge in a plasma contained in a torus, have contributed greatly to an understanding of the constricted discharge.

7.50. The general experimental arrangement is shown in Fig. 7.8. The power source is a bank of capacitors which is discharged through a spark

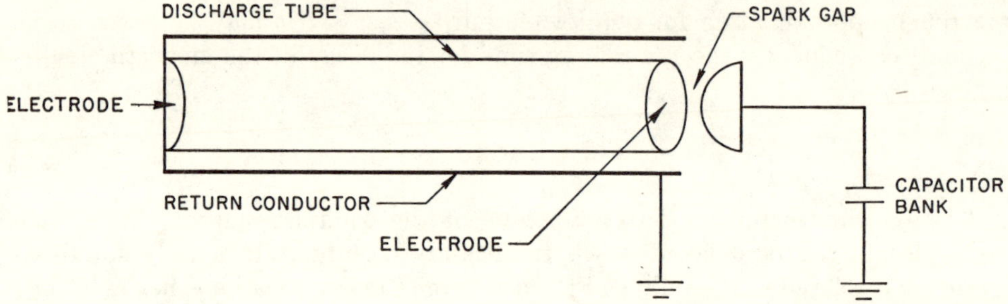

Fig. 7.8. Schematic arrangement for linear pinch experiments.

gap or an ignitron (or a number of spark gaps or ignitrons in parallel). The discharge passes through the tube between electrodes of stainless steel, copper, or aluminum. The return conductor is a metal cylinder which closely surrounds the discharge tube. In various experiments, the length of the tube has varied from a few inches to several feet, with diameters up to 2 feet or more. In most studies, the tube walls have been made of an insulating material, e.g., Pyrex, silica, or porcelain, but some work has also been done with metal tubes [36, 49]. Satisfactory pinches cannot be obtained in discharge tubes made of organic plastics. This is presumably due to evaporation and subsequent ionization of material from the walls at a rate greater than it is being removed by the azimuthal magnetic field of the discharge.

7.51. Much of the exploratory work on constricted discharges has been made with gases heavier than deuterium, e.g., xenon, argon, helium, and air. In the first place, the instability break-up of the pinched plasma occurs more slowly in these gases, because of the slower speed of sound at any given temperature. Furthermore, their spectra can be observed when that of deuterium can not, and so the pinches can be more easily studied by optical and photographic methods. It may be mentioned that, as far as the basic physical behavior of the discharge is concerned, the results obtained with heavier gases are essentially the same as with deuterium.

7.52. In the following section there will be described experimental results for "dynamic," i.e., unstabilized, pinches obtained with electric fields in the range from a few tens to a few hundreds of volts per centimeter. The current rise rates are generally 10^{10} to 10^{11} amp/sec and the initial gas pressures, usually of deuterium, are from about 2×10^{-3} to 1 mm of mercury. This operating range has been quite actively explored, in part, at least, because apparatus can be constructed with readily available components and useful diagnostic techniques are in many instances simple to perform.

7.53. Experiments have also been made on discharges with much lower electric fields ("slow" pinches) as well as with higher fields ("fast" and "superfast" pinches). The results obtained in these cases will be considered in later sections.

EXPERIMENTAL OBSERVATIONS

7.54. Apart from some minor differences, the observations made by various investigators are in general agreement, in spite of variations in the experimental conditions. The data presented here may therefore be regarded as fairly typical for linear discharges at moderate field strengths.

7.55. It appears that, if a satisfactory pinch is to form, the field strength (or the initial rate of increase of the discharge current, dI/dt) must be neither too small nor too large. If it is small, the rate at which the plasma contracts is of the same order as (or less than) the speed of sound. In this event, the amplitude of an instability may become large enough to destroy the inward coordinated motion of the plasma before contraction is complete. It has been found experimentally that a fairly rapid increase in the discharge current is necessary in order to obtain a well-defined pinch.

7.56. If, however, the rate at which the current is increased is too large, constriction of the discharge may be delayed. This seems to be a result of the pressure of the azimuthal magnetic field becoming greater than the kinetic pressure of the plasma before the latter is fully ionized. Neutral gas is then left behind when the plasma starts to contract and this gas ionizes more slowly than would have been the case had not the ions and electrons been removed by the discharge. In other words, the rapid increase in the magnetic field strength tends to sweep the charged particles inward before they have time to build up to a sufficiently high density in the plasma to cause rapid ionization of the neutral particles. The consequent delay in ionization means that the current is kept flowing near the wall until all the gas is ionized and then pinching will occur. If the current reaches its maximum value before ionization is essentially complete, the discharge may never entirely leave the wall [30].

7.57. Another necessary condition for a pinched discharge is that the gas pressure should not be too low. This requirement also appears to be associated with the presence of neutral gas at the walls of the discharge tube. At low densities, when the rate of ionization is relatively small, because of the infrequency of collisions between neutral and charged particles, the latter may be swept inward, by the increasing azimuthal magnetic field, before complete ionization can take place. The result is similar to that described above in connection with a very rapid increase in the discharge current. When the pressure is small, the rate at which instability break-up occurs may be greater than for a normal pinch, because of the smaller amount of mass present in the discharge. Thus, the discharge can become completely unstable even before all the gas is ionized.

7.58. If the conditions are appropriate for a normal pinched discharge, the plasma will move away from the walls of the tube when the current parallel to the axis has reached such a value that the pressure of the resulting azimuthal magnetic field exceeds the kinetic pressure of the particles in the plasma, i.e., $B_\theta^2/8\pi > nkT$. The dependence of the contraction time on the radius of the discharge tube, the density of the gas, and the initial rate of increase of the discharge current is observed to agree with equation (7.25) [10, 18, 37, 38, 51].

7.59. Following the first contraction of the plasma, there is usually a succession of pinches, e.g., two, three, or more, before break-up due to instability occurs. Thus, the formation of the first pinch, to which the calculations in §7.12 *et seq.* refer, is followed by a partial expansion of the plasma and a subsequent contraction, leading to a second pinch. Further expansion and con-

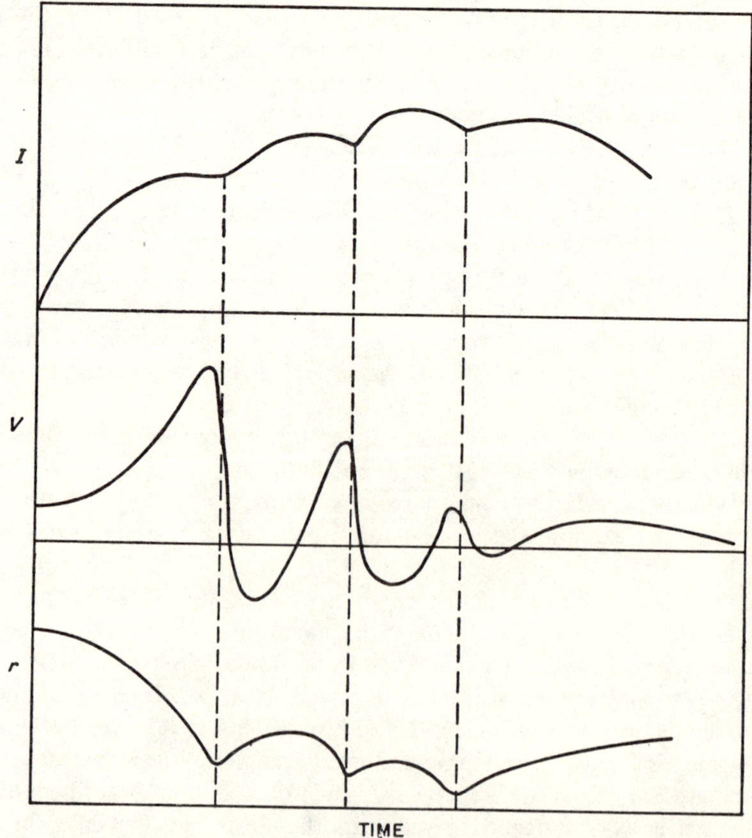

FIG. 7.9. Variations of current, voltage, and plasma column radius.

traction may lead to a third or a fourth pinch, the radius of the discharge generally decreasing with successive pinches.

7.60. The general explanation of this "bouncing" of the plasma that has been proposed independently by a number of workers is as follows. The rapid initial inward motion of the current sheath, when the discharge is first applied, causes a shock wave to travel radially through the plasma ahead of the sheath. When this reaches the center, a reflected shock is sent back in the outward direction, and this slows down and eventually reverses the inward motion of the plasma, so that expansion occurs. As the strength of the reflected shock wave dies down, the compressing effect of the radial self-magnetic field again becomes dominant and contraction takes place once more. As a result, another shock wave will move toward the axis, producing another reflected shock, leading to expansion of the discharge, followed by a further contraction and so on. The succession of contractions and expansions will continue until instabilities lead to complete break-up of the discharge.

7.61. The contractions and expansions of the plasma column produce characteristic oscillograms of voltage and current measured at the input terminals of the discharge tube. A typical set of V and I traces, as functions of time, together with a plot of the pinch radius, derived from V and I in the manner described below, is shown in Fig. 7.9. The results obtained can be interpreted as follows.

7.62. It will be recalled (§6.5) that the voltage on the tube terminals is given, in mks units, by

$$V = \frac{d\phi}{dt} + IR$$

$$= \frac{d}{dt}(LI) + IR. \tag{7.42}$$

For the conditions being considered here, the plasma resistance drops quickly to such a value that the IR term may be neglected [36], at least in a qualitative analysis of the system behavior, so that

$$V \approx L\frac{dI}{dt} + I\frac{dL}{dt}. \tag{7.43}$$

In any coaxial system, of which the linear pinched plasma is an example, the inductance in mks units is

$$L = \frac{\mu_0}{2\pi} l \ln \frac{r_0}{r} \tag{7.44}$$

where r is the pinch radius and r_0 is that of the outer return conductor; l is the tube length in meters. If l is expressed in centimeters, as is usually done, equation (7.44) becomes

$$L = 2 \times 10^{-9} l \ln \frac{r_0}{r} \tag{7.45}$$

in henrys. Differentiation with respect to time then gives

$$\frac{dL}{dt} = -2 \times 10^{-9} \frac{1}{r} \cdot \frac{dr}{dt}$$

and hence equation (7.43) can be written as

$$V = 2 \times 10^{-9} l \left[\frac{dI}{dt} \ln \frac{r_0}{r} - \frac{I}{r} \cdot \frac{dr}{dt} \right]. \tag{7.46}$$

7.63. When the discharge current just begins to increase, both the radial pinch velocity and the current are small; the second terms in equations (7.43) and (7.46), which are equivalent, can be neglected and the $L(dI/dt)$ term dominates the voltage. As soon as the pinch has contracted to some extent, the second term becomes important, since I and dr/dt have both become numerically larger with the decrease in the plasma radius. Finally, when the pinch radius is small, the voltage rises to a very high value, since r is small and dr/dt is fairly large and, of course, negative. As a result, the voltage across the discharge tube frequently becomes several times the capacitor bank voltage. If the latter is exceeded, dI/dt changes sign and there is a sharp, short dip in the current trace.

7.64. At the time the pinch reaches its minimum radius and begins to expand, the voltage takes an extremely sharp drop, often changing sign, in a small fraction of a microsecond. This is due to the fact that, whereas the second term in equation (7.46), equivalent to $I(dL/dt)$, is still numerically large, because r is small, it now has a negative sign, since dr/dt is positive when the pinch is expanding. The plasma is recompressed after its outward bounce and the whole process is repeated one or more times before instability sets in.

7.65. It has been observed that in some cases the current changes sharply when the pinch bounces while the voltage change is small; in other cases, on the other hand, the reverse occurs. The situation is determined by the circuitry between the capacitor bank and the discharge tube. This may be explained with the aid of Fig. 7.10, where C is the capacitor bank, L is the discharge inductance, and L_e represents the external parasitic inductance between the capacitors and the actual discharge. Over a short interval of time, such as that required for voltage reversal, an insignificant amount of charge leaves the capacitor bank, so that its voltage V_c remains essentially constant; hence,

$$V_c = L_e \frac{dI}{dt} + L \frac{dI}{dt} + I \frac{dL}{dt} = \text{constant} \tag{7.47}$$

$$= L_e \frac{dI}{dt} + V,$$

where V is the discharge voltage.

7.66. Near the time of maximum pinching, $I(dL/dt)$ is always large, as seen above, and if L_e is fairly small then $L(dI/dt)$ must be large and of opposite sign to $I(dL/dt)$ if V_c is to remain constant. This means that dI/dt must become large since, according to equation (7.45), L varies only as $\ln r_0/r$. The current trace will then show abrupt, deep dips at the time when pinching occurs. If L_e is sufficiently small, of course, V will not be greatly different from V_c and will stay almost constant. Hence, the changes in the voltage will be relatively small.

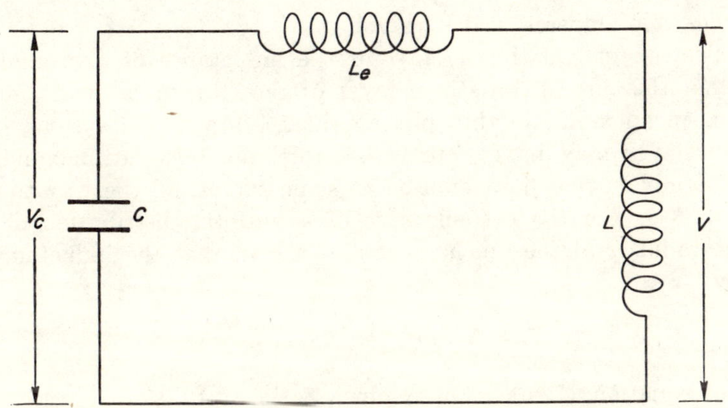

FIG. 7.10. Equivalent circuit diagram of capacitor bank and discharge.

7.67. If, however, L_e is much greater than L, the second term on the right of equation (7.47) can be neglected in comparison with the others, and then

$$V_c \approx L_e \frac{dI}{dt} + I \frac{dL}{dt}$$

$$= (V_c - V) + I \frac{dL}{dt}$$

so that

$$V \approx I \frac{dL}{dt}.$$

In these circumstances, the voltage makes wide excursions when pinching occurs, whereas I does not change very much. In the majority of experimental arrangements, L_e is of the same order of magnitude as the initial inductance of the discharge, and then the varations in V and I are similar to those depicted in Fig. 7.9.

7.68. The pinch radius at any instant can be derived from the observed values of I and V. If the IR term is neglected, as stated above, equation (7.43) can be written as

$$V \approx \frac{d}{dt} (LI),$$

so that the discharge inductance at any time t, when the current is I, is given by

$$L(t) = \frac{1}{I} \int_0^t V(t') \, dt'.$$

By making use of equation (7.45), it follows that

$$r(t) = r_0 \exp\left[-\frac{10^9}{2lI} \int_0^t V(t') \, dt' \right],$$

with l in cm, I in amperes, and V in volts.

7.69. The use of equation (7.45) for the inductance of a coaxial system assumes that the current flows in a layer of zero thickness in the inner conductor, i.e., in an infinitely thin plasma sheath. In a real system, however, the current layer may have appreciable thickness, and at maximum compression there is current flow, at least to some extent, all the way to the axis (see Fig. 7.11). For the extreme case of a uniform distribution of current within the radius r of the pinch, it can be shown that the inductance is expressed by

$$L = 5 \times 10^{-10} l \left(1 + 4 \ln \frac{r_0}{r} \right)$$

in place of equation (7.45). For a pinch ratio of 10, the inductance based on this expression is about 10 per cent more than for the thin-sheath value.

7.70. By using magnetic probes to measure the strength of the azimuthal magnetic field as a function of the radial distance from the discharge tube axis, at a given time after the initiation of the discharge, it is possible to derive the distribution of the plasma current density (cf. §6.111 *et seq.*) [29, 38, 52]. This distribution in space and time has been represented in the form of three-dimensional figures and contour diagrams. However, the general conclusions can be indicated more simply by the curves in Fig. 7.11; they show the current density distribution at four successive times during the first contraction of the plasma. The dotted line indicates the position of the

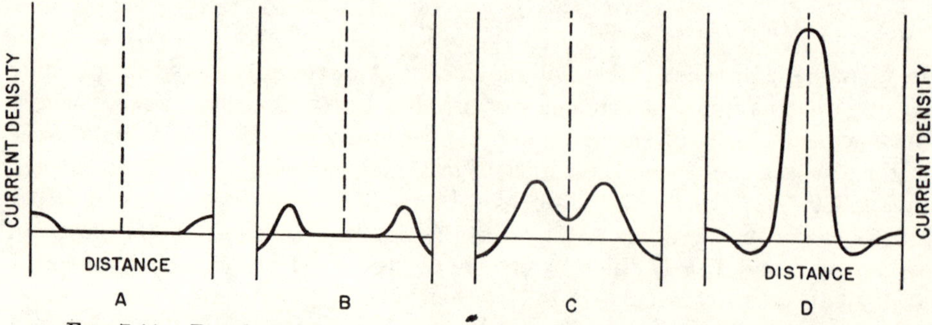

FIG. 7.11. Development of current density distribution in pinched discharge.

axis of the discharge tube.   It is seen from Fig. 7.11A that immediately after breakdown a current sheath forms near the walls, due to the skin effect of the plasma.   The sheath moves inward (Fig. 7.11B) as the current increases, and eventually, when the discharge is fully contracted, nearly all the current is carried by a column of plasma at the axis of the tube (Fig. 7.11D), in general, but not exact, accord with the Bennett distribution expressed by equation (7.10).

7.71.   An examination of Fig. 7.11C shows that, even while the plasma is only partially constricted, there is a current, near the axis, within the external current sheath.   It is evident, therefore, that the simple theory of §7.13 cannot be quantitatively correct.   Furthermore, it will be observed from Figs. 7.11B, C, and D that a region of negative current, i.e., flowing in the direction opposite to that of the main discharge current, exists between the positive sheath and the wall of the tube.   Some possible explanations of this negative current have been proposed, but the situation is not clear.

FAST LINEAR PINCHES

7.72.   Attempts have been made to produce magnetic-piston type heating of plasmas by increasing the applied electric fields to the order of 10^3 volts/cm. The purpose is to produce a fast pinch in which advantage is taken of the relationship between the rate of rise of the current, the gas density, and the contraction velocity (§7.20) to impart large radial kinetic energies to the ions. It is hoped that thermalization will occur near the axis of the tube before the pinched discharge is destroyed by instabilities.

7.73.   In general the construction of the equipment for the production of fast pinches has made necessary the use of more advanced engineering techniques than have been employed on discharges with lower electric fields [45, 53-58].   In this respect, reduction of parasitic circuit inductance has been the primary problem.   Many components are involved; capacitors, switches, transmission lines, and circuit connections all contribute series inductance between the energy storage and the discharge.   As a result, energy is diverted from the discharge and a limit is set on the rate of current rise and on the attainable peak current.

7.74.   The simplest and most commonly used procedure for reducing inductance is to divide the capacitor bank into a large number of parallel units, each connected to the load, e.g., the discharge tube, through a separate switch and transmission line, usually coaxial cable.   In this way, the total source inductance is decreased to $1/n$ of the inductance of a single unit, where n is the number of such equal units in parallel.   In a particularly successful application of this technique, a 100-microfarad capacitor bank was subdivided into 200 sections, each with its own spark-gap switch and cable.   The system inductance was 3×10^{-3} microhenry [56].

7.75.   In addition to the capacitor bank subdivision procedure, much work has been done, especially by industrial organizations, with the objective of reducing the inductances of individual components. For example, capacitors with configurations having very low inductances have become available. Coaxial transmission lines with very close spacings are now being manufactured; their inductances are lower, by a factor of about 5, than the common radiofrequency types previously used. Much study has also gone into decreasing the insulation thickness in transmission lines and switch assemblies. In this connection, successful results have been obtained by the use of insulating gases, e.g., sulfur hexafluoride.

Fig. 7.12. Low-inductance capacitor bank and parallel plate transmission lines for Columbus-II. (Courtesy of Los Alamos Scientific Laboratory)

7.76. The experimental data obtained with two particular fast-pinch devices are of interest; although there are some differences in construction, their behavior is quite similar. One of these utilizes the 200-section subdivided capacitor bank described above and achieves an initial current rise rate of about 2×10^{12} amp/sec in the pinched discharge [37]. The other, while employing less subdivision in the bank, uses special capacitors of low-inductance design together with a circular parallel plate transmission line to carry current from the capacitors, arranged in a circular array, to the discharge tube at the center (Fig. 7.12) [28]. The rate of increase of current is approximately 10^{12} amp/sec.

7.77. The principal diagnostic measurements made with these pinch systems were determinations of voltage and current, as well as of neutron emission. The dependence of the pinch times on gas density was in general agreement with theory (§7.17), although the experimental values were somewhat larger than predicted. An interesting observation made with both the devices referred to here is that, above a certain value of the applied electric field, the voltage and current traces did not give clear evidence of pinching.

7.78. The only indication that was available at the time concerning the temperatures reached in these experiments was based on the rate of neutron emission (§6.140) and the data were carefully analyzed to determine whether thermonuclear processes were indeed taking place. The evidence, however, was obscure. On the one hand an angular anisotropy was found in the neutron energy spectrum, this being characteristic of what is thought to be an instability process. At the same time, however, the neutron yield was not very sensitive to the presence of an axial (stabilizing) magnetic field within the pinched discharge, a result that is at variance with the idea that the neutrons owe their origin to plasma instability. The problem of neutron emission in pinched discharges will be discussed more fully later in this chapter.

THE SUPERFAST PINCH

7.79. As pointed out earlier (§7.20), the use of the fast dynamic pinch to attain temperatures in the region of even 10 kev presents serious problems, since electric fields of about 10^5 volts/cm and dI/dt values of 10^{13} to 10^{14} amp/sec would be required. Obviously, any conceivable experiments under these conditions can only be done with very short discharge tubes; otherwise the total applied voltage would be extremely high [59-62]. In order that end effects in the short tube may not become troublesome, the diameter should also be scaled down. The compression time can then be short enough that contaminants are not able to flow in very far from the electrodes. However, a very short compression time introduces a new complication: no stored energy can be utilized in the pinch if it is physically farther away than the pinch time multiplied by the velocity of light in the transmission line dielectric.

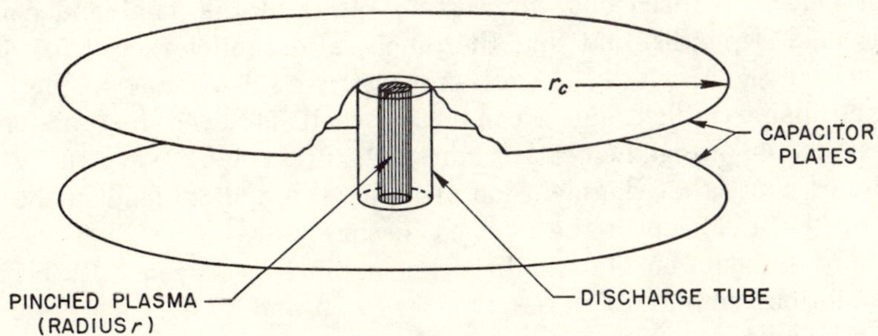

Fig. 7.13. Schematic arrangement for superfast pinch.

7.80. Probably the best that can be done in the conceptual design of a very fast (or *superfast*) pinch is shown in Fig. 7.13, in which the discharge tube is located at the center of two circular capacitor plates [61]. The only inductance in the system is that of the pinch; all the return current is displacement current between the plates. Since the pinch travels to its minimum radius in the same time that a wave travels from the center to the periphery of the largest useful plate and back again, it follows that

$$\frac{2r_c}{c} \leqslant \frac{r}{\bar{v}},$$

where r and r_c are the radii of the pinch and of the capacitor plate, respectively, and $\bar{v}$ is the average pinch velocity; c is the velocity of light. For $\bar{v}$ equal to 10^8 cm/sec, for example, this would require that

$$r_c \leqslant 150r.$$

7.81. The foregoing considerations have led to the construction of a pinch apparatus with a number of unusual parameters. The energy storage capacitor consists of a pair of concentric steel cylinders separated by Mylar insulation. The discharge tube is made of sapphire; it has a radius of 6 mm and a length of 4 cm and is inserted directly into the outer conducting cylinder of the capacitor. The 0.003-microfarad capacitor is charged to about 300 kilovolts, the stored energy being roughly 150 joules; the system when fired rings at a frequency of 18 to 20 megacycles/sec. The pinch discharge occurs through a highly ionized plasma produced by means of an electromagnetic shock tube.

7.82. Preliminary observations with the superfast pinch device have been made with a streak camera and current-measuring loops. Fairly clear pictures have been obtained of a rapidly contracting plasma; the average velocity observed was 3×10^7 cm/sec and the maximum was believed to be at least three times as large. The formation of a current sheath was indicated by probe signals [60, 62].

STABILIZED LINEAR PINCHED DISCHARGE*

7.83. It was seen in §7.36 *et seq.* that the so-called fully stabilized pinched discharge can be realized only for certain ratios of the axial and azimuthal magnetic field strengths, and that the pinch ratio cannot exceed 5.0 and will generally be less. Much of the work on constricted discharges in the presence of an axial magnetic field and a conducting shell has been done under conditions that do not altogether satisfy these requirements. Nevertheless, there is definite evidence that the inclusion of an axial magnetic field in the plasma does delay the break-up of the pinched discharge.

7.84. Experiments on the stabilized linear pinch have generally been made under conditions similar to those given in §7.50 and §7.51. The return con-

---

* Many of the references given in §7.48 include some work on the stabilized linear pinch.

ductor surrounding the cylindrical discharge tube serves as the conducting shell, and the axial magnetic field is provided by passage of a separate current through an insulated solenoid outside the conductor. Field strengths ranging from several hundred to a few thousand gauss have been commonly used. In making observations, the axial magnetic field is first established and when this reaches a maximum the main discharge is triggered.

7.85. Magnetic probe measurements made on B_z pinches show good agreement with the elementary treatment of sheath motion given in §7.31. The principal departure from the idealized assumptions is that the current sheath has appreciable thickness. But since it is, in many instances, thin compared with the tube radius, equations (7.33) and (7.34) still predict the radial behavior of the pinch quite well. It is usually observed, however, that even the stabilized pinch tends to oscillate slightly in radius after it has reached its maximum compression [41]. This is a direct consequence of the inertial term in equation (7.32).

7.86. By treating the current sheath as a mass moving back and forth between two springs, represented by the internal and external magnetic fields, it has been shown that the oscillation frequency, in radians/sec, is given by

$$\omega^2 = \frac{B^2}{4\pi r \rho_s}, \tag{7.48}$$

where r is the equilibrium pinch radius in centimeters, B is the magnitude in gauss of either B_θ or B_z, since they are equal, as pointed out in §7.32, when the particle pressure and inertial terms are small; ρ_s is the surface density of the sheath in g/cm². This oscillation, which is usually discernible, even though small, in stabilized pinches is quite different in origin from the bouncing observed in dynamic pinches. In the stabilized pinch, shocks are usually not present, since the radial velocities of the plasma particles do not reach large values. As indicated above, the current sheath simply behaves like a mass held between two springs.

7.87. The oscillations just described have proved useful as a diagnostic tool. Since ω, r, and B can be determined from experimental observations, by means of magnetic probes, it is possible to calculate ρ_s with the aid of equation (7.48). The results show to what extent the contracting sheath has swept the plasma particles inward. As a general rule, the sweeping-up process appears to be complete.

OBSERVATIONS OF PINCH STABILITY

7.88. In the earliest experiments on the B_z pinch it was established that instabilities of the $m = 0$ (sausage-type) mode were suppressed by the inclusion of quite weak axial fields. This is shown, for example, by the series of pictures in Fig. 6.21. It was found, however, that the pinched discharge was still unstable, presumably owing to the $m = 1$ (kink) mode. The existence of the kink distortion was, in fact, demonstrated by image-converter photographs of the pinch made through perforations in the return conductor of a discharge tube [46].

7.89. The kink instability has also been studied with the magnetic probe described in §6.110, which consists of a loop of wire wound in the azimuthal direction around the insulating wall of the discharge tube and inside the return conductor [46, 63-65]. The integrated voltage signal from this loop is proportional to the change in total B_z flux threading it during the progress of the discharge. This signal is normally small when the plasma is contracting, since the inflowing of the axial field lines originally between the loop and the return conductor results in only a slight increase in the total flux through the loop. However, at the time of the pinch instability, a large step is observed in the voltage signal, corresponding to a much greater total flux than could be accounted for by a normal compression. A reasonable explanation of this anomaly is that the pinch is distorted into a helix, with the resulting generation of a new B_z flux by the axial current through the pinch. In order to produce a signal in the loop, these new B_z lines must close on themselves by passing outside the loop between the insulating tube and the return conductor. The magnetic field system at the time of instability would then be of the form represented in Fig. 7.14.

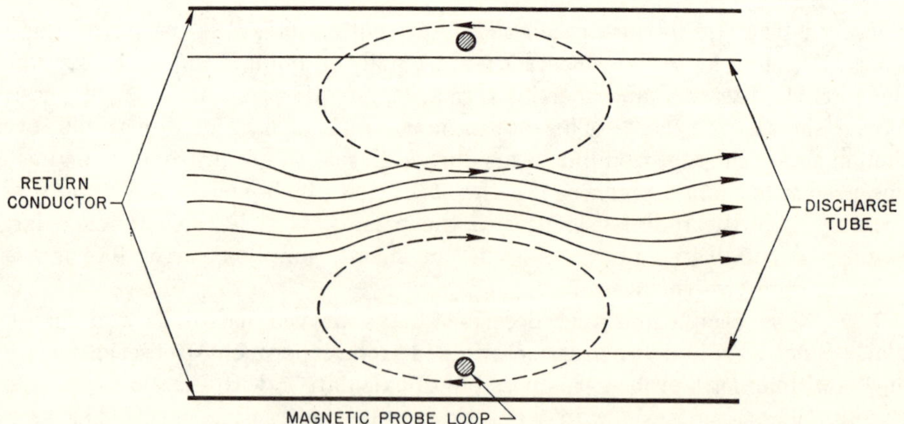

FIG. 7.14.   Magnetic field configuration accompanying kink instability.

7.90. The identification of the instability as a helical deformation has been further strengthened by observations made with a pair of small magnetic probe loops, located just inside the return conductor at opposite ends of a tube diameter. These probes give output voltages proportional to dB_θ/dt. From the simultaneous signals, displayed on a multibeam oscilloscope, from these two probes and the azimuthal loop, it was clear that the increase in B_z shown by the latter coincided with a local jump of the plasma to one side of the tube, as would be expected with a helical $m = 1$ deformation. In agreement with anticipation, the sign of the helix is such as to increase the total B_z flux [65].

7.91. As the axial field was increased to the point where stability against the $m = 1$ mode should have been expected theoretically, it was hoped that the abrupt step in the large-loop signal would suddenly disappear. What was actually observed was a continuous decrease in the size of the step as the axial, stabilizing field was increased, but the irregularity did not disappear completely. The pair of small opposed B_θ probes similarly show less and less evidence of a sidewise motion of the pinched plasma. But even at high values of the axial field, there always remains an uncorrelated fluttering of these small-loop signals.

7.92. The residual irregularity in the magnetic probe signals, though the $m = 1$ instability mode has apparently been suppressed, has become a characteristic of the B_z-stabilized pinch. It is a general experience that oscillograph signals from magnetic probes placed at any depth in the plasma are strikingly reproducible up to a certain time, after which the traces display

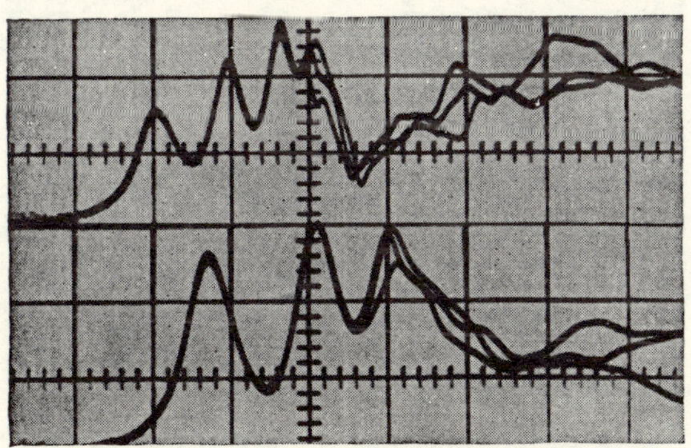

FIG. 7.15. Magnetic probe signals showing development of "fluttering" instability.

the variable raggedness. This is apparent, for example, in Fig. 7.15, which is a photograph of two sets of three superposed signals from a magnetic probe in a pinched discharge in deuterium [41].

7.93. Since the experimental conditions are such that both $m = 0$ and $m = 1$ instabilities do not occur, it seems possible that the irregularities in the probe signals are caused by small local instabilities rather than by gross movements of the discharge. That such is the case was shown in an experiment using a set of four parallel magnetic probe coils, 1.5 cm apart, oriented so as to couple the azimuthal field. The narrow tube containing the coils was placed at such a distance (3.8 cm) from the axis of the discharge tube that the probe was just outside the fully pinched plasma (equilibrium radius of

3 cm). If a large-scale motion of the discharge occurred, then a response should have been observed from all of the coils, whereas a local instability would be indicated by signals from one or two only.

7.94. When the axial field strength was too low (1000 gauss) for $m = 1$ stabilization, the results indicated that a gross instability did, in fact, develop, but when the field strength was increased (2000 gauss), simultaneous fluctuations were generally restricted to two adjacent coils. Furthermore, a search for correlated fluctuations on opposite sides of the discharge column indicated that no such correlation exists. It was concluded that, although the gross instabilities of a pinched discharge have probably been eliminated, local surface instabilities, which represent a kind of turbulent motion of the plasma, still persist. Because of the manner of its detection, the phenomenon is referred to as "fluttering" instability.

7.95. When experiments were made with tubes of different widths it was found that the time of onset of the fluttering instability increased with the tube radius for the same initial value of the axial magnetic field. Further investigation showed, however, that the development of the instability is determined only by B^2/ρ_s, where B and ρ_s have the same significance as in equation (7.48), and is otherwise independent of tube radius.

THE REVERSED AXIAL FIELD PINCH

7.96. The persistence of instability in the so-called fully stabilized B_z pinch, which is based on the postulate of an infinitely thin sheath, together with the oscillations which can be explained in terms of a sheath of appreciable mass, indicated that study should be made of the stability properties of a slightly interdiffused B_z-B_θ boundary layer. As a result of theoretical investigations of the diffuse current layer or "thick-sheath" pinch, as it is sometimes called, it has been concluded that it is indeed unstable, and special means have been proposed to correct the instability.

7.97. A particularly simple criterion for the stability of an arbitrary radial distribution of current has been derived. In a linear discharge, in which all properties are independent of the θ and z coordinates, it is found that a necessary, although not sufficient, condition for stability is that the inequality

$$\frac{r}{4}\left(\frac{1}{\mu} \cdot \frac{d\mu}{dr}\right)^2 + \frac{8\pi}{B_z{}^2} \cdot \frac{dp}{dr} > 0$$

must be satisfied; here p is the plasma pressure at a radial distance r and μ, defined by

$$\mu \equiv \frac{B_\theta}{rB_z},$$

is a measure of pitch of the helix which a local magnetic field line follows. More precisely, μ may be regarded as the number of rotations of a field line per unit increase in the z coordinate [66].

7.98. The quantity $(1/\mu)\,(d\mu/dr)$ is the rate of change of pitch angle with the radial distance and is called the "shear" of the field. Since this appears squared in the stability condition, it is evident that a large shear of either sign should be a stabilizing influence. It remains then for the second term, containing the pressure gradient dp/dr, to represent the cause of instability. The plasma pressure gradient in any conceivable pinched discharge system will certainly be negative, i.e., from the current sheath outward, since the heating of the plasma will inevitably result in a higher interior pressure than at the container walls. This situation should contribute toward the instability of the discharge, and for large enough negative gradients the necessary stability condition will be violated.

7.99. The limited stability of the conventional B_z-stabilized pinches produced in the laboratory suggests that the inequality given above may not be satisfied. Clearly, what is needed is an increase in the magnetic field shear so as to overcome the pressure gradient effect. One proposal for achieving this result is to apply an axial magnetic field outside the current sheath in a direction opposite to that of the stabilizing axial field trapped by the plasma. In passing through the boundary layer, the magnetic vector would then rotate more than the 90° that is the angle of rotation in an ideal, conventional stabilized pinch in which there is no axial magnetic field outside the plasma.

7.100. An attempt to test the stability of the "reversed-B_z" pinch, as it is called, was made by switching on an axial field, in a direction opposite to that of the conventional stabilizing field, immediately after the latter had been trapped by the formation of the current sheath in a pinch discharge [67]. The results in deuterium were inconclusive because the increased magnetic field shear was accompanied by an increase in the negative pressure gradient, thus leaving the inequality unsatisfied. The reason for this will be apparent from the second equation in §6.112. The increase in $\partial B_z/\partial r$ resulting from the presence of the external axial field means that the j_θ current flowing in the boundary layer will be greater than in the absence of this field. Hence, the plasma pressure in the sheath due to ohmic heating is also increased, thereby making dp/dr outside the shell more negative. In the deuterium system, therefore, the reversed-B_z field did not remove the instability.

7.101. A similar experiment with argon, at the same mass density as the deuterium in the work referred to above, led to ample satisfaction of the necessary stability criterion. The reason for this difference in behavior is that the argon, being incompletely stripped in this case, lost much of its energy as optical radiation, so that the pressure in the current sheath was reduced to a relatively low value. Significantly, however, there was no observed increase in the stability of the discharge. The instability indicated by the "fluttering" of the magnetic probe traces persisted in this case.

7.102. The inequality given in §7.95 is a necessary condition for stability, but it is evidently not sufficient. It is possible that other necessary conditions may exist which have not yet been elucidated and which were not satisfied in the experiment. However, it is conceivable that the factors responsible for the persistent instability, observed with and without the external, reversed axial field, may be outside the theoretical considerations upon which the stability inequality was based. In this connection, some attention is being given to problems of hydromagnetic turbulence and viscosity effects.

PLASMA PRESSURE DISTRIBUTIONS

7.103. From a knowledge of the radial distributions of the azimuthal (B_θ) and axial (B_z) magnetic fields at the same instant, the plasma pressure distribution in a constricted discharge can be determined by the method described in §6.111 *et seq.* [41, 44, 46]. The results shown in Fig. 7.16A were obtained with an apparatus (Columbus S-4) which is fairly typical of equipment used in stabilized pinch studies [41]. The discharge tube was of porcelain (Mullite) with an internal radius of 6.4 cm and a spacing of 60 cm between the electrodes. The discharge current rose to a peak value of about 2×10^5 amp in 6 microsec. The initial value of the stabilizing axial (B_z) field was 1800 gauss. The data in Fig. 7.16 apply to a time of 2.6 microsec after the initiation of the discharge when the plasma velocity was essentially constant and the correction for inertial forces, mentioned in §6.121, is not required.

7.104. The B_z and B_θ radial distribution curves show that the axial and azimuthal currents j_z and j_θ are flowing almost entirely in a hollow shell. For example, since j_θ is determined by $\partial B_z / \partial r_0$ (§6.112), it is clear from the shape of the B_z curve in Fig. 7.16A that j_θ has a peak value at a radius of about 2 cm, as seen in Fig. 7.16B. The axial current density j_z, as derived from the B_θ distribution, has a maximum at roughly the same radial distance. In addition, the plasma pressure distribution, calculated by means of equation (6.49), also peaks quite sharply at a radius of slightly more than 2 cm, as shown in Fig. 7.16A.

7.105. The close agreement of the pressure distribution, both in shape and position, with the current distribution suggests strongly that ohmic heating is responsible for the observed pressure increase. This view is supported by the following considerations. The maximum pressure in the plasma at 2.6 microsec, i.e., approximately 3×10^6 dynes/cm^2, is about 3×10^4 times as large as the initial gas pressure of 8×10^{-2} mm of mercury. Since the plasma has been pinched from an initial radius of 6.4 cm to a little over 2 cm, the volume has been decreased by a factor of no more than 10 or so. It is apparent, therefore, that adiabatic compression has contributed very little to the heating, and the pressure (and temperature) increase must be due to another cause, namely, ohmic heating.

7.106. In order to determine the plasma temperature from the observed pressure, a knowledge of the gas density distribution is required (§6.120). Since this cannot be measured directly, reliance must be placed on estimates based on indirect evidence.   The results suggest that the snowplow model, described in §7.14, is applicable; in other words, it is postulated that all the particles encountered by a thin current sheath in its inward motion are retained. The particle density at a radius of 2 cm would thus be somewhat more than 10 times the initial ionized gas density of 5.5×10^{15} particles/cm³.   On this basis,

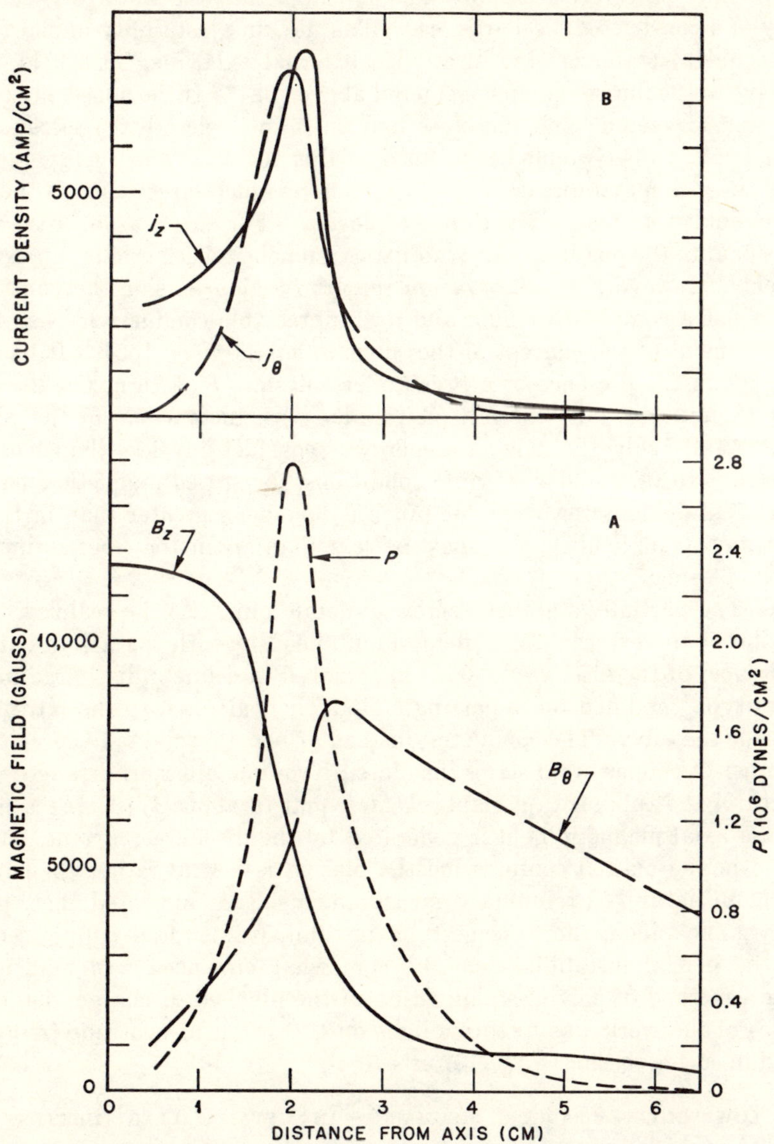

Fig. 7.16. Magnetic field, current density, and pressure distributions in pinched discharge.

the sheath temperature is estimated to be in the neighborhood of 30 ev, i.e., slightly more than 3×10^5 °K. This is in good agreement with the value derived from the plasma resistivity by the method outlined in §6.117.

THE SCREW DYNAMIC (LINEAR) PINCH

7.107.   Early theoretical studies had indicated that the sausage-type of instability of a constricted discharge, as well as the kink instability of short wave length, could be stabilized by an applied external axial magnetic field, as an alternative to the internal field considered above [68].   If the space outside the plasma were a vacuum and therefore had an infinite electrical resistance, the external applied field would be uniform in this region.   The energy required to maintain such a stabilizing field would then be much larger than if the axial magnetic field were trapped within the plasma.   For this reason, little attention was paid to the possibility of stabilizing a pinched discharge by an external axial field.   However, the observation that the region outside the constricted plasma is not a complete vacuum and has appreciable conductivity has led to the development of the concept of the *screw dynamic pinch* [69].   It has been shown that, in the presence of a pressureless plasma, B_z external to the pinch will not be uniform, but will fall off rapidly with distance from the plasma surface.   As a result, the magnetic energy is considerably less than had been anticipated.   In fact, under suitable conditions, the stored magnetic energy in the axial field of the screw dynamic pinch is not much greater than in the absence of any axial field, and it may be less than when the field is included within the plasma.

7.108.   The partially stabilized screw dynamic pinch can be realized by the application of the external axial field simultaneously with the azimuthal self-magnetic field of the discharge.   This is achieved in a linear discharge tube of the usual type, modified by imparting a slight helical pitch to the return current outside the tube.   The conductor, instead of being a solid cylinder, consists of a number of copper braid strips, insulated from one another, arranged along the length of the tube, with a slant $rd\theta/dz$ equal to about 1/10.   As a consequence, an axial magnetic field is generated by the discharge current, and the total field acting on and compressing the plasma is helical rather than purely azimuthal in nature.   Preliminary measurements have indicated that, in the screw dynamic pinch, the sausage-type instability is indeed suppressed, although the helical instability may be somewhat enhanced.   In addition to studying a method of partial stabilization of the pinched discharge, one of the objectives of the work was to throw light on the origin of neutrons frequently produced in such a discharge (§7.137 *et seq.*).

LOW VOLTAGE-GRADIENT DISCHARGES IN METAL-WALLED TUBES

7.109.   The effectiveness of eddy currents in the tube walls on the confinement of partially stabilized linear pinched discharges has been studied by means of fairly thick metal-walled tubes [49]. Such tubes are frequently made of a number of metal rings separated from one another by insulating gaskets. In order to prevent breakdown (or arcing) across the narrow insulator gaps, it is necessary to operate the discharge tube at appreciably lower voltage gradients than are generally employed in other pinch work. In the observations described below, the tube consisted of aluminum rings, 5 cm long and of 1.6-cm wall thickness, with 0.2-cm rings of polyethylene between them; the total length of the tube was 600 cm and its internal radius was 7.5 cm. The voltage gradients were in the region of 1 to 5 volts/cm, and the deuterium gas pressure ranged from 10^{-4} to 10^{-2} mm of mercury. An insulated solenoidal winding around the tube provided axial magnetic fields up to 500 gauss; this was not sufficient to produce a "fully stabilized" pinch.

7.110. When the current through the tube exceeded 10^4 amp, the constricted discharge developed the familiar type of kink instability, such as has been observed under other conditions. However, it was not possible to determine from streak camera photographs whether the discharge consisted of a tight pinch wound into an irregular helix or of a somewhat wider pinch with rapidly moving disturbances of large amplitude and short wave length. A distinction between the two possibilities could have been made, in principle, by means of magnetic probes, but owing to the lack of good reproducibility of the observations, it would have been necessary to make simultaneous measurements in a single discharge with a large number of probes. Apart from technical difficulties, it was felt that the introduction of several probes might so seriously perturb the discharge that the results would not be significant.

7.111. Studies with single magnetic probes showed that when the discharge current fell below 10^4 amp or if the current was initially less than this value, the plasma developed a regular oscillation. This is considered to be a form of hydromagnetic (or Alfvén) wave, i.e., one with displacements perpendicular to its direction of propagation (§4.115 *et seq.*). The frequency was found to be inversely proportional to the square root of the gas pressure (or density), and the product of the frequency and the measured wave length, which gives the velocity of propagation, was constant at 3×10^5 cm/sec.

7.112. From observations of the signals given by magnetic probes, the oscillations have been shown to be related to a rotating, helical disturbance of the plasma. The pitch of the helix was approximately half the diameter of the metal-walled discharge tube. The sense was such as to produce a magnetic field which enhanced the applied axial field (cf. §7.90), and the rotation of the helix caused it to move forward in the direction of the discharge current through the plasma.

7.113. It is of interest to mention that a regular helical oscillation similar to that just described has been observed under equivalent experimental conditions in a metal-walled tube of "race track" form, i.e., straight lengths connected by curved corner pieces to make an endless tube. The gas in the tube served as the secondary of an air-core transformer, the current being induced in the gas by a discharge through an external brass tube acting as the primary conductor (cf. §7.115). The results showed that the observed phenomena were not due to electrode effects. This conclusion is supported by the fact that, in the linear discharge tube, no spectroscopic lines were detected at the center of the discharge from either the tantalum anode or the nickel trigger electrode (cathode), although impurity lines of the wall materials, i.e., aluminum and carbon, were quite prominent [49].

TOROIDAL PINCHED DISCHARGES

EXPERIMENTAL TECHNIQUES

7.114. Although most of the earliest of the more recent experimental studies of the pinch phenomenon were made in endless tubes of either toroidal or race-track form [70-73], a great deal of the subsequent work was done with straight tubes, as described in the preceding sections. Nevertheless, from the beginning of pinch-effect research, it has been realized that a topologically toroidal discharge is the only kind which completely avoids contact with its surroundings [35, 43, 46, 74-86]. The incentive for working with a configuration which is more difficult mechanically, electrically, and analytically than the linear pinch

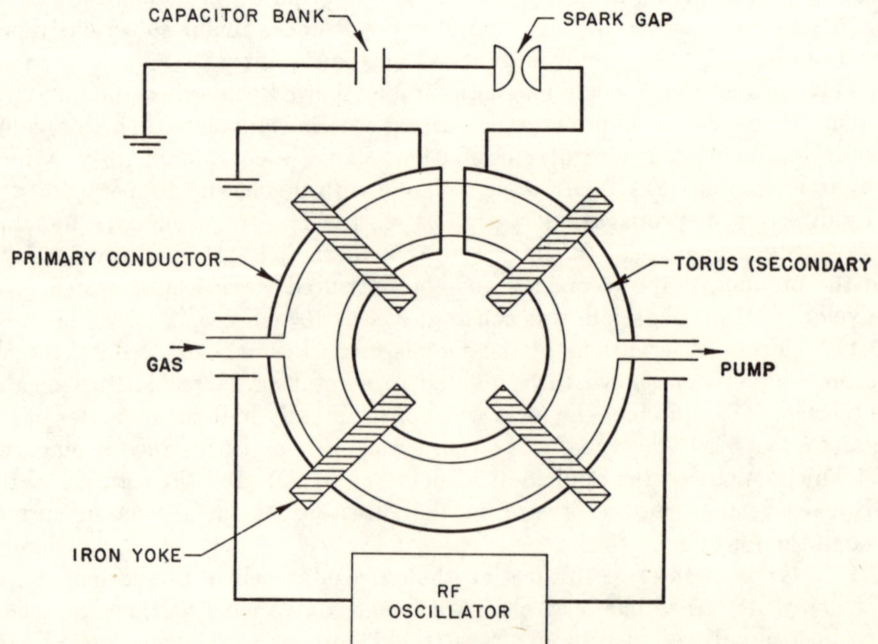

FIG. 7.17. Schematic arrangement for toroidal pinch experiments.

is the certainty that, as soon as sufficiently high temperatures are attained, electrode cooling effects will become intolerable.

7.115. Since a direct connection to the plasma circuit in a torus does not exist, the necessary electric field in the gas must be induced. This is accomplished by threading a time-varying magnetic field through the toroidal ring in some way or other. A particularly efficient procedure, from the viewpoint of circuit coupling, is to fit a conductor, usually a metallic shell or a closely spaced array of wires in parallel, over the outside of the toroidal vacuum chamber. Current is fed into this primary circuit from a high-voltage capacitor bank at a conveniently placed interruption. The system then becomes an air-core transformer with a one-turn primary and a one-turn secondary. In order to improve the coupling between primary and secondary circuits, as explained below, it is the common practice to include a number of iron yokes, as shown in Fig. 7.17 (see also Fig. 7.19), so that the arrangement is an iron-cored transformer. The radiofrequency oscillator indicated in the figure is used to produce preionization of the gas before passage of the main discharge.

7.116. An examination of the equivalent circuit of a toroidal pinch system has shown it to resemble that of a linear coaxial pinch of the same length, except that a parasitic inductance L_1, equal to the inductance of the toroidal primary, is shunted across the terminals, thereby diverting a fraction of the current. If I_1 is the primary (or input) current and I_2 is the secondary (or pinch) current, then

$$\frac{I_1}{I_2} = 1 + \frac{L_2 - L_1}{L_1} = 1 + \frac{L_c}{L_1}$$

for a single-turn primary, where L_2 is the inductance of the secondary and L_c, the coaxial inductance, is equal to $L_2 - L_1$. For toroidal aspect ratios, i.e., ratio of major to minor radius, in common use, L_1 is about three times L_c; the coupling coefficient, i.e., I_2/I_1, for an air-core transformer will then be 0.75. By passing an iron core through the primary circuit, however, L_1 may be made very large without affecting L_c. The coupling efficiency then approaches unity, and the pinch current is essentially equal to the primary (input) current.

7.117. When the primary consists of a single turn or its equivalent, e.g., several turns in parallel, around the torus, it is a property of this geometry that the voltage on the input terminals is equal to that across the secondary, regardless of the coupling efficiency. In some cases it is required to apply a voltage to the secondary that is less than across the primary, i.e., the capacitor voltage (§7.129). Alternatively, it may be desirable to increase the pinch current with respect to the primary current. Both these ends are achieved by the use of a multiturn primary, the voltage across the secondary being decreased and the current increased in proportion to the number of turns. One way of increasing the turns in the primary is to use several turns of wire in series around the torus. In practice, several wires in parallel make up each of the series turns.

7.118.   Another method for increasing the number of primary turns is to wind them around an iron transformer core.   The torus then threads the core, as shown in Fig. 7.18.   Electrically, the circuit is equivalent to that in Fig.

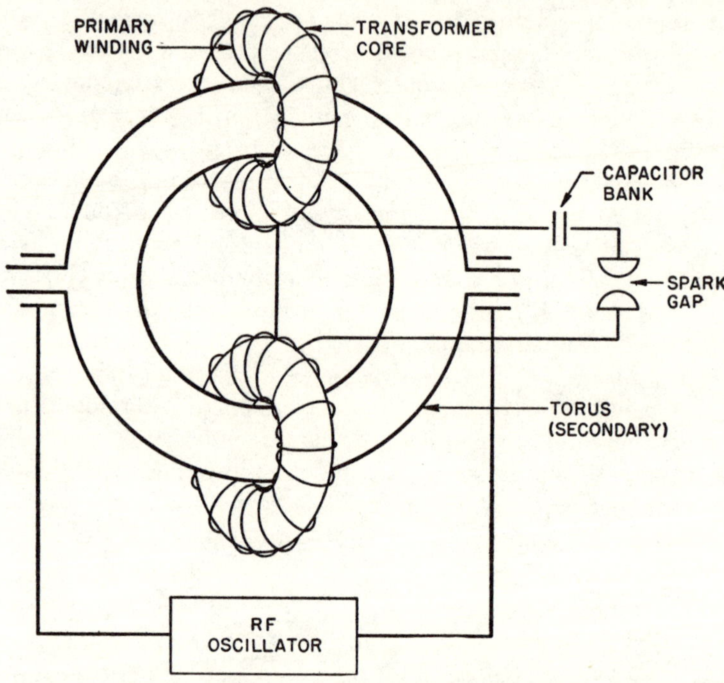

Fig. 7.18.   Toroidal system with multi-turn primary.

7.17 with several primary turns in series.   In devices with iron cores, it is a common practice to make use of what is known as "flux biasing" to increase the strength of the secondary current.   Prior to passage of the main discharge, a low-frequency or direct current is passed in the opposite direction through an auxiliary (bias) winding around the core (see Fig. 7.20), so as to produce a nearly saturated flux within it.   The total available flux change in the core during the main discharge is about twice that under unbiased conditions.   This flux change is roughly equal to LI_{max}, and so there is a corresponding increase in the peak value I_{max} of the main discharge current before the core becomes saturated again.

7.119.   If an axial magnetic field is to be used to stabilize the pinched discharge, a solenoid is wound around the torus, either inside or outside the primary conductor if it lies along the torus.   In any event, a conducting shell, for stabilizing purposes, is required inside the solenoid.   This shell, which may also be the primary circuit, must have gaps to prevent eddy currents from excluding the axial field but at the same time must not produce any appreciable asymmetry in the discharge current in the torus.

7.120. In order to prevent the electrical energy in the torus from returning to the capacitor bank, the primary winding is often short-circuited (or "crow-barred")* when the voltage in the secondary drops to zero. This occurs near the maximum of the pinch current if the resistance of the secondary circuit is small. Not only is the capacitor bank thus protected from receiving a reverse charge, which could be harmful, but the discharge current in the torus decays with a characteristic time of L/R. As a result, the current generally persists for a longer period than would otherwise have been the case.

7.121. The data in the following sections have been obtained with both insulating- and metal-walled toruses in a large variety of sizes, the smallest having a minor radius of about 4 cm [77] and the largest some 50 cm [75, 79]. Major radii have ranged from 15 to 150 cm. Much of the earlier work on the pinch effect was done with moderately heavy gases, such as argon, in order to improve the luminosity of the discharge and to delay the development of instabilities. In more recent studies, however, deuterium has almost invariably been used, at pressures ranging from about 10^{-4} to 0.1 mm of mercury. The electric fields employed fall roughly into two classes: a few tens to about 100 volts/cm in the medium-gradient devices and roughly 1 to 10 volts/cm in the low-gradient, metal-walled toruses. The descriptions given below of the experimental observations on toroidal pinches conveniently fall into these two categories.

MEDIUM VOLTAGE-GRADIENT TOROIDAL PINCHES

7.122. The work with electric fields of medium gradient, of the order of 10 to 100 volts/cm, has usually been done with fairly small toruses made of glass, quartz, or ceramic (Fig. 7.19). Qualitatively, the results are similar to those obtained with linear pinch discharges [35, 43, 46, 83]. Current is found to flow in a hollow sheath during the initial stages of the contraction and in the nonstabilized discharges the pinch breaks up after one or more bounces. If an axial, stabilizing field is included, the separation between B_z and B_θ is fairly distinct during the first contraction, after which interdiffusion proceeds rapidly. The plasma exhibits the same residual flutter as occurs in the otherwise grossly stabilized linear pinch (§7.91).

7.123. A phenomenon characteristic of toroidal discharges is the emission of a burst of hard X-rays shortly after application of voltage to the primary. The energy of the photons extends in some cases up to several hundred kilo-electron volts, indicating that electron runaway (§5.10), accompanied by a betatron type of acceleration, is taking place. A flux of lower-energy X-rays, persisting for a longer time, has also been detected. These radiations also

---

* In the British literature the term "clamping" is often used in this connection.

FIG. 7.19. Toroidal pinch system (Perhapsatron S-5) showing iron yokes and axial field coils. (Courtesy of Los Alamos Scientific Laboratory)

appear, however, in devices with metal walls using lower voltage gradients, as described below.

7.124. Comprehensive surveys of neutron production have been made on B_z-stabilized toroidal pinch discharges. The emission appears to originate from within the plasma column, but displays an energy anisotropy suggestive of streaming along the tube [82, 83, 85]. The experimental techniques employed were described in §6.133 *et seq.*, and the significance of the results will be discussed below (§7.147 *et seq.*).

7.125. At a variable interval after the formation of a stabilized pinch, large energy losses occur. These losses can be estimated from voltage, current, and magnetic probe data, as outlined in Chapter 6, by determining by how much the integrated power ($\int V \cdot I\, dt$) input exceeds that derived from adding the magnetic field energy to the kinetic pressure of the particles. In one experiment, a torus was connected to a subsidiary capacitor bank of low voltage and very high capacitance when the discharge voltage was nearly zero ("power crowbar"), so that, for several tens of microseconds, the current in the plasma remained almost constant. During this interval the voltage was also approximately constant but, significantly, the plasma energy did not appear to be changing. The power being fed into the discharge from the capacitor bank, as determined from V and I, was about 10^9 watts, and since the plasma was in a nearly steady state, this also represented the energy loss rate [83].

7.126. Investigation of both particle and electromagnetic radiation emission showed that the energy emitted in the vacuum ultraviolet region was of the same order of magnitude as the loss rate from the discharge determined from the voltage and current, as mentioned above. A few lines from highly ionized states of oxygen (O V and O VI), in the wave-length range of 100 to 200 A, appeared to account for most of the radiated energy [87]. The oxygen could either be an irreducible contaminant from the quartz tube wall or it could be the result of inadequate pumping. The latter was evidently significant, since the time at which the radiation loss occurred could be changed by varying the pumping speed in the gas-flow system. In fact, by the use of refined pumping techniques, it has been found possible to eliminate the ultraviolet radiation loss altogether during the first half-cycle—about 25 microsec—of the uncrowbarred pinch current [88].

7.127. It is possible that, in some circumstances, loss of electrons to the tube walls, with an associated emission of soft X-rays, can account for the major loss of energy from a pinched discharge. In one series of experiments, a scintillation probe inserted into a toroidal pinch near the wall showed the presence of an X-ray flux, above 1-kev energy, which was within an order of magnitude of the total power loss from the discharge. These X-rays were probably produced by runaway electrons striking the tube walls. The fact that the electrons apparently moved freely across the magnetic field lines suggests that they were driven by some form of plasma instability [46].

METAL-WALLED TORUSES

7.128. Several groups of investigators have constructed toroidal pinch tubes, covering a range of sizes, with metal walls which are exposed to the discharge [43, 79-82, 85]. A special advantage of such tubes is that, because of the larger thermal conductivity as compared with those made from insulating material, energy absorbed from the plasma can be removed more easily and the surface temperature kept lower. As a result, it should be possible to effect a considerable reduction in plasma contamination from wall material [89, 90]. In order that the metal wall may not act as a short-circuit and shield the gas from the electric field, it is interrupted at one or more places with insulation. A common practice is to shield each insulating break from the discharge by an overlapping extension of the metal. An alternative scheme is to employ a continuous metal liner which, by reason of its high resistance achieved through thinning, corrugation, and the use of a relatively resistant alloy, allows the electric field to enter in spite of its continuity.

7.129. The presence of insulating gaps in the metal tube makes the use of low electric fields imperative, so as to prevent arcing at the gaps. Since energy storage at low voltage is generally inconvenient, the metal-walled devices employ multiturn primary windings which are fitted closely to the vacuum torus. Capacitor banks with convenient storage voltages, e.g., 20 to 50 kilovolts, are then possible, with the added advantage that the switches need be capable of handling only currents of moderate strength. The axial, stabilizing magnetic field is provided in all the metal-walled systems by means of a solenoid wound about the outside of the torus (see, for example, Fig. 7.20).

7.130. The low electric fields (1 to 10 volts/cm) used in the discharges[*] mean that the dI/dt values are smaller and the current duration periods are longer than those usually encountered in the insulated-wall toruses. Consequently, pinching continues over a longer time and this permits intermixing of the magnetic fields to take place; hence, a discrete separation of the B_z and B_θ does not occur and a sheath of current is not observed to form. This interdiffusion does not, however, interfere with the stabilizing property of the axial field against wriggling, since any kinking of the plasma stretches the B_z lines independently of their radial distribution. Bouncing and shock phenomena are, of course, completely absent from these low voltage-gradient discharges, as also are dynamic effects in the gross pinch motion. This is not necessarily true, however, when instabilities occur.

7.131. As in the case of the medium voltage-gradient systems, the behavior of all the low voltage-gradient devices is qualitatively similar. The results obtained with the largest metal-walled torus, constructed in Great Britain, will be described here. The device, known as Zeta[†] [75, 79, 84], consists of an aluminum torus of 50-cm internal radius and an average major radius of 1.5 meters, with walls 2.5 cm thick. It is constructed in two halves, joined together by means of insulating and vacuum-tight gaskets around the minor circumference. To prevent the main discharge from reaching the gaps and causing them to break down, a closed-liner system is mounted in the interior of the torus; the space between the liner and the torus is about 3 cm in depth. In the original form of Zeta, the liner system consisted of 48 water-cooled metal sections, insulated from each other and from the walls of the torus. In a modification, known as Zeta 1A, it was replaced by continuous liner, made of thin, corrugated, stainless steel sheet [91]. This reduces problems of maintenance and makes possible the use of higher discharge currents without damage to the liner.

7.132. The gas in the torus, preionized by a radiofrequency discharge, constitutes the secondary of a large iron-cored pulse transformer in the manner represented in Fig. 7.20 [79]. By discharging a capacitor bank into the primary of the transformer, a unidirectional pulse is produced. The peak current

---

[*] Since several of the metal-walled toruses are large, the electric field gradient is small because of the large circumference, apart from the necessity to avoid arcing at gaps.

[†] From the initial letters of Zero Energy Thermonuclear Assembly. A similar apparatus, called Alpha, has been constructed in the U.S.S.R.

Fɪɢ. 7.20.   Cutaway drawing of part of Zeta 1A.   (Courtesy of United Kingdom Atomic Energy Research Establishment, Harwell).   (1) Corrugated liner; (2) axial field windings; (3) water cooling pipes; (4) bias windings; (5) primary pulse windings; (6) pump manifold; (7) refrigerator; (8) cold trap.

in most of the earlier work with Zeta was 1.6×10^5 amp, but this was later increased to 9×10^5 amp by using a larger capacitor bank.   With appropriate crowbarring (§7.120), the total duration period of the discharge current in the plasma is 3 or 4 millisec and the pulse is repeated every 10 sec.   A steady axial magnetic field, up to 320 gauss, for stabilizing the constricted discharge, is generated by a solenoidal coil around the torus.

7.133.   The deuterium gas pressures in Zeta are lower than those employed in other pinch devices; they range from 8×10^{-5} to 10^{-3} mm of mercury, with 10^{-4} mm a common value.   It is seen from equation (7.16) that the temperature in a constricted discharge in a state of quasi-equilibrium is related to I^2/N, where I is the current and N is the number of particles per centimeter length of the torus.   Since the magnitude of the current is limited, the attainment of a high temperature requires that N be relatively low.   In a torus of large internal diameter, this means that the particle density, and hence the gas pressure, must be exceptionally low.   The maximum electron temperatures attained were probably about 100 ev and the apparent ion temperatures were somewhat higher.

7.134.   Magnetic probe and streak camera studies show that while the secondary current is increasing, the plasma contracts to form a channel of approximately 15 to 20 cm radius. In deuterium, the pinched discharge becomes un-stable within about 100 microsec, so that it touches the walls. As a result, the plasma is cooled and impurities from the walls pass into the gas, as indicated by the spectral lines. In the absence of an axial field, the instability of the pinched discharge is greatly increased.

7.135. Magnetic probe measurements show that the axial (stabilizing) mag-netic field is partially trapped by the plasma and is drawn toward the axis as the discharge contracts. However, the general behavior, as indicated by the magnetic field and current density distributions in the radial direction, is dif-

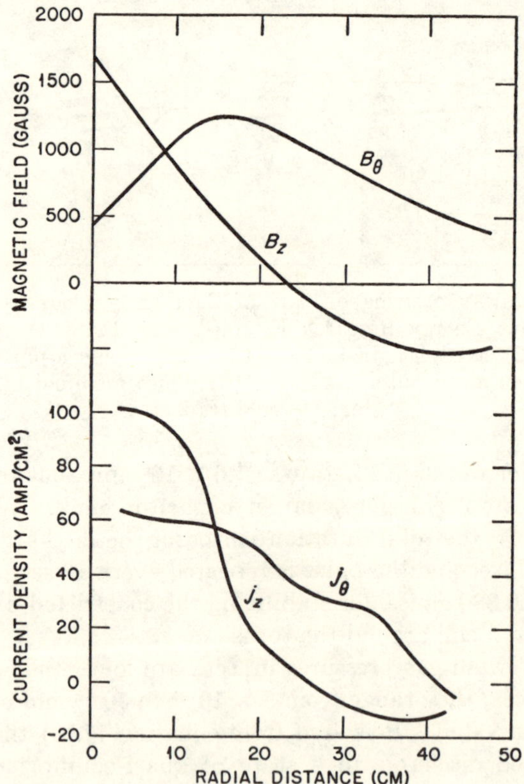

FIG. 7.21. Magnetic field and current density distributions in Zeta dis-charge.

ferent from that observed in other pinch work at higher voltage gradients (Fig. 7.21). It will be noted, in particular, that the axial and azimuthal currents are not restricted to a relatively thin sheath of plasma but are distributed fairly uniformly throughout the gas, as mentioned earlier.

7.136. An extensive series of measurements of Doppler broadening of im-purity lines was made with Zeta in order to determine a mean ion energy which, in the absence of rapid gross motions of the plasma, would provide an indication of the ion temperature (§6.59 *et seq.*). Known amounts of nitrogen, oxygen, or air—usually 5 per cent or less of the total gas pressure—were admitted to the discharge tube, and spectroscopic observations were made over a wide range of pinch currents and axial magnetic fields. It was found that certain impurity lines, e.g., of O V, which are brightest in the region of the pinch have Doppler widths corresponding to a temperature of 500 ev. However, it has not been established if mass motion due, for example, to instability or turbulence is contributing to the broadening of the spectral lines [79, 84].

NEUTRON PRODUCTION IN PINCHED DISCHARGES

GENERAL CHARACTERISTICS

7.137. The production of neutrons from a linear pinched discharge in deuterium was originally observed in 1952 and 1953, and since that time the results have been fully confirmed [28]. However, it soon appeared that these neutrons were not of thermonuclear origin; that is to say, they were not produced by the interaction of deuterons having a Maxwellian velocity distribution, moving in random directions in the plasma. One important reason for this conclusion was that the number of neutrons emitted from the discharge, generally in the range of 10^7 to 10^9 neutrons, in a pulse lasting only a few tenths of a microsecond in a dynamic pinch, was much greater than could be expected at the probable deuteron temperatures and densities. These neutrons have been called "false" neutrons, but a better term is "instability" neutrons because, as will be seen shortly, their production appears to be related to the development of instabilities in the pinched plasma.

7.138. As a general rule, especially in nonstabilized discharges, the neutrons have been observed immediately following the second or third contraction of the "bouncing" plasma. In some instances, however, especially when the potential gradient of the discharge was high, neutron emission occurred after the first contraction. It has been suggested that, after one or two "bounces," the inherent instabilities of the plasma have grown to such an extent as to provide conditions for the acceleration of deuterons. As a result a considerable number acquire energies greatly in excess of that corresponding to the kinetic temperature of the plasma or of the voltage across the pinched discharge. The observed neutrons then arise from the interaction of these accelerated deuterons with others in the plasma.

7.139. In some work with stabilized discharges, in both linear and toroidal tubes, the estimated plasma temperatures and pressures were such that the neutrons produced might possibly have been of true thermonuclear origin. However, even in these cases caution must be exercised in interpreting the results, and so far there is no conclusive evidence that such neutrons have been obtained in a pinched discharge [92].

7.140. It appears that generally the neutron yield increases at first and then decreases with increasing initial pressure of the deuterium gas, at a given discharge voltage. It also increases with increasing applied voltage at a particular gas pressure.

7.141. The effect of an included axial magnetic field on neutron emission is different in linear and toroidal pinches. In a linear system, the neutron yield has its maximum value when there is no B_z field and then decreases steadily with increasing field [12, 29, 37]. In a torus, on the other hand, there are no neutrons when B_z is zero, but the emission increases to a maximum and then decreases again as the axial stabilizing field is increased [77, 79, 82, 83]. Whether these differences would be observed with linear and toroidal pinches operated under otherwise identical conditions is uncertain. However, the nature of the results reported are represented qualitatively in Fig. 7.22. The rate

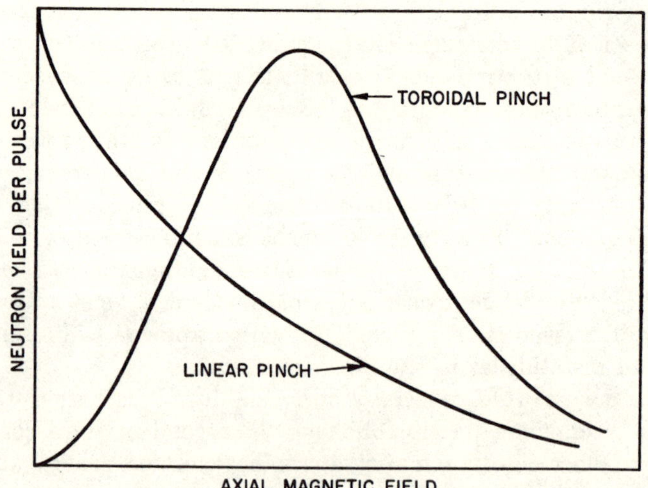

FIG. 7.22. Effect of axial magnetic field on neutron emission.

at which neutron emission falls off with increasing axial field, in the case of a linear pinch, seems to be related to the external circuit inductance; the higher the inductance the more rapidly does the neutron yield decrease.

7.142. At moderate values of the B_z field the time dependence of neutron emission is similar in both linear and toroidal pinches. The burst is fairly sustained and may persist for an appreciable fraction of the current duration. When the axial field strength is decreased, however, a very intense, short burst of neutrons occurs in a linear discharge at about the time of the first contraction. But no such emission is observed in a toroidal discharge. These results, together with those described above, suggest the possibility that two different neutron production mechanisms may operate in the linear pinch but only one in a toroidal system.

7.143. By means of collimated neutron detectors it has been shown that, except near the electrodes in a linear discharge, the neutrons are produced with reasonable uniformity over the length (or toroidal circumference) of the discharge tube, and that they appear to come from a narrow region near the central axis, i.e., from the constricted plasma [12, 83]. This observation disposes of the possibility that the neutrons might arise from interaction of fast deuterons with others adsorbed on the walls of the tube. The foregoing results are all compatible with a thermonuclear origin of the neutrons, but they have also been obtained with neutrons which are undoubtedly not thermonuclear. Small amounts, of the order of a few per cent or so, of certain gases, notably argon and air, have been found to suppress neutron emission almost completely; but helium is less effective in this respect.

7.144. If, as seems possible, the nonthermonuclear (or instability) neutrons are produced by deuterons which have acquired high energies due to extraneous conditions, e.g., plasma instabilities developed in the pinched discharge, then a directional effect might be expected in the energies of the neutrons. The anisotropic distribution of energy should be detectable by studying the proton recoils in nuclear cmulsion plates exposed at different angles to the direction of the discharge (§6.143 *et seq.*). Where this test has been applied, a definite anisotropy has been detected, showing that at least some of the neutrons are not of thermonuclear origin [12, 28, 36, 82, 83, 85, 92]. The energy distribution is such as to indicate that there is a high-velocity mass motion of the deuterons toward the cathode, i.e., in the direction of the applied electric field. In one experiment the mean energy of the accelerated deuterons was estimated to be about 50 kev, with some as high as 200 kev, although the maximum voltage across the discharge tube was only 35 kilovolts [12].

7.145. The fact that the neutron energy distribution is anisotropic does not necessarily mean that all the neutrons observed are nonthermonuclear in origin. It is difficult to prove, however, that thermonuclear neutrons are present, although they may be. This difficulty is compounded by the possibility that one proposed mechanism for deuteron acceleration might give an essentially isotropic distribution of neutron energies (§7.149).

7.146. It might be appropriate to inquire at this point if it matters, as far as energy release by fusion of deuterons is concerned, whether the neutrons resulting from D-D reactions in a pinched discharge are thermonuclear or not. The answer is that the distinction is important. If the reactions involve accelerated deuterons, the situation is somewhat similar to that described in §3.53 *et seq.* The large cross section for the scattering reaction between energetic deuterons and electrons of lower energy in the plasma means that there is a considerable irreversible loss of deuteron energy. The difficulty of attaining a self-sustaining system, in which the total energy produced by the D-D reactions is equal to that expended in the discharge, is thus greater than it would be for a strictly thermonuclear process, where on the average no energy is lost in collisions.

ORIGIN OF NONTHERMONUCLEAR NEUTRONS

7.147. Several mechanisms have been proposed to account for the acceleration of deuterons which can lead to the emission of nonthermonuclear neutrons from a pinched discharge. The first is based on a consideration of the rapid decrease in radius of a nonstabilized pinched discharge accompanying the development of the sausage-type instability [12]. In a discharge occurring in a tube surrounded by a conductor, the inductance increases when the discharge contracts (§7.62). Consequently, the fast growth of a sausage-type instability, accompanied by the narrowing (or "necking off") of the plasma, means that there will be a rapid increase of the inductance in the region of the neck (Fig. 7.23). This increase will result in the production of a voltage V across

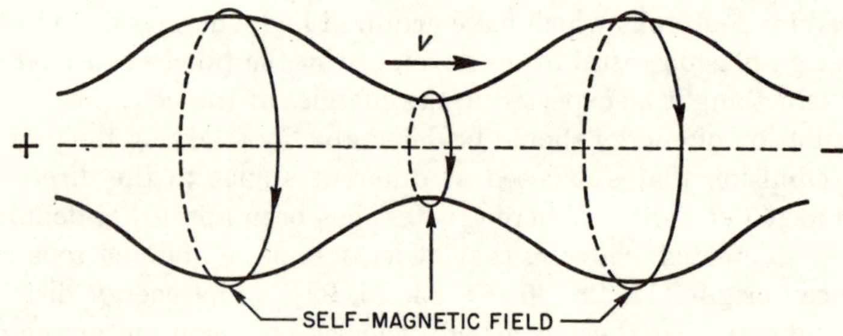

FIG. 7.23. Possible mechanism for production of accelerating potential.

the neck, equal to $d(LI)/dt$, in accordance with equation (7.43). The current I undergoes little change, but since dL/dt is large, the value of V will be correspondingly large. Because the self-magnetic field of the discharge is azimuthal, the voltage produced will be in the direction shown in Fig. 7.23. The interaction of a deuteron accelerated by this voltage, occurring at the neck of the sausage-type instability, with a deuteron in the bulk of the plasma will result in the emission of neutrons having the observed anisotropic energy distribution, i.e., corresponding to the preferred acceleration of deuterons in the direction of the discharge current.

7.148. Another possibility arises from the expected behavior of deuterons in the plasma when an $m = 0$ instability forms. Normally, when the deuterons strike the current sheath they are reflected and accelerated inward, as indicated in Fig. 7.24A. However, when the diameter of the plasma at the bottom of the neck is comparable with the thickness of the current sheath, theory indicates that the accelerated deuteron is turned, under the influence of the electric and magnetic fields, so as to move in the direction of the cathode (Fig. 7.24B), thus giving the observed anistropic distribution of neutron energy [93].

7.149. It may be noted, in relation to the foregoing mechanisms for deuteron acceleration, that the duration of the neutron pulses is much longer than would

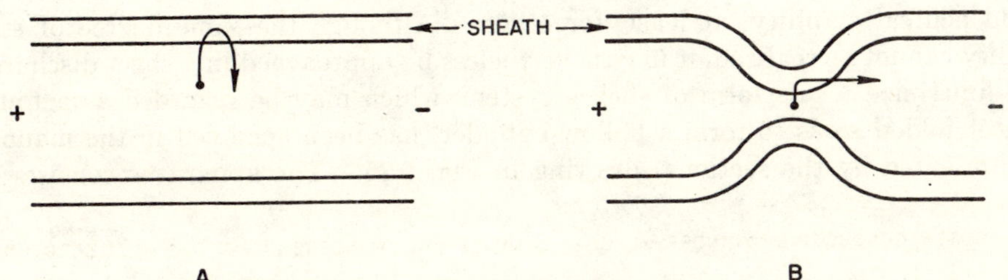

Fig. 7.24. Possible mechanism for acceleration of deuterons.

be anticipated from the rate of the necking-off process of a single sausage-type instability. It has been suggested that many such processes occur randomly throughout the length of the pinched discharge during the emission period. This would not only account for the comparatively long duration of the pulse, but would also explain the fairly uniform distribution of neutrons along the discharge.

7.150. In a so-called fully stabilized pinch, the evidence is very clear that the sausage-type instability is suppressed; nevertheless, neutron emission, probably of nonthermonuclear origin, still occurs. The suggestion has been made [17] that deuterons in a pinched plasma might be accelerated by a mechanism similar to one proposed to account for the high energy of the particles in cosmic rays [94]. A relatively few reflections of deuterons in the plasma by magnetic fields associated with hydromagnetic disturbances, moving with the Alfvén speed as given by equation (4.106), may result in the attainment of energies sufficiently high to produce nonthermonuclear neutrons. This is referred to as the Fermi-type mechanism for particle acceleration.

7.151. It is significant that, in both stabilized and nonstabilized discharges, the emission of neutrons always (or nearly always) appears to coincide with the onset of some kind of plasma instability, e.g., of the sausage, kink (helical), or fluttering type. In each case the rapid movement of the plasma must produce a time and space change in the magnetic field and will consequently be associated with the development of a local electric field capable of accelerating the deuterons. It is possible that, at the same time, there occurs a Fermi-type process of deuteron acceleration, by repeated reflection from moving magnetic fields. The conditions in the low-density plasma just outside the pinch proper, where the Alfvén velocity is high and the ion mean free paths are long, are particularly suited to the operation of the Fermi acceleration mechanism.

CONSTRICTED DISCHARGES OF SPECIAL TYPES

SHEET PINCH DEVICES

7.152. It has been shown theoretically that a plasma sheet of infinite extent carrying a current possesses positive stability for some types of perturbation and neutral stability, at least, for others. Although the same degree of stability cannot be realized, it may nevertheless be approached in a sheet discharge of finite size. One form of such a system, which may be regarded as a finite sheet folded so as to form a hollow cylinder, has been achieved in the manner represented by the sectional drawing in Fig. 7.25. The apparatus consists of

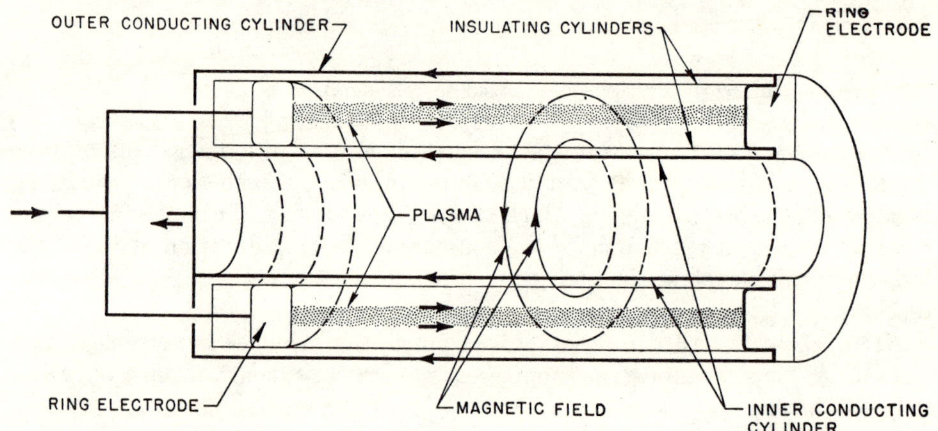

Fig. 7.25. Schematic representation of hollow cylindrical (Triax) discharge.

two coaxial insulating tubes, e.g., Pyrex or silica, with the gas in the annular space between them. Cylindrical metal conductors are placed outside the outer and inside the inner insulating tubes, respectively, as indicated. The discharge current, from a capacitor bank, enters through a ring electrode at the left end, travels through the gas (plasma) to the ring electrode at the right end, and returns by the conducting cylinders, as shown by the thick arrows. Magnetic probe and other measurements demonstrate that, with the passage of an electrical discharge, the plasma detaches itself from the containing walls and becomes constricted [95].

7.153. The pinched system just described has been called the *Triaxial* (or *Triax*) *pinch*. This name was used because of the general similarity to a three-conductor coaxial transmission line, known commercially as a triaxial cable. The Triax pinched discharge may be expected to exhibit something analogous to the sausage-type instability of the conventional channel (or filament) discharge. However, in a hollow cylinder of fairly large diameter, such as occurs in the Triax discharge, the rate of growth of this instability should be quite slow.

7.154. A possible disadvantage of the pinched discharge in a sheetlike plasma is the smaller compression obtainable, for a given current, than for an unstabilized pinch of the usual type. But this same objection applies to a filamentary pinch stabilized by an included axial magnetic field. As seen in

§7.36, the maximum theoretical value of the radial compression in a fully stabilized pinch is 5.0, and in practice it is probably considerably less. As far as compression is concerned, therefore, there is little difference between a sheet pinch, which does not require a magnetic field for stabilization, and the so-called fully stabilized pinched discharge.

7.155. The principal advantage claimed for the Triax pinch over the stabilized linear pinch is that the former offers the simplicity of a linear system without the drawback of heat losses to the electrodes. The reason is that, in the stabilized linear pinch, the axial magnetic field lines intersect the electrodes, so that energetic ions (and electrons), moving in spirals along these lines, are led to the electrodes where losses occur. This situation does not arise in the Triax system and experiments with thermocouples have shown that heat transfer to the electrodes is negligible.

7.156. The Triax pinch has been studied in tubes of 50-cm and 1-meter length; in the early experiments the diameter of the outer return conductor was 10 cm and that of the inner conductor ranged from 1.25 to 5 cm. Later work is being done with larger tubes. The initial pressure of the deuterium gas varied from 5×10^{-2} to 1 mm of mercury, the lower pressures being used in an effort to attain higher temperatures (cf. §7.133). When operating a Triax tube at low gas pressures and high discharge currents, satisfactory results were obtained only by preionizing and preheating the deuterium by means of a relatively low-current discharge, insufficient to cause pinching to occur. Measurements of the voltage across the discharge and also of dI/dt indicate that the contracting plasma may undergo as many as eight or ten bounces, lasting for several microseconds, but with no signs of instability.

7.157. In experiments made with large currents and low gas pressures, a burst of 10^4 to 10^5 neutrons has been observed at a microsecond or two after initiating the discharge. Simultaneously there occurs a transient increase of 2 or 3 kilovolts in the voltage across the tube, indicating a temporary increase in the plasma impedance. The cause of this increase has not been ascertained, and so the origin of the neutrons, with which it is evidently associated, is uncertain.

7.158. The transient increase in the voltage might suggest the development of an instability, but this seems to be quite different in nature from those occurring in ordinary linear or toroidal discharges. Acceleration of the deuterons in a longitudinal direction seems to be ruled out by the fact that insertion of dummy annular electrodes, supported by friction between the insulating tubes and floating electrically, had no effect on voltage, current, or neutron production. Moreover, analysis of the neutron energy showed that the production was isotropic within the limits of accuracy (± 4 kev) of the experiment. It is not impossible, therefore, that the neutrons are of thermonuclear origin.

7.159. It is of interest, and perhaps of significance, that if the neutrons are thermonuclear, the yields are consistent with the conditions existing in the

Triax pinch. As noted earlier, this is not the case for the instability neutrons produced from linear pinched discharges. The observed rate of neutron emission (2×10^4 neutrons per pulse) would indicate an ion temperature of about 300 ev in the Triax pinch, using a heating current of 1.3×10^6 amp and an initial deuterium gas pressure of about 0.1 mm of mercury. This is roughly the temperature that might be expected from ohmic and shock heating and adiabatic compression. The observed resistivity, however, points to a much lower electron temperature.

7.160. In addition to the hollow cylindrical pinch described above, a disk-shaped sheet pinch has also been established. The discharge is passed between an inner (central) electrode and an outer ring electrode located between two flat circular metal plates, as represented in Fig. 7.26. A radial current flows

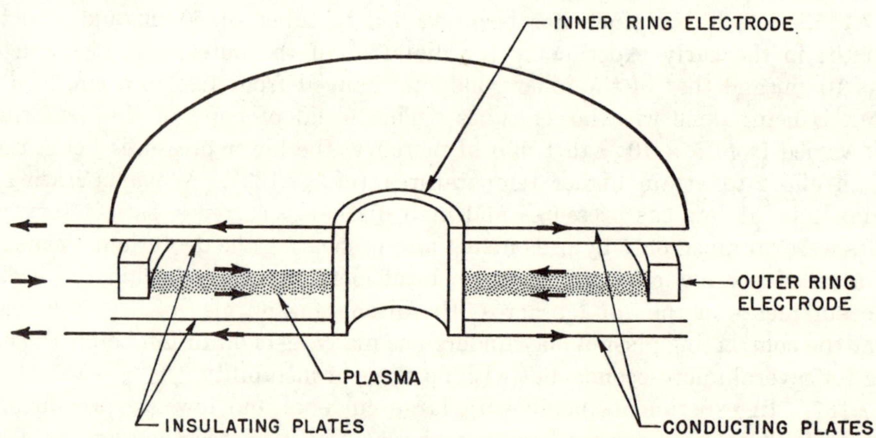

Fig. 7.26. Schematic representation of circular sheet discharge.

through the sheet plasma, the direction being indicated by the arrows. As a result there is some movement of the plasma away from the walls, so that the effect is similar to a pinched discharge having a small compression.

THE "HARD-CORE" DISCHARGE

7.161. Theory indicates that, when confined by a magnetic field, a plasma should be hydromagnetically stable if the plasma-field interface is everywhere convex toward the plasma (§13.7). This has led to the development of a special type of discharge configuration called the *hard-core system*, similar to the conventional stabilized pinch but having the positions of the azimuthal and axial fields reversed [96-100]. The discharge, although constricted, is different from a conventional pinched discharge in the respect that the constriction is due to a self-magnetic field produced inside the plasma rather than outside it.

7.162. In its simplest form the apparatus used is depicted in Fig. 7.27. The

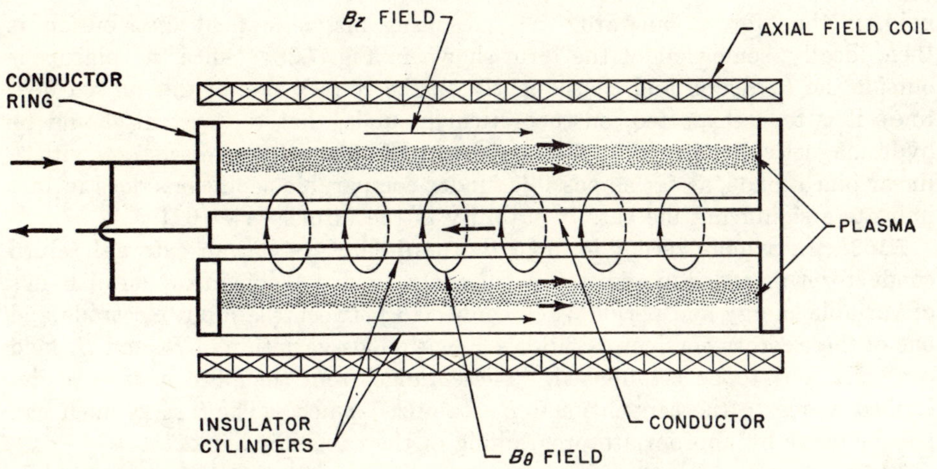

Fig. 7.27. Schematic representation of hard-core pinch.

discharge current passes through the gas and returns through a metallic conductor at the center of the tube; it is to this feature that the name "hard-core" owes its origin. The current first flows in a thin sheath next to the insulator which separates the center rod from the gas. The azimuthal (B_θ) field, produced by the current in the rod, which is now between the plasma sheath and the central conductor, exerts an outward pressure on the sheath, expanding it in radius. The B_z field, previously established by means of an external sole-

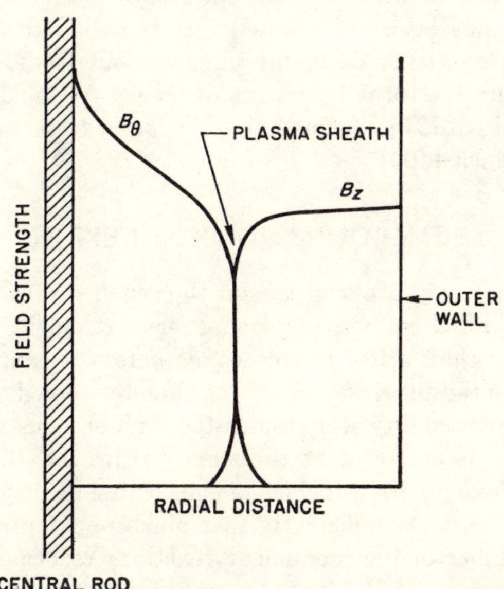

Fig. 7.28. Idealized magnetic field distribution in hard-core pinch.

noid, is thus forced outward. The resulting magnetic field distribution is then, ideally, somewhat of the form shown in Fig. 7.28. Since the plasma is outside the B_θ region and consequently on the convex side of the curved field lines, it is to be expected, on theoretical grounds, that the system should be hydromagnetically stable.* A comparison of a hard-core discharge with a linear pinch made, as far as possible, under comparable conditions does in fact indicate a significant increase in stability in the former case [97].

7.163. In a more recent form of the hard-core system, an external return conductor is provided in addition to the central rod. A pair of capacitor banks of variable energy and period, each connected between the input electrode and one of these return conductors, allows a continuous variation of B_θ and B_z field configurations to be established. These range from the pure hard-core described above to the conventional B_z-stabilized pinch. The Triax pinch can also be made by appropriate programing of the capacitor banks.

7.164. A systematic survey of various configurations has shown that, while a certain amount of fluttering (§7.91) still appears in the simple hard-core distribution of Fig. 7.28, the discharge is nearly quiescent when B_θ is kept greater than zero within the plasma. The experimental results show that the magnetic field configurations which provide the maximum stability of the plasma are indeed those to be expected from hydromagnetic theory. This suggests that the small, irreducible flutter which still remains is not hydromagnetic in origin [98].

7.165. Because of the interesting results which have been obtained in the preliminary experiments with a linear hard-core discharge a toroidal system of the same type has been constructed. It is called the *Levitron* since the hard core consists of a shorted metal ring, without leads, which is supported ("levitated") within the torus by means of pulsed AC fields. The constricted discharge obtained with this device has been found to be more stable than the conventional B_z pinch [100].

SUMMARY AND CONCLUSIONS

7.166. From the early experiments on the pinched discharge it was apparent that the constricted column of plasma was extremely unstable and could persist for no more than a few microseconds before it came into contact with the walls of the containing vessel and was thereby cooled. Although some of the reasons for the instability were understood, no obvious method for stabilizing the discharge was apparent at the time. However, theoretical studies of the dynamic behavior of the pinched plasma subsequently raised hopes that, in spite of its instability, a sufficiently fast pinch might provide conditions for a considerable number of thermonuclear reactions to occur. The inward ve-

* It should be noted that in the linear Triax pinch, the magnetic field lines are convex on one side of the plasma (inner radius) but concave on the other side (outer radius).

locity of the plasma sheath, which theory indicates to be proportional to $(dI/dt)^{1/2}$, i.e., to the square root of the rate of current rise, determines the radial kinetic energy of the particles swept up by the sheath in its motion. Hence, by increasing the discharge current at a sufficiently high rate, which would require an appropriately high voltage, it should be possible to heat the ions to temperatures of thermonuclear interest, provided they were able to make a few scattering collisions with one another after reaching the vicinity of the axis of the containing tube.

7.167. Observations on the fast pinched discharge showed that neutrons were indeed being produced at appreciable rates. But it soon appeared that the neutron flux from such a "dynamic" pinch is anisotropic in its energy distribution, indicating that the neutrons probably arise from the interaction of deuterons which are accelerated to high velocities by spurious electric fields. There is little doubt that these fields originate from instabilities which develop in the pinched plasma. Consequently, if any neutrons are also produced by true thermonuclear reactions, their presence is completely masked by the large flux of instability neutrons.

7.168. Although the possibilities of the fast (or superfast) dynamic pinched discharge may not have been completely exhausted, interest has largely shifted to the study of the so-called fully stabilized pinch. Theoretical work has indicated that the combination of an external conductor with a trapped axial magnetic field should, in the limit of infinite conductivity, provide complete hydromagnetic stability of a pinched discharge. In its simplest conceptual form, the B_z-stabilized pinch is an infinitely thin cylindrical shell of current containing within it the whole of the axial (B_z) flux which originally filled the tube and surrounded on the outside by the azimuthal field set up by its own current. Provided the kinetic pressure of the plasma is relatively small, so that the $8\pi nkT$ in equation (7.33) can be neglected, equation (7.34) may be rearranged into

$$r = \frac{5B_{z0}r_0^2}{I},$$

since the pinch ratio κ is equal to r_0/r; in this expression r and r_0 are in centimeters, B_{z0} in gauss, and I in amperes. It follows that, for a B_z-stabilized pinch, the radius of the discharge is governed by the instantaneous value of the total pinch current, for a given value B_{z0} of the initial axial field and of r_0, the tube radius.

7.169. Experimental attempts to produce a stabilized pinch have been only partly successful. The inclusion of an axial magnetic field in the plasma, together with the use of an external conductor, has produced some stabilization, but the life of the pinched discharge is still relatively short. One reason for this situation is that the current shells produced in actual plasma discharges have appreciable thickness. It has been shown that this may be due not only to the low conductivity of the plasma in the early stages of the discharge, but also to the influx of gas from the tube walls as the sheath is forming.

7.170. More serious, however, is the persistence of instabilities when the requirements for stabilization of a very thin current sheath are essentially satisfied. Although gross motions of the discharge column, such as formation of local constrictions, kinking, spiraling, and general wriggling, are suppressed in the B_z-stabilized pinch, an irregular (fluttering) motion of the sheath surface, similar to a fine-grained turbulence, develops after a short time. This time appears to be related directly to the pressure of the gas in the discharge tube and inversely to the strength of the axial magnetic field.

7.171. It has not yet been established whether these residual instabilities are harmful in themselves. Certainly, magnetic probe and other measurements do not show that there is any tendency at this stage for the pinched plasma to be driven to the walls. Some evidence has been reported that the onset of the fluttering instability could be correlated with the appearance of the spectra of contaminants from the walls as well as an increase in the plasma resistivity, which would indicate a decrease of temperature. Later work, however, has cast some doubt on this correlation, since it has been found possible to create a discharge free from contaminant spectra but nevertheless exhibiting the turbulent flutter.

7.172. The future of the pinch effect as a means of generating thermonuclear power appears to depend on finding satisfactory solutions to three main problems; they are concerned with (a) the elimination (or retardation) of the residual instability of the plasma, (b) the heating of the plasma to thermonuclear temperatures, and (c) the reduction of energy losses. These will be considered in turn.

7.173. For a current sheath of appreciable thickness, theory indicates that shear of the magnetic field across the plasma, such as should result from axial fields in opposite directions inside and outside the sheath, might be expected to give enhanced stability. While the results obtained with a conventional linear pinch were not too promising, it is possible that the reversed-B_z field, as it is called, may be effective in a hard-core discharge. Such a discharge is superficially similar to a pinch, but it differs in the respect that the return current is carried by a central conductor and the sheath moves outward from the core. One of the significant features of the hard-core discharge is that the interface between the magnetic field and the plasma is everywhere convex toward the plasma, and such a configuration should be intrinsically stable against hydromagnetic disturbances. It is possible, therefore, that the study of hard-core discharges, especially in toroidal geometry, may prove of interest in determining whether the flutter instability is in fact hydromagnetic.

7.174. In the dynamic pinch, it is conceivable that high kinetic temperatures could be attained as a result of both adiabatic and nonadiabatic heating of the plasma during compression. But in the B_z-stabilized pinch, the high radial velocities required for nonadiabatic compression are not easily attained. Furthermore, the requirement that the pinch ratio be less than 5, with practical values in the vicinity of 3, precludes any appreciable heating as a result of adiabatic compression. In short, therefore, compression alone can contribute relatively little to the heating of a B_z-stabilized pinch.

7.175. Although the rate of direct ohmic heating drops to low values, even for the highest plasma currents, for temperatures above 0.5 kev or so, the variation known as intermixing heating (§§5.8, 7.41) presents possibilities. In this procedure, a B_z pinch configuration is set up with as thin a current sheath as possible and crowbarred at the input terminals. The B_θ and B_z fields then diffuse into one another and the energy decrease of the field configuration appears as heat. While the B_z-stabilized pinch is being formed, this intermixing must be avoided, but after pinch formation is complete, it can contribute to the heating of the plasma.

7.176. In order to facilitate the intermixing of the fields, equation (7.41) indicates that the plasma resistivity should be high and the sheath radius small. Unfortunately, it is just when intermixing heating is most desirable, at temperatures above about 0.5 kev, that the resistivity becomes negligible. Hence, the interdiffusion of the magnetic fields will inevitably be slow. A detailed analysis of the problem shows that it may take a time interval of the order of 0.1 sec to heat the plasma in a fairly small tube, of 6-cm radius, to a temperature of a few kilovolts by this means [101]. The maintenance of the pinched discharge for this period presents a problem for which no solution is yet in sight. The life of the discharge, before it reaches the walls of the containing tube, can be extended by increasing the tube radius. But this would mean a corresponding increase in the discharge radius and consequently in the intermixing time.

7.177. Finally, reference will be made to the subject of energy losses from the discharge. The inevitable losses such as bremsstrahlung and, possibly, electron cyclotron radiation, discussed in Chapter 2, have not been a problem in experimental work, because, at the temperatures attained so far, they have been negligible. Apart from these, however, the existence of serious sources of energy loss has been repeatedly demonstrated in experimental pinch studies. At least two processes have been shown to be important loss channels.

7.178. First, the presence of high-intensity X-rays, with energies of 1 kev or more, suggests that much energy is lost from the plasma in the form of runaway electrons driven by an instability. The electrons produce thick-target bremsstrahlung when they strike the walls of the containing tube. It is probable that the energy loss due to runaway electrons can be decreased by increasing the particle pressure in the plasma. The resulting increase in the number of collisions with ions will tend to prevent the steady increase in the energy of the electrons (§5.11).

7.179. It has been found, in another series of experiments, that a considerable amount of energy is lost from a pinched discharge as radiation in the far ultraviolet region of the spectrum, mainly in the lines of highly ionized oxygen. The dependence of the emission of this radiation on the pumping speed of the gas inlet system suggests that at least part of the oxygen impurity entered with the deuterium. However, the wall materials of the discharge tubes used in the experiments contained oxygen, in the form of silica (or silicates), and so it is possible that oxygen is ejected from the wall. If this is the case, more efficient pumping methods may reduce but not prevent the energy losses. It is evident, therefore, that the development of materials for containing tubes, which would keep the impurities in the discharge to a minimum, would represent an important problem in the design of a thermonuclear reactor based on the pinch effect (cf. §3.4).

7.180. There is another possible means by which a pinched plasma may dissipate its energy, although its importance has not yet been confirmed [102]. If the conductors which establish the electric and magnetic field configurations, e.g., primary circuit and B_z-field windings in a toroidal system, have irregularities or gaps, as they ordinarily do in practice, a particle following a line of force will be transferred to one of slightly larger radius each time it makes a circuit of the torus and passes the irregularity in the field. As a result, the particle can ultimately be led into the wall. Computations of a situation of this kind indicate that small field perturbations, such as are encountered in typical pinch devices, could produce large particle (and energy) losses for certain operating parameters.

REFERENCES FOR CHAPTER 7

1. W. H. Bennett, *Phys. Rev.*, **45**, 90 (1934).
2. L. Tonks, *Trans. Electrochem. Soc.*, **72**, 167 (1937); *Phys. Rev.*, **56**, 360 (1939).
3. C. L. Northrup, *Phys. Rev.*, **24**, 474 (1907).
4. A. Schlüter, *Z. Naturforsch.*, **5a**, 72 (1950).
5. P. C. Thonemann and W. T. Cowhig, *Proc. Phys. Soc. (London)*, **B64**, 345 (1951).
6. M. Blackman, *Proc. Phys. Soc. (London)*, **B64**, 1039 (1951).
7. P. C. Thonemann, UKAEA Report AERE P-17/P-6 (1955), Chapter 2.
7a. R. S. Pease, *Proc. Phys. Soc. (London)*, **B70**, 11 (1957).
8. M. Rosenbluth, R. Garwin, and A. Rosenbluth, USAEC Report LA-1850 (1954).
9. M. Rosenbluth, in R. K. M. Landshoff, Ed., *Magnetohydrodynamics*, Stanford University Press, 1957, p. 57.
10. M. Leontovich and S. M. Osovets, *J. Nuclear Energy*, **4**, 209 (1957).
11. J. E. Allen, *Proc. Phys. Soc. (London)*, **B70**, 24 (1957).
12. O. A. Anderson, W. R. Baker, S. A. Colgate, H. P. Furth, J. Ise, R. V. Pyle, and R. E. Wright, *Phys. Rev.*, **110**, 1375 (1958).
13. S. I. Braginsky and A. B. Migdal, *Plasma Physics and the Problem of Controlled Thermonuclear Reactions*, Pergamon Press, Inc., Vol. II, 1959, p. 28.
14. M. Rosenbluth, USAEC Report LA-2030 (1956).
14a. M. A. Levine, USAEC Report TID-7503 (1955), p. 195.
15. R. J. Tayler, *Proc. Phys. Soc. (London)*, **B70**, 31, 1049 (1957).
16. V. D. Shafranov, *J. Nuclear Energy*, **5**, 86 (1957).

17. I. V. Kurchatov, *J. Nuclear Energy*, **4**, 193 (1957).
18. L. A. Artsimovich, A. M. Andrianov, O. A. Bazilevskaya, Y. G. Prokhorov, and N. V. Filippov, *J. Nuclear Energy*, **4**, 213 (1957).
19. L. A. Artsimovich et al., *J. Nuclear Energy*, **4**, 213 (1957).
20. S. Y. Lukyanov and V. Sinitsyn, *J. Nuclear Energy*, **4**, 216 (1957).
21. S. Y. Lukyanov and I. M. Podgornyi, *J. Nuclear Energy*, **4**, 224 (1957).
22. A. L. Bezbatchenko, J. N. Golovin, D. P. Ivanov, V. D. Kirilov, and N. A. Yavalinsky, *J. Nuclear Energy*, **5**, 71 (1957).
23. L. C. Burkhardt, R. E. Dunaway, J. W. Mather, J. A. Phillips, G. A. Sawyer, T. F. Stratton, E. J. Stovall, and J. L. Tuck, *J. Appl. Phys.*, **28**, 519 (1957).
24. S. Berglund, R. Nilsson, P. Ohlin, K. Siegbahn, T. Sundström, and S. Svennerstedt, *Nucl. Instr.*, **1**, 223 (1957).
25. P. Ohlin, K. Siegbahn, T. Sundström, and S. Svennerstedt, *Nucl. Inst.*, **3**, 237 (1958).
26. H. A. B. Bodin and J. A. Reynolds, *Engineering*, **84**, 538 (1957).
27. L. C. Burkhardt, R. H. Lovberg, and J. A. Phillips, *Nature*, **181**, 224 (1958).
28. D. C. Hagerman and J. W. Mather, *Nature*, **181**, 226 (1958).
29. L. C. Burkhardt and R. H. Lovberg, *Nature*, **181**, 228 (1958).
30. S. A. Colgate, USAEC Report UCRL-4895 (1957).
31. L. C. Burkhardt, R. H. Lovberg, G. A. Sawyer, and T. F. Stratton, *J. Appl. Phys.*, **29**, 964 (1958).
32. R. Dunaway and J. A. Phillips, *J. Appl. Phys.*, **29**, 1137 (1958).
33. E. Fünfer, *Z. Naturforsch.* 13a, 524 (1958).
34. L. A. Artsimovich, *Proc. Second U.N. Conf. on Peaceful Uses of Atomic Energy*, **31**, 6 (1958).
35. L. Biermann, *Proc. Second U.N. Conf. on Peaceful Uses of Atomic Energy*, **31**, 21 (1958).
36. A. M. Andrianov, O. A. Bazilevskaya, S. I. Braginskii, B. G. Brezhnev, S. Khavaschevski, V. A. Khrabrov, N. G. Kovalski, N. V. Filippov, T. I. Filippova, V. E. Palchikov, I. M. Podgornyi, Y. G. Prokhorov, and M. M. Sulkovskaya, *Proc. Second U.N. Conf. on Peaceful Uses of Atomic Energy*, **31**, 348 (1958).
37. S. C. Curran, K. W. Allen, H. A. B. Bodin, R. A. Fitch, N. J. Peacock, and J. A. Reynolds, *Proc. Second U.N. Conf. on Peaceful Uses of Atomic Energy*, **31**, 365 (1958).
38. V. S. Komelkov, U. V. Skvortsov, and S. S. Tserevitinov, *Proc. Second U.N. Conf. on Peaceful Uses of Atomic Energy*, **31**, 374 (1958).
39. J. L. Tuck, *Proc. Second U.N. Conf. on Peaceful Uses of Atomic Energy*, **32**, 3 (1958).
40. J. W. Mather and A. H. Williams, *Proc. Second U.N. Conf. on Peaceful Uses of Atomic Energy*, **32**, 26 (1958).
41. L. C. Burkhardt and R. H. Lovberg, *Proc. Second U.N. Conf. on Peaceful Uses of Atomic Energy*, **32**, 29 (1958).
42. I. N. Golovin, D. P. Ivanov, V. D. Kirilov, D. P. Petrov, K. A. Razumova, and N. A. Yavlinsky, *Proc. Second U.N. Conf. on Peaceful Uses of Atomic Energy*, **32**, 72 (1958).
43. R. Aymar, E. Etievant, P. Hubert, A. Samain, B. Taquet, and A. Torossian, *Proc. Second U.N. Conf. on Peaceful Uses of Atomic Energy*, **32**, 92 (1958).
44. D. F. Brower, R. E. Dunaway, J. H. Malmberg, C. L. Oxley, M. Stearns, D. W. Kerst, F. R. Scott, S. P. Cunningham, and R. G. Tuckfield, *Proc. Second U.N. Conf. on Peaceful Uses of Atomic Energy*, **32**, 110 (1958).
45. K. Siegbahn and P. Ohlin, *Proc. Second U.N. Conf. on Peaceful Uses of Atomic Energy*, **32**, 113 (1958).

46. S. A. Colgate, J. P. Ferguson, and H. P. Furth, *Proc. Second U.N. Conf. on Peaceful Uses of Atomic Energy*, **32**, 129 (1958).
47. T. Sundström and S. Svennerstedt, *Nucl. Instr.*, **4**, 213 (1959).
48. R. Latham and J. A. Nation, *Nucl. Instr.*, **4**, 261 (1959).
49. D. A. Baker, G. A. Sawyer, and T. F. Stratton, *Proc. Second U.N. Conf. on Peaceful Uses of Atomic Energy*, **32**, 34 (1958).
50. O. A. Anderson and W. R. Baker, USAEC Report TID-7520 (1956), p. 134.
51. J. W. Mather, USAEC Report TID-7503 (1956), p. 180.
52. L. C. Burkhardt, R. H. Lovberg, and J. A. Phillips, USAEC Report LA-2131 (1956).
53. *Proc. Conf. on Controlled Thermonuclear Reactions*, UKAEA Report AERE GP/R-2371 (1957), Chapter 12 (S. C. Curran); Chapter 22 (R. A. Fitch).
54. R. A. Fitch, USAEC Report TID-7526 (1957), p. 301.
55. D. L. Smart, *Proc. Inst. Elec. Eng. (London)*, **106A**, Suppl. No. 2, 107 (1959).
56. R. A. Fitch and N. R. McCormick, *Proc. Inst. Elec. Eng. (London)*, **106A**, Suppl. No. 2, 117 (1959).
57. H. de B. Knight, L. Herbert, and R. C. Maddison, *Proc. Inst. Elec. Eng. (London)*, **106A**, Suppl. No. 2, 131 (1959).
58. R. A. Fitch, *Proc. Inst. Elec. Eng. (London)*, **106A**, Suppl. No. 2, 177 (1959).
59. D. B. Fried, USAEC Report TID-7558 (1958), p. 486.
60. V. Josephson, USAEC Report TID-7558 (1958), p. 507; *Space Technology Laboratories Report* TR-59-0000-09880 (1959).
61. M. V. Clauser and E. S. Weibel, *Proc. Second U.N. Conf. on Peaceful Uses of Atomic Energy*, **32**, 161 (1958).
62. L. O. Heflinger and S. L. Leonard, Division of Plasma Physics APS, Monterey Conf., J 3 (1959).
63. O. A. Anderson and W. R. Baker, USAEC Report TID-7520 (1956), p. 134.
64. S. A. Colgate and J. P. Ferguson, USAEC Report TID-7520 (1956), p. 279.
65. R. H. Lovberg, L. C. Burkhardt, G. A. Sawyer, and T. F. Stratton, USAEC Report TID-7520 (1956), p. 308.
66. B. R. Suydam, *Proc. Second U.N. Conf. on Peaceful Uses of Atomic Energy*, **31**, 157 (1958).
67. L. C. Burkhardt and R. H. Lovberg, *Bull. Am. Phys. Soc.*, II, **4**, 365 (1959).
68. M. Kruskal and J. L. Tuck, *Proc. Roy. Soc.*, **A245**, 222 (1958); also in USAEC Report LA-1716 (1953).
69. S. A. Colgate, J. P. Ferguson, H. D. Furth, and R. E. Wright, *Proc. Second U.N. Conf. on Peaceful Uses of Atomic Energy*, **32**, 140 (1958).
70. A. A. Ware, *Trans. Roy. Soc.*, **A243**, 197 (1951).
71. S. W. Cousins and A. A. Ware, *Proc. Phys. Soc. (London)*, **B64**, 159 (1951).
72. W. H. Bostick, USAEC Report WASH-115 (1952), p. 101.
73. J. L. Tuck, USAEC Report WASH-146 (1953), p. 46.
74. R. Carruthers and P. A. Davenpert, *Proc. Phys. Soc. (London)*, **B70**, 49 (1957).
75. P. C. Thonemann, R. Carruthers, D. W. Fry, R. S. Pease, G. N. Harding, S. A. Ramsden, E. P. Butt, R McWhirter, P. J. Lees, A. Dellis, S. Ward, and A. Gibson, *Nature*, **181**, 217 (1958).
76. N. L. Allen, T. E. Allibone, D. R. Chick, R. F. Hemmings, T. P. Hughes, S. Kaufman, B. S. Liley, J. Mack, H. T. Miles, R. M. Payne, J. Read, A. A. Ware, J. Wesson, and R. V. Williams, *Nature*, **181**, 222 (1958).
77. J. L. Honsaker, H. Karr, J. Osher, J. A. Phillips, and J. L. Tuck, *Nature*, **181**, 231 (1958).

78. P. C. Thonemann, *Proc. Second U.N. Conf. on Peaceful Uses of Atomic Energy*, **31**, 4 (1958).

79. E. P. Butt, R. Carruthers, J. T. D. Mitchell, R. S. Pease, P. C. Thonemann, M. A. Bird, J. Blears, and E. R. Hartill, *Proc. Second U.N. Conf. on Peaceful Uses of Atomic Energy*, **32**, 42 (1958).

80. G. G. Doglov-Saviliev, D. P. Ivanov, V. S. Mukhovatov, K. A. Razumova, V. S. Strelkov, M. N. Shepelyev, and N. A. Yavlinsky, *Proc. Second U.N. Conf. on Peaceful Uses of Atomic Energy*, **32,** 82 (1958).

81. J. Andreolotti, C. Breton, J. Charon, P. Hubert, P. Jourdan, and G. Vendryes, *Proc. Second U.N. Conf. on Peaceful Uses of Atomic Energy*, **32,** 100 (1958).

82. T. E. Allibone, D. R. Chick, G. P. Thomson, and A. A. Ware, *Proc. Second U.N. Conf. on Peaceful Uses of Atomic Energy*, **32,** 169 (1958).

83. J. P. Conner, D. C. Hagerman, J. L. Honsaker, H. J. Karr, J. P. Mize, J. E. Osher, J. A. Phillips, and E. J. Stovall, *Proc. Second U.N. Conf. on Peaceful Uses of Atomic Energy*, **32,** 297 (1958).

84. G. N. Harding, A. N. Dellis, A. Gibson, B. Jones, D. J. Lees, R. W. P. McWhirter, S. A. Ramsden, and S. Ward, *Proc. Second U.N. Conf. on Peaceful Uses of Atomic Energy*, **32,** 365 (1958).

85. A. A. Ware, *Proc. Inst. Elec. Eng. (London)*, **106A,** Suppl. No. 2, 30 (1959).

86. R. J. Bickerton, *Proc. Inst. Elec. Eng. (London)*, **106A,** Suppl. No. 2, 148 (1959); also in *Nucl. Instr.*, **4,** 273 (1959).

87. H. Karr, J. Osher, and E. Knapp, Division of Plasma Physics APS Monterey Conf., J 2 (1959).

88. J. P. Mize, D. Hagerman, and D. Lier, Division of Plasma Physics APS, Monterey Conf., J 2 (1959).

89. J. L. Craston, R. Hancox, A. E. Robson, S. Kaufman, H. T. Miles, A. A. Ware, and J. A. Wesson, *Proc. Second U.N. Conf. on Peaceful Uses of Atomic Energy*, **32,** 414 (1958).

90. A. E. Robson and R. Hancox, *Proc. Inst. Elec. Eng. (London)*, **106A,** Suppl. No. 2, 47 (1958).

91. J. T. D. Mitchell, H. R. Whittle, E. M. Jackson, and P. R. Clarke, *Proc. Inst. Elec. Eng. (London)*, **106A,** Suppl. No. 2, 74 (1958).

92. B. Rose, A. E. Taylor, and E. Wood, *Nature*, **181,** 1630 (1958).

93. J. L. Tuck, USAEC Report TID-7520 (1956), p. 22.

94. E. Fermi, *Phys. Rev.*, **75,** 1169 (1949).

95. O. A. Anderson, W. R. Baker, J. Ise, W. B. Kunkel, R. V. Pyle, and J. M. Stone, *Proc. Second U.N. Conf. on Peaceful Uses of Atomic Energy*, **32,** 150 (1958).

96. D. H. Birdsall, S. A. Colgate, and H. P. Furth, *Fourth Int. Conf. on Ionization Phenomena in Gases*, pp. 888, 892 (1959).

97. K. L. Aitken, J. N. Burcham, and P. Reynolds, *Fourth Int. Conf. on Ionization Phenomena in Gases*, p. 896 (1959).

98. D. H. Birdsall, S. A. Colgate, and H. P. Furth, Division of Plasma Physics APS, Monterey Conf., J 4 (1959).

99. S. A. Colgate and H. P. Furth, USAEC Report UCRL-5392 (1958).

100. R. Spoerlin, S. A. Colgate, H. P. Furth, L. Sandy, and O. Twite, Division of Plasma Physics APS, Monterey Conf., J 5 (1959).

101. M. N. Rosenbluth, *Proc. Second U.N. Conf. on Peaceful Uses of Atomic Energy*, **31,** 85 (1958).

102. D. W. Kerst, unpublished.

Chapter 8

THE STELLARATOR

THE ROTATIONAL TRANSFORM

INTRODUCTION

8.1.   For the purpose of bringing about thermonuclear reactions in a heated deuterium or deuterium-tritium plasma, confinement in an endless (or toroidal) system is desirable in order to avoid the inevitable losses of energy at the ends of a straight tube, even when used in conjunction with a magnetic mirror. If an externally applied axial i.e., longitudinal, magnetic field is to be used for confining the plasma in a torus then, as seen in §4.33 *et seq.*, a field that is everywhere parallel to the tube (or minor) axis is not satisfactory. Since the magnetic field strength varies with distance from the major axis of the torus, there will be a tendency for the positively charged ions to drift in one direction while the negative electrons drift in the opposite direction perpendicular to the plane of the torus. As a result of this separation of charges, electric fields are produced which, in conjunction with the axial magnetic field, serve to drive the plasma particles to the walls of the containing tube.

8.2.   In a planar torus in which an axial magnetic field is produced by means of a simple solenoidal winding, such as is shown in Fig. 4.4, every magnetic line of force closes exactly upon itself after one circuit around the torus. Such a closed line of force is said to be *degenerate*. Because of this degeneracy, no path is available whereby appreciable mutual neutralization of the separated charges would be possible by the spiral motion of the charged particles along the magnetic lines of force. However, if the degeneracy could be removed so that, in making a single circuit around the torus, each line of force was somewhat displaced from its original position and so did not close upon itself, charge neutralization could occur. This would be especially the case if, after many circuits of the torus, a line of force generated a complete toroidal surface. By following a single line of force, an electron, for example, could travel from a region in which these particles are in excess, as a result of their drift in the inhomogeneous field, to another region where there is an excess of positive ions. Charge neutralization could then take place, and the

290

magnitude of the electric fields, which tend to counteract the confinement of plasma particles by the externally applied axial magnetic field, would be greatly reduced.

8.3.   Displacement of a magnetic line of force in a single circuit about a toroidal tube* so that it is no longer degenerate is achieved by means of a field having what is called a *rotational transform.* There are several methods, as will be seen shortly, of producing a magnetic field with a rotational transform, and the general name of *stellarator* is applied to any device which employs, for the purpose of confining a plasma, an axial field which has this property and which is large compared to any self-magnetic fields set up by currents in the plasma [1, 2].

PROPERTIES OF ROTATIONAL TRANSFORMS

8.4.   The properties of rotational transforms may be understood by follow-ing the lines of force in a toroidal tube containing an axial magnetic field. Consider any plane through the tube; for convenience, this plane will be taken to be perpendicular to the direction of the applied field. If P_0 is any point in the plane, then a line of force will pass through it. This line of force is now followed in the direction of the magnetic field until it has com-pleted a circuit of the toroidal tube. If the field is such that it possesses a rotational transform, the line of force will now intersect the plane at a point, such P_1 in Fig. 8.1, that is not coincident with P_0. With limited exceptions,

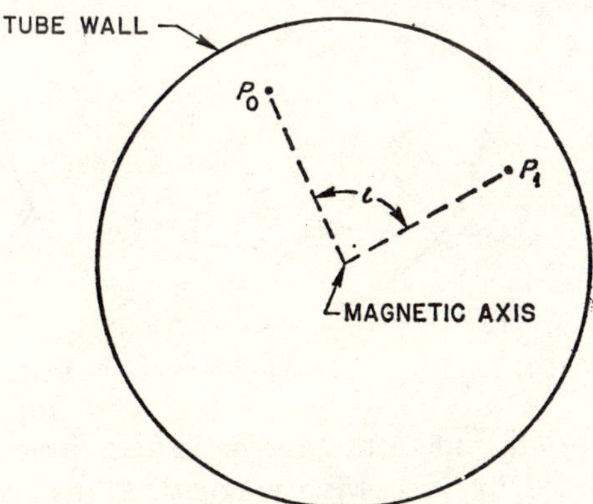

Fig. 8.1. Magnetic field with rotational transform.

which will be indicated below, the same is true for all points P_0 in the given plane. Although the lines of force passing through some of the points P_0 will intersect the walls, the system can be designed so that most of the lines remain in the tube. Thus, it may be concluded that any point, for example P_0, in

---

* The term "toroidal tube" will be used here in a general sense to describe any endless tube which, although it may be deformed, is topologically a torus.

the cross-sectional plane under consideration will be transformed into another point, namely P_1, in the same plane, as the result of following the line of force from P_0 once around the toroidal tube. The transformation of the set of points P_0 into the set of points P_1 is called a *magnetic transform* or (*H transform*) of the plane into itself [3].

8.5. It is now postulated that the magnetic transform is primarily rotational; that is to say, the points—at least those in the outer portions of the plane—all rotate in the same direction in a single circuit. Such is the case in actual stellarator systems. It has been shown that there is usually only one point in the plane that transforms into itself, and so there is only one line of force that closes upon itself after one circuit around the toroidal tube. This line is called the *magnetic axis*. The angle subtended by the arc P_0P_1 in Fig. 8.1 at the magnetic axis is known as the *rotational transform angle* and is represented by the symbol ι.

8.6. Apart from points lying on the magnetic axis, any point P_0 when followed through successive circuits of the toroidal tube will intersect the cross-sectional plane at a series of points falling on (or very near to) a single closed curve. The situation is illustrated in Fig. 8.2, which shows a plane

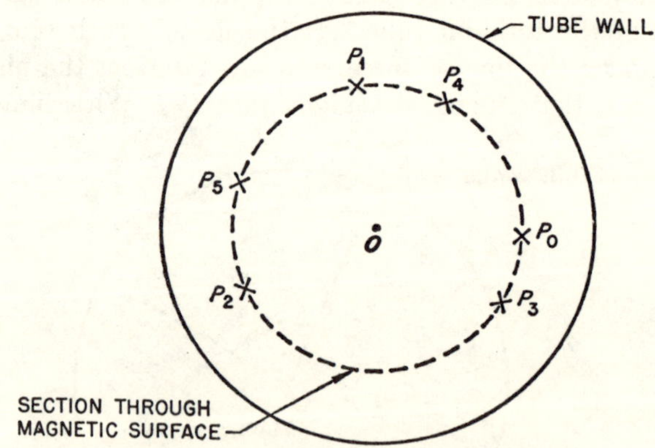

FIG. 8.2. Points generated in a plane by successive magnetic transforms.

through the tube perpendicular to the magnetic field direction; P_1, P_2, P_3, etc., are the points generated by successive magnetic transforms of the arbitrary point P_0. That is to say, the line of force starting from P_0 will intersect the cross-sectional plane at the points P_1, P_2, P_3, etc., upon successive circuits of the tube, provided the magnetic field possesses a rotational transform. If O represents the position of the magnetic axis and the magnetic field is uniform in this cross section, the angles, P_0OP_1, P_1OP_2, P_2OP_3, etc., are all equal to ι, the rotational transform angle. It has been shown that, after a sufficient number of magnetic transforms, the points P_0, P_1, P_2, etc., fall upon what is effectively a closed curve, so that, after several circuits, the line of force starting

from any point P_0 will ultimately return to that point or to a point very close
to P_0.   If ι is not a rational fraction (or multiple) of 2π, the return exactly to
P_0 is of course impossible, but the deviation can become arbitrarily small after
a sufficiently large number of circuits [4].

8.7.   An important consequence of the foregoing result is that, if followed
many times around the toroidal tube, a single line of force in a magnetic field
possessing a rotational transform will spiral about a central line of force, i.e.,
the magnetic axis, and will always remain at more or less the same distance from
this central line.   A single line of force thus generates a closed toroidal surface, referred to as a *magnetic surface,* in the course of a large number of circuits of the tube.   Since every line of force generates such a surface, there
will be a family of surfaces, one within the other, about the magnetic axis
of the system.   Ideally, the outermost magnetic surface would be made to
coincide with the inner surface of the toroidal tube, but in practice this is
rarely the case because of small variations in the axial magnetic field.

8.8.   One result of the existence of the magnetic surfaces is that a single
charged particle, in the absence of fluctuating fields, should be confined for
many circuits around the toroidal tube while it follows a line of force.   The
only requirement is that the coaxial magnetic surfaces (or most of them) do
not intersect the walls of the tube.   Furthermore, if the nonuniformity of the
magnetic field due to the curvature of the tube causes the ions and electrons
to drift in opposite directions, the magnetic surfaces provide a means for
bringing oppositely charged particles together and thus reducing the electric
field resulting from charge separation.

8.9.   The presence of magnetic surfaces will, in any event, decrease the net
drift of the ions and electrons in the toroidal tube.   The reason is that the
line of force which maps out a given surface in the course of its circuits around
the torus is sometimes above the magnetic axis and sometimes below it.
Suppose that, owing to the curvature of the magnetic lines of force, the ions
tend to drift upward and the electrons downward; then the situation when
the line of force is respectively above and below the axis can be represented by

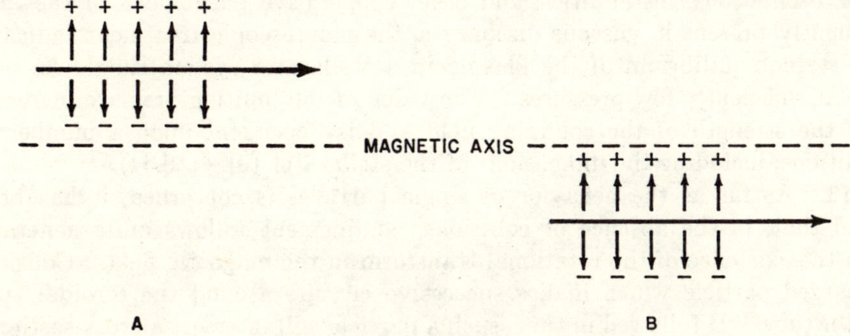

FIG. 8.3.   Particle drifts above and below magnetic axis.

Figs. 8.3A and 8.3B. In the former case, the positive ions tend to drift away from the magnetic axis, as in Fig. 8.3A; but as they follow the line of force over the magnetic surface the ions will enter a region where they tend to drift toward the axis, as in Fig. 8.3B. The electrons will, of course, behave in an analogous manner. It is seen, therefore, that in a magnetic field possessing a rotational transform and hence giving rise to magnetic surfaces, the net radial drift of the charged particles in a plasma will be decreased, if not eliminated.

CONFINEMENT BY MAGNETIC FIELD WITH ROTATIONAL TRANSFORM

8.10. The arguments presented above show that in a stellarator system, i.e., in the presence of an axial magnetic field with a rotational transform, the electrostatic fields due to the oppositely directed drifts of positive and negative particles can be largely eliminated. It remains to be demonstrated, however, that a stellarator system will provide adequate confinement of a plasma. To be rigorous, this would require a detailed proof that the number of charged particles striking the walls of the toroidal containing tube is negligibly small. Because of the complex configuration of the magnetic field, solution of the appropriate equations would be very difficult, and so an approximate precedure has been adopted [2, 5].

8.11. The problem is divided into two parts; first, the macroscopic behavior of the plasma is considered and it is demonstrated that an equilibrium situation is possible in which the plasma as a whole is confined. However, although the plasma as a whole may be confined, particles with velocities in certain ranges of magnitude and direction might conceivably escape. Hence, in the second part of the problem, a microscopic treatment is employed which examines the motion of individual particles in the electric and magnetic fields derived from the macroscopic approach in order to show that the loss is not significant.

8.12. By making certain postulates, the most important being that a steady state exists in which the plasma is quiescent, to the complete exclusion of turbulence, oscillations, instabilities, and other cooperative phenomena of the kind frequently present in gaseous discharges, the macroscopic treatment indicates that stable equilibrium of the plasma in a stellarator system should be possible at sufficiently low pressures. The value of the limiting pressure increases with the strength of the confining field and is dependent upon a number of conditions, including the dimensions of the stellarator tube (§8.34).

8.13. As far as the behavior of single particles is concerned, it has been found that, in the absence of collisions, confinement follows quite generally from the existence of the rotational transform in the magnetic field. Consider a charged particle which makes successive circuits around the toroidal stellarator tube. If followed in time, such a particle will intersect a cross-sectional plane similar to that considered in §8.4, at a large number of points. Successive intersections of a group of similar particles with this plane produce a transformation of the plane into itself; this is called a *particle transform* and it satisfies the same conditions as the magnetic transform described above. Hence, such single particles may be expected to be confined by the magnetic field to the same approximation as the lines of force remain within the stellarator tube.

8.14. For particles that do not make successive circuits of the toroidal path, either because they are trapped between two regions of relatively high magnetic field or because they are moving so slowly that the velocity component parallel to the magnetic axis is virtually zero, the situation is more complicated. In these cases, however, adequate confinement is to be expected because of the rotation of the particles about the magnetic axis. Two mechanisms are operative in causing this rotation. First, the diamagnetic effect of the plasma produces a radial gradient of the magnetic field and this results in a rotational drift of the guiding centers about the magnetic axis. Second, the presence of a radial electric field produces a similar rotational motion of the charged particles. Since the macroscopic velocity of the plasma as a whole is taken to be zero, such a radial field must be present in order to compensate, in the steady state, for the velocity associated with the radial pressure gradient.

PRODUCTION OF A ROTATIONAL TRANSFORM

8.15. To produce a rotational transform in a vacuum magnetic field, resulting from the passage of current through a solenoid wound around a toroidal tube, so that the field is everywhere parallel to the tube axis, it is sufficient to twist the torus out of a single plane. Almost any such distortion will remove the degeneracy of the lines of force in a flat torus, so that the magnetic field has a rotational transform. Historically, the first distortion proposed for a stellarator system was to twist the torus into a figure-8 geometry [1]. The side (A) and end (B) views of the resulting shape are shown in Fig. 8.4.

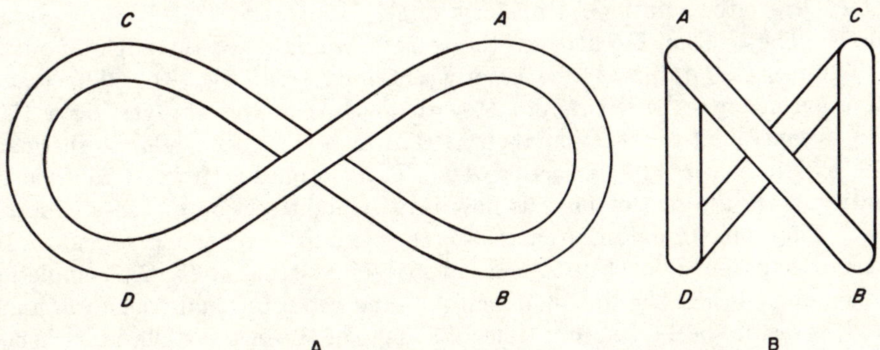

FIG. 8.4. Side (A) and end (B) views of figure-8 stellarator.

The two curving end sections AB and CD, which are both bent through an angle of 180°, are in separate planes, each tilted at an angle to the parallel planes in which are located the straight sections AD and BC.

8.16. The existence of a rotational transform may be seen from an examination of Fig. 8.5, which shows the cross-sectional planes at A, B, C, and

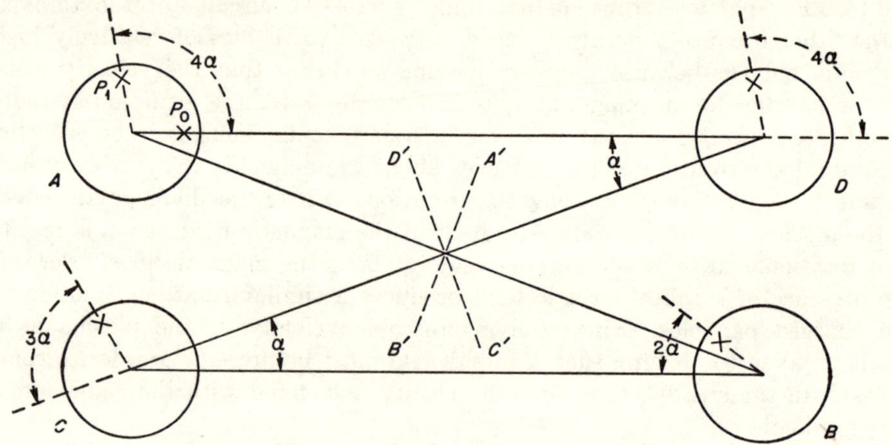

Fig. 8.5. Angle of tilt and rotational transform.

D as seen from the end of the figure-8 stellarator tube. The angle between the planes containing AB and AD, and that between the corresponding planes containing CD and BC, is represented by α; it is called the *angle of tilt* of the stellarator. The lines connecting the four centers mark the path of the magnetic axis. The cross at P_0 indicates an arbitrary point where a single line of force intersects the plane at A; the intersections of this line of force with the planes B, C, and D are also shown by crosses. The complete circuit of the stellarator tube brings the line of force back to P_1 in the plane A, so that P_0OP_1 is the rotational transform angle.

8.17. The relationship between the transform angle and the angle of tilt of the figure-8 stellarator can be readily obtained from Fig. 8.5. The transformation of cross section A into B, with the line of force traversing a 180° bend, is equivalent to reflection about $A'B'$, which is perpendicular to the magnetic axis in the middle of section AB. Transformation from B to C is an identity, since the line of force is now in a straight section of the stellarator tube. The transformation from C to D is again a reflection, about $C'D'$, whereas that from D back to A is an identity. Starting at P_0, the completion of the circuit brings the line of force to P_1 and the rotational transform angle P_0OP_1 is seen to be 4α, i.e., four times the angle of tilt of the stellarator; hence,

$$\iota = 4\alpha.$$

8.18. A modification of the geometry described above is achieved by making all the bends of the figure-8 into right angles.* The torus consists of two parallel tubes in a horizontal plane joined at the ends by V-shaped sections in two vertical planes (see Fig. 8.17). The lengths of the legs of the two V's are different and it is this asymmetry that leads to a rotational transform in a magnetic field which is everywhere parallel to the tube axis. The transform angle can range from 0° to 180°, according to the degree of asymmetry of the end sections. A further description of this type of stellarator and the reasons underlying its design will be given later (§8.84).

8.19. Another method for producing a rotational transform, which is of considerable interest for several reasons, including its expected stabilizing influence on the confined plasma, is to superimpose on the axial confining field a magnetic transverse field whose direction rotates with distance along the magnetic axis. This transverse field is capable of producing a rotational transform in the magnetic field in an undistorted, planar torus.

8.20. In order to treat the situation resulting from the superposition on the axial field of a transverse magnetic field which rotates with distance, consider, for simplicity, an infinite cylindrical tube within which the coordinates of any point are represented, in the usual manner, by r, θ, and z. The changes in the coordinates of a point moving along a single line of force, as z increases, are then related by

$$dr = \frac{B_r}{B_z}\, dz \quad \text{and} \quad rd\theta = \frac{B_\theta}{B_z}\, dz, \tag{8.1}$$

where B_z, B_r, and B_θ are the components of the superposed magnetic fields. Suppose that the fields B_r and B_θ are produced by $2l$ wires wound in the form of a helix on the outside of the cylinder, such that current flows in opposite directions in adjacent wires (Fig. 8.6). Let the pitch of the helix be $2\pi l/h$,

FIG. 8.6. Currents producing helical transverse magnetic field.

so that $2\pi/h$ is the wave length (or period) of the helix; in other words, as z increases by $2\pi/h$, the field direction rotates through an angle of 2π. If the radius of the cylindrical tube is r_t, then if hr_t is small in comparison with

---

* A device based on this concept has been called the Model B-64 stellarator, because its design "starts with a figure-8 and makes all corners square, and $8^2 = 64$." The letter B refers to the series of experimental studies (see §8.84).

unity, the appropriate solutions of Laplace's equation for small values of r/r_t are

$$B_r = Ar^{l-1} \sin (l\theta - hz) \tag{8.2}$$

and

$$B_\theta = Ar^{l-1} \cos (l\theta - hz), \tag{8.3}$$

where A is a constant dependent upon the strength of the helical, transverse magnetic field. There is also a component B_z arising from the current in the helical wires, but its magnitude is small and so the dominant axial field may be taken as that produced by the usual solenoidal winding.

8.21. In order to follow a line of force whose initial coordinates, i.e., in the absence of the transverse magnetic field, are r_0 and θ_0, equation (8.1) is integrated by means of a power-series expansion in the coefficient A which appears in the expressions used for B_r and B_θ, i.e., equations (8.2) and (8.3), respectively. To the first order in A of the expansion, it is found that $r - r_0$ and $\theta - \theta_0$ vary as $\cos (l\theta - hz)$ and $\sin (l\theta - hz)$, respectively, so that the line of force is a helix and its intersections with a plane moving along the z axis is a circle. To higher orders in A, it is necessary to take into account the fact that for values of l greater than unity B_θ is larger on the outside of the circle, i.e., for $r > r_0$, where B_θ is positive, than on the inside, i.e., for $r < r_0$, where B_θ is negative. As a result, the positive values of $d\theta/dz$, derived from equation (8.1), exceed the negative ones, so that on a line of force θ increases steadily with increasing z. In other words, the magnetic field possesses a rotational transform. A detailed integration [6, 7] shows that, for values of $l > 1$, the rotational transform ι_h *for each period* $2\pi/h$ of the helical field windings is given by

$$\iota_h \approx \frac{\pi A^2 r^{2l-4}}{h B_z^{\,2}} \{2(l - 1)\}. \tag{8.4}$$

8.22. Since the transverse magnetic field alone can produce a rotational transform, it is no longer necessary to distort the torus in any way. Hence, either a planar, circular torus or a tube of race-track form, i.e., one having semicircular (or equivalent) ends connected by straight pieces, may be used. Furthermore, the radial variation of ι_h, as indicated by equation (8.4), provided $l > 2$, inhibits certain instabilities which are characteristic of the simple figure-8 stellarator. This aspect of transverse magnetic fields will be considered later (§8.44).

8.23. The foregoing methods for obtaining a rotational transform apply to a vacuum magnetic field; that is to say, they do not depend upon the presence of a plasma. However, if the toroidal tube contains a plasma, a rotational transform of the axial magnetic field may be obtained, even in an undistorted torus, by passage of a current having a component parallel to the magnetic axis. Consider the case of a simple toroidal system in which the lines of force are circular; let R be the radius of curvature of any one of these lines of force. The position angle of a point P on a line of force around the z axis of rotational symmetry is defined by ϕ; as before θ is the azimuthal angle of the point in a given plane about the magnetic axis, and r is the radial distance from

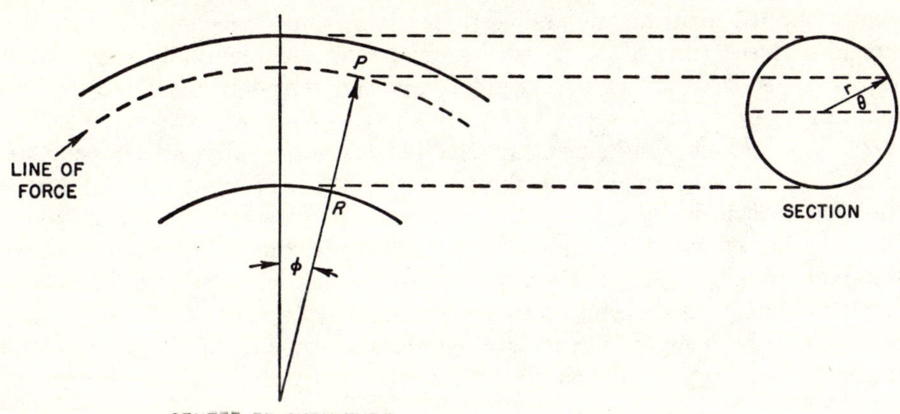

Fig. 8.7. Rotational transform and axial current.

this axis (Fig. 8.7). For small changes in the angles θ and ϕ, it follows (cf. equation 8.1), since $Rd\phi$ is equal to dz and B_ϕ is equivalent to B_z, that

$$\frac{rd\theta}{B_\theta} = \frac{Rd\phi}{B_\phi},$$

where B_θ and B_ϕ are the radial and axial (or longitudinal) components of the magnetic field at the given point. Hence,

$$d\theta = \frac{R}{r} \cdot \frac{B_\theta}{B_z} d\phi,$$

where the more usual symbol B_z has been written in place of B_ϕ.

8.24.   The total rotation of the angle θ for a point on the line of force in a circuit of the torus, which is equal to the rotational transform angle, is then given by

$$\iota = \int_{\phi=0}^{\phi=2\pi} d\theta = \frac{R}{r} \cdot \frac{B_\theta}{B_z} 2\pi,$$

where R is taken as the major axis of the torus. If I amp is the current parallel to the magnetic axis and r is expressed in cm, then B_θ in gauss is equal to $I/5r$, by equation (3.23). Since $2\pi R$ can be replaced by L, the total axial length around the middle of the torus, it follows that

$$\iota = \frac{LI}{5B_z r^2}, \tag{8.5}$$

with L in centimeters. This equation gives the rotational transform angle in radians.

8.25. Ohmic heating of a plasma (§5.6) can be achieved by passing a current through it. If the direction of the current is parallel to the axis of the confining field, there will be produced a rotational transform in the externally applied axial magnetic field. However, since the ohmic heating currents in a closed tube are induced currents, they are inevitably transient in nature, and so they are of no practical value for long-term confinement in a stellarator system.

8.26. To permit confinement of the plasma, especially at temperatures higher than these attainable by ohmic heating, it is preferable to produce a rotational transform by means of one of the other methods described above, since they are effective in the absence of plasma currents. Nevertheless, the rotational transform produced by such currents has an important connection with the stability of the plasma in a stellarator system, as will be seen later. Use has also been made of the rotational transform arising from the ohmic heating current to provide confinement in experiments with a planar toroidal tube (§8.87).

CORRUGATED MAGNETIC FIELDS

8.27. A different approach to the problem of compensating for the particle drifts in the curved region of a race-track type of toroidal tube is based on the proposal to make use of what have been called "corrugated" magnetic fields. Such fields can be realized by a suitable spacing of the windings of a number of coils which, when energized, produce the axial magnetic field, as shown in Fig. 8.8. A charged particle moving in an inhomogeneous field

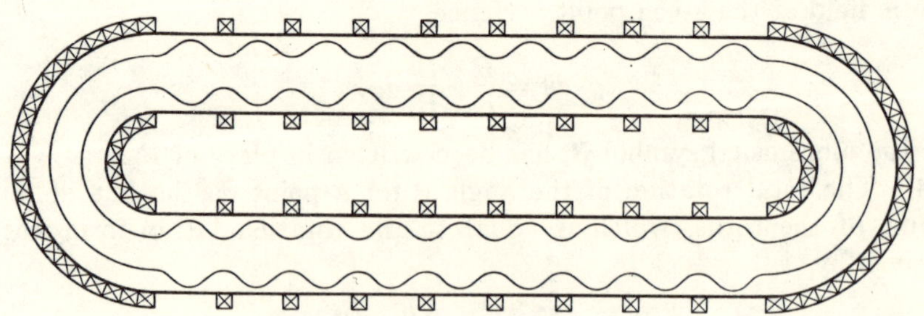

Fig. 8.8. Corrugated magnetic field.

of this kind will be subjected to a circular (or rotational) drift, so that its trajectory will rotate about the axis of the magnetic field. If the length of a straight piece of the race-track tube is appreciably greater than that of the curved portions, the length of a peripheral line of force is determined by the corrugated regions and the contribution of the bends is not significant. In these circumstances, the forced rotational drift of the particles in the corrugated field will compensate for the up and down drifts in the curved sections, provided $\Delta > \pi r$, where Δ is the difference in length between a peripheral (corrugated) field line and one in the center of the tube, and r is the tube radius [8]. It may be remarked, however, that, until an experimental test is made, there may be some question concerning the stable confinement of a plasma by a corrugated field.

8.28. A proposal, somewhat similar in principle to that just described, for inhibiting the particle drifts in a toroidal system, has been called the "bumpy torus" [9]. It consists of a number of circular current loops around a torus, as indicated in Fig. 8.9. These will produce what is effectively a continuous

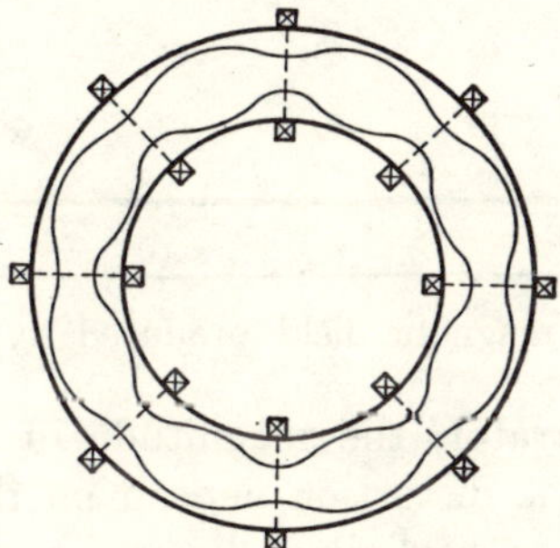

FIG. 8.9.   Bumpy torus.

corrugated magnetic field as shown in the figure. In this geometry, there should be formed a series of closed precessional magnetic surfaces which should not intersect the walls. Losses could result from the scattering of particles in collisions, but calculations suggest that the time constant will be much greater than the normal relaxation time for scattering.

CONFINEMENT LIMITATIONS IN THE STELLARATOR

PLASMA PRESSURE LIMITATIONS IN EQUILIBRIUM CONFINEMENT

8.29. The limitations on the plasma particle pressure for which equilibrium confinement is theoretically possible, for a given value of the confining magnetic field, arises from the presence of longitudinal plasma currents, i.e., currents parallel to the field direction. In one curved section of a stellarator tube, the particles of a given sign tend to drift in one direction (cf. §8.9), whereas in the other curved section, separated from it by straight pieces, they drift in the opposite direction. When the particles move into the straight sections along the lines of force, the separation of charges, resulting from the drifts, produces secondary, longitudinal currents. These flow in one direction in the upper parts of the straight sections and in the opposite direction in the lower parts. The longitudinal currents produce a weak secondary magnetic field which is approximately at right angles to the confining field (Fig. 8.10). Consequently, the magnetic surfaces of the vacuum field become partially distorted when a plasma is present in the stellarator tube, the distortion increasing with the length of the straight section. If, as a result of this distortion, some magnetic surfaces intersect the tube walls, equilibrium confinement of the plasma is reduced to a smaller region near the magnetic axis; that is to say, the effective aperture of the tube is decreased.

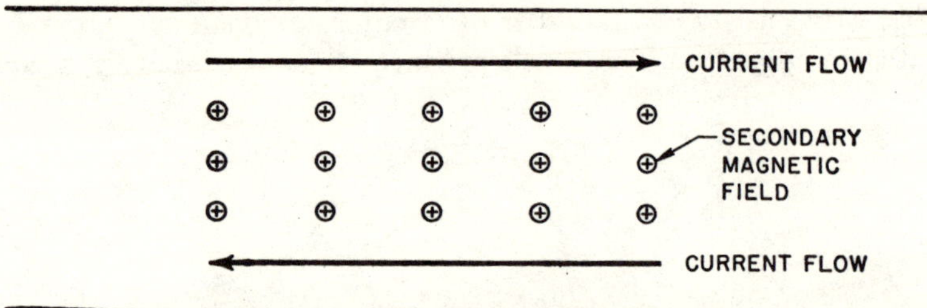

FIG. 8.10. Secondary magnetic field produced by longitudinal currents.

8.30. In a given stellarator, the magnitude of the longitudinal currents and hence the extent of the distortion depend on the particle density in the plasma. The condition for negligible distortion of the magnetic surfaces, which should give the condition for equilibrium plasma confinement, may be taken to be that the deviation of the magnetic axis shall be small in comparison with the tube radius. It is to be expected, therefore, that there will be a relationship between the limiting particle density which can be confined in a stellarator tube and its dimensions. An approximate derivation of such a relationship is given below [10, 11].

8.31. The magnitude of the drift current density in a curved section (or U bend) of the stellarator, i.e., j_{bend}, in cgs-esu, is

$$j_{\text{bend}} = 2n_i e v_d,$$

where v_d cm/sec is the particle drift velocity (up or down), n_i is the number density of either the ions or electrons, since they are equal, and e statcoulomb is the electronic charge; the factor 2 arises from the fact that the ions and electrons drift in opposite directions with equal velocities. The current in the U bend is therefore represented approximately by

$$I_{\text{bend}} \approx \pi R \times 2r \times 2n_i e v_d, \tag{8.6}$$

where R is the radius of curvature of the bend and r is the tube radius. For a Maxwellian distribution of the particle energies, v_d is given by equation (4.19) as

$$v_d = \frac{2kTc}{eBR},$$

where B gauss is the confining (axial) field strength. Upon substituting this result into equation (8.6), it follows that

$$I_{\text{bend}} \approx \frac{8\pi n_i kTrc}{B} = \frac{4\pi nkTrc}{B},$$

where n, equal to $2n_i$, is the total particle density. In this expression, I_{bend} is in statamperes and upon conversion to practical amperes, the result is

$$I_{\text{bend}} = \frac{40\pi nkTr}{B}.$$

8.32. The secondary, longitudinal currents flowing in the straight sections of the stellarator (Fig. 8.10) have the same magnitude I in both upper and lower parts, and each is equal to half the total drift current in the bends, i.e.,

$$I \approx \frac{20\pi nkTr}{B}.$$

The resultant magnetic field $B_\perp$ gauss in the direction perpendicular to the straight sections, produced by the two secondary currents, is

$$B_\perp \approx 2 \times \frac{I}{5r} = \frac{8\pi nkT}{B}.$$

Upon introducing the definition of β given in §3.20, i.e., the ratio of the kinetic pressure of the plasma to the external (or confining) magnetic pressure, it follows that

$$\beta \equiv \frac{nkT}{B^2/8\pi} \approx \frac{B_\perp}{B}. \tag{8.7}$$

8.33. The extent to which a line of force parallel to the tube axis will be distorted, i.e., moved toward the wall, over a given distance is roughly equal to the product of the distance and the ratio $B_\perp/B$. Over a distance equal to $\frac{1}{4}L$, where L is the total length around the magnetic axis of the stellarator, the distortion Δr will be

$$\Delta r \approx \frac{L}{4} \cdot \frac{B_\perp}{B}.$$

This is the maximum distortion which will occur, since from symmetry considerations the distortion will start to reverse at the midpoint of a bend. As implied above, the condition for plasma confinement is that the deviation Δr shall be small in comparison with the tube radius r; that is,

$$\frac{\Delta r}{r} \approx \frac{L}{4} \cdot \frac{B_\perp}{Br} \ll 1$$

or, by equation (8.7),

$$\beta \ll 4\frac{r}{L}.$$

A more rigorous treatment [10, 12] leads to the required condition as

$$\beta \ll \frac{16}{\pi} \cdot \frac{r}{L}.$$

8.34. If the stellarator geometry is of the figure-8 form, it is doubtful whether the ratio r/L, i.e., the ratio of the tube radius to the total axial length, can exceed 0.02, so that the condition for equilibrium confinement is that β shall be much less than 0.1. The small value of β, probably not more than 0.01, for which equilibrium plasma confinement in a figure-8 stellarator should be theoretically possible, places a very serious limitation on the particle density and thus on the rate of generation of thermonuclear energy.

8.35. There appear to be two ways, at least, in which the maximum attainable value of β for equilibrium confinement of the plasma in a stellarator system could be increased. The first is applicable when the rotational transform in the confining magnetic field is produced by distortion of the toroidal tube, as in the figure-8 stellarator. It has been suggested [13] that the net particle drift, which is the basic cause of the distortion of the magnetic surface and consequently of the pressure limit for equilibrium confinement, could be reduced if each curved end section were replaced by a series of short alternating curved pieces, called *scallops*, in which the magnetic field lines have opposite but equal curvatures, as shown in Fig. 8.11. Since the particle

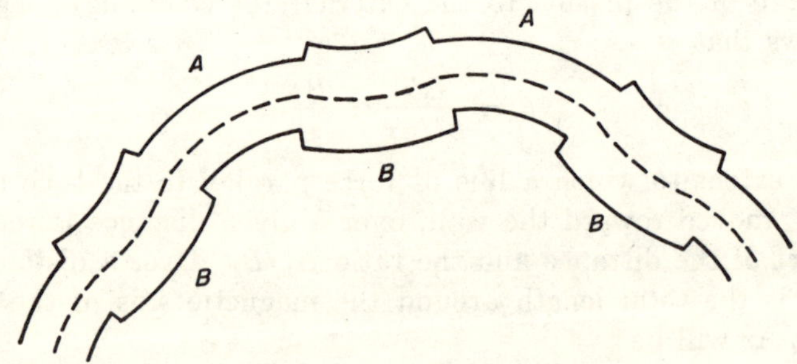

FIG. 8.11. Scallops in bend of stellarator tube.

drifts in alternate scallops will be in opposite directions, the resultant drift will be small and the longitudinal current in the straight sections will be greatly reduced.

8.36. In order to obtain a net bending of the lines of force, that is, for the overall curvature of the scalloped section to be the same as that of the end section it replaces, the scallops A, which curve in the required direction, must be longer than those marked B, having equal curvature in the opposite direction. However, to equalize the particle drifts in adjacent scallops, the magnetic field strength in the shorter sections must be decreased relative to those in the longer sections, in proportion to their lengths, assuming the radii of the curvatures to be equal. This can be achieved by making the shorter scallops wider than the longer ones, as indicated in Fig. 8.11.

8.37. In a stellarator in which the rotational transform is produced by means of a transverse, helical magnetic field, the axial distance required for the direction of particle drift to be reversed is that in which the lines of force are twisted through an angle of π radians. If this distance is small in comparison with the length around the magnetic axis of the stellarator tube, the theoretical value of β for equilibrium confinement of the plasma may approach unity. However, in these circumstances the limiting value of β is probably determined by plasma instabilities (§8.43) and not by deformation of the magnetic surfaces produced by the secondary, longitudinal currents.

PLASMA INSTABILITIES: THE LIMITING CURRENT

8.38. The instabilities which can occur in a plasma confined by magnetic fields fall into at least two broad categories, namely, hydromagnetic and electrostatic. The former are somewhat similar to those considered in Chapter 7, in which the magnetic field and the confined plasma move together to form constrictions, kinks, etc. The electrostatic instabilities, on the other hand, arise from the development of extraneous electric fields in the plasma which tend to drive the charged particles across the magnetic field lines so that they are no longer confined.

8.39. The character of the electrostatic instabilities is not yet clear, but there are two types of hydromagnetic instabilities arising in a stellarator discharge that are fairly well understood. One is similar to the kink instability of a pinched plasma (§7.23); magnetic energy is converted into kinetic energy of mass motion of the plasma as the instability develops. As a result of a detailed theoretical treatment [14, 15], it has been concluded that the $m = 0$ (or sausage-type) of instability is not to be expected in a stellarator, but that the $m = 1$ (or kink-type) of instability should occur when the longitudinal current in the plasma, e.g., that used for ohmic heating, exceeds a certain critical (or limiting) value. This critical current is often referred to as the *Kruskal limit*. Other instabilities of higher modes, i.e., with $m > 1$, are theoretically possible [16].

8.40. The interesting fact has emerged from the hydromagnetic calculations that the Kruskal limiting current is equal in magnitude to the longitudinal plasma current which would produce a rotational transform angle of either ι or $2\pi - \iota$, due to the effect described in §8.25, where ι is the rotational angle of the axial confining field. In other words, the kink instability of the plasma sets in when the superposition of the rotational transform due to the longitudinal current upon that of the confining magnetic field results in a net rotational angle of either zero or 2π. The limiting current is such that it cancels the rotational transform of the magnetic field, since a rotational angle of 2π is effectively the same as one of zero. However, the agreement between the results of the hydromagnetic calculations and the rotational transform derived geometrically should not be given too much significance. For currents considerably in excess of the Kruskal limiting value, where a net rotational transform has again been established, the plasma is still unstable.

8.41. The value of the rotational angle produced by a longitudinal current in the plasma in a circular torus is given by equation (8.5), and this expression may be taken as being approximately valid for toroidal geometries in general. According to the results described above, it follows, therefore, that the value of I_c, the limiting (or Kruskal) current in amperes, is related to other properties of the stellarator system by

$$\frac{LI_c}{5Br^2} \approx \iota \quad \text{or} \quad 2\pi - \iota,$$

where B gauss is the strength of the axial magnetic field, r cm may be taken as the radius of the tube containing the plasma, and L cm is its total axial length; hence,

$$I_c \text{ (amp)} \approx \frac{5Br^2}{L}\iota \quad \text{or} \quad \frac{5Br^2}{L}(2\pi - \iota), \tag{8.8}$$

with ι in radians.

8.42. The two values of the critical current for kink instability correspond to the two directions of flow of the longitudinal current. If the current flow is such as to produce a rotational transform in the opposite direction to that of the confining magnetic field, then the first value in equation (8.8) will apply. On the other hand, the second value will be observed if the current flows in the other direction and the rotational transforms are both in the same sense. The existence of these two possible magnitudes for the critical current has been verified experimentally (§8.91).

8.43. Another type of hydromagnetic instability, called *interchange* (or *flute*) *instability* [7, 17, 18], which may arise in a wide variety of magnetically confined plasma systems, is due to the presence of ionized matter and may occur at fairly low values of β, i.e., when the particle pressure in the plasma

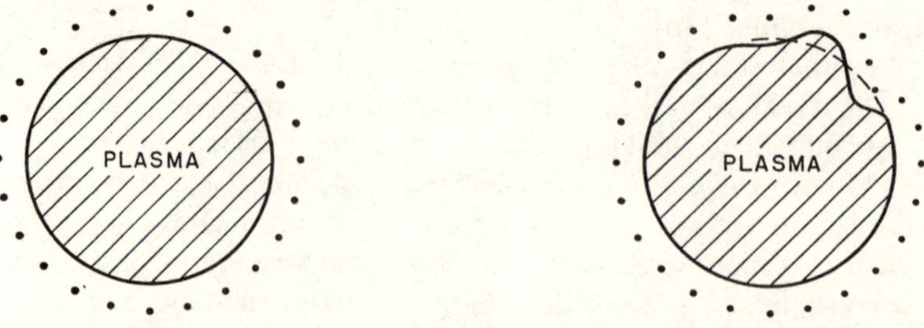

Fig. 8.12. Formation of interchange (or flute) perturbation.

is low relative to the magnetic pressure of the confining field. Consider a steady-state system in which the plasma and the magnetic field are imbedded in each other. A displacement of the plasma is then possible which results in an interchange of the lines of force in such a way as to leave the magnetic field strength and the magnetic energy essentially unaffected (cf. Fig. 8.12). The perturbation along the length of the plasma then resembles the fluting of a column. This occurs most readily if the lines of force can change position with little or no bending. If, as a consequence of such an interchange, the material energy of the plasma would be increased, the system is stable. On the other hand, if the energy were to decrease, the plasma could be unstable. In the latter event, the excess material energy is converted into kinetic energy of the plasma as a whole. Interchange (or flute) instability may be contrasted in this respect with the kink instability, where the kinetic energy of the moving plasma is provided by the magnetic energy of the system. Calculations show that in a figure-8 (or topologically similar) stellarator, in which the rotational transform angle is constant over the cross section of the tube, interchange of lines of force leads to a decrease of material energy and so interchange instability may be expected to occur.

PLASMA STABILIZATION

8.44. A method of stabilizing the plasma against interchange instability was originally proposed based on the use of transverse magnetic fields produced by pole pieces placed at right angles to the plasma with alternate pole pieces rotated in space [19]. However, an examination of the procedure for obtaining a rotational transform by the use of helical windings around the stellarator tube showed that this automatically introduces a similar stabilizing effect due to the variation of the rotational transform angle with distance from the magnetic axis. It will be noted from equation (8.4) that the rotational transform angle varies as r^{2l-4}; it follows that for $l = 3$ or more, the pitch angle of the lines on a given magnetic surface will be different from those on surfaces inside or outside it. As a consequence, the interchange of lines of force at different distances from the axis cannot be achieved without considerable distortion. This will generally result in the interchange being energetically unfavorable, thereby making the system more stable. Since the lines of force of the magnetic field under consideration change direction with distance from the axis, it has the property of shear. As seen in §7.96, the stabilizing effect of shear in the magnetic field extends beyond the particular case discussed here.

8.45. When $l = 2$, it is evident from equation (8.4) that the situation is a singular one in which there is no shear. However, this value of l is particularly favorable for the cancellation of particle drifts since it can yield large rotational transform angles. If l were greater than 3, the result would be increased shear, but in order to obtain sufficiently large transverse fields near the axis, very large fields must be generated near the external windings, and this could require the prohibitive expenditure of power in the coils. An optimum arrangement has been found to be a combination of $l = 2$ and $l = 3$, one on each curved section of the containing tube (see §8.123).

8.46. The stabilizing effect of the transverse, helical magnetic field is opposed by the kinetic pressure of the plasma particles. Hence, if a certain pressure is exceeded, the system becomes inherently unstable. There is, consequently, a critical value of β associated with a given helical field above which stability is not possible. Neglecting the effect of curvature of the stellarator tube, calculations indicate that, when a transverse, helical field for which $l = 3$ is superimposed on an axial confining field, the critical value of β may be as large as 0.1 or possibly 0.2. However, the theory upon which the computations are based is admittedly approximate, and in any event the conditions for maximum β may be outside the range of validity of the theory. It is felt at present, therefore, that the critical value of β for which a stellarator plasma is stable can probably be determined only by experiment.

8.47. The influence of the helical field on the kink instability is quite complex, but a simple approach can be used to indicate the conditions that might be expected to maximize the value of the limiting current. It can be seen from equation (8.8) that I_c may be increased by increasing either ι or $2\pi - \iota$. Since the latter alternative requires ι to be small, the choice is between making ι as large as possible, but less than 2π, or as small as possible, but greater than zero. According to equation (8.4), the value of ι_h is proportional to A^2, the other parameters remaining constant, and so the rotational transform angle can be changed by varying the strength of the helical magnetic field. Hence ι can be made either large or small, to provide the conditions for the maximum value of the Kruskal limiting current.

8.48. For practical purposes, in a stellarator system for the production of thermonuclear energy, it is preferable to increase the critical current by increasing ι, that is, by increasing the strength of the transverse, helical field. As indicated earlier, the critical value of β, above which the plasma is unstable, is related to the helical field strength. It would therefore appear to be advantageous to operate under such conditions as to provide a large rotational transform angle.

STELLARATOR OPERATIONS

PLASMA HEATING

8.49. After initial breakdown of the gas, possibly by means of a radio-frequency pulse discharge (§5.2), the remainder of the ionization and preliminary heating of the deuterium or deuterium-tritium plasma will be produced by an axial electric field, applied in the manner described in §5.6

et seq. In other words, ohmic heating is proposed for the first stage of raising the temperature of the thermonuclear reactants. The voltage gradient in the discharge tube would be of the order of 0.1 volt/cm, which is at least an order of magnitude less than that used to constrict and heat the plasma in the pinch effect.

8.50. In experimental stellarator devices, which will be described shortly, the ohmic heating voltage pulse, produced by discharge of a capacitor bank through the primary of a transformer, has an approximately rectangular wave form; this is referred to as *constant voltage* operation (cf. Fig. 8.15). As ionization proceeds, the electron density generally increases, apart from the effect of certain instabilities, and the plasma current begins to rise rapidly. When ionization is complete, the increase in temperature causes the resistance of the plasma to decrease, and the current increases correspondingly.

8.51. Since there is a possibility that, in its rapid initial rise, the current may exceed the Kruskal limit and the plasma may consequently develop the kink instability, an alternative mode of ohmic heating, at essentially *constant current*, has been employed. A resistance is inserted in the primary circuit, thereby increasing its impedance, so that there is an upper limit to the heating current. The current in the plasma is thus maintained approximately constant, after an initial rise, during the course of the heating pulse. By the use of a suitable resistance, the constant current can be kept at a value below that at which the onset of the kink instability may be expected.

8.52. Because the maximum plasma current is limited and approximately constant during the so-called constant current discharge, the voltage across the discharge decreases as ionization proceeds. Since the ohmic heating current is transformer coupled, there is a maximum product of voltage and time that is available. As a result, the heating current at constant current lasts longer than at constant voltage for the same stored energy. Hence, constant current operation is sometimes referred to as slow heating, in contrast to the faster heating at constant voltage.

8.53. The maximum electron temperatures that can be reasonably expected from ohmic heating in a stellarator are around 100 ev and the ion temperatures may be somewhat lower. To heat the ions to temperatures of thermonuclear interest, e.g., 10 kev or above, it has been suggested that either magnetic pumping (§5.51) or ion cyclotron resonance heating (§5.72) could be employed. As regards magnetic pumping, the conditions would at first be best suited to collisional heating; and although in principle this method could be used to heat the ions in a fully ionized plasma to very high temperatures, its application would actually be limited. The reason is that the rate of collisional heating varies approximately as $T^{-\frac{1}{2}}$, so that it falls off with increasing temperature. It is of interest to record that, since the rate of energy absorption is greatest at low temperatures, collisional heating by magnetic pumping was at one time considered as an alternative to ohmic heating in a stellarator, but the idea was discarded because, according to simple theory, it is not an effective method for ionizing the gas.

8.54. The final stage of heating, to reach the thermonuclear ignition temperature, might be performed by magnetic pumping with the frequency of the field oscillations adjusted to give transit time heating. Since the optimum frequency increases with temperature, the magnetic pumping frequency should be changed correspondingly as the temperature rises. If this can be achieved, transit time heating might be an effective method for attaining very high ion temperatures, particularly as the rate of energy absorption varies as $T^{3/2}$.

8.55. Ion cyclotron resonance heating appears to be an attractive alternative to magnetic pumping as a means of raising ion temperatures from about 20 ev to the kilo-electron volt range. There are reasons for believing [20] that at moderate β values, such as would probably exist in a stellarator system in order for the plasma to be stable, the ion cyclotron resonance method of heating, especially in a system of small volume, may have the advantage over magnetic pumping in the respect that coupling between the external field and the plasma may be better. In other words, a larger proportion of the energy of the radiofrequency magnetic field is utilized for heating the plasma in the former case.

THE DIVERTOR

8.56. The *divertor* was conceived as a basic component of a stellarator reactor system to serve three main purposes [1]. In the first place, it was intended to be a device for reducing the number of energetic ions, heated by the thermonuclear reaction, that would impinge on the walls of the discharge tube. In an operating reactor, the heat generated in the walls by bremsstrahlung and neutron absorption alone would represent a difficult cooling problem without the additional heat produced by the charged particles. The situation becomes worse with increasing radius of the stellarator discharge tube, as is desirable from other considerations (§8.34), since the rate of energy input to the walls increases as the square of the radius whereas the surface area available for heat removal is proportional only to its first power. One objective of a divertor would thus be to transfer the energetic charged particles to a region where a large surface area facilitates transfer of heat.

8.57. Even if the heat generated in the walls of the discharge tube were not excessively large, it would still be necessary, since it operates as a steady-state device, to remove continuously from the stellarator the products of the thermonuclear reactions, namely, helium and ordinary hydrogen ions. This represents the second function of a divertor and can, of course, be fulfilled by the transfer of the ions to a region where their energy is removed. The atoms resulting from recombination of the cold ions with electrons can then be extracted by vacuum pumps.

8.58. The third purpose of a divertor is to reduce the influx of impurities into the plasma and thus minimize the resulting energy loss.   In the absence of a divertor, the ions, after transferring much of their energy to the walls, will diffuse back into the discharge and absorb energy from the hotter plasma. Moreover, the impact of the energetic ions (and photons) upon the walls results in the sputtering of ions or atoms of fairly high atomic number into the plasma. The loss of energy from the discharge at high temperatures, in the form of bremsstrahlung, would thus be greatly increased (§2.74).   The presence of heavy atomic or ionic impurities also hinders the early stages of the plasma heating process because the occurrence of several excited electronic states facilitates the loss of energy as radiation (§12.70).

8.59.   The basic principle of the divertor is that, by the use of a reversed magnetic field, the thin outer cylindrical shell of magnetic flux, near the stellarator walls, is brought out locally from the main discharge tube and spread out into a wider chamber.   Charged particles diffusing outward from the body of the plasma will enter the magnetic flux shell and follow the lines of force into the chamber where they strike collector plates to which they transfer their energy as heat.   In essence, the effect of the divertor is to surround the main discharge by a protective sheath or scrape-off layer (cf. §8.62) which leads to an auxiliary chamber where heat can be removed and the resulting cooled products of the thermonuclear reaction, as well as impurities, can be pumped off.

8.60.   The divertor chamber is cylindrically symmetrical, but narrower at the end which has an opening into the stellarator discharge tube, as represented in the section in Fig. 8.13.   The diversion of the flux tube into the chamber is achieved by means of the divertor coil in which flows the same current that produces the confining magnetic field, but in the opposite direction.   The reversal of current flow is indicated by the dots and crosses in the figure, in which the bending of the lines of force by the divertor coil is also shown. By means of a vacuum pump, the particles diverted into the chamber are continuously removed.

8.61.   The design of the coil system for a divertor is a difficult matter, but it has been facilitated by the development of a resistance analogue capable of solving a variety of magnetic field problems in which there is axial symmetry [21.]   By means of the analogue, the configuration of the magnetic field in the divertor and in the adjacent region of the discharge tube can be plotted for various coil designs and divertor dimensions.   The characteristics of the divertor constructed for the Model B-65 stellarator (§8.89) are indicated in Fig. 8.14.   Since the divertor has cylindrical symmetry, only one quarter section is shown; the other three are similar.

8.62.   The effective *magnetic aperture* of the stellarator tube with a divertor is determined by the radius, measured in the region where the magnetic field has not been affected, of the flux tube which represents the boundary between the tubes passing by the divertor and those guided into it.   The intersection

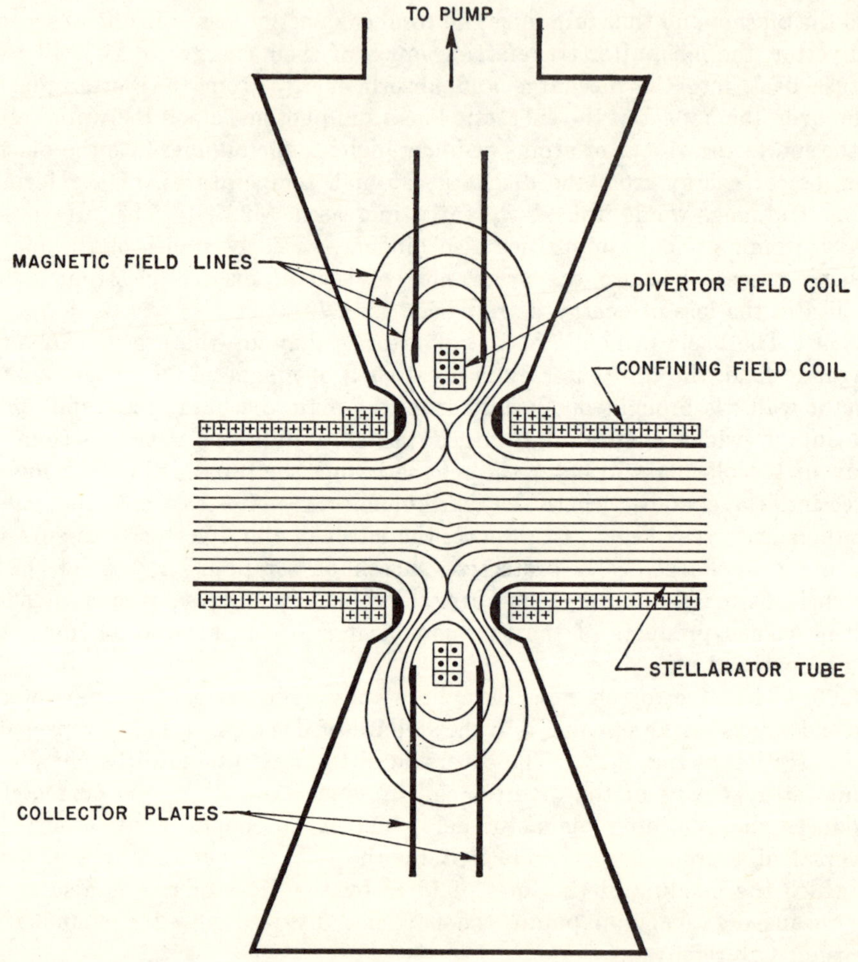

FIG. 8.13. Section through divertor chamber.

of this boundary flux tube with a plane through the axis of the coil system is called the *diverted line*. A similar line associated with the smallest flux tube which intersects, or is tangential to, the wall of the discharge tube at some point is referred to as the *limiting line*. The region between the boundary and tangent flux tubes has been termed the *sheath* or *scrape-off layer* and the distance between the diverted and limiting lines is known as the *scrape-off distance*. In Fig. 8.14 the aperture is seen to be 2.64 cm, and the scrape-off distance is 1.40 cm.

8.63. The particular device to which Fig. 8.14 refers was intended to test the effectiveness of a divertor in maintaining a low concentration of impurities in a plasma. Any material sputtered from the wall of the discharge tube will

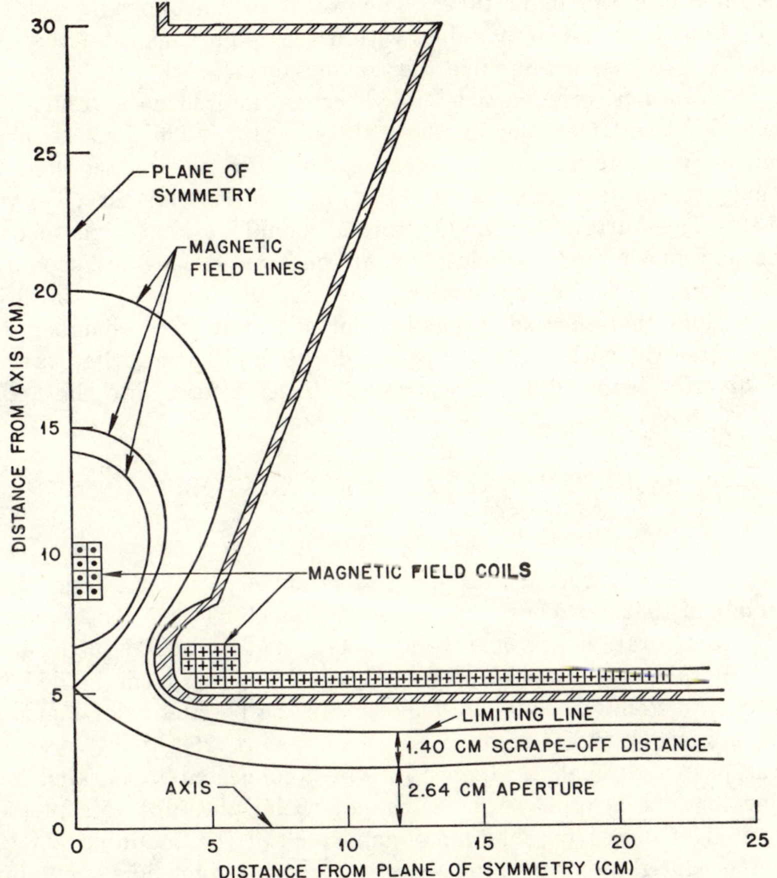

FIG. 8.14. Magnetic field configuration in divertor region.

first enter the scrape-off layer, and if it is in ionized form, it will tend to follow the magnetic lines of force and so be led into the divertor. If these impurities are to be kept out of the plasma, the back flow of neutral particles from the divertor chamber to the main discharge tube must be minimized. The same requirement would apply, of course, to the thermonuclear reaction products which would have to be removed from an operating reactor.

8.64. It happens that there are a number of circumstances which help to reduce the back flow of diverted particles. Most of these particles will be released from the collector plates in the neutral form, and it is possible, at least in principle, to remove a large proportion by pumping. If the rate of removal does not keep pace with the rate of release, the back flow of neutral atoms into the main discharge tube is then determined by the pressure attained in the divertor chamber and the gas-kinetic conductance of the opening from this chamber into the main tube. The width of this opening (or throat) should be kept as small as possible, within the requirements of the divertor magnetic field, so as to minimize the conductance.

8.65. Released neutral particles may become ionized as a result of collisions with ions and electrons entering the divertor. The ions so produced will tend to pass into the main discharge, but may well be reflected by the strong magnetic mirror existing at the throat of the divertor chamber (cf. Fig. 8.13). The mirror ratio of the throat should be kept high to decrease the back flow of charged particles; but at the same time, there should be no magnetic mirror effect for ions entering the divertor from the main discharge tube. Any ions that succeed in passing out of the divertor chamber through the mirror are still within the scrape-off sheath and so may be drawn back into the divertor before diffusing across the lines of force into the main discharge.

EXPERIMENTAL STELLARATORS*

PRELIMINARY STELLARATOR MODELS

8.66. The first experimental stellarator device, called Model A, was operated late in 1952. It consisted of a 5-cm-diameter Pyrex glass tube having a figure-8 configuration and an overall axial length of about 350 cm. The radius of curvature of the end sections was just over 15 cm. The purpose of the Model A stellarator was to demonstrate the possibility of ionizing and confining a gas. A radiofrequency current, from a 250-kilocycle/sec oscillator, was passed through a number of wires, connected in parallel, outside the stellarator tube in order to ionize the air within the tube at a pressure of 2×10^{-4} mm of mercury. The maximum value of the confining axial magnetic field, produced by means of a steady current in a solenoid surrounding the tube, was 1400 gauss. It was observed that, to produce a measurable discharge, the strength of the magnetic field required for the stellarator was an order of magnitude less than when the same Pyrex glass tubing was assembled in the form of a race track. This result was taken as indicating an improved confinement of the plasma by a magnetic field in the figure-8 geometry, as compared with that in an analogous system possessing no rotational transform.

8.67. The axial magnetic field in the Model A device was weak and confinement was poor, believed to be due to plasma oscillations and turbulence which caused diffusion of the particles to the walls. Nevertheless, the results obtained were considered to be of sufficient promise to justify further investigation of plasma confinement in a stellarator system. A new experimental apparatus, known as Model B-1, was therefore built, with the figure-8 configuration, in order to make further tests of confinement by the use of much

---

*This section is based on information contained in a number of papers presented at USAEC Project Sherwood Conferences [22, 23].

stronger magnetic fields than in the Model A. Another objective was to study ohmic heating of the plasma by means of an induced longitudinal current pulse. Preliminary operation of the B-1 device started in 1954, and full operation, with an ultra-high vacuum system, began two years later.

8.68. The Model B-1 stellarator tube was made of flexible, stainless steel hose, about 5 cm in diameter and 450 cm axial length. The radius of curvature of the bends was increased to roughly 25 cm, in order to provide greater uniformity of the magnetic field across the tube diameter without much increase in length. The angle of tilt of the figure-8 was 49°, and so the rotational transform angle was 196°, i.e., 1.09π radians. Helium was selected as the experimental gas, rather than hydrogen or deuterium, because it lends itself particularly to spectroscopic diagnostic measurement, e.g., of electron and ion temperatures (cf. §6.41 *et seq.*). The base pressure to which the stellarator tube could be evacuated was in the vicinity of 10^{-5} mm, and the discharges were made at helium gas pressures in the range of 5×10^{-5} to 5×10^{-3} mm of mercury.

8.69. The 250-kilocycle/sec radiofrequency discharge for initial breakdown was applied across an insulating gap in the metal tube and the ohmic heating current was induced in the plasma by the discharge of capacitors through the primary windings of a Hypersil transformer, the plasma forming the secondary circuit (cf. Fig. 5.5). The energy for exciting the axial magnetic field was supplied by another bank of capacitors which was "crowbarred," i.e., short-circuited, shortly after the peak current was reached so as to maintain the magnetic field strength (cf. §7.120). The magnetizing current then fell exponentially toward zero with a time constant of some 40 millisec. The energy of the capacitor bank was sufficient to produce an axial field of 50,000 gauss around the stellarator, but because of the displacement of the coils which occurred when high fields were used, the system was usually operated at 10,000 to 15,000 gauss maximum.

8.70. To study the configuration of the magnetic field, a small electron gun, capable of emitting a narrow beam of 200-volt electrons parallel to the field, was inserted into the stellarator tube. With a moderate pressure of air, e.g., about 10^{-4} mm of mercury, in the tube the electron beam produced a visible track which mapped out part of a magnetic surface. The formation of several tracks, representing successive passages of the electron beam past a viewing port, provided visual demonstration of the existence of a rotational transform in the magnetic field. By moving the electron gun radially outward until one of the tracks began to decrease, indicating that the electron beam was striking the tube wall, it was possible to determine the magnetic aperture of the stellarator. In the absence of divertor action this is the cross-sectional area of the first magnetic surface which just touches the wall.

8.71. Electron gun experiments, sometimes made in conjunction with a fluorescent screen mounted diagonally across the discharge tube in such a manner that the luminosity can be seen from an observation port, have been used for various purposes. One of these is to adjust the solenoid coils so as to make the axial field as uniform as possible and give a maximum magnetic aperture. During preliminary operation of the Model B-1 stellarator, for example, the diameter of the aperture was only 2 cm, although the total tube diameter was 5 cm. Later, the aperture was increased to 3 cm.

8.72. A typical operating sequence of the Model B-1 and other stellarators is depicted in Fig. 8.15. The capacitor bank is first discharged through the

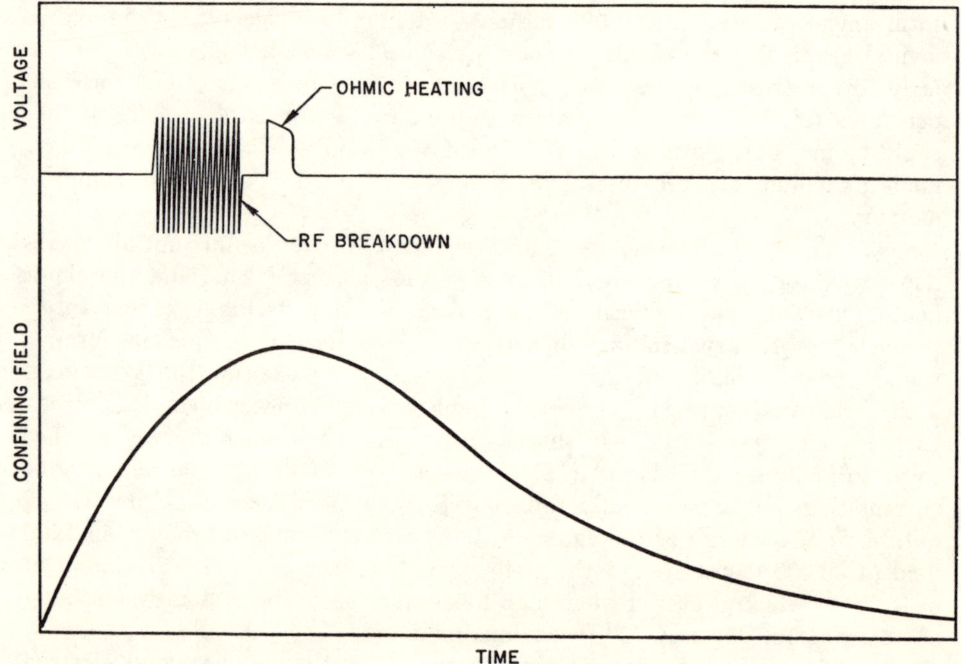

Fig. 8.15. Stellarator operating sequence.

solenoidal coils to provide the axial magnetic field. As the approximately constant field strength is approached, the radiofrequency current is applied. Following a further delay, the ohmic heating (constant voltage) pulse is initiated. This coincides roughly with the attainment of the peak strength of the magnetic field.

8.73. In the experiments with the Model B-1, measurements were made of the time variation of the loop voltage and of the plasma current during the discharge, in addition to spectroscopic, X-ray, and microwave observations. The results showed that the plasma ionizes and heats more rapidly the greater the strength of the axial magnetic field. This was interpreted as implying that confinement improves with the field strength. From the electrical resistivity of the plasma and the intensities of the lines in the helium spectrum, it appeared that an electron temperature of 10 to 100 ev was attained for a very short time during the ohmic heating pulse, although Doppler broadening of spectral lines indicated an appreciably lower ion temperature. Spectroscopic observations also showed that impurities were present in the plasma; these probably resulted from the bombardment of the walls of the stellarator tube by energetic charged particles.

8.74. One of the most interesting observations made with the Model B-1 stellarator was in connection with the X-rays emitted from the discharge tube quite early in the heating pulse. The photon energies covered a considerable range, but the maximum number had energies of about 150 kev. These were undoubtedly bremsstrahlung arising from the deceleration of runaway electrons (§5.10) with at least this energy. Since the heating voltage, which accelerates the electrons, was only 100 volts per turn, it appears that the electrons producing the X-rays must have traversed the stellarator tube at least 1500 times. This result was taken as clear proof that single particle confinement is provided by the geometry of the figure-8 stellarator.*

8.75. In a modified form of the Model B-1, made of Pyrex glass, the detection of X-rays produced by runaway electrons was improved by the use of an aperture limiter; this consisted of a tantalum or tungsten disc, with a hole about 2.5 cm in diameter in the center, placed across the stellarator discharge tube. Essentially all the runaway electrons eventually ran into the limiter and the thick-target bremsstrahlung was thus produced at a definite location in a known material, instead of being widely distributed over the tube walls. The X-rays penetrated a thin "window" near to the limiter and reached the detector without having to pass through the magnetic field coils surrounding the discharge tube. X-ray photons with energies up to 1.4 Mev were observed, indicating that the runaway electrons made as many as 20,000 circuits of the tube before striking the aperture limiter. For lower confining fields the maximum energy of the X-rays was less, implying shorter particle confinement times, in general agreement with expectation. Incidentally, microwave measurements of electron density also indicated that confinement of the plasma became less effective with decreasing strength of the axial magnetic field.

8.76. When the B-1 device was first operated, it was observed that the plasma current rose rapidly when the ohmic heating voltage was first applied and then fell to a much lower value, instead of continuing to rise throughout the heating pulse. This result was taken to indicate that heating of the plasma was being limited by contamination of the discharge by impurities. That such was the case was confirmed by the appearance of spectroscopic lines of various

---

* There is a theoretical limit to the energy of runaway electrons in a stellarator, since the electrons cannot be accelerated beyond the point at which the drift in going around a bend in the tube is greater than the radius of the aperture.

species other than helium. It was concluded, therefore, that a cleaner system was desirable if the attainment of high temperatures, as well as improved confinement, were to be achieved.

LATER STELLARATOR MODELS

8.77. A result of the effort in the direction of reducing impurities was the development of the bakable version of the Model B-1 stellarator. Like the original form, this was made of stainless steel tubing and had the same dimensions. Demountable, hard-seal joints, together with gold wire seals, were used to insure vacuum tightness even after several heating and cooling cycles [24]. A short section of alumina ceramic was welded into the figure-8 tube to prevent the metal walls from short-circuiting the ohmic heating transformer and to provide a point for applying the 250-kilocycle/sec radiofrequency pulse for initial breakdown of the gas in the tube. An aperture limiter of tungsten, with a hole 3.2 cm in diameter centered on the magnetic axis, served to define the size of the discharge column. It also provided a reference potential for Langmuir probe measurements (§6.82),* as well as a localized bremsstrahlung target for runaway electrons. The field coil mountings permitted the use of axial magnetic fields up to about 30,000 gauss.

8.78. Apart from the special vacuum-tight seals, the essential difference between this stellarator model and previous designs is that it could be thoroughly "baked out" so as to remove adsorbed impurities. By passing a low-voltage current of about 600 amp through the metal walls, the tube, which was well insulated thermally, could be heated to 450°C. As a result of baking the stellarator tube and the stainless steel connecting line to the pump in this manner for some 12 hours, and using an improved vacuum system, base pressures of 3×10^{-10} mm of mercury were attained, in comparison with normal values of 10^{-5} mm in unbaked systems.

8.79. The bakable Model B-1 stellarator commenced operation in October 1956, and it was soon evident that the influx of impurities into the plasma was greatly decreased. Systematic observations of the discharge phenomena were then possible because of the improved reproducibility of the data. Nevertheless, it was concluded that the influx of material from the tube wall still had a significant effect on the behavior of the plasma under most operating conditions.

8.80. Stellarator Model B-2, with a figure-8 geometry and a rotational transform angle of 196°, was completed in April 1956, before the development of the bakable vacuum system. It was intended, in particular, for the study of heating by magnetic pumping. As in the Pyrex form of Model B-1, the glass tube had a diameter of 5 cm, but in order to accommodate a magnetic pumping

---

* Neither Langmuir probes nor magnetic probes were found useful in the active region of the discharge of this stellarator. Although probe studies were possible in earlier devices with higher impurity levels, the high plasma temperatures and abundant runaway electrons of the baked system soon destroyed the probe.

bulge, about 50 cm long and nearly 15 cm in diameter, the axial length was increased to 600 cm.   A seven-turn coil around the bulge, connected to a 240-kilocycle/sec oscillator, provided the magnetic pumping field.   The maximum strength of the axial magnetic field in the main discharge tube was 18,000 gauss and, without the magnetic pumping field, the field strength in the bulge was approximately one-sixth of this value.

8.81.   Some preliminary measurements were made with the Model B-2, but the results were greatly influenced by impurities from the walls of the tube. The entry of these impurities into the plasma appeared to be enhanced by the magnetic pumping field.   In order to reduce this source of disturbance, the Model B-2 stellarator was rebuilt in bakable form, using stainless steel tubing instead of Pyrex glass.

8.82.   A further improvement in the Model B-2 stellarator is represented by Model B-3, which also has the figure-8 geometry (Fig. 8.16).   The angle of twist, however, is 41°, so that the rotational transform angle is 164°, i.e., 0.81π radians.   The model B-3 is made of stainless steel tubing, about 5 cm in diameter, and has a total axial length of 640 cm.   As in earlier steel models, a ring of alumina ceramic was inserted in the tube to serve as an insulator.   The inside diameter of this ring, approximately 4.4 cm, defines the magnetic aperture

FIG. 8.16.   The Model B-3 stellarator.   (Courtesy Project Matterhorn, Princeton University)

of the plasma.   All parts of the vacuum system and the discharge tube can be baked at 450°C, so as to give the low base pressure characteristic of bakable systems.   In addition, it is planned to include a divertor in the Model B-3, to reduce possible effects due to residual impurities.

8.83.   The water-cooled copper conductors constituting the axial magnetic field coils can carry currents up to 9000 amp, to produce a field of 50,000 gauss. The supports of the coil sections are designed to withstand normal operating stresses at fields up to 100,000 gauss.   Although the axial confining field has a rotational transform, due to the figure-8 geometry, the Model B-3 stellarator also has helical windings to take advantage of the anticipated contribution of the transverse, helical field to stabilization of the plasma.

FLEXIBLE STELLARATOR MODELS

8.84.   The Model B-64 stellarator, to which reference was made in §8.18, was designed primarily as a flexible experimental tool.   For this reason, it was constructed from two basic units, namely, 10-cm diameter straight tubes and 90° elbows of stainless steel, so that its dimensions and rotational transform could be changed and other components added as desired.   Each structural unit has flanges at its ends and the flanges of adjacent pieces are clamped together, with a rubber ring providing a vacuum seal.   Where electrical insulation is desired, a Teflon ring, protected by ceramic, is inserted between the flanges at a coupling.   The magnetic field coils are wound directly on the tube sections.   This greatly simplifies the mounting problem but precludes a high-temperature bake-out.

8.85.   By making the lengths of the legs in one of the V-shaped ends of the distorted toroid in the ratio of 2 to 1, a rotational transform angle of 74°, i.e., 0.41π radians, was obtained.   Viewing ports at the ends of the straight pieces make it possible to look down the whole length of each horizontal tube, thereby improving the intensities of the spectral lines observed during the discharge.*   As originally constructed, the exterior length of the Model B-64 stellarator was roughly 105 cm and its width was 74 cm.   Since, as stated above, the device does not lend itself to baking, the lowest base pressure attainable is about 10^{-6} mm of mercury.

8.86.   Because of its flexibility, the Model B-64 was readily adapted to include, first, a divertor and, subsequently, both a divertor and facilities for magnetic pumping.   For the latter purpose, a Pyrex glass bulge was added to one of the horizontal tubes and a stainless steel straight section of the same length was included in the other tube.   A diagrammatic representation of the Model B-64 stellarator with divertor but no magnetic pumping system is given in Fig. 8.17.

8.87.   Two other devices have been built using the same type of flexible construction as the Model B-64, although the rotational transforms are obtained

---

*This feature has also been incorporated into other stellarator models.

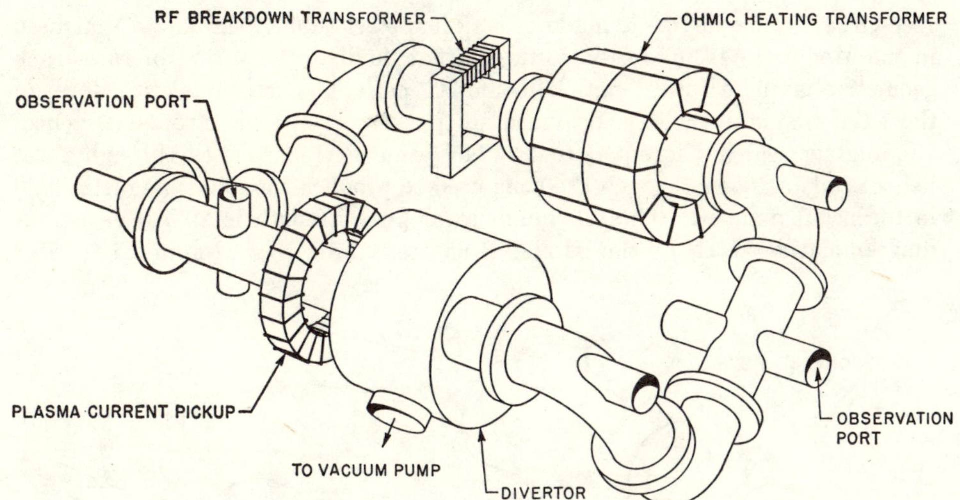

Fig. 8.17.   The Model B-64 stellarator.

in different ways.   In the Model B-64-0, so named because the rotational transform angle in the absence of a longitudinal plasma current is 0°, the 10-cm-
diameter, stainless steel straight tubes and elbows are assembled in the geometry of a flat torus (or race track) with straight sides and 90° bends.   The axial
magnetic field is generated by discharge of a capacitor bank through water-
cooled coils in the usual manner.   The rotational transform is provided by the
ohmic heating current which can be varied within limits.   Under conditions of
zero current, the plasma cannot be in an equilibrium state since there is then
no rotational transform of the magnetic field.

8.88.   The main purpose of the Model B-64-0 stellarator was to study the
ohmic heating behavior of a plasma using relatively large currents.   Since the
rotational transform angle of the axial magnetic field itself is zero, the critical
current for the onset of the kink instability is given by the second form of
equation (8.8) as $10\pi Br^2/L$ amp.   By making the axial length small in comparison with other stellarator models, large heating currents are possible without the danger of kink instability of the plasma.   For example, with an axial
field of 19,000 gauss and an ohmic heating field of 1.24 volts/cm, which is much
higher than is generally used in stellarator studies, the current of 8000 amp
used for heating the plasma was only about half the Kruskal limit.   Another
objective of the Model B-64-0 was to observe the effect of zero rotational transform on the initiation of the discharge current as well as on the confinement
properties of the device when the ohmic heating current has returned to zero.

8.89.   The second modification, known as Model B-65, was built primarily
to investigate heating of a plasma by ion cyclotron resonance (§5.72) and to
make a quantitative study of the effect of a divertor in reducing impurities

(see §8.61).   The device is made of stainless steel tubing, similar to that used in the Model B-64, but in the form of a flat, undistorted toroid of race-track geometry having a total length of some 207 cm.   The rotational transform of the axial magnetic field is provided by passing a current through a helical winding for which l is equal to 3.   The same current passes through these helical coils as flows through the coils used to produce the axial magnetic field in the usual manner.   The maximum axial field strength is 20,000 gauss.   A diagrammatic sketch of the Model B-65 stellarator is shown in Fig. 8.18.*

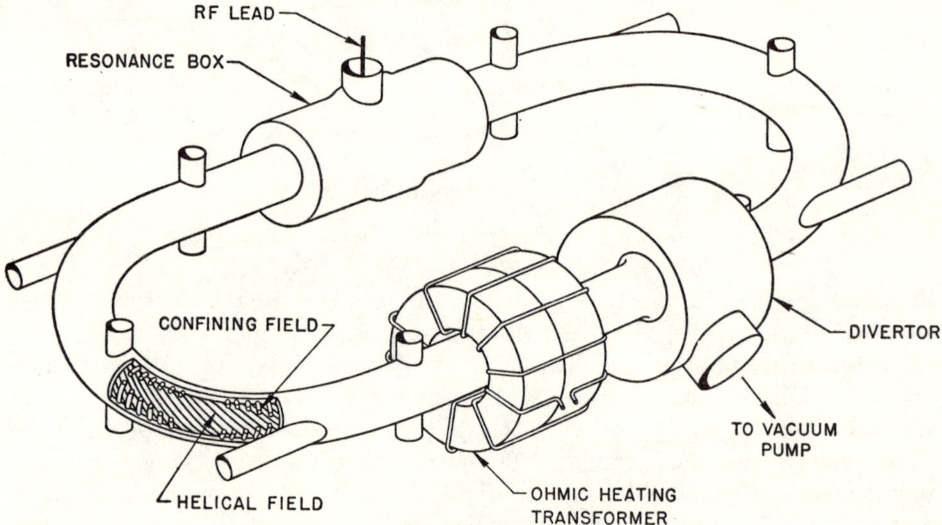

FIG. 8.18.   The Model B-65 stellarator.

The ohmic heating field is 0.24 volt/cm and the plasma current is usually about 2000 amp, which is considerably less than the kink instability limit.   The vacuum system is unbaked, so that the base pressure is about 1×10^{-6} mm of mercury.

RESULTS OF EXPERIMENTAL STUDIES

OHMIC HEATING

8.90.   Among the significant results obtained with the baked Model B-1 stellarator are those connected with ohmic heating and the Kruskal limiting current [15, 25].   Most of the experiments were made with helium in the discharge tube, but the phenomena observed with hydrogen and deuterium, which behaved very much alike, were generally similar, although there were also

* A description of the resonance box for studying ion cyclotron resonance heating is given in §8.108.

some differences.   However, the information summarized below appears to be applicable, in general, to both helium and the isotopic forms of hydrogen.

8.91.   For moderately high ohmic heating fields, under so-called constant voltage conditions, the plasma current rises fairly rapidly to a peak value which is essentially independent of the applied voltage and the gas pressure. The results obtained with helium in the discharge tube are given in Fig. 8.19;

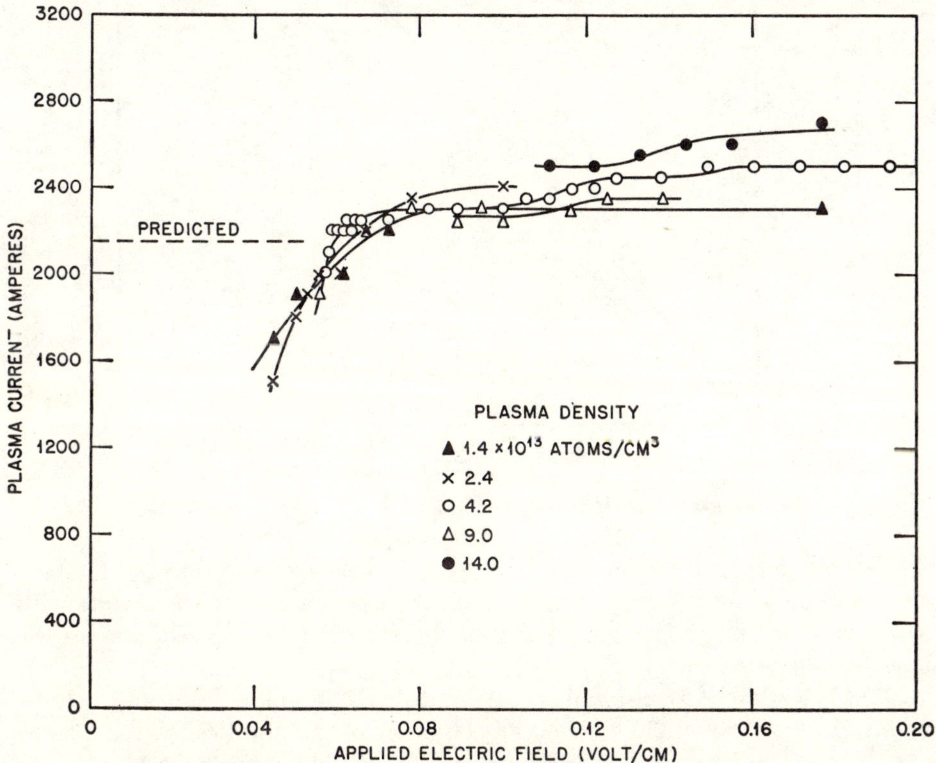

FIG. 8.19.   Development of kink instability in stellarator system.

the maximum plasma current attained is seen to be close to the calculated Kruskal limit.   The dependence of the observed peak current on the strength of the axial, confining magnetic field, in the range of 10,000 to 27,000 gauss, is indicated in Fig. 8.20.   The two sets of points are for the longitudinal plasma currents flowing in opposite directions.   The linear nature of the plots of the peak current versus the confining field strength is in harmony with equation (8.8) for the Kruskal limit.

8.92.   The rotational transform angle of the Model B-1 stellarator is 1.09π radians; hence, it follows from the two alternative forms of equation (8.8) that the slopes of the lines in Fig. 8.20 should be in the ratio of $\iota/(2\pi - \iota)$, i.e., 1.19.

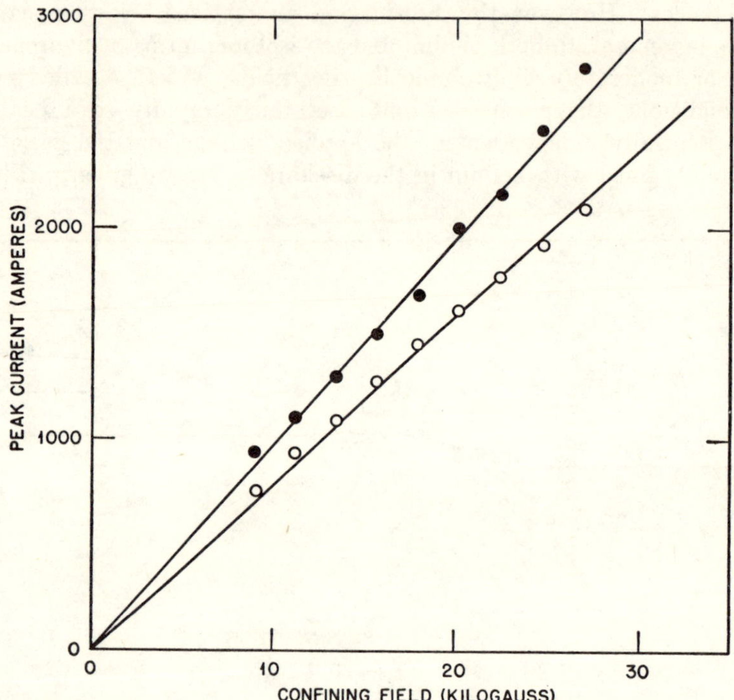

FIG. 8.20. Dependence of limiting current on strength of confining field.

The ratio found from the figure is 1.22, the difference between this and the theoretical value being within the limits of experimental error. Experiments with hydrogen show that under equivalent conditions the limiting currents are the same as in helium.

8.93. When the Kruskal limit is attained, both plasma current and voltage exhibit violent fluctuations, indicating the onset of instabilities. At the same time the spectral lines of impurities, principally carbon, oxygen, and hydrogen, are very intense in a helium discharge. Because the plasma is carried to the walls by the kink instability, currents greater than the Kruskal limit may be expected to be difficult to attain. However, it can be seen from Fig. 8.19 that if the electric field strength is high enough, the current does exceed the theoretical limit. Similar results have been reported for the Models B-2 and B-3, with axial fields up to 53,000 gauss in the latter, but in every case there is a break in the plot of peak current versus ohmic heating field which corresponds closely to the theoretical critical current. It may be mentioned that, in experiments with the Model B-64, currents as large as five times the Kruskal limit were observed. It is probable that the steel wall of the discharge tube contributed to the conduction of the current as a result of contact with the unstable plasma.

8.94. By operating at approximately constant current conditions (§8.51), the current may be prevented from reaching the Kruskal limit. Electrical resistivity measurements indicated that, in the bakable Model B-1 stellarator, an electron temperature of 90 ev has been attained by ohmic heating in helium gas at an initial pressure of 5×10^{-4} mm of mercury. The result is subject to some uncertainty because a significant fraction of the plasma current may be carried by runaway electrons and even the other, apparently thermal, electrons probably do not have a Maxwellian distribution of velocities. Electron temperatures derived from the relative intensities of singlet and triplet lines of helium (§6.72) are also unreliable, because the velocity distribution may not be Maxwellian, as well as for other reasons, but they are in general agreement with those determined from the electrical resistivity of the plasma.

8.95. With the high ohmic heating fields and large currents used in the Model B-64-0, the Doppler broadening of the helium spectral lines indicated that an ion temperature as high as 100 ev had been attained. Calculations show that the actual rate of heating of the ions was greater than the maximum estimated from classical electron collisions. It has been concluded, therefore, that under the experimental conditions some type of cooperative phenomenon (or instability) was responsible for the enhanced rate of ion heating.

8.96. Measurements by the microwave method, made in the bakable Models B-1 and B-3, show that, after the initial increase due to ionization of the gas, the electron density falls during the heating pulse [26, 27]. About 200 microsec after the electron density has reached its peak value it has decreased by a factor of 2. This loss of particles, frequently referred to as *pump-out*, occurs at plasma currents well below those required to produce the kink (or $m = 1$) instability. Observations made with hydrogen gas have been interpreted as implying that the average time spent by an electron-ion pair in the discharge is only about 0.1 millisec. The reason for the brief confinement period of the plasma during the ohmic heating pulse is not understood. It is due to some kind of instability, but there is evidence against its being hydromagnetic in nature [27]. After a short time, the electron density may increase again as a result of re-emission of helium or other atoms from the wall and their subsequent ionization in the discharge.

8.97. There is other evidence, in addition to pump-out, that the plasma is not quiescent during the ohmic heating phase, even at moderate electric field strengths when the current rises smoothly and rapidly to the Kruskal limit [28]. The emission of X-rays from the aperture limiter during the rise of the heating current suggests that some of the electrons are confined by the axial magnetic field of the stellarator for periods of not more than 0.1 millisec or so, depending upon the field strength. The fact that the thick-target bremsstrahlung emission starts abruptly, rather than gradually, implies that some kind of internal disturbance in the discharge suddenly begins to carry runaway electrons to the limiter. Other evidence supports the view that the instability develops fairly abruptly.

8.98. When the current reaches or slightly exceeds the Kruskal limit, the emission of X-rays ceases. This result is to be expected, since it is not likely that electrons will be confined and accelerated to sufficiently high energies to produce observable X-rays during the time that the kink instability dominates the discharge. Further, runaway electrons already present in the plasma can presumably no longer be confined, once the instability sets in.

PLASMA PHENOMENA AFTER HEATING PULSE

8.99. In contrast to the poor confinement while the heating pulse is being applied, the plasma can exhibit a long decay time under certain conditions. If the constant-voltage heating field is cut off (or crowbarred) within less than about 0.2 millisec from its initiation, that is, before much gas has been lost by pump-out or the walls have been heavily bombarded, the electron density falls rapidly at first and then more slowly. In the latter phase, when the plasma is undoubtedly quiescent, the decay time constant or $1/e$-folding time, i.e., the time for the electron density to change by a factor of $1/e$, may be as long as 8 millisec for a confining field of nearly 30,000 gauss. The appearance of a large burst of X-rays some 10 millisec after crowbar time indicates that many runaway electrons have been confined for this period.

8.100. In experiments in which the voltage was crowbarred at a later stage in the ohmic heating pulse, the electron density decay time constant was decreased to about 1 millisec. Presumably the impurity ions which enter the plasma during the discharge facilitate the removal of the electrons by recombination.

8.101. If the constant heating voltage is crowbarred after the Kruskal limit has been reached, the current decays rapidly to zero and no X-rays are emitted during the decay period. However, if the crowbar is applied while the current is still rising, X-ray emission occurs and may continue for several milliseconds after the heating current has ceased. In general, the earlier the crowbar time during the period of initial X-ray emission, the greater the total X-ray intensity after the heating voltage is cut off. More X-rays are produced in the crowbarred discharge than when the heating current, in a constant voltage pulse, is allowed to decay in the normal manner. This result implies that, in the latter situation, electrons are lost from the discharge without producing detectable X-rays. A possible explanation of this fact, which receives support from other observations to be described below (§8.103), is that, when the voltage is not crowbarred but is allowed to decay normally, some of the energy gained by runaway electrons can later be absorbed into the plasma as a result of non-radiative interactions.

8.102. After the voltage is crowbarred the residual current has been observed to decay in one or other of two different ways [28]. In the first case, the plasma current falls immediately to a low constant value, some 5 to 10 per cent of its initial value, where it remains for several milliseconds, and then drops abruptly almost to zero. This step in the current coincides with the emission of a burst of X-rays and an increase, instead of the previous decrease, in the electron density, as indicated by microwave measurements. If the discharge period before the voltage is crowbarred is increased, the second type of current decrease is observed after crowbar time. It consists of a rapid succession of thee or four steps, each being accompanied by a much more intense X-ray emission than is associated with the single, larger step in the other case. Moreover, at the small steps there is no evidence of any increase in the electron density.

8.103. When a single step only is observed, that is, when the discharge period before crowbar time is relatively short, it is possible that a cooperative process decelerates many of the runaway electrons and they are suddenly reabsorbed into the main body of the plasma, as suggested above. The energy thereby gained causes reionization to occur and this accounts for the observed increase in electron density. Only a small proportion of the unconfined runaway electrons strike the aperture limiter and produce bremsstrahlung. In such circumstances that current decay occurs in several small steps, each accompanied by strong X-ray emission, it appears that groups of runaway electrons suddenly lose confinement and are led out of the discharge to the limiter.

8.104. Experiments on the rate of decay of the electron density in the Model B-64-0, as determined by microwave and Langmuir probe measurements, after the end of the ohmic heating pulse, makes possible a comparison of confinement by magnetic fields with and without rotational transforms. The decay time in the Model B-64-0, which had zero transform angle when the heating current ceased, was found to be appreciably less than in the Model B-64. Nevertheless, it was still greater than might have been expected. It appears that some factors are operative in the Model B-64-0, which has a divertor, that counteract the drift of electrons and ions in opposite directions in the planar toroidal tube. One possibility is that charge neutralization may occur at the so-called stagnation point of the divertor where the magnetic field strength is zero and charged particles can flow across the lines of force. In addition, transfer of electrons to regions where the ions are in excess may occur by flow along the magnetic lines of force to the conducting wall of the discharge tube.

HEATING BY MAGNETIC PUMPING

8.105. The experiments on heating by magnetic pumping [29] made with the Model B-2 stellarator were largely vitiated by impurities which entered the plasma during the discharge. The frequency of the pumping oscillator, namely, 240 kilocycles/sec, was such as to make heating most efficient for ions having temperatures of at least 30 ev. But owing to the influx of impurities, the ion temperatures attained by preliminary ohmic heating probably did not exceed about 5 to 8 ev. Magnetic pumping thus produced no detectable heating, although it did seem to increase the amount of impurities in the plasma.

8.106.   An unexpected effect was observed when the magnetic field modulation (§5.57) was relatively high, e.g., about 0.7.   Even without previous ohmic heating, the magnetic pumping field caused very rapid ionization and heating of the helium gas in the stellarator tube.   The electrons were being heated very effectively and it is estimated that temperatures of about 100 ev were attained. The ion temperatures were believed to be in the vicinity of 15 to 20 ev.   In view of the very short time in which these temperatures were reached, it appears probable that the ions were being heated directly rather than by collisions with energetic electrons as is the case in ohmic heating.

8.107.   Since ion temperatures of about 50 ev had been indicated in the Model B-64 stellarator with a divertor, it was hoped that heating by magnetic pumping would be more effective in this system [30].   However, when the device was modified to include a magnetic pumping facility, it was not possible to attain such high ion temperatures.   This result was attributed to an increase in the impurities in the plasma when the magnetic pumping oscillator was operating.   The presence of a 35-kilocycle/sec "hash" in the discharge voltage indicated plasma instabilities which may have contributed to wall bombardment and the consequent release of impurities.   There was some indication that magnetic pumping did cause ion heating, but the observed effect was not large.

HEATING BY ION CYCLOTRON RESONANCE

8.108.   Direct observation of heating by ion cyclotron resonance would require the use of a radiofrequency oscillator of high power.   However, measurements made with an induction coil, coaxial with the plasma, and energized with radiofrequency power provide definite indication that the expected energy absorption by the plasma does occur (cf. §5.76).   The Model B-65 stellarator, with divertor, which was described in §8.89, was used for the experiments on ion cyclotron resonance [31].   The resonance box, which is shown in greater detail in Fig. 8.21, contains a ceramic (Mullite) tube, 53 cm long, around which is

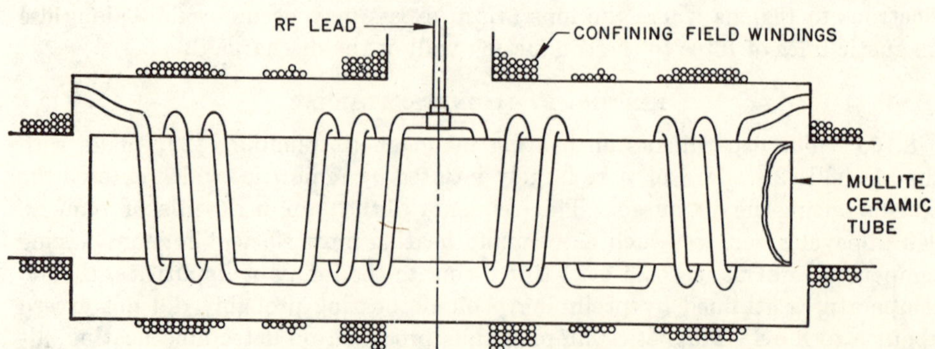

Fig. 8.21.   Resonance box for generation of ion cyclotron waves.

wound the induction coil. The box is inserted in a section of the stellarator tube where specially designed field windings provide an essentially uniform magnetic field. The induction coil is seen to consist of four sections arranged so that the radiofrequency current, which is brought in at the center of the coil while the two ends are grounded, flows in opposite directions in adjacent sections. The coil is thus two wave lengths long and this establishes the wave length of the ion cyclotron waves which should be generated.

8.109. After initial breakdown of the deuterium gas in the discharge tube at a pressure of 10^{-3} mm, an ohmic heating field of 0.24 volt/cm was applied for about 0.3 millisec at the time when the axial magnetic field strength had become approximately constant. The ohmic heating field was then turned off and the ion cyclotron resonance coil was energized at a frequency of 11.5 megacycles/sec for 2 millisec. At this time, the input voltage and current were measured and from these the impedance of the resonant circuit could be determined; this is related to the radiofrequency power absorbed by the plasma, i.e., the plasma loading. By repeating the observations over a range of axial magnetic field strengths, results of the type described in §5.77 were obtained. These indicated that ion cyclotron waves were indeed formed and that they absorbed energy from the radiofrequency field.

EXPERIMENTS WITH DIVERTOR

8.110. The first tests of the effectiveness of a divertor in decreasing the influx of impurities into the plasma during a discharge were made with the Model B-64 stellarator. The results were encouraging from the outset. Whereas in the (unbaked) Model B-1, the electron temperature fell after a few tenths of a millisecond of heating, owing to the cooling by impurities, in the Model B-64 with a divertor it rose to 50 ev or more and remained there for several milliseconds. In fact, both electron and ion temperatures as high as 90 ev were reported. In the Model B-64 without the divertor the maximum electron temperatures were almost independent of the axial magnetic field strength, from 3000 to 18,000 gauss, indicating that the effect of impurities dominates that of the confining field. With the divertor in action, however, the electron temperature attained increased with the field strength.

8.111. A more detailed investigation of divertor action was made in the Model B-65 [32]. This device has helical coils for providing a rotational transform to the axial magnetic field. For most of the experiments, however, they were not used, since they decreased the thickness of the scrape-off layer (§8.62) from 1.40 cm to 0.66 cm. The rotational transform was then that arising from the longitudinal current of the ohmic heating field. In order to make a reliable comparison between the behavior with and without the divertor, an aperture limiter was inserted in the discharge tube to make the magnetic apertures equal in both cases. Observations were made alternately, first with the divertor in action and the limiter retracted, and then with the divertor inactive and the limiter inserted.   The axial field strength was 20,000 gauss and the peak current of 3500 amp was somewhat less than half of the Kruskal limiting value.   The base pressure in the discharge tube was 2×10^{-6} mm and the experiments were made with helium at a pressure of 1×10^{-3} mm of mercury.

8.112.   From the curves in Fig. 8.22, which represent the variations in inten-

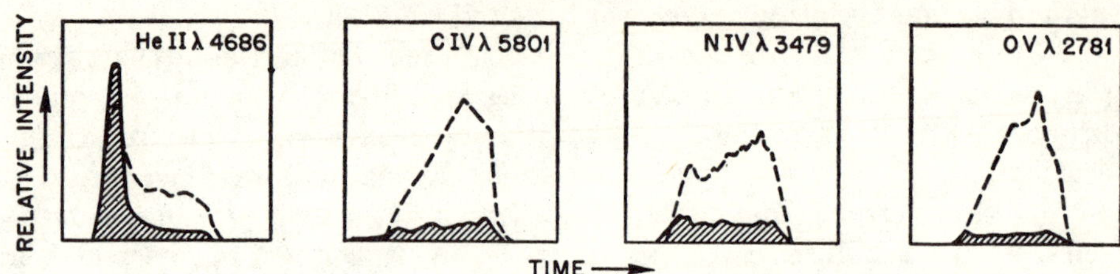

FIG. 8.22.   Effect of divertor in decreasing impurities.

sity of several spectral lines as a function of time during the course of the approximately 1.5-millisec ohmic heating pulse, it is immediately apparent that the divertor markedly decreased the impurities in the discharge.   The full lines indicate the intensities with the divertor in operation, and the broken lines show the corresponding intensities without the divertor.   The increased intensity of the helium (He II) line at 4686 A suggests that higher temperatures are attained with the divertor, and this has been confirmed by determination of the ion temperature from the Doppler broadening.   With the divertor, a maximum temperature of 60 ev was estimated for the helium ions, compared to 40 ev without the divertor.   A temperature as high as 130 ev was indicated from the O V lines with the divertor in action; this may have been due to spatial variations of temperature in the discharge.   The ability to attain higher ion temperatures when a divertor is used is undoubtedly associated with the reduction in the amount of impurities entering the plasma from the walls of the discharge tube.

8.113.   Increased heating accompanying the use of a divertor was also observed when the helical windings of the Model B-65 stellarator were energized. However, the ion temperatures attained were not as high as those given above, as was to be expected from the larger scrape-off distance of the system when the rotational transform was only that provided by the ohmic heating current.

SUMMARY AND CONCLUSIONS

RESULTS OF EXPERIMENTS

8.114.   The results obtained with the various experimental stellarators have shown that it is remarkably easy to bring about virtually complete ionization of a gas and to attain electron and ion temperatures of 50 to 100 ev by ohmic heating. In some cases the observed ion temperatures have been higher than expected and it appears that processes other than transfer of energy from accelerated electrons are responsible. The emission of X-rays, produced by runaway electrons, even during the early stages of the ohmic heating pulse, suggests that the plasma is by no means quiescent. The pump-out action which occurs at about the same time is also indicative of some type of plasma instability. In addition to these instabilities of uncertain origin, the occurrence of the catastrophic, hydromagnetic kink instability, at current values in excellent agreement with theory, has been definitely established.

8.115. Although the temperatures achieved by ohmic heating may be regarded as very satisfactory, the inexplicable instabilities make plasma confinement during the heating pulse disappointingly poor. The runaway electrons may possibly produce oscillations which facilitate diffusion of charged particles across the lines of force so that they reach the walls. There is also a possibility that interchange instability or higher modes $(m > 1)$ of the kink instability may exist during the ohmic heating pulse. No matter what the cause, the very short confinement time represents a serious problem that must be solved if ohmic heating is to be used in a practical stellarator reactor for power production.

8.116. Detailed theoretical calculations have been made on ohmic heating of hydrogen and helium in stellarators in which the electron and ion temperatures are followed [33-35]. The observed behavior through the ionization phase is generally what was anticipated, but because of the simplifications made in the theoretical treatment, quantitative agreement with experiment is not to be expected. The calculations postulate that the energy distribution remains Maxwellian throughout the ionization and heating processes. But the existence of runaway electrons and the probable distortion of the electron energy distribution by the large heating fields show that this cannot be the case in practice.

8.117. A fact of some interest is that, in spite of poor plasma confinement, single-particle confinement in the stellarator system is quite marked. The emission of X-rays for some time after the termination of the heating pulse shows that the runaway electrons may be confined for several milliseconds, even though the strength of the axial magnetic field has then decreased appreciably. These electrons may make as many as half a million circuits of the discharge tube while the plasma is quiescent; such a situation would not be possible in a simple torus, where the particle drifts are of the order of one radius of gyration in each circuit of the tube.

8.118. Assuming adequate confinement can be achieved, methods must be developed for raising the temperature of the plasma above 100 ev, since this is about the practical limit for ohmic heating. The procedures being considered, namely, magnetic pumping and ion cyclotron resonance, both depend upon the use of radiofrequency, oscillating electromagnetic fields. No definite conclusions concerning heating by magnetic pumping can be drawn from the tests made so far. Owing to the presence of impurities, it was not possible to raise the plasma temperature by ohmic heating to the point where magnetic pumping is expected to be effective. Furthermore, the radiofrequency power available has not been sufficient to offset the losses from the plasma resulting from poor confinement and excessive amounts of impurities.

8.119. Direct evidence of plasma heating by ion cyclotron resonance has been obtained. The measurements made so far indicate definite resonant loading of the plasma for frequencies of the external circuit at and also somewhat below those corresponding to ion resonance in the axial magnetic field. A prerequisite for efficient heating is that the energy fed into the plasma exceeds that dissipated in the external circuit, and the data indicate that this is the case in the experiments at low power. At low values of β, i.e., the ratio of the material pressure to the external magnetic pressure, and in small systems, theory indicates that ion cyclotron resonance should have an advantage over magnetic pumping as a method of heating, since the coupling between the external circuits and the plasma is much better in the former case. For a large system, however, and at moderate β values, it is not apparent from theory which method of heating would be the more effective.

8.120. The early experimental studies with the stellarator have brought to light the difficulty of keeping impurities out of the discharge. If high plasma temperatures are to be attained, it is mandatory that the influx of impurities be kept at a minimum. Two techniques show promise of providing at least a partial solution to this problem. Baking of the vacuum system, including the discharge tube, at 450°C for several hours removes some of the impurities adsorbed on the stainless steel walls and makes possible base pressures of the order of 10^{-10} mm of mercury. Further, by means of a divertor, the impurities produced as a result of bombardment of the walls by energetic charged particles and photons can be largely prevented from entering the main body of the discharge. The effectiveness of the divertor in the experiments described above appears to have been limited by the diffusion of the plasma across the field lines, combined with a thin scrape-off distance. Provided the influx of impurities from the walls is due to ion bombardment, and not to bombardment by photons, it is probable that the divertor action can be improved by better magnetic alignment.

THE MODEL C STELLARATOR

8.121. In reviewing the foregoing summary, it will be evident that further development of the stellarator as a device for attaining thermonuclear reactions depends on the possibility of improving the confinement of the plasma, especially during the ohmic heating pulse. There are two fairly obvious ways in which this can be achieved. The first is to increase the strength of the axial magnetic field. This would decrease both the drift velocity of the charged particles across the lines of force and the radius of gyration about the lines of force, as may be seen from equations (4.11) and (4.4), respectively. Apart from the effects of instabilities, an increase in the field strength, therefore, may be expected to decrease the rate at which charged particles strike the walls of the discharge tube.

8.122. Secondly, if cooperative phenomena drive the plasma particles to the wall by a process of diffusion, the confinement time, for a given field strength and particle density, should increase as the square of the diameter of the tube. Qualitatively, the reason for this is as follows: the rate at which charged particles escape across the magnetic field lines is proportional to the pressure gradient from the interior to the exterior of the plasma, and this is related to n/r, where n is the particle density and r is the tube radius. The distance a particle has to travel in order to escape from confinement is proportional to r, and so the confinement time is determined, roughly, by r^2/n. Hence, an improvement in confinement should result from the use of wider stellarator tubes.

8.123. In the event, however, that the difficulties associated with ohmic heating should persist, there remains the possibility that high-modulation magnetic pumping may be used for the initial ionization and heating of the plasma, thus eliminating altogether the need for net longitudinal currents.

8.124. With these and other objectives in mind, the Model C stellarator has been designed as a large but flexible experimental facility [36]. It is made of stainless steel tubing, 20.4 cm internal diameter, in the form of a planar race track having a total axial length of 1200 cm (nearly 40 ft).* The rotational transform, 0.95π radian, is provided by an $l = 3$ transverse, helical magnetic field at one U bend and an $l = 2$ field at the other. These fields are produced by conductors wound helically around the discharge tube with current flowing in opposite directions in adjacent conductors. The axial magnetic field strength is 55,000 gauss for intermittent operation. By using DC generators as the source of current, the axial field can be held constant for periods up to 1 sec or so. In addition to being bakable, the Model C will have a divertor, with the possibility of including another later, to minimize the impurities in the plasma. Some benefit in this respect is also anticipated from the large size of the discharge tube. The ratio of plasma volume to wall surface area increases with the tube radius, and so it is hoped that the effect of impurities coming from the wall will be less than in earlier stellarator models.

8.125. The current for the primary circuit of the ohmic heating transformer is supplied by a capacitor bank in the usual manner. However, in order to insure rapid heating the current in the discharge tube secondary must be increased as rapidly as possible to a value just below the Kruskal limit, which is about 23,000 amp, and kept constant until the desired temperature, in the vicinity of 100 ev, is reached. To do this, a system has been developed for

---

* An earlier design was based on a figure-8 geometry, but it was discarded in favor of one using transverse, helical fields because of the expected improvement in plasma stability [37].

modified constant-voltage ohmic heating while the current is increasing.   The voltage across the plasma is maintained approximately constant at a high value by means of a feedback circuit which supplies sufficient voltage to compensate for the L (dI/dt) drop of the plasma loop.   In the Model C stellarator, both the self-inductance of the plasma and the current are large, as compared with other models, and so a high constant voltage is necessary to insure a rapid increase in the plasma current.*   When the desired current is attained, the voltage is dropped to maintain it essentially constant for a period of time.

8.126.   In order to raise the plasma temperature in the Model C stellarator above 100 ev, heating by both magnetic pumping and ion cyclotron resonance will be tried.   It should then be possible to determine which of these procedures is more effective in a large system.

REFERENCES FOR CHAPTER 8

1. L. Spitzer, USAEC Report NYO-993 (1951).
2. L. Spitzer, *Phys. Fluids*, **1**, 253 (1958); also in *Proc. Second U.N. Conf. on Peaceful Uses of Atomic Energy*, **32**, 181 (1958).
3. L. Spitzer, USAEC Report NYO-995 (1951).
4. M. D. Kruskal, USAEC Report NYO-998 (1952).
5. L. Spitzer, USAEC Report NYO-7316 (1957).
6. J. L. Johnson, C. R. Oberman, R. M. Kulsrud, and E. A. Frieman, USAEC Report NYO-7904 (1958).
7. J. L. Johnson, C. R. Oberman, R. M. Kulsrud, and E. A. Frieman, *Phys. Fluids*, **1**, 281 (1958); also in *Proc. Second U.N. Conf. on Peaceful Uses of Atomic Energy*, **31**, 198 (1958).
8. B. B. Kadmotsev and S. I. Braginsky, *Proc. Second U.N. Conf. on Peaceful Uses of Atomic Energy*, **32**, 233 (1958).
9. G. Gibson, W. C. Jordan, and E. J. Lauer, *Phys. Rev. Letters*, **4**, 217 (1960).
10. L. Spitzer, USAEC Report NYO-997 (1952).
11. A. Simon, *An Introduction to Thermonuclear Research*, Pergamon Press, Inc., 1959, p. 57.
12. M. D. Kruskal and R. M. Kulsrud, *Phys. Fluids*, **1**, 265 (1958); also in *Proc. Second U.N. Conf. on Peaceful Uses of Atomic Energy*, **31**, 213 (1958).
13. L. Spitzer, D. J. Grove, W. E. Johnson, L. Tonks, and W. F. Westendorp, USAEC Report NYO-6047 (1954).   This report contains much interesting discussion of the stellarator as a useful source of power.
14. M. D. Kruskal, USAEC Report NYO-6045 (1954).
15. M. D. Kruskal, J. L. Johnson, M. B. Gottlieb, and L. M. Goldman, *Phys. Fluids*, **1**, 421 (1958); also in *Proc. Second U.N. Conf. on Peaceful Uses of Atomic Energy*, **32**, 217 (1958).
16. W. Bernstein, A. Z. Kranz, and F. Tenney, Division of Plasma Physics APS, Monterey Conf., D 15 (1959).
17. E. Teller, USAEC Report TID-7520 (1956), p. 10.
18. M. N. Rosenbluth and C. L. Longmire, *Ann. Physics*, **1**, 120 (1957).

---

*Since $V \approx L$ (dI/dt) if the plasma resistance is low, the time taken to attain a current I is approximately equal to LI/V, where L is the self-inductance and V the applied voltage.

19. L. Spitzer, in R. M. Kulsrud, E. A. Frieman, and J. L. Johnson, USAEC Report TID-7503 (1955), p. 232.
20. T. H. Stix, USAEC Report TID-7558 (1958), p. 389.
21. K. E. Wakefield, USAEC Report TID-7503 (1955), p. 75.
22. USAEC Reports WASH-115 (1952); WASH-146 (1953); WASH-184 (1954); WASH-289 (1955); TID-7503 (1955); TID-7520 (1956); TID-7536 (1957); TID-7558 (1958).
23. A. S. Bishop, *Project Sherwood*, Addison-Wesley Publishing Co., Inc., 1958, Chapters 4 and 11.
24. D. J. Grove, *Trans. Fifth Natl. Symp. on Vacuum Techniques* (1958), p. 9.
25. T. Coor, S. P. Cunningham, R. A. Ellis, M. A. Heald, and A. Z. Kranz, *Phys. Fluids*, **1**, 411 (1958); also in *Proc. Second U.N. Conf. on Peaceful Uses of Atomic Energy*, **32**, 201 (1958).
26. W. Bernstein, M. A. Heald, and A. Z. Kranz, USAEC Report NYO-7901 (1957).
27. R. A. Ellis, L. P. Goldberg, and J. G. Gorman, USAEC Report MATT-27 (1960); *Phys. Fluids*, **3**, 468 (1960).
28. W. Bernstein, F. F. Chen, M. A. Heald, and A. Z. Kranz, *Phys. Fluids*, **1**, 430 (1958); also in *Proc. Second U.N. Conf. on Peaceful Uses of Atomic Energy*, **32**, 210 (1958).
29. E. B. Meservey, M. S. Jones, S. P. Cunningham, and L. M. Goldman, USAEC Report TID-7520 (1956), p. 224.
30. M. S. Jones, E. B. Meservey, R. W. Palladino, and T. H. Stix, USAEC Report TID-7536 (1957), p. 326.
31. T. H. Stix and R. W. Palladino, *Phys. Fluids*, **1**, 446 (1958); also in *Proc. Second U.N. Conf. on Peaceful Uses of Atomic Energy*, **31**, 282 (1958).
32. C. R. Burnett, D. J. Grove, R. W. Palladino, T. H. Stix, and K. E. Wakefield, *Phys. Fluids*, **1**, 438 (1958); also in *Proc. Second U.N. Conf. on Peaceful Uses of Atomic Energy*, **32**, 225 (1958).
33. J. M. Berger and E. A. Frieman, USAEC Report NYO-6043 (1953).
34. J. M. Berger et al., USAEC Reports NYO-7311 and 7312 (1956).
35. J. M. Berger, I. R. Bernstein, L. M. Goldman, and R. M. Kulsrud, *Phys. Fluids* **1**, 297 (1958); also in *Proc. Second U.N. Conf. on Peaceful Uses of Atomic Energy*, **32**, 197 (1958).
36. Members of the Project Matterhorn Staff, USAEC Report NYO-7899 (1957).
37. USAEC Report NYO-3709 (1956).

Chapter 9

MAGNETIC MIRROR SYSTEMS

===

THEORY OF MAGNETIC MIRRORS *

INTRODUCTION

9.1.   The primary purpose of magnetic mirror systems in controlled thermonuclear research is to minimize the loss of charged particles from the ends when
a plasma is confined, by means of an externally applied axial magnetic field,
in a straight cylindrical tube [1-4].   A charged particle moving into a region
where the lines of force converge, owing to an increase in the field strength,
tends to have its direction of motion reversed (§3.44).   In other words, the
stronger field can act as a magnetic mirror and reflect the particle back into
the region where the field is weaker.   Hence, if the axial magnetic field at both
ends of a straight tube is stronger than it is in the middle, it is to be expected
that the escape of charged particles from the ends will be greatly decreased.

9.2.   In addition to its confinement function, a magnetic mirror system can
be used to compress a plasma, both radially and axially, thereby increasing its
density and also raising its temperature (§5.18).   Since magnetic mirrors,
which are produced by passing a current through solenoidal coils, can be moved
either mechanically or electrically, they can also be utilized to transfer a
plasma from one region to another.   Thus, magnetic mirrors can serve several
important purposes in the study of controlled thermonuclear processes and,
possibly, in the realization of thermonuclear power.

9.3.   In the subsequent theoretical treatment it will be supposed that the
plasma is contained in a straight tube.   The axial confining field is produced
by current passing through coils wound around the tube.   At the ends the
magnetic fields are stronger than in the central region, thus causing the required mirror effect (Fig. 3.4).   A satisfactory magnetic mirror configuration
can also be obtained by means of solenoids at the ends of the tube only.   The
length of the central region, where the magnetic field is taken to be uniform,
will usually not be specified.   Provided the plasma density is not too high, this

---

* In translations from the Russian literature, such terms as magnetic "plugs," "stoppers,"
and "traps" are used.

336

length does not affect the essential ability of the mirror system to confine charged particles (cf. §9.32). If the strengths of the mirror fields are not the same at both ends of the containing tube, escape of the particles will be determined by the weaker of the two fields. As a rule, therefore, the mirror field strengths will be assumed to be equal.

9.4. The general principle of charged-particle confinement by magnetic mirrors does not require the magnetic field to be symmetrical about the axis of the tube. However, such symmetry will be postulated, since it is found to be desirable for a number of reasons. Perhaps the most important is the following. Because of the higher field strength in the mirror region, the magnetic field is inhomogeneous and the lines of force are curved, as indicated in Fig. 3.5. Consequently, the charged particles of the confined plasma will tend to drift in opposite directions across the field. As explained earlier (§4.18), the charge separation resulting from these drifts will produce an electric field which, in conjunction with the magnetic field, will cause a general movement of the plasma toward the walls of the containing tube. But if the confining magnetic field has axial symmetry, the particle drifts will also possess this symmetry. The particles will thus exhibit a rotational drift around the axis, the ions and electrons traveling in opposite directions. Since each flow then closes upon itself, there will be little or no tendency for charge separation to occur, and so the drift of the plasma to the walls due to the inhomogeneous nature of the field will not be significant.

ADIABATIC INVARIANTS: MAGNETIC MOMENT

9.5. The confinement of a plasma by a magnetic mirror system can be most simply explained in terms of the individual particle. In this case, use is made of certain properties called *adiabatic invariants* of the motion of a charged particle in a magnetic field. In the present sense, the term adiabatic invariant implies a quantity which remains essentially constant for slow changes, in both time and space, of the magnetic field.

9.6. The *magnetic moment* μ, associated with the rotational component of the helical motion of a charged particle in a magnetic field, is expressed in Gaussian-cgs units by

$$\mu = \frac{IA}{c}, \tag{9.1}$$

where I is the current produced by the particle moving in a circular orbit of area A. For the singly-charged particles in a hydrogen isotope plasma, the gyromagnetic radius of the orbit is given by equation (4.5), so that

$$\mu = I \times \frac{\pi}{c}\left(\frac{mv_\perp c}{eB}\right)^2, \tag{9.2}$$

the symbols having the significance given in §4.4. In the present case, the current is equal to e/τ_g, where τ_g is the gyromagnetic period, i.e., the time for the particle to make a single orbital rotation. Using equation (4.7) for the gyromagnetic period, it follows that

$$I = \frac{e^2 B}{2\pi mc}.$$

Upon substituting this value for I in equation (9.2), the result is

$$\mu = \frac{\tfrac{1}{2}mv_\perp^2}{B} = \frac{W_\perp}{B}, \tag{9.3}$$

where $\tfrac{1}{2}mv_\perp^2$, represented by $W_\perp$, is the component of the kinetic energy in the direction perpendicular to that of the magnetic field lines and B is the field strength at the line of force on which the guiding center of the particle is moving (§4.10).

9.7.   It is now required to show that the magnetic moment remains essentially constant when the magnetic field varies slowly in space and time. As far as the space variation is concerned, it will be postulated that the change in the magnetic field is small over a distance equal to the gyromagnetic radius r_g of the particle.   Since the magnetic force is exerted perpendicular to the circular motion of the particle, its angular momentum about the center of gyration will remain virtually constant.   Upon inserting the expression for r_g from equation (4.5), it is seen that

$$\frac{m^2 v_\perp^2 c}{eB} \approx \text{constant}.$$

Since c and e are constants, it follows immediately that

$$\frac{W_\perp}{B} = \mu \approx \text{constant}$$

under the conditions postulated.

9.8.   When a charged particle moves through a time-varying magnetic field, an electromotive force is induced which changes the energy of the particle. According to Faraday's electromagnetic induction equation, the emf, represented by V, induced by the field is given in Gaussian units by

$$V = -\frac{1}{c} \cdot \frac{d\phi}{dt},$$

where ϕ, equal to $\pi r_g^2 B$, is the flux at any instant of time through a circular path of radius r_g cm of a particle moving in a field of B gauss. The energy gain $\Delta W_\perp$ by the particle during one turn in the field is equal to $-eV$, where e is the charge on the particle, i.e., the electronic charge for a hydrogen isotope plasma.   The negative sign arises from the fact that $\Delta W_\perp$ is negative for an ion and positive for an electron.   It follows, therefore, that

$$\Delta W_\perp = \frac{\pi r_g^2 e}{c} \cdot \frac{dB}{dt}. \tag{9.4}$$

9.9. If the magnetic field in which the particle is moving remains essentially constant during one gyromagnetic period, $\Delta W_\perp$ will be relatively small. It is then permissible to set $dW_\perp/dt$ as approximately equal to $\Delta W_\perp/t$, where t is the time required for the particle to make a single turn, i.e., the gyromagnetic period given by equation (4.7); hence,

$$\frac{dW_\perp}{dt} \approx \Delta W_\perp \frac{eB}{2\pi mc}.$$

Upon introducing the expression in equation (9.4) for $\Delta W_\perp$, using equation (4.5) to eliminate r_g, and writing $W_\perp$ for $\frac{1}{2}mv_\perp^2$, it is found that

$$\frac{1}{W_\perp} \cdot \frac{dW_\perp}{dt} \approx \frac{1}{B} \cdot \frac{dB}{dt},$$

and so

$$\frac{W_\perp}{B} = \mu \approx \text{constant}.$$

Hence, the magnetic moment is also an adiabatic invariant for slow variation of the magnetic field with time.

9.10. Since $W_\perp/B$ is essentially constant in space and time for slowly varying fields, it follows that $v_\perp^2 r_g^2$ and $v_\perp^2/B$ are also constant, because $v_\perp^2$ is proportional to $W_\perp$ and r_g to $v_\perp/B$; hence,

$$r_g^2 B = \text{constant} \tag{9.5}$$

or

$$\pi r_g^2 B = \text{constant}.$$

The quantity $\pi r_g^2 B$ represents the magnetic flux through the gyromagnetic orbit circle of the particle; the flux thus remains constant in an adiabatic process of the kind described. In other words, the motion of the particle is such as to conserve the magnetic flux through the orbit circle.

9.11. The theory of particle trapping in a magnetic mirror configuration depends upon the constancy of the magnetic moment, and so it is important to examine the validity of this conclusion. In the arguments presented above, it was postulated that the variation in the magnetic field is small over a distance equal to the radius of a particle orbit. It is to be expected, therefore, on general grounds, that the magnetic moment might not be truly an adiabatic invariant in circumstances such that the radius of gyration of the charged particle is not infinitesimal in comparison with the dimensions of the magnetic field.

9.12. Detailed theoretical treatments have indicated that fluctuations in the magnetic moment associated with nonadiabatic effects are of importance only for relatively large particle orbits and decrease approximately exponentially with the reciprocal of the orbit radius [5, 6]. It should thus always be possible to scale the experimental conditions in such a manner as to make the nonadiabatic effects negligible. In any event, the calculations show that the deviations of the magnetic moment from a constant value, resulting from successive periods of back-and-forth motion of the trapped particles between the magnetic mirrors, are cyclic, rather than cumulative, in nature. The maximum fractional change in the magnetic moment is related to the magnetic field strength and the distance between the mirrors and by increasing these two quantities the deviation from strictly adiabatic behavior can be made negligible, provided that disturbing effects, such as collisions, are absent.

PARTICLE REFLECTION IN A MAGNETIC FIELD

9.13. For a charged particle moving freely in a magnetic field and undergoing no collisions with other particles, another invariant of the motion is the total kinetic energy (§4.11). That is to say, in moving along a line of force, the sum of the energy components of a particle perpendicular and parallel to the field axis, i.e., $W_\perp + W_\parallel$, should be a constant. Hence, for any two points on a given line of force, such as those indicated by a and b in Fig. 9.1,

$$W_{\perp(a)} + W_{\parallel(a)} = W_{\perp(b)} + W_{\parallel(b)}. \tag{9.6}$$

If B_a and B_b are the corresponding magnetic field strengths, then the invariance of the magnetic moment means that

$$\frac{W_{\perp(a)}}{B_a} = \frac{W_{\perp(b)}}{B_b},$$

and consequently equation (9.6) yields

$$W_{\parallel(b)} = W_{\parallel(a)} - W_{\perp(a)}\left(\frac{B_b}{B_a} - 1\right) \tag{9.7}$$

$$= W_{\parallel(a)} - W_{\perp(a)}(R - 1), \tag{9.8}$$

where R, defined by B_b/B_a, is the ratio of the magnetic field strengths at the two points.

9.14. When R is greater than unity, i.e., the magnetic field strength at b is greater than at a, as is the case in Fig. 9.1, then it follows from equation (9.8) that $W_{\parallel(b)}$ will be less than $W_{\parallel(a)}$. Hence, in moving into a stronger field, the

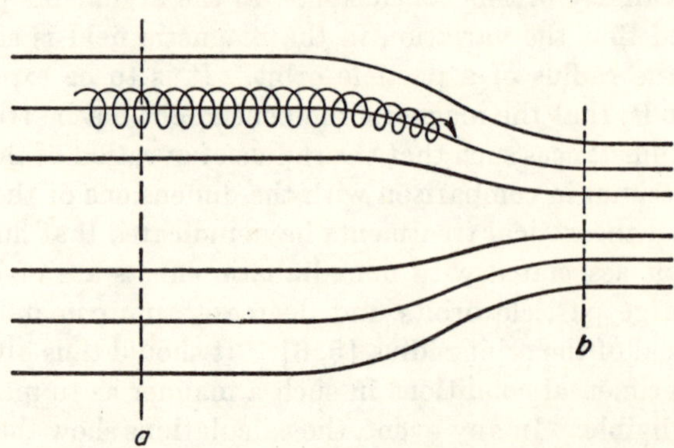

FIG. 9.1. Constancy of particle kinetic energy along a line of force.

velocity component of the particle along the field lines is decreased, so that it is slowed down. If the value of $W_{\parallel (a)}$ is not too large, the second term on the right of equation (9.8) may predominate. In this event, $W_{\parallel (b)}$ would be negative, which means that the motion of the particle along the field lines is reversed. In other words, the region of higher magnetic field can act as a *magnetic mirror* and reflect the particle back into the region of lower field.

CONDITIONS FOR REFLECTION BY A MAGNETIC MIRROR

9.15. The condition for reflection can be obtained from equation (9.8) by setting $W_{\parallel (b)}$ equal to (or less than) zero; the charged particle is then just stopped (or reversed) by the stronger magnetic field. Slight rearrangement gives this condition as

$$\frac{W_{\parallel (a)}}{W_{\perp (a)}} \leqslant R - 1, \tag{9.9}$$

which relates the magnetic field ratio for reflection to the initial ratio of the parallel and perpendicular components of the particle energy. Alternatively, since $W_{\perp (a)} + W_{\parallel (a)}$ is equal to the total kinetic energy W, the condition may be written as

$$\frac{W}{W_{\perp (a)}} \leqslant R. \tag{9.10}$$

9.16. In magnetic mirror systems of interest for plasma confinement, the magnetic field strength generally varies qualitatively in the manner shown in Fig. 9.2. There is a fairly uniform central region in which the field strength is

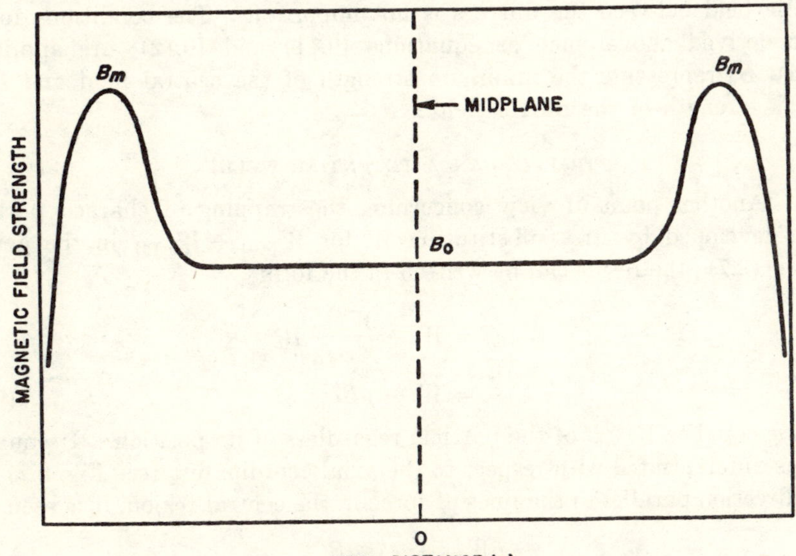

FIG. 9.2. Variation of field strength in magnetic mirror system.

B_0 and it rises to a maximum of B_m at the mirror peaks. The *mirror ratio* is then defined by

$$R_m \equiv \frac{B_m}{B_0}.$$

It follows, therefore, from equations (9.9) and (9.10) that the condition for a particle to be confined by the mirrors, either by being brought to rest or reflected back, is

$$\frac{W_{\parallel (0)}}{W_{\perp (0)}} \leqslant R_m - 1 \tag{9.11}$$

or

$$\frac{W}{W_{\perp (0)}} \leqslant R_m, \tag{9.12}$$

that is,

$$\frac{W_{\perp (0)}}{W} \geqslant \frac{1}{R_m}.$$

It is seen, therefore, that only particles having a component $W_\perp$ of energy perpendicular to the field lines that is a sufficiently large proportion of the total energy W will be reflected; all other particles will escape through the mirror fields.

9.17. It should be noted that the condition for confinement of a particle by a magnetic mirror system is determined by the *ratios* of energy components and of magnetic field strengths, and not by their absolute values. Furthermore, the results derived above are independent of the mass and charge of the particles. Another matter of interest is that the detailed configuration of the magnetic field between the mirrors is not important. The conditions for reflection derived above, such as equations (9.11) and (9.12), are applicable provided B_0 represents the minimum strength of the central field and B_m is the peak strength of the mirror field.

TRAPPING IN A POTENTIAL WELL

9.18. Another point of view concerning the trapping of charged particles can be developed by first substituting W for $W_{\perp (a)} + W_{\parallel (a)}$ in the general equation (9.7); the result can be written in the form

$$W_{\parallel (b)} = W - \frac{W_{\perp (a)}}{B_a} B_b$$

$$= W - \mu B_b, \tag{9.13}$$

since μ is equal to $W_\perp / B$ of the particle regardless of its position. If equation (9.13) is differentiated with respect to the axial coordinate z (see Fig. 9.2), i.e., in the direction parallel to the lines of force in the central region, it is seen that

$$\frac{\partial W_\parallel}{\partial z} = -\mu \frac{\partial B}{\partial z},$$

since W is independent of position on a given field line. But $\partial W_{\parallel}/\partial z$ is equal to the force, in the direction parallel to the field lines, exerted on the particle; hence, if a charged particle is moving into a region of magnetic force, it is acted upon by a retarding force F_z given by

$$F_z = -\mu \frac{\partial B}{\partial z}. \tag{9.14}$$

Because the force is the negative gradient of potential, this result may be expressed in the form

$$F_z = -\nabla_z(\mu B),$$

where the quantity μB is equivalent to a potential.

9.19. In view of the definition of R_m as B_m/B_0, the reflection condition in a magnetic mirror system may be written as

$$W \leqslant \frac{W_{\perp(0)}}{B_0} B_m$$

$$\leqslant \mu B_m. \tag{9.15}$$

It is evident, therefore, that a charged particle located in the region between two magnetic mirror fields of peak strength B_m may be regarded as being in a potential well between two potential maxima of height μB_m. Particles with total energy less than μB_m will be bound between the mirrors, but others can escape.

PARTICLE ENERGY AND PITCH ANGLE TRANSFORMATIONS

9.20. The manner in which the energy components of a particle change in its motion in a magnetic field of varying strength can be derived from an alternative form of equation (9.6), namely,

$$W_{\perp(a)} \left(\frac{W_{\parallel(a)}}{W_{\perp(a)}} + 1 \right) = W_{\perp(b)} \left(\frac{W_{\parallel(b)}}{W_{\perp(b)}} + 1 \right). \tag{9.16}$$

Because of the constancy of the magnetic moment, the ratio $W_{\perp(b)}/W_{\perp(a)}$ is equal to the ratio of the field strengths B_b/B_a, which is represented by R. Hence, equation (9.16) can be transformed into

$$\frac{W_{\parallel(a)}}{W_{\perp(a)}} = R \left[\frac{W_{\parallel(b)}}{W_{\perp(b)}} + 1 \right] - 1. \tag{9.17}$$

This expression relates the energy component ratio $W_{\parallel}/W_{\perp}$ at any point b to its value at another point a on the same line of force in terms of the ratio R of the field strengths at the two points.

9.21. If $v_{\perp}$ and $v_{\parallel}$ are the velocity components of a particle in the directions perpendicular and parallel, respectively, to the magnetic field axis, then

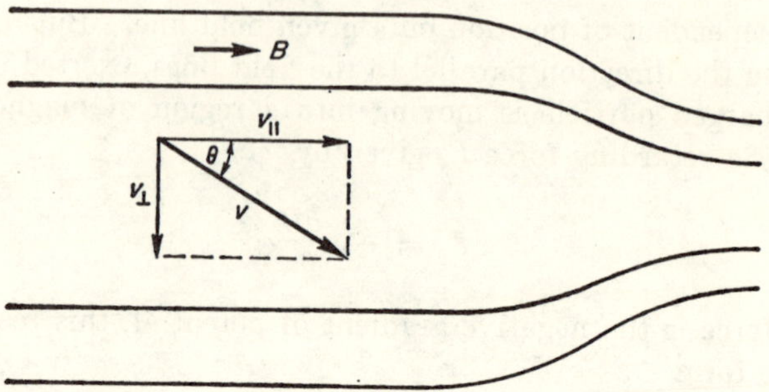

FIG. 9.3.   Pitch angle θ of particle in magnetic field.

the *pitch angle* θ, the angle the velocity vector of the particle makes with this axis (Fig. 9.3), is defined by

$$\cot \theta = \frac{v_\parallel}{v_\perp}. \tag{9.18}$$

Since $W_\parallel/W_\perp$, which is equal to $(v_\parallel/v_\perp)^2$, changes with the magnetic field strength, the magnitude of the pitch angle changes also.   Replacing the ratios $W_\parallel/W_\perp$ in equation (9.17) by the corresponding values of $\cot^2 \theta$, it is found that

$$\cot^2 \theta_a = R(\cot^2 \theta_b + 1) - 1.$$

Because $\cot^2 \theta + 1 = 1/\sin^2 \theta$, it follows, upon rearrangement, that

$$\sin^2 \theta_b = R \sin^2 \theta_a$$

or

$$\sin \theta_b = R^{1/2} \sin \theta_a, \tag{9.19}$$

which shows how the pitch angle varies with the field strength.

9.22.   If the pitch angles in the central region and at the peak of the mirror field are represented by θ_0 and θ_m, respectively, then from equation (9.19)

$$\sin \theta_m = R_m^{1/2} \sin \theta_0.$$

The condition for reflection is that $v_\parallel$ is zero or negative at the mirror peak, so that $\sin \theta_m$ is equal to or less than unity (cf. Fig. 9.3); thus, for the particle to be bound between the mirrors, the requirement is

$$\sin \theta_0 \geqslant \frac{1}{R_m^{1/2}}. \tag{9.20}$$

LOSS CONES AND LOSS PROBABILITY

9.23.   It is apparent that in a plasma in which the motion of the charged particles is isotropic, that is to say, is completely random and all instantaneous directions of motion are about equally probable, some of the particles will escape through the mirrors because they do not satisfy the conditions derived above, e.g., equation (9.20). The escape probability can be considered in terms of the concept of *loss cones*. If reflection is to occur, the pitch angle of a particle in the central region between the mirrors must exceed a critical value θ_c, and by equation (9.20)

$$\sin \theta_c = \frac{1}{R_m^{1/2}}.\tag{9.21}$$

The loss cone, of solid angle Ω_c, is then shown in Fig. 9.4. Thus, all particles traveling toward the mirror region within the solid angle Ω_c will not be reflected, but will escape. It should be understood, of course, that the mirror loss cone refers to the velocity space of the charged particles rather than the configuration space of the confining magnetic field.

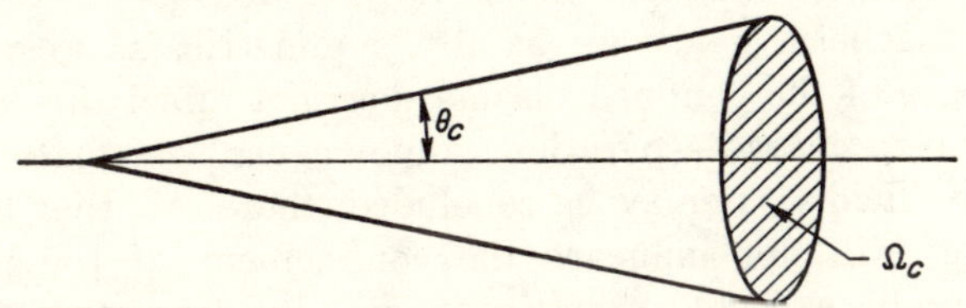

FIG. 9.4.   Loss cone in magnetic mirror system.

9.24.   Since all the particles are moving within a solid angle of 2π, the probability P that a particle will be lost, by passing through the mirror, is given by

$$P = \frac{\Omega_c}{2\pi} = \int_0^{\theta_c} \frac{2\pi \sin \theta \, d\theta}{2\pi}$$
$$= 1 - \cos \theta_c.\tag{9.22}$$

From trigonometrical considerations, $\cos \theta = (1 - \sin^2 \theta)^{1/2}$ and so, utilizing equation (9.21), it follows that

$$P = 1 - \left(\frac{R_m - 1}{R_m}\right)^{1/2}.$$

If the mirror ratio R_m is moderately large,* then

$$\left(\frac{R_m - 1}{R_m}\right)^{1/2} \approx 1 - \frac{1}{2R_m},$$

so that

$$P \approx \frac{1}{2R_m} = \frac{B_0}{2B_m}.\tag{9.23}$$

Hence, the higher the mirror ratio R_m, the smaller the probability that charged particles will escape through the mirror region.

---

* If R is very large, the conditions of adiabaticity may not be applicable (§9.31).

PLASMA CONFINEMENT AND STABILITY

9.25. It has been seen that, provided the velocity vector of a charged particle lies outside the loss cone for the given mirror system, it will not escape from the confining field. Hence, from the individual particle standpoint, it appears that magnetic mirrors can serve to provide confinement and trapping in a cylindrical tube. However, collisions among the particles can cause some of those which previously were outside the loss cone to enter this cone, so that they will escape. In addition, confinement of the plasma as a whole depends on cooperative phenomena. Such phenomena can influence the confinement as a result of either static or time-varying electric fields arising from charge separation, and by diamagnetic effects which alter the strength of the confining magnetic fields.

9.26. If the magnetic fields are axially symmetric, as postulated above, and if the presence of the confined plasma does not affect this symmetry, the electric fields due to charge separation cannot cause systematic drifts of the plasma across the fields. It may be concluded, therefore, that this particular cooperative effect will not influence the confinement of the plasma. Furthermore, it is to be expected that, in an axially symmetric field, the diamagnetic effect of the plasma will lead to a decrease of the axial magnetic field only in the central region of the mirror configuration. Since the plasma particle density is low at the peaks of the mirrors, where the charged particles either escape or are reflected back, the field strengths at the mirrors will remain substantially unchanged. Symmetric diamagnetic effects would thus tend to increase the mirror ratio above the value in a vacuum field and so would decrease the loss of particles by escape through the mirrors.

9.27. Because the lines of force in the central region of a magnetic mirror system are concave toward the axis, the plasma-magnetic field configuration should be intrinsically unstable (§13.6). Specifically, with the magnetic field lines coplanar with the axis of the plasma column, instability of the interchange (or flute) type, such as occurs in simple stellarator systems, is to be expected. If such perturbations occurred and they traveled approximately with the speed of sound, stable confinement of the plasma for periods exceeding a few microseconds would not be possible at elevated temperatures. However, in experiments to be described later (§9.136), confinement times up to 30 millisec have been observed, the typical half-life at peak compression being about 2 millisec. Moreover, scintillation counter measurements indicate that the rate of outward radial motion of the electrons in a plasma confined by a mirror field is relatively slow, in general agreement with collisional diffusion theory. If there had been any gross instability of the plasma, the movement to the walls of the containing vessel would have been much faster.

9.28. The demonstrated stability of the plasma has been attributed to two factors operative in the mirror geometry. First, the magnetic field lines are "tied down" outside the plasma by conducting coatings on the chamber walls. During the development of the flute instability there is a charge separation which produces a difference of potential between the crest and through of the flute and this tends to drive the instability [6]. The fact that charges can move freely along the lines of force to (and from) the conducting ends means that a path is available for the charges to be neutralized and so remove the potential difference. The second factor is the absence of any applied electric field parallel to the main magnetic field, thus avoiding the kink instability which might otherwise have been expected for the column of plasma. Whether these two special circumstances will be sufficient to provide stability in a magnetic mirror geometry at higher temperatures and particle densities is, of course, not known [7].

9.29. By considering a general magnetic mirror configuration with conducting end-plates, but making a number of simplifying postulates, it has been shown theoretically that hydromagnetic stability should be possible under certain conditions [8]. The maximum permissible value of β, the ratio of the plasma particle pressure to the magnetic pressure of the confining field (§3.20), appears to be dependent mainly on the aspect ratio, i.e., the ratio of the midplane diameter of the tube to the distance between the mirrors, and on the peak mirror ratio. For a fixed aspect ratio, a decrease in the mirror ratio results in an increase of the maximum value of β for which hydromagnetic stability can be realized. On the other hand, for a given mirror ratio, there is an optimum value of the aspect ratio which increases as the mirror ratio decreases. By suitable choice of the aspect and mirror ratios, theory indicates that it should be possible to obtain hydromagnetic stability in plasmas with relatively large β values, e.g., 0.3 or more, confined in a magnetic mirror field. It should be noted, however, that in view of the simplifications made in treating the highly complex problem of stability of a plasma in a mirror system, the quantitative results of the theoretical calculations are probably somewhat uncertain.

LOSSES FROM MAGNETIC MIRROR SYSTEMS

LOSS BY COLLISIONS

9.30. The treatment of charged particle trapping in a magnetic mirror configuration given above, based on the use of adiabatic invariants, implies that particles which did not satisfy the conditions for reflection would soon escape through the mirrors but the others would be confined indefinitely. In the absence of collisions, the ratio $W/W_{\perp(0)}$ would remain unchanged as a result of reflection and the particle would be repeatedly reflected back and forth between the mirrors. There are, however, various mechanisms whereby the initially trapped particles may gradually escape through the mirrors [1, 9-12]. For example, it is apparent that, since the conclusions reached depend on the constancy of the magnetic moment, trapping will be less effective, because of the changing $W/W_{\perp(0)}$ ratio, if μ is not strictly an adiabatic invariant of the magnetic field [1].

9.31. If the magnetic mirror ratio is very large, e.g., of the order of 100 or more, particles reflected near the mirror peak will have a large $W_{\parallel}$ between the mirrors, and the transit time will be less than the gyromagnetic period. In these circumstances, the magnetic moment will not be constant and the particles which were originally trapped may escape after a few reflections [9]. However, for the much smaller mirror ratios being considered here, the departure from adiabatic behavior has little effect on the losses through the ends of the mirror system.

9.32. The most significant mechanisms for the escape of particles arise from collisions. As a result of scattering, the ratio $W/W_{\perp(0)}$, which previously satisfied the condition for trapping, can be changed in such a manner as to cause the velocity vector of a particle to enter the loss cone. Consequently, when this particle reaches one of the mirrors it will escape. In plasmas of relatively high density, especially at low temperatures, in which the relaxation distance (or scattering mean free path) of the ions is less than the distance between the mirrors, collisions are so frequent that the loss cone is kept filled all the time. That is to say, particles are scattered (in velocity space) into the loss cone as fast as they are lost by escape. The confinement time of the particles is then determined by the rate at which they reach one or other of the mirrors; this is of the order of $\frac{1}{2}l/v$, where l is the distance between the mirrors and v is the average particle velocity [11, 12].

9.33. At lower densities, particularly at higher temperatures, when the relaxation length is much greater than the distance between the mirrors, the loss of particles will be dependent on the rate at which scattering occurs as a result of Coulomb collisions. The confinement time of a particle may then be identified with the average time required for a particle to be scattered into the mirror region. An approximate method for estimating this time is then as follows.

9.34. Let $(\overline{\sigma v})_s$ represent an appropriate average scattering rate parameter (cf. §2.16); then the rate of loss of particles from the ends of a magnetic mirror system, due to binary collisions in the plasma, can be written as

$$-\frac{dn}{dt} = n^2(\overline{\sigma v})_s f(R_m), \tag{9.24}$$

where n is the particle density at any instant and $f(R_m)$ is an escape factor which is a function of the peak mirror ratio. If $(\overline{\sigma v})_s$ remains approximately constant during the time the plasma is confined, integration of equation (9.24) gives the relationship between n at any time t and the initial particle density n_0 as

$$n = n_0 \left(\frac{\tau}{t + \tau} \right), \tag{9.25}$$

where

$$\tau = \frac{1}{n_0 (\overline{\sigma v})_s f(R_m)}. \tag{9.26}$$

From the arguments in §4.65, it can be seen that $1/n_0 (\overline{\sigma v})_s$ is the average time elapsing before the required particle scattering into the mirror field occurs; if this is represented by t_s, then

$$\tau = \frac{t_s}{f(R_m)}. \tag{9.27}$$

9.35. In order to obtain an expression for τ, two approximations are made. First, $f(R_m)$ is set equal to unity; this overestimates the loss through the mirrors, since it implies that every particle that is scattered through a large angle escapes. Second, t_s is identified with the relaxation time for large-angle, distant ion-ion scattering, as given by equation (4.49); this underestimates the loss, since it does not take into account all possible mechanisms which could lead to scattering of the ions into the loss cone.* Hence, to a reasonably good approximation, τ may be replaced by t_{ii}; thus equation (9.25) becomes

$$n \approx n_0 \left(\frac{t_{ii}}{t + t_{ii}} \right), \tag{9.28}$$

so that t_{ii} is approximately equal to the time for one-half of the plasma originally present to escape through the mirrors as a result of particle collisions [1].

9.36. The foregoing treatment of particle losses due to scattering resulting from long-range collisions has been based on the assumption that $(\overline{\sigma v})_s$ remains constant. But if the plasma is being heated by compression in the mirror system, allowance must be made for the fact that t_{ii}, which is here taken to be equal to $1/n_0 (\overline{\sigma v})_s$, increases with increasing temperature. Moreover, consideration should be given to the possibility that the ratio β of the particle pressure to the confining magnetic field pressure may be less than unity and may change during the course of the compression process (cf. §5.27). The effect of these factors is to make the losses much less than calculated above. Thus, it has been estimated that, instead of the loss being 50 per cent during the scattering (or relaxation) time t_{ii}, as given by equation (9.28), it is only 5 per cent for β equal to 0.1 [12].

9.37. It will be seen shortly that the particles which escape through the mirrors tend to have smaller energies than those remaining. Since t_{ii} is pro-

---

* Although the calculations are made here for the rate of loss of ions, that for electrons will be essentially the same because of space-charge effects (see, however, §9.42 *et seq.*).

portional to $W^{3/2}/n$ and W increases while n decreases, the particle loss itself will tend to cause t_{ii} to increase. It is evident, therefore, that during compression in a mirror system the rate of loss of particles will be much less, and the confinement time considerably longer, than is derived from the treatment leading to equation (9.28).

9.38. A more exact calculation of the losses through the ends of a magnetic mirror configuration due to scattering in a plasma at constant temperature has been based on the Boltzmann transport equation [5, 13]. The rate of escape of particles can be written as

$$-\frac{dn}{dt} = n^2 \left[\frac{4\pi e^4}{3m^2} \, (\overline{v^{-1}})(\overline{v^{-2}}) \ln \Lambda \right] f(R_m), \tag{9.29}$$

where m is the mass of the ion and Λ is the ratio defined in §4.39; $(\overline{v^{-1}})$ and $(\overline{v^{-2}})$ are the mean values of the reciprocal velocity and of the reciprocal of the square velocity, respectively, of the particle distribution.

9.39. By comparing equation (9.29) with equations (9.24) and (9.26), it is seen that the characteristic (scattering) time τ may be represented by

$$\tau = \left[\frac{3}{4} \cdot \frac{m^2}{\pi e^4 n_0} \cdot \frac{1}{\ln \Lambda} \right] \left[(\overline{v^{-1}})(\overline{v^{-2}}) \right]^{-1} \frac{1}{f(R_m)}. \tag{9.30}$$

If $\overline{v^2}$ is the mean square velocity of the particles, so that $\frac{1}{2}m\overline{v^2}$ may be replaced by the mean kinetic energy W, it can be readily shown that

$$\tau = \left[\frac{3}{2} \cdot \frac{(2m)^{1/2}W^{3/2}}{\pi e^4 n_0} \cdot \frac{1}{\ln \Lambda} \right] \left[(\overline{v^{-1}})(\overline{v^{-2}})(\overline{v^2})^{3/2} \right]^{-1} \frac{1}{f(R_m)}. \tag{9.31}$$

Apart from a numerical factor, the quantity in the first brackets is identical in form with equation (4.49) for the relaxation time for large-angle, ion-ion scattering. The expression in the second brackets will not differ greatly from unity, and so the simple relaxation time treatment given above, which makes τ equal to t_{ii}, is a reasonably good approximation. However, it has been found that the more exact value of τ, derived from equation (9.31), may be about half the scattering time obtained from equation (4.49), so that simple calculations based on relaxation time considerations tend to overestimate the confinement times by a factor of 2 or so. On the other hand, it should be noted that if the plasma is being heated by compression, the effective scattering time is much greater than that given by equation (9.31), for the reasons stated in §9.36. Hence, the use of this equation, like that of equation (9.28), underestimates the confinement time under these conditions.

9.40. If W_s is the mean energy of the escaping particles, the rate of energy loss is equal to $-W_s(dn/dt)$; by use of the Boltzmann transport equation it can be shown that

$$-W_s \frac{dn}{dt} = n^2 \left[\frac{2\pi e^4}{3m} \left(\frac{T}{v} \right) \ln \Lambda \right] f(R_m), \tag{9.32}$$

and upon dividing this by equation (9.29), the result is

$$W_s = \tfrac{1}{2}m(\overline{v^{-2}})^{-1}. \tag{9.33}$$

If the energy distribution of the confined particles is assumed to be Maxwellian, it is found that equation (9.33) becomes

$$W_s = \tfrac{1}{2}kT, \tag{9.34}$$

which may be compared with the value $\tfrac{3}{2}kT$ for the mean (kinetic) energy per particle in a Maxwellian distribution. Hence, for this distribution, the average energy of the particles which escape through the magnetic mirrors as a result of collisions is only about one-third that of the confined particles. In actual practice the energy distribution will probably not be Maxwellian; neverthless, the general conclusion is correct, namely, that the average energy of the escaping particles is appreciably less than that of those which remain confined. Physically, the reason for this situation is that particles of low energy are more readily scattered than those of high energy and hence have a greater probability of escaping through the magnetic mirrors.

9.41. It should be understood that the conclusions reached here apply only if the confinement time of the particles is determined by the scattering time, i.e., when the relaxation length of the ions greatly exceeds the distance between the mirrors. If the reverse is true, and the confinement time is essentially equal to the rate at which the particles can reach the mirrors, those which escape are the ones having the largest velocity, and hence energy, component in the direction parallel to the field lines. If the distribution is Maxwellian, the average energy of the particles escaping through the mirrors will be larger than the value for the trapped particles.

LOSS DUE TO AMBIPOLAR EFFECTS

9.42. The relaxation times for Coulomb electron-electron scattering is considerably less than for ions of equal energy (§4.67); hence, it is to be expected that the initial loss rate of electrons from the ends of a mirror configuration will exceed that of ions having the same energy. In an isolated plasma, the resulting separation of charges will lead to the establishment of a positive space-charge potential which will tend to equalize the rates of escape of electrons and ions. For plasmas in general, the common rate of "ambipolar" diffusion, i.e., of charged particles of opposite sign, is determined by the diffusion rate of the slower moving species, namely, the ions (§12.25). This situation appears to apply also to the particles escaping from the ends of a magnetic mirror system. The result of the ambipolar effect is to introduce a correction to the loss rate, the magnitude of which depends upon the relative values of the electron and ion temperatures.

9.43. If the electron temperature is appreciably smaller than that of the ions, as is often the case in experiments made in a magnetic mirror geometry, it has been shown that the ambipolar phenomenon establishes a potential which effectively changes the mirror ratio for the ions to a somewhat smaller value [14]. The effective mirror ratio R_{eff} can be related, approximately, to the theoretical value R_m, based on the actual magnetic field strengths, by

$$R_{\text{eff}} \approx \frac{R_m}{1 + A(T_e/T_i)}, \tag{9.35}$$

where the constant A is of the order of unity and T_e and T_i are the electron and ion temperatures, respectively.

9.44. When the electron temperature exceeds that of the ions, the fact that the electrons, which tend to escape more readily, have higher energy than the ions lost through the mirrors may possibly have important ambipolar effects. The consequences of the tendency for ions and electrons to have different diffusion rates, and of the electrical potentials developed in the plasma due to charge separation resulting from this tendency, are not fully understood. There may even be some effect on plasma stability, but this aspect of the problem requires further study, both theoretical and experimental.

ENERGY LOSS DUE TO IMPURITIES

9.45. As with plasmas in general, the presence of impurities in a mirror system, such as those originating from the walls of the containing tube, will lead to energy loss in the form of radiation. The equilibrium concentration of impurities in any magnetically confined plasma will be determined by the difference between their rate of influx and rate of removal in various ways. It is of special interest in this connection to note that, in a magnetic mirror configuration, the trapping of impurity ions, which have higher atomic numbers, is much less effective than for hydrogen-isotope ions. A mechanism therefore exists for discriminating against impurity ions by facilitating their escape through the mirrors. The preferential loss of the heavier ions is due to three main causes, which will be considered in turn.

9.46. First, the relaxation time for Coulomb scattering collisions involving stripped ions of higher atomic number is much less than for those between hydrogen-isotope ions of the same energy alone. This arises from the fact that the scattering cross section, for a given energy, increases with the charge on the ion. Consequently, there is a decrease in the relaxation time, and hence also in the confinement time, of the nuclei of higher atomic number compared with those of hydrogen isotopes.

9.47. In the second place, the energy of the impurity ions which enter the plasma from the walls of the containing vessel will invariably be lower than the mean energy of the hydrogen-isotope ions in the plasma. Since the relaxation time is proportional to the three-halves power of the particle energy, the confinement time of the impurity ions will be less than that of the plasma ions, even apart from the effect of charge considered above.

9.48. Third, for the reason given in §9.42, and also because, under the operating conditions to be expected for a high-temperature plasma confined by a magnetic mirror geometry, the average ion energy will probably be larger than that of the electrons, the plasma will acquire a positive space-charge potential. As a result, there will be a tendency for the rate of loss of ions of all kinds to be increased. Only ions of sufficiently high energy can be confined and so the impurity ions, because of their lower energy, will escape.

PLASMA INJECTION IN MAGNETIC MIRROR SYSTEMS

INTRODUCTION

9.49. In both the stellarator and pinched discharge systems, the plasma is produced by passage of electrical discharges through an initially cold gas. When magnetic mirrors are used for confinement, however, other methods have been proposed, involving the injection of a plasma or of a beam of highly energetic charged or neutral particles into the region between the mirrors. The problem of injection is a difficult one, since it is generally not possible to trap charged particles, originating from outside, within a system of static magnetic fields under conservative conditions. Several methods have been proposed for overcoming the difficulty by making a suitable change in the system during injection, so that the magnetic field is not static or the conditions are not conservative.

9.50. Three general types of injection procedures will be described here. They are (*a*) the use of time-varying magnetic fields, (*b*) order-disorder mechanisms, and (*c*) change of charge-to-mass ratio of the injected particles. In the first category there will be considered the methods of external injection, internal injection, traveling mirror injection, and radiofrequency trapping. The choice between the various possible methods must be made on the basis of efficiency or of adaptability to the particular situation of interest.

TIME-VARYING FIELDS: EXTERNAL INJECTION

9.51. In the method of *external injection*, also called *external adiabatic trapping*, the plasma is trapped by injecting it through the mirror (or mirrors) into a rapidly increasing magnetic field [1, 2]. The procedure is similar to that employed for injection of electrons into a betatron. The general principle can be understood by considering the condition for particle binding in a mirror configuration given by inequality (9.15), i.e., $W \leqslant \mu B_m$. If the plasma is to be injected through a magnetic mirror from outside the confinement zone, the essential condition is the reverse of that for trapping, namely, that the particle energy must exceed μB_m, where B_m is the initial peak value of the mirror field. Obviously, since $W > \mu B_m$, these particles will immediately tend to escape. Suppose that, during passage of the plasma through the confinement region, the strength of the mirror field is increased to B_m', so that $\mu B_m'$ is larger than the energy of the injected particles. Under the condition $W < \mu B_m'$, which is now applicable, many of the injected particles will be trapped between the mirrors. External adiabatic trapping, as just described, may be achieved either by increasing the entire magnetic field, i.e., both at the mirrors and in the central region, or by strengthening the mirror fields only.

9.52. In order to derive the conditions under which a particle injected through a mirror may be trapped in a time-rising magnetic field, consider a symmetrical mirror system with a source located just outside the mirror at the left end. For simplicity, it will be supposed that the mirror regions are very narrow compared to the distance between the peaks, as indicated in Fig. 9.5, so that all trapped particles, irrespective of their energy, travel

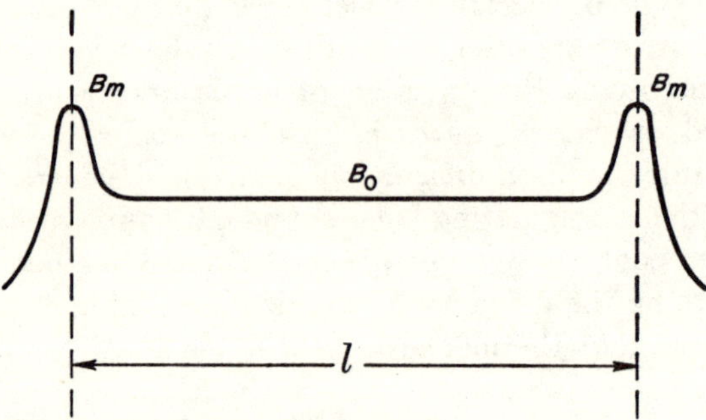

Fig. 9.5. Idealized magnetic mirror configuration.

the same net axial distance l between successive reflections at the two mirrors. In actual practice, the magnetic field strength will increase more gradually as the mirror is approached (cf. Fig. 9.2), so that trapped particles of lower energy will be reflected at points where the field strength is lower. The axial distance traveled between successive reflections will thus be less than for particles of higher energy (cf. §9.111). However, provided the distance between the mirror peaks is fairly large, the general conclusions reached from a consideration of the type of field shown in Fig. 9.5 will still be applicable.

9.53. Suppose v is the velocity with which a charged particle is injected through the mirror and let ϵ be the injection angle with respect to the plane of the magnetic mirror. The initial components parallel and perpendicular to the field direction may then be written as $v_{\parallel(m)}$ and $v_{\perp(m)}$, respectively (Fig. 9.6), so that

$$v^2 = v_{\perp(m)}{}^2 + v_{\parallel(m)}{}^2 \tag{9.36}$$

and

$$v_{\parallel(m)} = v_{\perp(m)} \tan \epsilon. \tag{9.37}$$

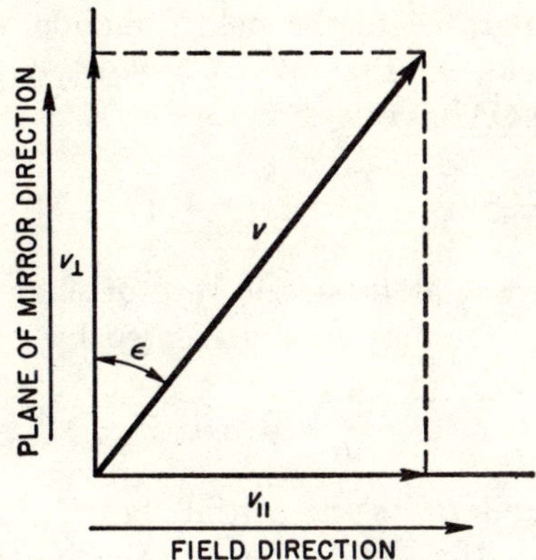

Fig. 9.6.   Definition of injection angle ϵ.

It will be seen shortly that successful external injection requires the angle ϵ to be small, so that $\tan \epsilon \approx \epsilon$; hence, equation (9.37) may be written as

$$v_{\parallel (m)} \approx v_{\perp (m)}\epsilon. \tag{9.38}$$

9.54.   The energy transformation as the injected particle leaves the mirror region and enters that part of the system where the field strength is B_0 can be derived from equation (9.17); thus, replacing the initial and final locations a and b by m and 0, respectively, it is seen that

$$\frac{W_{\parallel (0)}}{W_{\perp (0)}} = R_m\left[\frac{W_{\parallel (m)}}{W_{\perp (m)}} + 1\right] - 1, \tag{9.39}$$

where R_m has been written for B_m/B_0. Since $W_{\parallel}/W_{\perp} = v_{\parallel}^2/v_{\perp}^2$, it follows from equation (9.38) that

$$\frac{W_{\parallel (m)}}{W_{\perp (m)}} = \epsilon^2,$$

and so equation (9.39) reduces to

$$\frac{W_{\parallel (0)}}{W_{\perp (0)}} = R_m(\epsilon^2 + 1) - 1. \tag{9.40}$$

EXTERNAL INJECTION: CONSTANT MIRROR RATIO

9.55.   Suppose that, as the particles are injected, the magnetic fields are increasing with time in such a manner that the mirror ratio remains constant [1, 2]. By the time a particle reaches the mirror at the right of the system, the magnetic field B_0 will have been increased to B_0', but if the magnetic moment is adiabatically invariant,

$$\frac{W_{\perp (0)}}{B_0} = \frac{W_{\perp (0)}'}{B_0'}. \tag{9.41}$$

The energy component parallel to the field direction will not be affected by the time-rising axial field, so that $W_{\parallel(0)} = W_{\parallel(0)}'$; hence, combination of equations (9.40) and (9.41) gives

$$\frac{W_{\parallel(0)}'}{W_{\perp(0)}'} = \frac{B_0}{B_0'}\,[R_m(\epsilon^2 + 1) - 1]. \tag{9.42}$$

9.56. For short intervals, at least, the rate of change of the central magnetic field strength with time may be represented by

$$\frac{dB_0(t)}{dt} = aB_0, \tag{9.43}$$

where a is a constant and B_0 is the initial value of the field strength. If t_t is the transit time, i.e., the time taken for the particle to travel from the left end, where it is injected, to the right end, then integration of equation (9.43) over this interval gives

$$B_0' = B_0(1 + at_t), \tag{9.44}$$

Hence, from equations (9.42) and (9.44), it follows that, at time t_t after injection,

$$\frac{W_{\parallel(0)}'}{W_{\perp(0)}'} = \frac{R_m(\epsilon^2 + 1) - 1}{1 + at_t}. \tag{9.45}$$

9.57. The condition for the injected particle to be trapped when it reaches the right-end magnetic mirror can be obtained from equation (9.11). It is that either side of equation (9.45) shall be equal to or less than $R_m - 1$; thus, if the injected particle is not to escape,

$$\frac{R_m(\epsilon^2 + 1) - 1}{1 + at_t} \leqslant R_m - 1$$

or

$$\epsilon^2 \leqslant at_t\left(\frac{R_m - 1}{R_m}\right). \tag{9.46}$$

This result is in agreement with expectation since it shows that ϵ, the angle of injection with respect to the plane of the mirror, must be less than a certain value if the particle is to be trapped. In other words, trapping of an injected particle in a time-rising magnetic field requires that the direction of injection should not diverge greatly from that normal to the direction of the main (axial) confining field.

9.58. If both sides of equation (9.40) are divided by R_m, i.e., by B_m/B_0, and the constancy of the magnetic moment is taken into account, the expression

$$W_{\parallel(0)} = W_{\perp(m)}\left[(\epsilon^2 + 1) - \frac{1}{R_m}\right] \tag{9.47}$$

is obtained. Under the conditions of interest, that is, when the injected particles are trapped, ϵ^2 may be neglected in comparison with unity and it is a reasonably good approximation to write equation (9.47) as

$$W_{\parallel(0)} \approx W_{\perp(m)} \left(\frac{R_m - 1}{R_m} \right)$$

and then

$$v_{\parallel(0)}^2 \approx v_{\perp(m)}^2 \left(\frac{R_m - 1}{R_m} \right). \tag{9.48}$$

Furthermore, if ϵ is small, $v_{\parallel(m)}$ is also small, by equation (9.38), and so, from equation (9.36), $v_{\perp(m)}^2 \approx v^2$, so that equation (9.48) becomes

$$v_{\parallel(0)}^2 \approx v^2 \left(\frac{R_m - 1}{R_m} \right). \tag{9.49}$$

9.59. If l is the distance between the mirror peaks (Fig. 9.5), then the time t_t required for a particle to travel from the left-end mirror to the one at the right end is $l/v_{\parallel(0)}$, where $v_{\parallel(0)}$ remains constant according to equation (9.49), in spite of the time-rising magnetic field, because R_m does not change. Upon substituting this value for t_t in equation (9.46) and utilizing equation (9.49) for $v_{\parallel(0)}$, the condition for trapping an injected particle is seen to be

$$\epsilon^2 \leqslant \frac{al}{v} \left(\frac{R_m - 1}{R_m} \right)^{\frac{1}{2}}. \tag{9.50}$$

Finally, if the expression for a obtained from equation (9.43) is introduced, the result, after rearrangement, is

$$\frac{dB_0(t)}{dt} \geqslant \epsilon^2 B_0 \left(\frac{R_m}{R_m - 1} \right)^{\frac{1}{2}} \frac{v}{l}. \tag{9.51}$$

Thus, the required condition is that the time-rising magnetic field should increase at a rate greater than a particular value determined by the magnetic field strengths, the injection angle, the velocity of injection, and the distance between the mirrors.

EXTERNAL INJECTION: INCREASING MIRROR RATIO

9.60. The case will now be considered in which the mirror field only is increased in order to trap the injected particles, while the central confining field remains constant [1, 2]. If, as before, the mirror field region is assumed to be very narrow, so that the change in the mirror field strength is negligible while the injected particles are traversing it, equation (9.40) is applicable, with R_m representing the initial value of the mirror ratio. However, by the time the injected particles have reached the right-end magnetic mirror, the mirror ratio has increased to R_m' and so the condition for trapping, as given by equation (9.11), is

$$\frac{W_{\parallel(0)}}{W_{\perp(0)}} \leqslant R_m' - 1.$$

If this result is combined with equation (9.40), it is found that for trapping

$$R_m' \geqslant R_m(\epsilon^2 + 1). \tag{9.52}$$

9.61. Suppose that the time rate of increase of the magnetic mirror field is represented by

$$\frac{dR_m(t)}{dt} = bR_m, \tag{9.53}$$

so that

$$R_m' = R_m(1 + bt_t),$$

where t_t is again the transit time of the particle. Equation (9.52) can now be rearranged to give

$$\epsilon^2 \leqslant bt_t \tag{9.54}$$

as the condition for trapping. As before, $t_t = l/v_{\parallel(0)}$, but since R_m now changes with time, it is seen from equation (9.48) that $v_{\parallel(0)}$ will be a function of the time of injection while the mirror field increases. Thus, in the present case, equation (9.49) should be written as

$$v_{\parallel(0)}{}^2 \approx v^2 \left[\frac{R_m(t) - 1}{R_m(t)} \right],$$

where $R_m(t)$ is the mirror ratio at any time t after the mirror field has started to increase, i.e.,

$$R_m(t) = R_m(1 + bt).$$

By using equation (9.53) to eliminate b, equation (9.54) becomes, after rearrangement,

$$\frac{dR_m(t)}{dt} \geqslant \epsilon^2 R_m \left[\frac{R_m(t) - 1}{R_m(t)} \right]^{\frac{1}{2}} \frac{v}{l}.$$

Finally, since $R_m(t) = B_m(t)/B_0$ and $R_m = B_m/B_0$, it follows that

$$\frac{dB_m(t)}{dt} \geqslant \epsilon^2 B_m \left[\frac{R_m(t) - 1}{R_m(t)} \right]^{\frac{1}{2}} \frac{v}{l} \tag{9.55}$$

is the condition for trapping. This expression is somewhat similar to equation (9.51)* except that the term in the brackets on the right of equation (9.55) is a slowly varying function of time, whereas the corresponding term in equation (9.51) is constant.

TIME-VARYING FIELDS: INTERNAL INJECTION

9.62. Another method of injection using time-varying magnetic fields for trapping is that of *internal injection* or *internal adiabatic trapping* [1]. In this case, the plasma source is located just inside the peak of the mirror region

---

*It should be noted that since equation (9.51) is based on a constant mirror ratio, dB_0/dt and B_0 can be replaced by dB_m/dt and B_m, respectively, thereby bringing out more clearly the similarity to equation (9.55).

where the magnetic field has a gradient in the axial direction. The particles are injected, as far as possible, in a direction perpendicular to that of the axial field lines, so as to provide maximum trapping. Since the injected ions spiral about the lines of force, an essential requirement is that they miss the source on their first turn. This is achieved by having the field gradient large enough at the point of injection to insure sufficient displacement of the particle orbit in the axial direction.

9.63. To determine the magnitude of the displacement of an ion in a single rotation, consider equation (9.14); thus, it is possible to write

$$F_z = m \frac{\partial^2 z}{\partial t^2} = -\mu \frac{\partial B}{\partial z}, \qquad (9.56)$$

where m is the mass of the injected particle. If the origin of the axial co-ordinate z is taken at the plasma source, it may be assumed that z and $\partial z/\partial t$ are both zero at $t = 0$. Hence, integration of equation (9.56) gives

$$\Delta z = - \frac{\mu t^2}{2m} \cdot \frac{\partial B}{\partial z} \qquad (9.57)$$

for the displacement at any time t after injection, provided $\partial B/\partial z$ remains constant during this time interval. The time required for a single rotation is $2\pi r_g/v_\perp$, where r_g is the gyromagnetic radius of the particle, i.e., the radius of curvature of its helical path in the field. If this is substituted for t in equation (9.57) and μ is replaced by $\frac{1}{2}mv_\perp^2/B$, the result is the displacement in a single turn, namely,

$$\Delta z = -\pi^2 r_g^2 \frac{1}{B} \cdot \frac{dB}{dz}. \qquad (9.58)$$

9.64. If a characteristic distance λ associated with the magnetic field gradient in the source region is defined by

$$\frac{1}{\lambda} \equiv - \frac{1}{B} \cdot \frac{\partial B}{\partial z}, \qquad (9.59)$$

equation (9.58) can be written as

$$\Delta z = \frac{\pi^2 r_g^2}{\lambda}. \qquad (9.60)$$

This result is based on the supposition that the particles are emitted from the source in a direction perpendicular to the axial field lines; if ϵ is the angular deviation from the perpendicular direction, an additional term $2\pi r_g \epsilon$ per rotation must be included, so that equation (9.60) becomes

$$\Delta z = \frac{\pi^2 r_g^2}{\lambda} \pm 2\pi r_g \epsilon, \qquad (9.61)$$

where the plus or minus signs represent forward or backward deviation, respectively. The first condition for internal injection, therefore, is that Δz, as given by equation (9.61), shall exceed the distance s, which is the axial dimension of the source that must be cleared by the returning particle (Fig. 9.7).

9.65. Another requirement is that, having missed the source on their first rotation, the ions may travel through the confinement region to the other mirror where they are reflected back. Upon returning to the injection end of the system, they must again miss striking the source. Actually, because the magnetic field between the mirrors varies somewhat with the radial distance, the guiding center of an ion will drift in the azimuthal direction about the axis (cf. §4.27). The extent of precession is small for each reflection, since

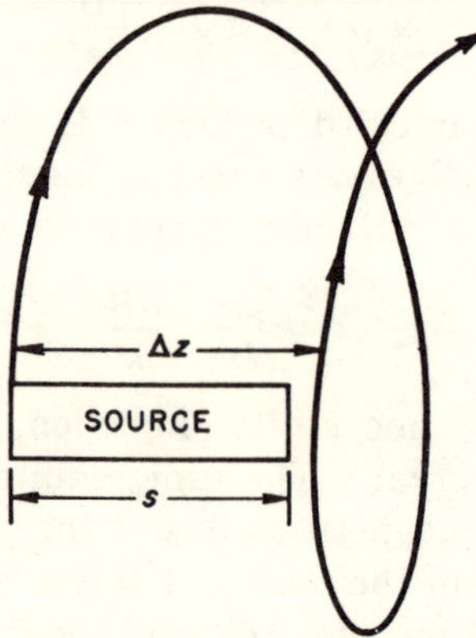

Fig. 9.7.   Method of internal injection.

the inhomogeneity of the field in the central region is generally not considerable. However, it is possible, by suitable adjustment of the conditions, to make the precession large enough that the ions do not hit the source upon their first return to the injection end, after reflection by the mirror at the other end. If this is the case, the ions may continue to be reflected back and forth, as many as fifty or a hundred times, until they have precessed completely around the magnetic field axis. When this occurs, there is a large probability that the ions will hit the plasma source and thus be lost. The period of injection and confinement would then be limited to approximately the time required for the particle to precess completely around the axis, and this is too short to be of practical interest.

9.66. The time of useful injection may be greatly increased over that for a single precession period by increasing the magnetic field strength while the particle is moving back and forth during this period. As a result the particle will, in general, be reflected at some distance short of the point of injection and so may miss the source completely (§9.111). In this way, the length of time over which injection can be accomplished may be extended to something like a thousand times the single transit time, i.e., of the order of ten to a hundred times as long as in a constant magnetic field.

9.67. Assuming a time-rising magnetic field, let δz be the distance in front of its starting point to which an ion returns after the time δt required for a complete precession about the axis of the confinement system. Suppose the direction of emission is such that the particles have energy components both parallel ($W_\parallel$) and perpendicular ($W_\perp$) to the central field lines. Upon their return, after time δt, the magnetic field strength will have increased to such an extent that the $W_\parallel$ component is negligible. Hence, if z and t represent the space and time coordinates for ion emission, so that $z + \delta z$ and $t + \delta t$, are the corresponding values upon the return of the particle to the source, the energy balance requirement is

$$W_\parallel(z, t) + W_\perp(z, t) \approx W_\perp(z + \delta z, t + \delta t).$$

Making use of the adiabatic condition that $W_\perp/B$ remains unchanged, this result can be expressed as

$$B(z, t)\,\frac{W_\parallel(z, t)}{W_\perp(z, t)} \approx B(z + \delta z, t + \delta t) - B(z, t)$$

$$\approx \left(\frac{\partial B}{\partial z}\right)\delta z + \left(\frac{\partial B}{\partial t}\right)\delta t. \tag{9.62}$$

9.68. If ϵ is the angle of injection with respect to the mirror plane, as in §9.53, then

$$\frac{W_\parallel(z, t)}{W_\perp(z, t)} = \tan \epsilon^2 \approx \epsilon^2,$$

since ϵ will be small. Furthermore, if λ is defined by equation (9.59), and a time parameter τ by

$$\frac{1}{\tau} \equiv \frac{1}{B} \cdot \frac{\partial B}{\partial t},$$

equation (9.62) yields

$$\delta z = \lambda \left(\frac{\delta t}{\tau} - \epsilon^2\right). \tag{9.63}$$

This gives the amount by which the returning particle will miss its point of origin when it strikes the source after a complete precession. A large value of δz is seen to require large λ, i.e., small relative magnetic field gradient in the axial direction, and small τ, i.e., rapid relative increase of the field strength with time during injection. If the field increases exponentially with time, τ is constant and hence δz will be independent of time. On the other hand, for a field increasing linearly with time, τ increases steadily and δz decreases correspondingly. Hence, at a certain time after injection, the value of δz will become too small to permit the particle to miss the source.

9.69. For injection to be successful, both Δz and δz, as expressed by equations (9.61) and (9.63), respectively, must simultaneously be larger than the source dimension s. The maximum value of the injection time can be estimated by setting Δz and δz equal, both being greater than s, and taking the angle ϵ as zero; it is then found that

$$\tau_{\max} = \left(\frac{\lambda}{\pi r_g}\right)^2 \delta t, \tag{9.64}$$

where δt, as before, is the duration of one precession period. For typical values of λ and r_g, which would be compatible with the conditions that Δz and δz exceed the source length, it appears that the maximum injection time can be of the order of a millisecond for a magnetic mirror system of moderate dimensions.

TRAVELING MIRROR INJECTION

9.70. Suppose that, instead of having a single mirror at each end, the confining field has, in addition, a number of peaks of lower strength between the end mirrors, so that the variation of the field strength along the containing tube is of the form shown in Fig. 9.8. This situation can by achieved by varia-

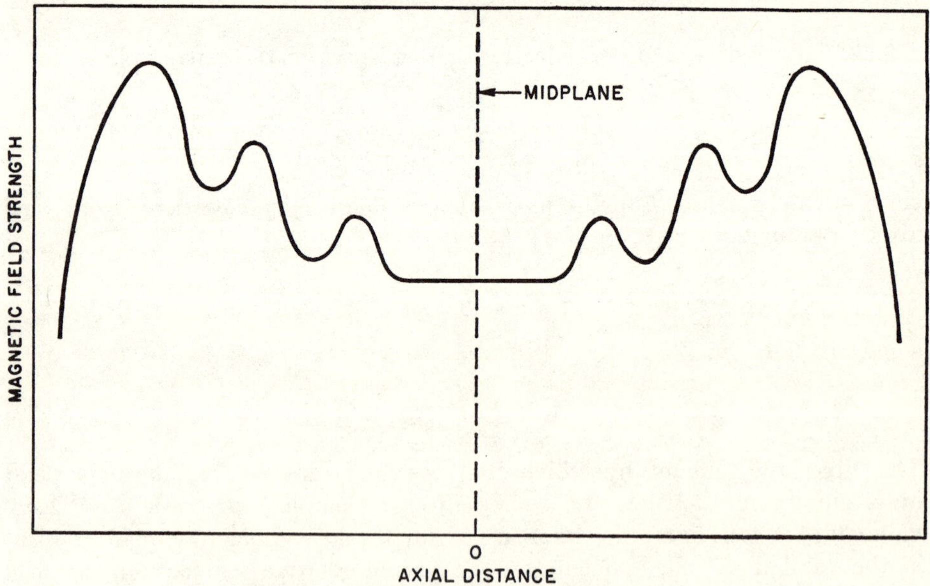

Fig. 9.8. Injection by traveling mirrors.

tions in the strength of the current passing through individual coils surrounding the tube. Particles of different energy groups will then be trapped in the "valleys" which constitute potential wells between each pair of mirrors.

9.71. By means of time-varying currents of the proper amplitude and phase, it is possible to make the mirrors move inward from both ends toward the middle.   As they approach the central region, the mirrors decrease in strength and disappear.   The disappearance of each wave in the magnetic field strength, i.e., of each mirror, is accompanied by the formation of a new one in the outer (or end) regions.   In this manner there can be produced a continuous train of magnetic mirrors of decreasing strength moving inward.   As a result, the particles trapped in the potential wells between the mirrors will be carried into the central region and deposited there.   It would appear, therefore, that if a plasma were introduced through the end mirrors, by either external or internal injection, as described above, the charged particles would be carried by the traveling mirror system into the central confinement zone where a high concentration could be built up.

9.72.   It can be shown, however, that the number of particles which can be trapped in the central region cannot be large.   The reason is that, although moving magnetic mirrors serve to bring in particles, they also provide a means for escape.   The particles already trapped undergo axial compression as the mirrors move inward and so their energy increases (§§5.30, 9.108).   Ultimately the energy reaches a point where the particles are no longer bound in the potential well of the mirror system and so they are able to escape [15, 16, 17].

RADIOFREQUENCY TRAPPING

9.73.   The principle of radiofrequency trapping of a plasma is based on the use of time-varying fields of a different kind from those considered above. The plasma is injected axially into a static magnetic mirror system and the ions are subjected to a radiofrequency field in resonance with their gyromagnetic (or cyclotron) frequency (§5.61).   The irreversible energy gain in $W_\perp$, i.e., in the direction perpendicular to the magnetic field axis, should result in trapping between the mirrors.   This method of trapping involves problems of penetration of the radiofrequency field into the plasma, especially at relatively high densities, but it is nevertheless considered to be of potential interest.

ORDER-DISORDER TRAPPING

CAPTURE BY COLLISIONAL INTERACTIONS

9.74.   In principle, any cooperative (or nonconservative) mechanism may be expected to lead to capture of part of an injected plasma, and the conditions must be chosen to make this part as large as possible.   It was seen in §9.32 that collisions between particles can result in loss by escape through the mirrors, but collisions can also lead to the trapping of an injected plasma. It is possible that, after a particle is injected and before it has time to reach a mirror, it may suffer one or more collisions with other particles which change its energy in such a way that it is reflected when it does enter the mirror field.

Thus, if a relatively dense plasma at low temperature is suddenly discharged into the region between magnetic mirrors, there is an appreciable probability that some of the plasma will be captured as a result of mutual collisional effects within the confinement volume.

9.75. The approximate conditions for capture by interparticle collisions can be derived from relaxation time considerations. By analogy with certain relationships developed in connection with stellar dynamics [6, 16, 18], the rate of growth of the pitch angle of an ion, subjected to scattering collisions by other ions of the same kind, can be represented by

$$\overline{\Delta\epsilon^2} \approx \frac{t}{t_{ii}}, \tag{9.65}$$

where $\overline{\Delta\epsilon^2}$ is the mean square of the change in the angle of injection with respect to the plane of the mirror at time t, and t_{ii} is the ion-ion relaxation time. This expression is in general agreement with the definition of the relaxation time (§4.68) as that required to produce a large-angle scattering; thus, if t in equation (9.65) is set equal to t_{ii}, the mean value of $\Delta\epsilon$ is seen to be approximately 1 radian, i.e., nearly 60°.

9.76. Suppose the injected ion has a value of the angle ϵ such that, without collisions, it would not be trapped. The requirement for trapping is then, very roughly, that the change in the pitch angle due to scattering shall be at least equal to ϵ in the time t_t in which the ion travels between the source and a mirror. Utilizing equation (9.65), the trapping condition may be stated in the approximate form

$$\frac{t_t}{t_{ii}} \geqslant \epsilon^2. \tag{9.66}$$

If l is the distance between source and mirror at the other end, which may be taken as the distance between the mirrors, then $t_t = l/v_{\parallel(0)}$, where the value of $v_{\parallel(0)}$ for a constant mirror ratio is given by equation (9.49). Upon substituting for t_t in equation (9.66), the trapping condition is seen to be

$$\frac{l}{v t_{ii}} \left(\frac{R_m}{R_m - 1} \right)^{1/2} \geqslant \epsilon^2.$$

Using equation (4.49) for t_{ii} and substituting $(2W_i/m_i)^{1/2}$ for v, this condition may be expressed as

$$\epsilon^2 \leqslant \frac{2\pi e^4 n_i l \ln \Lambda}{W_i^2} \left(\frac{R_m}{R_m - 1} \right)^{1/2}, \tag{9.67}$$

where n_i is the ion number density in the plasma. It appears, therefore, that trapping of charged particles as a result of collisions should be favored by high plasma densities, a large distance between the mirrors, a small mirror ratio, and low particle energies.

9.77. A different kind of trapping due to cooperative effects involves injection of plasma bursts (§5.85 *et seq*.) across the containing volume and so directed that they collide in the center of that volume. Although such plasma bursts can move across a magnetic field in an evacuated region, they are unable to do so if they enter a conducting medium, such as another plasma disposed along the field lines. Consequently, when the plasma bursts meet, they will tend to mix with one another and also with any plasma that may be already present. By carefully timing the confining field with respect to the energetic bursts, it is possible to capture much of the plasma when the beams intercept each other in the center of the magnetic mirror system.

ENTROPY TRAPPING OF CHARGED PARTICLES

9.78. Suppose that a collision-free, diamagnetic plasma is confined on all sides by a magnetic field of magnitude B, such that $B^2/8\pi$ is equal to the plasma pressure. If the plasma were to assume the highly improbable, but not impossible, configuration in which all the particles are moving parallel to each other toward a small region of the confining field, the pressure there would become greater than the magnetic pressure. The field would then be forced aside and the plasma would escape. By reversing the situation, that is, by directing a broad jet of plasma in which the particles are all moving in one direction toward a magnetic field, e.g., at one end of a magnetic mirror system, a possible method of injection becomes apparent. The problem still remains, however, of trapping the injected particles.

9.79. It has been proposed to employ a gun of the type described in Chapter 5 to produce a body of plasma which moves toward the magnetic mirror (or cusp, cf. §11.33) system along its axis at a velocity large compared with the average thermal velocity within the moving frame of the gas itself. Provided that the plasma density and velocity are sufficiently large, for the given field strength, the magnetic field will be forced aside and the plasma will enter the confining region of the mirror (or cusp) geometry. It should be noted that the assumption is made that the plasma is able to exclude all the field.

9.80. During the penetration and especially during the subsequent expansion of the plasma in the interior where the magnetic field is weaker, the fast ions undergo specular reflection at the interface between the field and the diamagnetic plasma. As a result, their originally ordered motion parallel to the axis becomes more and more disordered, since some of the energy is transferred into the transverse direction at each reflection (Fig. 9.9). The increased ran-

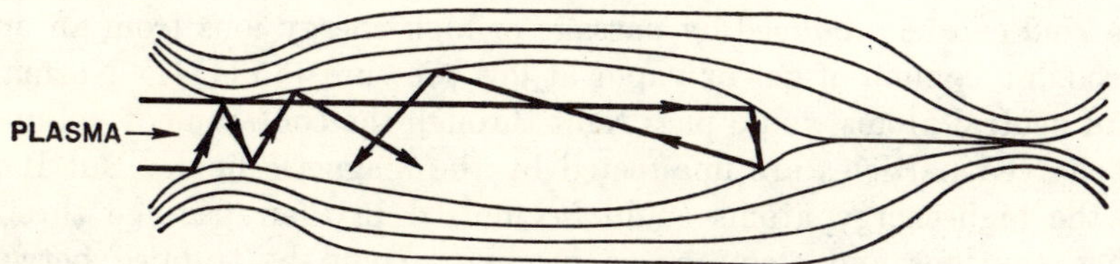

Fig. 9.9. Principle of entropy trapping.

domness of the ion motions corresponds to an increase in the entropy of the system, so that the penetration process is irreversible and the injected ions are unable to escape. It is for this reason that the process has been called *entropy trapping*. Once inside the mirror system, most of the ions lose the necessary directedness of motion to pry open the mirror at the other end, or to escape through the mirror at which they entered, and so they are trapped [19].

ION MAGNETRON TRAPPING

9.81. Another proposal for converting the directed motion of the particles in a plasma beam into a more random distribution in a mirror geometry is based on the principle of the ion magnetron, in which essentially steady electric and magnetic fields are at right angles to one another [20]. In the center of one of the mirrors there is inserted a plasma source which emits a cylindrical beam of small transverse dimensions along the axis of the confining magnetic field. Close to the other mirror, located symmetrically with respect to the source and connected with it electrically, is a "reflector" electrode. If a potential which is positive with respect to the walls of the containing vessel is applied to the source and reflector, the system is similar to a magnetron.

9.82. The charged particles emitted from the central plasma column are thus subjected to and accelerated by a radial electric field. As a result, they can acquire a considerable energy component in a direction perpendicular to the magnetic field lines. If the electric field strength is not too large, so that the plasma does not rotate as a whole, the ratio $W_\perp/W_\parallel$ (or $W_\perp/W$) for many of the particles should become sufficiently high to permit trapping between the magnetic mirrors. The arrangement described here has features in common with those used in the study of rotating plasmas, and a further discussion will be found in Chapter 11, where such plasmas are treated more fully.

TRAPPING BY CHANGE IN CHARGE-TO-MASS RATIO

INJECTION OF HIGH-ENERGY NEUTRAL ATOMS

9.83. Two general methods have been proposed for trapping a plasma in a mirror system, based on a change in the charge-to-mass ratio of the particles. One procedure is to start with a beam of neutral particles of high energy and then to ionize them in the region between magnetic mirrors. Neutral beams of this kind can be produced by passage of high-energy ions from an accelerator through a column of gas or vapor at low pressure (§5.111). Normally, the beam of neutral atoms would pass right through the containment volume, since the uncharged particles are unaffected by the magnetic field. But if part or all of the high-energy atoms could be ionized in their passage through the chamber, the ions and electrons so produced could be trapped between the mirror fields. Ionization may occur as a result of a collision between the highly energetic atom and a cold atom in the residual gas present in the containing vessel. In addition, the incoming, energetic atoms can be ionized by interaction with ions which may be present initially or which are produced by the former mechanism.

9.84. Ionization due to interaction between two neutral particles requires that the relative energy of the system shall be large. The best way to achieve this is to inject highly energetic atoms, and if the latter are introduced at the midplane of the system, in a direction perpendicular to the field lines, many of the ions produced will have large $W_\perp$ values. The probability of escape of such ions through the mirrors is thus relatively small. However, the production of a high-energy beam containing a large proportion of neutral atoms is difficult, especially by charge exchange, e.g., by the reaction

$$D^+ \text{ (fast) } + D \text{ (slow) } \rightarrow D^+ \text{ (slow) } + D \text{ (fast)},$$

because the cross section decreases sharply with increasing energy, at least for energies of interest, i.e., in the kilo-electron volt region (§12.68).

9.85. The scattering loss through the mirrors of the low-energy ions, which would result from the injection of atoms of moderate energy, could be offset by the increased ionization that would result from the charge exchange of these atoms with ions already present. A possibility which has been considered is to create an initial plasma, by means of one of the injection processes described above, e.g., one involving time-varying magnetic fields for trapping, and then to make use of this plasma to bring about the ionization of a beam of neutral particles.

9.86. Some of the conditions for the injection of high-energy atoms may be derived by considering the equilibrium which describes the time rate of change of the trapped ion density as the result of introducing a beam of energetic neutral particles into a partially ionzed gas [21]. By making a number of simplifications, e.g., assuming the volume of the atomic beam to be about equal to that of the containment system as a whole, it is found that

$$\frac{dn_i}{dt} = \phi_0 n_0 \sigma_{nn} + \phi_0 n_i \sigma_{ni} - n_i n_0 \sigma_{cx} v_1 - \tfrac{1}{2} n_i^2 \sigma_{ii} v_2, \tag{9.68}$$

where n_0 and n_i are the particle densities of cold neutral atoms and trapped ions, respectively, ϕ_0 is the flux of high-energy neutral particles in the injected beam,* and v_1 and v_2 are the velocities of a trapped ion relative to a cold neutral atom and to a trapped ion, respectively; σ_{nn} and σ_{ni} are ionization cross sections due to energetic neutral-cold neutral and to energetic neutral-trapped ion interactions, respectively; σ_{cx} is the charge exchange cross section for trapped ion-cold neutral interaction, and σ_{ii} is the large-angle Coulomb scattering cross section of trapped ions by other trapped ions. The

---

* Alternatively, ϕ_0 may be expressed as $4I_0/S$, where I_0 particles/sec is the current strength of the injected high-energy neutral beam, and S is the surface area of the plasma.

first two terms on the right of equation (9.68) give the rates of formation of ions by collision of energetic neutral particles with cold atoms and trapped ions, respectively; the third and fourth terms represent the loss of ions as a result of charge exchange with cold atoms and of ion-ion scattering which leads to escape through the mirrors, respectively. The loss due to scattering of ions by neutral atoms is small and has been omitted.

9.87. It is probable that the first and fourth terms on the right of equation (9.68) will be small, so that their algebraic sum may be neglected. Consequently, for the density of trapped ions to build up exponentially, the sum of the coefficients of n_i in the second and third terms on the right of equation (9.68) must be positive; thus,

$$\phi_0 \sigma_{ni} > n_0 \sigma_{cx} v_1.$$

In other words, the density of neutral atoms in the containment volume must be such that

$$n_0 < \frac{\phi_0 \sigma_{ni}}{\sigma_{cx} v_1}.$$

If the actual density n_0 is much less than the limiting value, i.e., $n_0 \sigma_{cx} v_1 \ll \phi_0 \sigma_{ni}$, the third term in equation (9.68) can be neglected, and this equation reduces to

$$\frac{1}{n_i} \cdot \frac{dn_i}{dt} \approx \phi_0 \sigma_{ni} \quad \text{or} \quad n_i(t) \approx n_i(0) \exp(\phi_0 \sigma_{ni} t).$$

The e-folding time for the build-up of the trapped ions is thus approximately equal to $1/\phi_0 \sigma_{ni}$. Finally, if the density of neutral particles in the hot plasma is negligible, after build-up of the trapped ions has occurred to a considerable extent, the steady-state density of these ions is obtained by taking the second and fourth terms on the right equation (9.68) to be equal; thus,

$$n_i \approx \frac{2\phi_0 \sigma_{ni}}{\sigma_{ii} v_2}. \tag{9.69}$$

9.88. It is apparent from the foregoing results that the desirable conditions of a short e-folding time for the build-up of the trapped ions and a high steady state density are both favored by having $\phi_0 \sigma_{ni}$ large. Since σ_{ni} is known to decrease with increasing temperature, it is evident that there must be a compromise in the choice of the energy of the injected beam of neutral atoms. If it is too high, the build-up time will be too long for confinement of the resulting plasma, but if the energy is too low, so also will be that of the trapped ions.

INJECTION BY ARC DISSOCIATION OF MOLECULAR IONS

9.89. Another approach to the method of injection into a magnetic mirror geometry by changing the ratio of charge to mass is to inject high-energy molecular deuterium ions and then to dissociate them by passage through an arc discharge in the manner described in §5.104. This represents a form of internal injection in which the ions are made to miss the source, in the course of their gyration in the magnetic field, by changing the radius of rotation. If the injection occurs at the median plane of the magnetic mirror system, in a direction normal to the field lines, it is to be expected that most of the ions will be trapped, at least until scattering collisions become significant.

9.90. One of the problems associated with this method of injection arises from the inevitable presence, particularly at start-up, of large numbers of neutral particles of low energy. As a result of charge-exchange reactions, the energetic ion becomes a fast neutral particle, which will not be confined by the magnetic field and so will soon strike the wall of the containing vessel and be lost. At the same time, the low-energy neutral will be converted into a slow ion which will readily escape through the mirrors. Thus, the presence of neutral atoms will lead to serious energy losses from the system. Although the charge-exchange cross section decreases sharply with energy, in the region of interest, the losses would be considerable even for ion energies as high as several hundred kiloelectron volts. Hence, the formation of a high-temperature plasma will be very difficult, if not impossible, unless the density of residual neutral particles is reduced in some manner.

9.91. There are two mechanisms by which the injected ions tend to remove neutral particles. One is the charge-exchange reaction just considered, and the other, which is generally more important, is the normal ionization, i.e., the removal of electrons from neutral particles with the formation of ion pairs, caused by the highly energetic ions. Thus, for deuterons of 300-kev energy, the charge-exchange cross section with deuterium atoms is about 2×10^{-18} cm^2, whereas the ionization cross section is roughly 4×10^{-17} cm^2, so that ionization is roughly twenty times more probable than charge exchange at the given energy. In view of the removal of neutral particles in these two ways, it seems probable that there will be a critical value of the input current of deuterons for which neutral particles in the plasma are removed by ionization as fast as they enter from outside. The point where this occurs has been called *burnout*. If the critical current for the burnout of neutral atoms is exceeded, the density of neutral particles will decrease, thus leading to an increase in the ion density and an even greater rate of removal of neutral particles. Consequently, once the burnout point has been passed, there will be a rapid decrease in the neutral particle density [22-24].

9.92. The theoretical treatment of neutral particle burnout is based on equations giving the time dependence of the densities of ions (n_i) and of neutral atoms (n_0), respectively. Thus, for the ions

$$\frac{dn_i}{dt} = \frac{I_i}{V} - n_i n_0 \sigma_{cx} v_1 - \tfrac{1}{2} n_i^2 \sigma_{ii} v_2, \tag{9.70}$$

where I_i is the strength, expressed in ions per second, of the injected ion current and V is the plasma volume; the other symbols have the same significance as in §9.86.   The first term on the right gives the rate at which ions are brought into unit volume of the system by a steady current, and the other two terms give the losses by charge exchange and by escape through the mirrors, respectively.

9.93.   Assuming that there is a steady stream of incoming neutral particles of flux ϕ_0, from all parts of the system, the rate of change of neutral particle density is given by

$$\frac{dn_0}{dt} = \frac{\phi_0}{4} \cdot \frac{S}{V} - \frac{n_0 v_0}{4} \cdot \frac{S}{V} - n_i n_0 (\sigma_{cx} + \sigma_{ni}) v_1, \qquad (9.71)$$

where v_0 is the thermal velocity of the neutral particles and S is the surface area of the plasma; the other quantities have the same meaning as before. The first two terms on the right represent the rates, as derived from kinetic theory, of streaming of neutral particles into and out of the plasma, respectively. The last term gives the loss of neutrals due to charge exchange and direct ionization.

9.94.   Theoretical calculations have shown that the injected ions should not lose much of their energy when they enter the plasma.   Hence, it is a good ap-

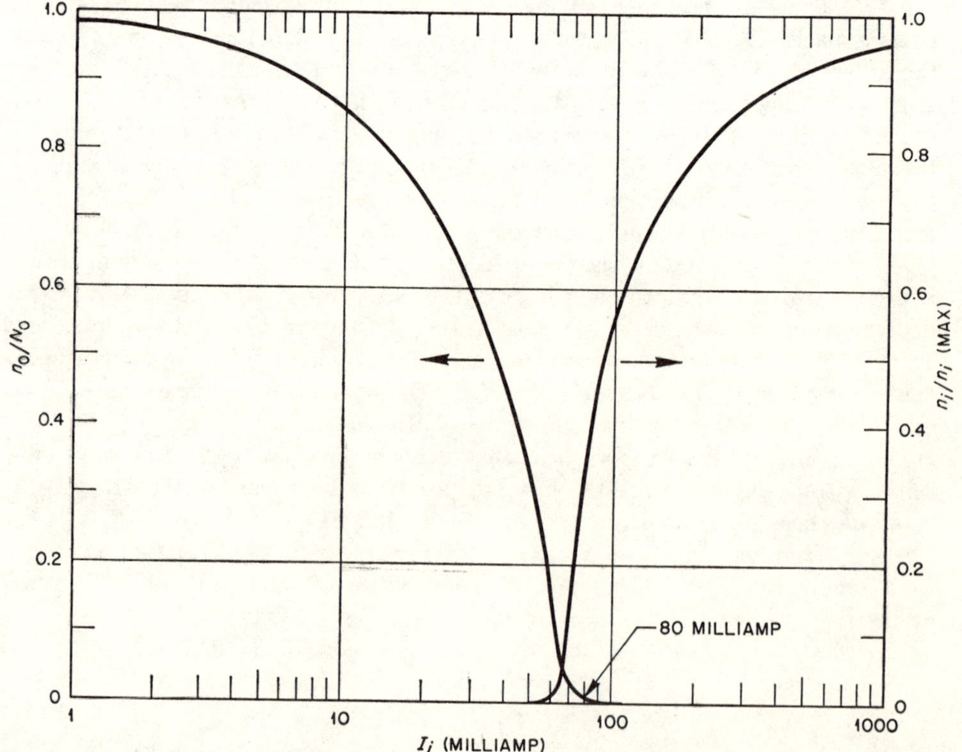

Fig. 9.10.   Determination of ion current for burnout.

proximation to take all cross sections and velocities, except v_0, in equations (9.70) and 9.71) as those applying to the same energy, which is close to that of the injected deuterons. The equations so obtained have been solved by means of an analog computer to determine the steady state values of n_i and n_0 for a range of 250-kev deuteron currents I_i and certain specified system pressures which determine ϕ_0. The nature of the results obtained, for a plasma of 4×10^4 cm³ volume and 6×10^3 cm² surface area, may be illustrated by means of the curves in Fig. 9.10. The pressure outside the plasma is taken to be 10^{-6} mm of mercury. The density of neutral particles is expressed relative to N_0, i.e., as n_0/N_0, where $N_0 v_0$ is equal to ϕ_0. For the given pressure, N_0 is about 3×10^{10} particles/cm³, and v_0 is assumed to be 1.9×10^5 cm/sec, corresponding to a temperature of 300°K for the neutral gas leaking into the plasma volume. The ion density is given as n_i/n_i (max), where n_i (max) is the maximum value consistent with losses due only to escape through the mirrors as a result of scattering. The injected deuteron current is expressed in milliamperes.*

9.95. It is seen that when the ion current exceeds about 80 milliamp, the density of neutral particles has dropped almost to zero; at the same time the ion density is increasing rapidly. Hence, this may be regarded as the region of neutral particle burnout. It should be noted, however, that burnout is not a sudden transition, but is rather a gradual phenomenon occurring over a range of ion currents.

9.96. A simple, approximate treatment of neutral burnout, which gives values for the critical current in good general agreement with those derived from equations (9.70) and (9.71), is based on the postulate that burnout occurs when the neutral particles in the plasma are being removed as fast as they are coming in from outside. The ratio of the number of neutral particles removed (by charge exchange and ionization) to the number of ions lost (by charge exchange) is $(\sigma_{cx} + \sigma_{ni})/\sigma_{cx}$. This represents the number of neutrals removed by an injected ion before being lost itself. The rate of removal of neutral atoms may thus be written as $I_i(\sigma_{cx} + \sigma_{ni})/\sigma_{cx}$ particles/sec, whereas the rate at which neutrals enter the plasma from outside is $\frac{1}{4}\phi_0 S$, i.e., $\frac{1}{4}N_0 v_0 S$. The critical ion current for burnout is obtained by equating these two rates, so that

$$I_i \text{ (burnout)} \approx \frac{\sigma_{cx}}{\sigma_{cx} + \sigma_{ni}} \cdot \frac{N_0 v_0 S}{4} \quad \text{ions/sec.†} \tag{9.72}$$

9.97. Using the values of N_0, v_0, and S in §9.94, which apply to Fig. 9.10, and taking σ_{cx} as 5.5×10^{-18} cm² and σ_{ni} as 1×10^{-16} cm² for 250-kev deuterons, it is found from equation (9.72) that the critical current for neutral

---

* One milliampere of current is equivalent to 6.2×10^{15} ions/sec.

† The same result may be derived from equations (9.70) and (9.71), by taking the first two terms on the right of equation (9.70) to be approximately equal before burnout occurs and the approximate equality of the first and third terms of equation (9.71) as the condition of burnout. Upon eliminating $n_0 n_i$ from the two resulting expressions, equation (9.72) is obtained.

burnout should be 76 milliamp. This result is in good agreement with that derived from the more exact calculations. It will be noted, too, that according to equation (9.72) the burnout current should increase in proportion to the density N_0 of incoming neutral particles, i.e., to the pressure in the region outside the plasma. The curves obtained by the solution of equations (9.70) and (9.71) lead to the same conclusion. It follows that burnout can be achieved more readily, i.e., with smaller ion currents, the lower the gas pressure in the mirror confinement system.

9.98. Once the burnout condition has been attained, the dominant energy loss mechanism of the injected ions will no longer be charge exchange but transfer of energy to the colder ions and electrons present. This process will continue until the energy of these particles approaches that of the injected ions, so that the rates of energy transfer are reduced. The cross section for energy transfer will then be less than that for Coulomb scattering and so the motion of the particles will tend to become more and more random in nature and a thermonuclear plasma, with an approximately Maxwellian distribution, should be produced.

9.99. The calculation of the energies in such a plasma can be made, in principle, by means of the Boltzmann transport equation, but a simple model can be used to provide moderately reliable results. In this model, the injected ions with energy E_i are assumed to lose energy smoothly and uniformly until they form a Maxwellian plasma at some average energy E_+. After the plasma is produced, the ions are lost through the mirrors in the usual manner. Similarly, cold electrons entering the system, at an energy E_e, are heated until they form a plasma of average energy E_-. The behavior of the system according to the model is illustrated schematically in Fig. 9.11, in which the solid lines

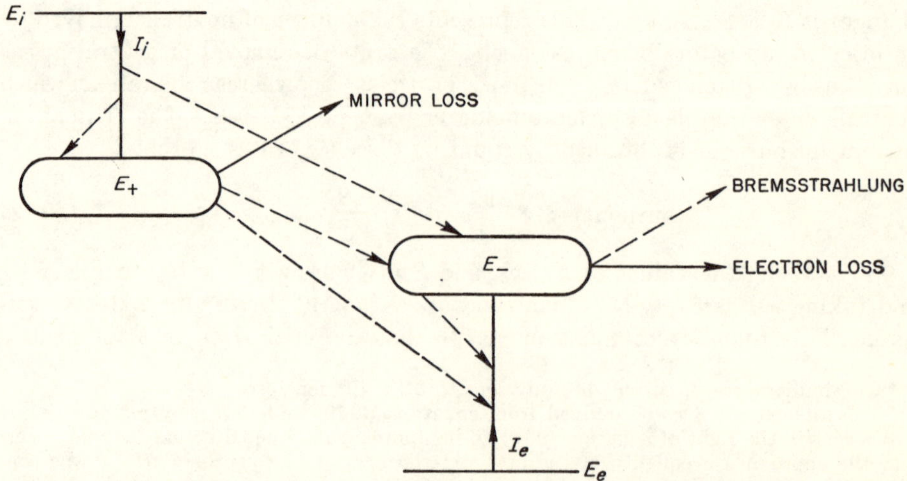

Fig. 9.11.   Particle energies in equilibrium plasma.

represent particle paths in energy space and the broken lines indicate the general directions of energy flow [22-24]. The rate of electron loss from the system is such as to satisfy the requirement of space-charge neutrality.

9.100. It is seen that the ion beam loses energy to both the ion and electron plasmas,* which have lower energies. Furthermore, the ion plasma transfers energy both to the electron plasma and to cold electrons which are being heated up. The electron plasma loses energy by bremsstrahlung, as well as by heating the incoming cold electrons. Direct transfer of energy from the injected ions to the cold electrons is neglected.

9.101. If the injected ion current and energy are specified, there are four unknowns of the system to be determined, namely, the ion and electron densities and the respective plasma energies. These should be obtainable by the solution of four equations resulting from the requirements of energy balance in both the ion and electron plasmas, ion conservation, and space-charge neutrality. The four equations can be further reduced to two coupled equations in the unknowns E_+ and E_- involving such parameters as the energies E_e and E_i of the entering electrons and ions and I_e/I_i, the ratio of their currents. The equations are, however, independent of the actual value of I_i and of the plasma volume. Some of the results obtained by numerical solution, for different values of I_e/I_i, are given in Table 9.1. They refer to the case in which E_i is 300 kev, E_e is 1

TABLE 9.1. CALCULATED EQUILIBRIUM PLASMA ENERGIES FOR
INJECTED 300-KEV IONS

I_e/I_i	E_+ (kev)	E_- (kev)
0	291	221
1	219	78
2	186	56
5	134	33

ev, and the mirror ratio is 2.0. It is apparent that, provided the influx of cold electrons is not too large, it should be possible to attain ion plasmas with an average Maxwellian energy that is a considerable fraction of the energy of the injected ions [22-24].

DISSOCIATION OF MOLECULAR IONS BY GAS COLLISIONS

9.102. In the large magnetic mirror device known as OGRA (§9.168), deuterium-molecule ions of 200-kev energy are injected transversely in such a manner that if they remained unchanged they would travel a distance of 1 kilometer before returning to the source [25, 26]. Because of their long mean free path, it is expected that dissociation of the molecular ions, and consequent trapping, will occur as a result of collisions with neutral molecules (or atoms)

---

* The terms ion and electron plasmas are used here in a somewhat loose sense; they refer to the ion and electron systems which constitute the hot plasma.

of the background gas, with previously trapped atomic ions, and with other molecular ions (§5.113). The situation is quite different from that of dissociation by an arc, considered in the preceding section; nevertheless, it is expected that a critical current for neutral burnout should exist in the present case. The main distinction between the two situations lies in the fact that, although in the gas-breakup scheme the onset of neutral burnout results in a reduction of the extent of dissociation due to collisions with neutral particles, there is an accompanying increase in collisions with trapped ions.

9.103. By neglecting the effect of molecular ions on dissociation and burnout, as compared with the contributions of neutral gas and trapped ions, it has been found possible to set up a steady state equation which can be solved [27]. The plot of the trapped ion density n_i against the strength of the injected molecular ion current I_i is of the form shown in Fig. 9.12. Three general regions

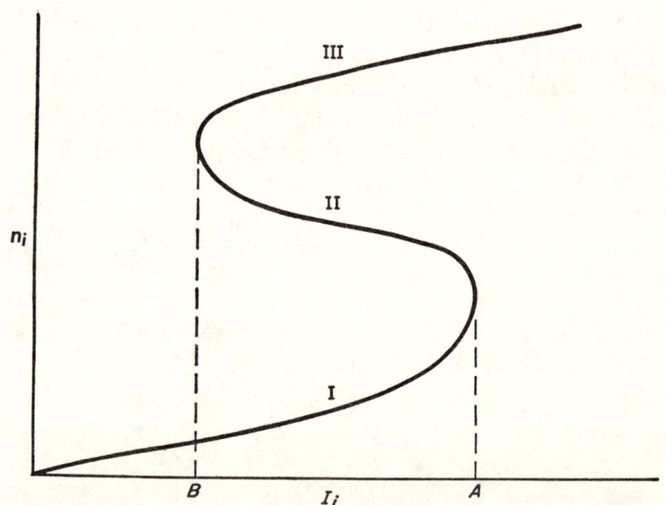

Fig. 9.12. Trapped ion density and molecular-ion current.

are distinguishable, as indicated by the numbers I, II, and III. In region I, neutral burnout has set in, and a steady state is achieved by a balance between the dissociation of the molecular ions resulting from collisions with neutral background particles, on the one hand, and charge-exchange loss of trapped ions, on the other hand. With increasing burnout, the neutral particle density falls while the trapped ion density increases, as indicated above. Hence, in region II, collision between molecular and trapped (atomic) ions is the main cause of dissociation, while charge exchange remains the chief source of loss. The latter is essentially constant in region II, because the rate of the charge-exchange process is proportional to $n_i n_0$, where n_0, the average density of neutral particles in the plasma region after burnout, is approximately proportional to $1/n_i$.

Finally, in region III, escape of ions through the mirrors becomes the dominant loss, since the trapping rate varies directly as the density of the trapped ions, but escape through the mirrors is dependent upon its square (cf. §9.34).

9.104. It will be noted from Fig. 9.12 that there are two critical currents, having the values A and B, corresponding to different trapped ion densities. The upper critical current A is important because this is the molecular-ion current which must be achieved in order to build up a high plasma density when starting initially from a low density. However, once a high ion density has been reached, it may be maintained by steady injection of molecular ions at a current strength somewhat larger than the lower critical current B.

PLASMA COMPRESSION IN MAGNETIC MIRROR SYSTEMS

RADIAL COMPRESSIONS

9.105. The plasma introduced into a magnetic mirror configuration may, in general, not have a sufficiently high density or temperature for the thermonuclear reactions to take place at the desired rate. Both density and temperature can, however, be increased by compression of the plasma. The subject has been treated in Chapter 5 and a further discussion of certain aspects will be given here [1, 2].

9.106. A *radial compression* of the confined plasma can be achieved by an increase in the strength of the confining axial magnetic field. If the process occurs under such conditions that very few particles escape during the compression and that collisions leading to an interchange of energy are not important, then it can be shown that the particles have two degrees of translational freedom. The increase in energy accompanying the radial compression thus occurs in the transverse component $W_\perp$ of the particle motion, i.e., in the direction perpendicular to the field lines. Hence, radial compression leads to more effective confinement of the particles by the magnetic mirrors. By taking into account the constancy of the magnetic moment, it is seen that

$$W_\perp(t) = W_\perp(0) \frac{B_0(t)}{B_0(0)}, \tag{9.73}$$

where $W_\perp(0)$ and $B_0(0)$ are the initial values, before compression, of the transverse energy of a particle and the field strength, respectively, in the central region between the mirrors and $W_\perp(t)$ and $B_0(t)$ are the corresponding values at any time t during the compression.

9.107. Because of the constant flux condition, derived in §4.24, the density of a plasma will vary directly as the field strength during the course of a radial magnetic compression (cf. §5.38) in a system of constant length. The radial compression factor α is defined as the ratio of the volumes before and after compression, or what is the same thing, the ratio of the particle density $n(t)$ at time t to the initial value $n(0)$. It follows, therefore, that

$$\alpha \equiv \frac{n(t)}{n(0)} = \frac{B_0(t)}{B_0(0)}, \tag{9.74}$$

and upon combining this result with equation (9.73), it is seen that

$$W_\perp(t) = W_\perp(0)\alpha. \tag{9.75}$$

The relative increase of transverse energy is thus equal to the radial compression factor.

AXIAL COMPRESSION

9.108. To achieve *axial compression*, the magnetic mirror fields are moved closer together, either mechanically or by increasing the current passing through the appropriate coils surrounding the tube containing the plasma. As a result, the longitudinal energy $W_\parallel$, in the direction parallel to the magnetic field lines, is increased. If $l(0)$ and $l(t)$ represent the distances between the mirrors initially and at a time t, respectively, the corresponding values of the energy $W_\perp$ for one degree of freedom are related, according to equation (5.8), by

$$W_\parallel(t) = W_\parallel(0)\left[\frac{l(0)}{l(t)}\right]^2, \tag{9.76}$$

since $W_\parallel$ is proportional to $v_\parallel^2$. If χ represents the *axial compression factor*, which may be defined in a manner analogous to the radial compression factor, it is seen that

$$\chi \equiv \frac{n(t)}{n(0)} = \frac{l(0)}{l(t)}. \tag{9.77}$$

Hence, equation (9.76) becomes

$$W_\parallel(t) = W_\parallel(0)\chi^2. \tag{9.78}$$

9.109. An alternative derivation of equation (9.76) is of interest since it makes use of an adiabatic invariant defined as *the action integral of the momentum of a particle along a line of force*, taken over the region in which the particle is bound between magnetic mirrors [1, 2]. Expressed mathematically, this is

$$\int_{z_1}^{z_2} p_\parallel \, dz = \text{constant}, \tag{9.79}$$

where $p_\parallel$ is the longitudinal momentum of the particle and z_1 and z_2 are the axial coordinates of the points at which the particles are turned back by the two magnetic mirrors. This means that if the path of the particle is represented in phase space, with momentum $p_\parallel$ and distance z from the midplane as the coordinates, as shown in Fig. 9.13, the area enclosed by the path remains constant. For a symmetrical system, z_1 and z_2 are equal numerically but of opposite sign; the lower limit of integration may then be taken as zero, i.e., at the midplane, and the result will still be constant. It is of interest to point

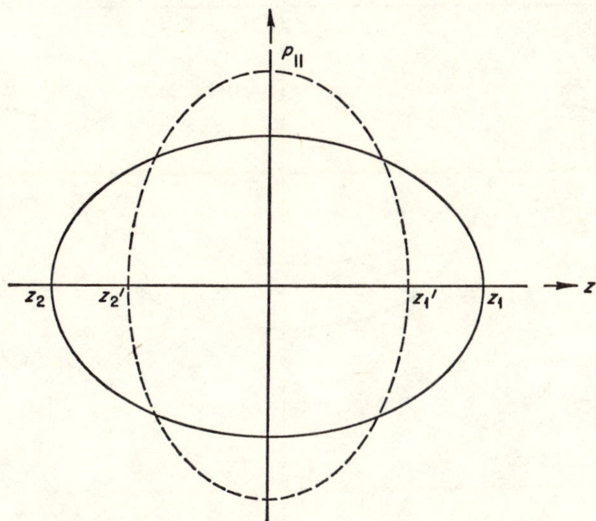

FIG. 9.13.   Action integral as adiabatic invariant.

out that equation (9.79) is a special case of Liouville's theorem in statistical mechanics, according to which the volume (or area) occupied in phase space by a single particle remains constant.

9.110.   If the mirrors are moved closer together, in an axial compression, so that the reflection points z_1 and z_2 are changed to z_1' and z_2', respectively, the path of the particle in phase space is changed from the full line to the broken line in Fig. 9.13.   However, the areas enclosed by the two orbits must be equal, so that, if the magnetic mirror system is symmerical, a good approximation is to treat $p_{||}l$, as constant, where l, the distance between the mirror peaks, is taken as being equal to the distance between the reflection points.   Since $p_{||}$ is proportional to $v_{||}$, it follows that

$$\frac{v_{||}(t)}{v_{||}(0)} = \frac{l(0)}{l(t)}$$

or

$$\frac{W_{||}(t)}{W_{||}(0)} = \left[\frac{l(0)}{l(t)}\right]^2,$$

which is indentical with equation (9.76).

EFFECTS OF FINITE DIMENSIONS OF A MIRROR FIELD

9.111.   As stated in §9.52, the assumption has generally been made in the preceding treatment that the magnetic field is uniform in the central region and then rises abruptly at the ends to provide the mirror system (see Fig. 9.5). The actual variation of the magnetic field strength is, however, much more

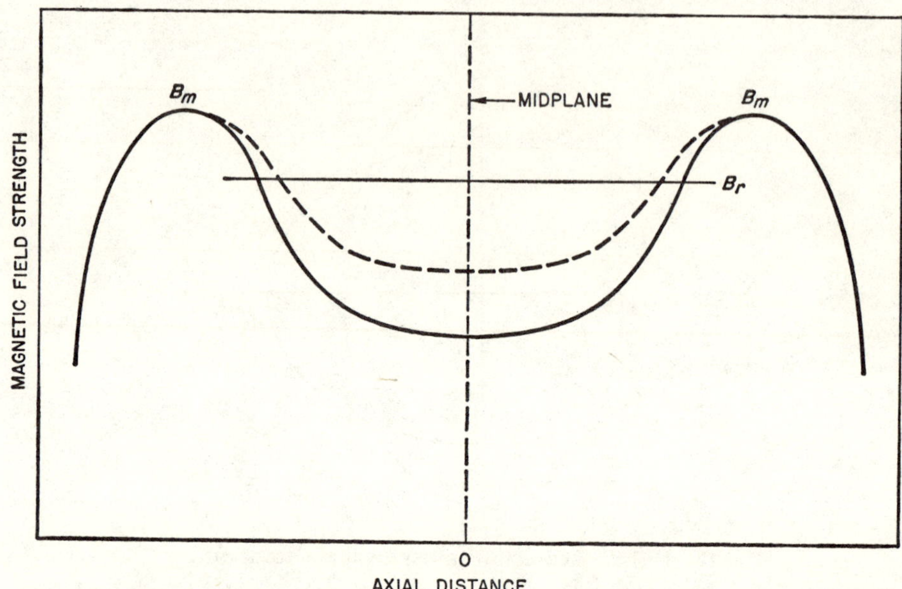

FIG. 9.14. Change in effective distance between magnetic mirrors.

like that indicated by the full line in Fig. 9.14. Consider a particle having such a value of $W_\parallel$ that is reflected at the points where the magnetic mirror field is equal to B_r, as shown on the figure. Suppose the plasma is compressed radially by increasing the strength of the central magnetic field, so that it is represented by the broken line. Since radial compression does not affect $W_\parallel$, the particle will still be reflected at the point where the mirror field strength is B_r. It is seen, therefore, that the particle does not now have to travel quite so far into the mirror before it is reflected. This means, in effect, that there is an indirect axial compression of the plasma in addition to the radial compression, even though the mirror field is not deliberately moved. In Fig. 9.14, the radial compression is not accompanied by a change in the peak strength of the mirror field, but even if it were increased at the same time as the central field, the general conclusion would remain unchanged.

9.112. The extent of the indirect axial compression can be derived by means of the adiabatic invariant defined by equation (9.79); thus, for a symmetrical system, since $p_\parallel$ is proportional to $W_\parallel{}^{1/2}$, it follows that

$$\int_0^{z_r} [W_\parallel(z)]^{1/2}\, dz = \text{constant}, \tag{9.80}$$

where z_r is the axial coordinate of the point at which the particle is reflected. At this point, $W_\parallel(z)$, which may be represented by $W_\parallel(z_r)$, is zero, so that the total energy W is equal to $W_\perp(z_r)$. Hence, for any value of z,

$$W_{\parallel}(z) = W - W_{\perp}(z)$$
$$= W_{\perp}(z_r) - W_{\perp}(z).$$

Making use of the constancy of the magnetic moment and the definition of the mirror ratio, this result can be transformed into

$$W_{\parallel}(z) = W_{\perp(0)}[R_r - R(z)],$$

where R_r and $R(z)$ are the mirror ratios at the reflection point z_r and at any arbitrary point z, respectively. Upon substitution for $W_{\parallel}(z)$ in equation (9.80) and noting that $W_{\perp(0)}/B_0$ is the magnetic moment and so is constant, it is seen that

$$B_0{}^{\frac{1}{2}} \int_0^{z_r} [R_r - R(z)]^{\frac{1}{2}} \, dz = \text{constant} \tag{9.81}$$

or, upon changing the variable to R,

$$B_0{}^{\frac{1}{2}} \int_1^{R_r} [R_r - R(z)]^{\frac{1}{2}} \frac{dz}{dR} \, dR = \text{constant}. \tag{9.82}$$

If the magnetic field strength (or mirror ratio) can be expressed as a function of the axial distance, it should be possible to solve equation (9.81) or (9.82) for z_r.

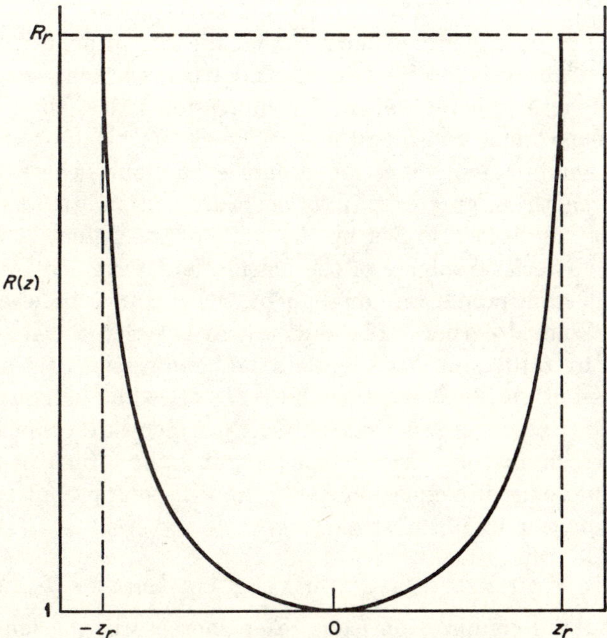

FIG. 9.15. Parabolic magnetic mirror field.

9.113. A simple case to consider is that of a field in which the variation of the field strength or the mirror ratio with the distance from the median plane is parabolic (Fig. 9.15) and can be represented by

$$R(z) = 1 + cz^2,$$

where c is a constant. Upon substituting for $R(z)$ in equation (9.81) and performing the integration, the result is

$$B_0^{\frac{1}{2}}z_r^2 = \text{constant}.$$

Hence, in a radial compression, utilizing equation (9.74),

$$\alpha = \frac{B_0(t)}{B_0(0)} = \left[\frac{z_r(0)}{z_r(t)}\right]^4. \tag{9.83}$$

The axial compression factor χ accompanying the radial compression is equal to $z_r(t)/z_r(0)$, and so by equation (9.83)

$$\chi = \left[\frac{B_0(t)}{B_0(0)}\right]^{\frac{1}{4}} = \alpha^{\frac{1}{4}}.$$

Thus, the indirect axial compression factor is equal, in this particular case, to the fourth root of the radial compression factor. The total compression factor, which is equal to $\alpha\chi$, is then $\alpha^{5/4}$ for the radial magnetic compression in a parabolic field [1].

9.114. Another consequence of the fact that the mirror field increases gradually, rather than abruptly, to its peak is that when the magnetic mirrors are brought closer together the actual axial compression is less than would be expected from geometrical considerations. The reason is that, since $W_{\parallel}$ increases in the compression, reflection occurs at a point where the magnetic mirror field strength is greater. In other words, after the axial compression the particle will ascend higher up the walls of the potential well before it falls back. The effective volume of the plasma is determined by the distance between the reflection points, rather than by the distance between the mirror peaks, as was assumed earlier. If the distances between the reflection points are represented by $D(0)$ and $D(t)$ before axial compression and at time t during the compression, respectively, then $D(0)/D(t)$, which is equal to the true axial compression factor, is less than $l(0)/l(t)$, which is the apparent or geometrical compression factor. Strictly speaking it is the former of these quantities which should be used in equation (9.76), as will be apparent from the derivation of this equation in §9.109 *et seq.*

COMBINED RADIAL AND AXIAL COMPRESSION

9.115. Since axial compression leads to an increase in the longitudinal energy of the plasma particles, the confinement by the mirrors becomes less effective. This drawback can be overcome, within certain limits, if both radial and axial compression are applied simultaneously. The total compression factor is then $\alpha\chi$ so that

$$\frac{n(t)}{n(0)} = \alpha\chi. \tag{9.84}$$

The condition that the plasma particles shall remain at least as well confined by the magnetic mirrors after a combined radial and axial compression as it was before can be derived as follows. If the injection pitch angle ϵ is small, as should be the case, equation (9.40) becomes

$$\frac{W_{\parallel}(0)}{W_{\perp}(0)} \approx R_m(0) - 1,$$

where the $W_{\parallel}(0)$ and $W_{\perp}(0)$ values are those in the central region between the mirrors before compression occurs, and $R_m(0)$ is the initial mirror ratio. As the result of a radial compression, $W_{\perp}$ is increased by a factor α, according to equation (9.75), but $W_{\parallel}$ remains unchanged; hence,

$$\frac{W_{\parallel}(t)}{W_{\perp}(t)} \approx \frac{1}{\alpha}[R_m(0) - 1].$$

At the same time as the central field is increased, to produce radial compression, the mirrors are moved closer together to achieve axial compression. Hence, by equation (9.78), the energy $W_{\parallel}$ is increased by a factor χ^2, and so for simultaneous radial and axial compression

$$\frac{W_{\parallel}(t)}{W_{\perp}(t)} = \frac{\chi^2}{\alpha}[R_m(0) - 1]. \tag{9.85}$$

9.116. The condition for reflection after the compressions is given by equation (9.11) as

$$\frac{W_{\parallel}(t)}{W_{\perp}(t)} \leqslant R_m(t) - 1,$$

and by combining this result with equation (9.85), the condition may be written in the form

$$\chi^2 \leqslant \alpha \frac{R_m(t) - 1}{R_m(0) - 1}.$$

If the mirror ratio remains unchanged during the compression, i.e., by increasing the mirror field strength in proportion to that of the central field, the condition for reflections becomes

$$\chi^2 \leqslant \alpha. \tag{9.86}$$

9.117. To determine the variation of the total energy W of a particle in a compression process, it may be noted, first, that at the point of reflection, i.e., when the mirror field strength is B_r, all the energy is transverse, so that W is equal to $W_{\perp(r)}$. Hence, as a result of the adiabatic constancy of the magnetic moment,

$$\frac{W_\perp}{B} = \frac{W}{B_r} = \text{constant.}$$

This result is completely general and so, in any compression process,

$$\frac{W(0)}{B_r(0)} = \frac{W(t)}{B_r(t)}$$

or, since $R = B_r/B_0$,

$$W(t) = W(0)\,\frac{R(t)}{R(0)} \cdot \frac{B_0(t)}{B_0(0)}. \tag{9.87}$$

This expression relates the final kinetic energy $W(t)$ after the compression to the initial value $W(0)$.

9.118.   Returning now to equation (9.86), the special case in which $\chi^2 = \alpha$ is of interest; in these circumstances equation (9.84) shows that

$$n(t) = n(0)\alpha^{3/2}, \tag{9.88}$$

which is equivalent to uniform compression of the plasma by a factor of $\alpha^{1/2}$ in three dimensions.   Further, when $\chi^2 = \alpha$, it is seen from equations (9.75) and (9.78) that $W_\parallel/W_\perp$ remains unchanged as a result of the compression.   Hence, the mirror ratio at which reflection occurs will also remain the same, that is, $R(t)$ and $R(0)$ in equation (9.87) are equal.   Upon introducing equation (9.74) it is seen that, in the case under consideration,

$$W(t) = W(0)\alpha. \tag{9.89}$$

From equations (9.88) and (9.89) it follows that the mean value W of the particle energy is proportional to $n^{2/3}$.   This is equivalent to saying that the average temperature is inversely proportional to $V^{2/3}$, where V is the plasma volume; hence, in the given circumstances

$$TV^{2/3} = \text{constant,}$$

which is identical with the familiar expression for the adiabatic heating of an ideal gas.   Thus, if a plasma is compressed radially and axially at the same time, under such conditions that $\alpha = \chi^2$, the general results are identical with those to be expected for equipartition of the energy among three degrees of freedom of translational motion.

LOSSES DURING RADIAL COMPRESSION

9.119.   The increase in the transverse energy of the particles and the accompanying improvement in confinement of the plasma that result from radial compression are offset by the disordering effect of collisions.   But since the cross section for Coulomb scattering decreases with increasing energy of the particles, it would appear to be possible to carry out the radial compression in such a manner as to overcome the losses through the mirrors, due to scattering collisions, during the whole of the heating period [1].   In order for this to be the case, it is necessary that the component of the particle energy perpendicular to the magnetic field direction shall be much greater than that in the parallel direction, i.e., $W_\perp(t) \gg W_\parallel(t)$, at all times during the compression operation. The treatment given below is based on the assumption that the plasma densities and temperatures are such that it is the scattering time which determines the escape of particles through the mirrors (§9.32). In addition, the effect of the ratio β and its variation during compression are neglected.

9.120. As a rough approximation, the average value of $W_\parallel$ may be taken as $\frac{1}{3}W$, so that the rate at which the longitudinal energy of the ions increases, as a result of collisions, may be written as

$$\frac{dW_\parallel}{dt} \approx \frac{1}{3} \cdot \frac{dW}{dt} \approx \frac{1}{3} \cdot \frac{W}{t_{ii}}, \tag{9.90}$$

where, as above, t_{ii} is the relaxation time for ion-ion collisions. From equation (4.49) it is seen that t_{ii} may be expressed in the general form

$$t_{ii} \approx A\,\frac{W^{3/2}}{n},$$

where n is the ion density, and A is a constant, so that equation (9.90) gives

$$dW_\parallel \approx \frac{1}{3A} \cdot \frac{n}{W^{1/2}}\,dt.$$

Thus, at any time t after the start of the radial compression,

$$W_\parallel(t) \approx \frac{1}{3A} \int_0^t \frac{n}{W^{1/2}}\,dt. \tag{9.91}$$

Since it is required that $W_\perp \gg W_\parallel$, it follows that $W_\perp \approx W$; hence, from equation (9.87),

$$W(t) \approx W(0)\alpha(t),$$

where $\alpha(t)$ is the radial compression factor, i.e., $n(t)/n(0)$, at any time t. Upon substituting these expressions for n and W into equation (9.91), the result is

$$W_\parallel(t) \approx \frac{n}{3A\,W^{1/2}} \int_0^t [\alpha(t)]^{1/2}\,dt, \tag{9.92}$$

where n and W are the values at the start of the compression process.

9.121. Suppose that the central magnetic field strength increases linearly with time, so that

$$\frac{dB_0(t)}{dt} = \frac{B_0(0)}{\tau},$$

where τ is a constant having the dimensions of time. Then the radial compression factor, which is equal to $B_0(t)/B_0(0)$ at any time t, is given by

$$\alpha(t) = 1 + \frac{t}{\tau}. \tag{9.93}$$

Hence, upon integration, equation (9.92) gives

$$W_{\parallel}(t) \approx \frac{2n\tau}{9AW^{1/2}} (\alpha^{3/2} - 1), \tag{9.94}$$

where, for simplicity, α has been written for $\alpha(t)$.

9.122. The change in the transverse energy in the radial compression is given by equation (9.75), but since $W_{\perp} \approx W$, it is permissible, in the present case, to write

$$W_{\perp}(t) \approx W(0)\alpha(t),$$

so that, utilizing equation (9.87), the condition that $W_{\perp} \gg W_{\parallel}$ at all times becomes

$$\frac{2n\tau}{9AW^{3/2}} \left(\alpha^{1/2} - \frac{1}{\alpha} \right) \ll 1. \tag{9.95}$$

If α is large, so that $\alpha - 1 \approx \alpha$, it follows from equation (9.93) that $\tau \approx t/\alpha$; upon making this substitution into equation (9.95) and negelcting $1/\alpha$ as being small in comparison with $\alpha^{1/2}$, it follows that

$$\frac{2nt}{9AW^{3/2}\alpha^{1/2}} \ll 1$$

is the required condition for particle confinement during the radial compression. This result may be expressed as a simple restriction on the time t_α within which the compression must be carried out if appreciable loss of particles through the mirrors, due to scattering, is to be avoided; thus,

$$t_\alpha \ll \frac{9AW^{3/2}\alpha^{1/2}}{2n}, \tag{9.96}$$

a requirement which can be most easily satisfied for plasmas of low density and high particle energy, particularly for large compression factors. Since not all the charged particles will have the same energy, it is probable that these of low energy will escape from the mirror confinement during the radial compression operation.

9.123. A general idea of the required compression time may be obtained by considering a fairly typical experimental case in which the energy $W_{\perp}$ of the injected particles, namely, deuterons, is 0.1 kev, so that $W \approx W_{\perp} = 0.1$ kev, and n is 10^{12} ions/cm³. Suppose that this plasma is to be compressed radially by a factor of 100, so as to make W about 10 kev and the particle density 10^{14} ions/cm³. According to equation (4.51), the value of A for deuteron-deuteron scattering is about 2×10^{10}, if W is expressed in kilo-electron volts. Hence, it follows from equation (9.96) that t_α should be less than about 0.03 sec; this does not represent a serious restriction upon the compression operation. The compression time is increased if the initial particle energy is greater than that assumed above, but it is decreased if the density is higher.

9.124. If the radial compression is carried out too slowly, so that the condition represented by equation (9.96) is not satisfied, most of the particles of high energy will be retained whereas those of lower energy will be lost before the compression is completed. The result is that the mean particle energy will be somewhat higher than if the compression had been more rapid, but the particle density will be lower than expected from the simple compression equation (9.74). It should be noted that the condition derived above for minimizing losses through the mirrors during compression will not apply if the radial compression is accompanied by axial compression. In addition to the fact that the density change is larger than that due to radial compression alone, the axial compression results in an increase in the longitudinal energy of the particles, thereby increasing the probability of escape.

9.125. Another source of loss is that the condition for adiabaticity may not be satisfied during radial compression. If the process is to be adiabatic, or essentially so, the change in the magnetic field strength must be small during the time a particle makes a single orbital turn, i.e., during the gyromagnetic period given by equation (4.7). Thus, the adiabatic condition may be represented by

$$\left(\frac{1}{B} \cdot \frac{dB}{dt}\right) \frac{2\pi mc}{eB} \ll 1.$$

If the magnetic field increases in a linear manner with time, so that $dB/dt = B(0)/\tau$, as in §9.121, this condition for adiabatic behavior may be readily transformed into

$$\frac{eB(0)t_\alpha \alpha^2}{2\pi mc(\alpha_{\max} - 1)} \gg 1,$$

where α is written for $\alpha(t)$ and $\alpha_{\max}$ is the maximum value of the radial compression factor attained at the end of the time t_α.

9.126. In the early stages of the compression, when α is small, the adiabatic condition may not be satisfied, especially if $\alpha_{\max}$ is large. However, as α increases, the process approaches adiabatic behavior. The fraction of the compression time during which the system is not adiabatic is decreased by a decrease in $\alpha_{\max}$; this can be achieved by increasing τ, by an increase in $B(0)$, the initial field strength, and by the factors which permit an increase in t_α, i.e., high particle energy and low density.

EXPERIMENTAL RESULTS

EARLY STUDIES OF MAGNETIC MIRROR SYSTEMS[*]

9.127. The first tests of confinement of a plasma in a magnetic mirror con-

---

[*] The work described in §§9.127-9.154 is part of what has been called the Pyrotron Program [1].

figuration were made with a vacuum chamber consisting of a glass tube on the interior of which a thin coating of stainless steel had been deposited by evaporation. The purpose of this conducting layer was to make sure that all points on the inside of the tube were at the same potential. Hydrogen gas at low pressure was ionized by means of a pulsed radiofrequency field [28]. The decay of the plasma was then traced by the behavior, as a function of time, of a microwave signal propagated through it. When magnetic mirror fields were energized at the ends of the tube, there was evidence of increased confinement of the plasma.

9.128. This work was followed by an experiment, called Albedo, on the use of a radiofrequency field for enhancing the confinement of charged particles in a mirror system [29]. A beam of low-energy ions (and electrons) was injected along the central axis of the magnetic field; as expected, the losses through the mirrors were considerable. However, when a radiofrequency field which was in resonance with the gyromagnetic frequency of ions was applied, so as to increase the energy component perpendicular to the axial field lines, appreciable trapping of the ions resulted. Although the work showed promise, it was abandoned at the time because of the problems arising from penetration of the radiofrequency field into the plasma at moderately high density.

9.129. A preliminary study of magnetic compression was made with a glass T-shaped electromagnetic shock tube (cf. §5.43) [30]. Two electrodes were mounted in the arms at the top of the T and coils for a magnetic mirror system were wound around the leg. The shock-heated plasma, resulting from passage of a discharge between the electrodes, passed down the leg of the tube where it could be trapped by the time-rising magnetic mirror fields. Within a few microseconds of the discharge, the fields were pulsed to a peak value of about 200,000 gauss in a period of about 0.1 millisec. Time-resolved photographs of the luminous plasma, obtained with a streak camera, showed definite indications of compression by the magnetic field with no signs of instability.

9.130. Although the experiments just described seemed to prove the basic feasibility of the plasma compression process, the magnetic field technology at the time did not permit its effective use with the high particle densities of the shock-produced plasmas. Consequently, further efforts were devoted to the development of occluded-gas ion sources, such as those described in §5.80 *et seq.,* which could be used to inject more highly ionized plasmas of lower density in a magnetic mirror geometry.

9.131. In the apparatus known as Table Top I,* the containing tube (vacuum chamber) was 15 cm in diameter and about 1 meter in length. Plasma injection was effected by means of a deuterium-loaded titanium spark source located at the peak of one of the mirrors. A fraction of the plasma produced was trapped, apparently owing to a combination of collisional and time-rising field effects between the pulsed magnetic mirror fields, which reached a peak

* The name Table Top has also been given to various modifications of this apparatus.

strength of 30,000 gauss. The energy of the injected deuterons was estimated to be 5 to 10 ev. The confinement of the plasma was observed by the microwave method and the time for the electron density to fall to one-third of its initial value, of about 3×10^{11} electrons/cm³, was of the order of 300 or 400 microsec; this is in rough agreement with the calculated loss rate for escape through the mirrors due to Coulomb scattering. Thus, it was concluded that no significant loss mechanism, other than interparticle collisions, appeared to be needed to account for the observed decay rate of the plasma [31, 32].

9.132. With the objective of obtaining higher magnetic fields, smaller devices were built, using Pyrex glass tubing either 2.5 or 5 cm in diameter, to which the name Toy Top was given [32-35]. In general, a weak DC magnetic field served to confine the plasma, which was introduced from a spark source of titanium wires loaded with deuterium; the mean energy of the injected ions was probably about 5 ev. A strong pulsed magnetic field, with a mirror ratio of approximately 2.0 and a peak strength up to 400,000 gauss at the mirrors, was applied to produce radial compression of the plasma. The most direct evidence that heating occurred was the production of X-rays of 10 to 200 kev energy, which presumably represented bremsstrahlung from interaction of high-energy electrons with the tube walls. The fact that the signals sometimes persisted for as long as 3 millisec indicated considerable confinement of the heated plasma. The confinement times appeared to be limited more by the decay of the pulsed magnetic field than by plasma instabilities.

9.133. Additional evidence for confinement and radial magnetic compression of a plasma in a magnetic mirror configuration was obtained later from the Saturn experiment in which an equatorial ring plasma source was used (§5.83) [36]. This source was located at the midplane of a 46-cm diameter tube between two solenoid coils which produced the pulsed mirror field. Luminosity and streak camera observations indicated that compression of the plasma could be achieved. From microwave determinations it appeared that a measurable electron density could be maintained for a period of approximately 1 millisec. The persistence of the emission of X-rays and of microwave noise from the plasma provide general support for this conclusion. The maximum ion temperature attained was estimated to be about 100 ev.

9.134. In order to study the problems of plasma stability and diffusion across the magnetic field in a mirror geometry, it was considered desirable to work with plasmas in which the particle pressure nkT represented as large a fraction β as possible of the external field pressure $B^2/8\pi$. The simplest way to achieve this result was to decrease the strength of the confining field, so that plasmas of low density and low temperature could be employed in the experiments. For this purpose, several large systems, referred to as Cucumber (or Q-cumber), with diameters up to 46 cm, were constructed [37, 38]. The central confining field was as low as 20 gauss and the DC mirror fields had a peak strength up to 6000 gauss. The plasma source, of the deuterium-loaded titanium wire type utilizing a spark discharge, was inserted into one of the mirror fields. Because of the constancy of the magnetic fields, trapping by collisions, which operated effectively, was relied upon exclusively. Plasma densities in the range of 10^{10} to 10^{12} particles/cm^3 with an estimated energy of 5 to 20 ev were obtained in this manner. The pressure in the vacuum chamber was usually 2 to 4×10^{-6} mm of mercury. At higher pressures the plasma was soon lost, presumably due to charge exchange and other interactions of the ions and electrons with neutral particles.

9.135. Both microwave measurements of electron density and Langmuir probe studies indicated that, with the mirror fields in operation, confinement times of at least 2 millisec could be realized, at β values up to 0.1, without any indication of plasma instability. The end losses appeared to vary approximately inversely as the mirror ratio over a considerable range, in general agreement with expectation. Furthermore, under the experimental conditions, there was no indication of an anomalously high rate of diffusion of the plasma in a direction perpendicular to the field lines (cf. §12.35).

HIGH-COMPRESSION AND HEATING EXPERIMENTS

9.136. In the later high-compression experiments, carried out in the device called Table Top II (or System III), a plasma of about 5-ev energy, from a titanium-deuterium spark source, was injected into an evacuated chamber in an initially weak DC magnetic mirror field. The field strength was then increased rapidly to a high value, in a period of 200 to 500 microsec, thereby causing trapping of the plasma and its radial compression by factors as large as 1000. The initial plasma bursts produced a density of 10^{11} to 10^{12} particles/cm^3 and so the final density might be 10^{14} ions/cm^3 or more. Correspondingly, ion and electron temperatures in the kilo-electron volt range should have been attained [39, 40].

9.137. In Table Top II, the main vacuum chamber, with a base pressure of 2×10^{-7} mm of mercury, is a Pyrex glass tube of 15-cm internal diameter, with a thin coating of stainless steel evaporated on to the inner surface. Two sets of magnetic field coils surround the tube and are coaxial with it (Fig. 9.16) [40]. Direct current passing through the larger outer coils provides the weak initial mirror field having a maximum strength of 400 gauss. The discharge through the inner coil of smaller diameter, of a 1.2-millifarad capacitor bank, which can be charged to 40 kilovolts, produces a pulsed, time-rising field with a peak value of 44,000 gauss and a mirror ratio of about 1.5. In a typical operation, the pulsed field is energized about 20 microsec after injection of the plasma. The field strength increases almost sinusoidally, reaching its maximum value in approximately 500 microsec. The circuit is then crowbarred in order to maintain a large current through the field coil for as long as possible. The decay of the field strength is exponential, with a

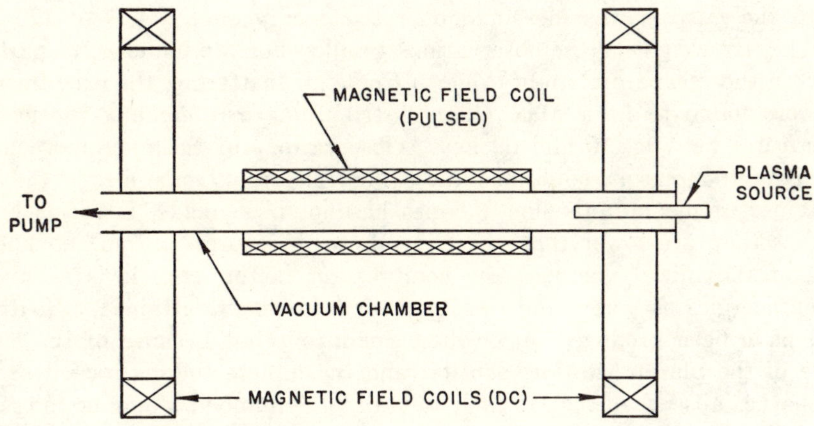

Fig. 9.16.   Schematic representation of Table Top II.

$1/e$-folding time of roughly 30 millisec determined from the L/R value of the circuit.

9.138. The compression of the plasma has been followed by means of Langmuir-type probes. A thin wire probe, whose radial position could be varied, was inserted at the midplane of the mirror system. At a given position, the signal was found to drop suddenly to zero at a definite time after the pulsed field was applied. This was taken to be the time the plasma boundary, in its inward motion due to radial compression, passed the end of the probe. In a series of observations it was thus possible to determine the radius of the plasma (or the plasma density) as a function of time. The results obtained in this manner were compared with those calculated from equation (9.74), using the known time-variation of the confining field strength. Satisfactory agreement was obtained between the experimental and calculated values. Such small discrepancies as were observed were attributed to the indirect axial compression which may be expected to accompany the radial compression (§9.111).

9.139. Further evidence of compression of the plasma by the magnetic field was provided by observations on the longitudinal distribution of X-rays outside the vacuum chamber beyond the mirrors. These X-rays are apparently bremsstrahlung produced at the walls by electrons which have escaped from the plasma by traveling along the magnetic field lines. Hence, by extrapolation to the interior of the vacuum chamber, the X-ray distribution can provide information concerning the radial distribution of the compressed plasma. The results indicated that the plasma had been compressed by the magnetic field into a narrow filament at the axis of the containment tube.

9.140. By using scintillation detectors for electrons, together with suitable absorbers placed on the tube axis outside the mirror at the end opposite to that where the source is located, the energy distribution of the electrons escaping through the mirrors was measured over the energy range of 3 to 120 kev [40, 41]. By weighting the observations to allow for the influence of particle energy on the escape probability due to Coulomb scattering, the experimental data were found to fit a Maxwellian distribution corresponding to electron temperatures between 10 and 20 kev. Observations of the microwave noise emitted by the plasma confirmed these high electron temperatures (§6.32). Direct measurement of the ion energies has not been made, but it is probable that they are lower than the electron temperatures. From the known values of the initial energies, the compression factor, etc., it is estimated that ion energies of 1 kev, and possibly 2 or 3 kev, were attained. Neutrons should have been produced under these conditions but because of the small volume of the plasma, its low density, and incomplete thermalization of the deutrons, i.e., absence of the Maxwellian "tail," the numbers would be too small to detect.

9.141. A striking feature of these high-compression studies is the relatively long confinement times and the absence of any indication of gross plasma instabilities. From the persistence of X-ray and electron emission and of microwave noise, it is concluded that the decay half-life of the compressed plasma is generally several milliseconds. In some cases the emission of highly energetic particles of radiation has been observed to continue for as long as 30 millisec. Estimates of the total kinetic energy of the plasma particles, using scintillator and thermocouple techniques, indicate that β values of 0.08 have been attained.

9.142. One of the unexpected results was concerned with the high-energy (above 100 kev) X-rays produced at the midplane of the mirror system. The peak X-ray signal was received 1 or 2 millisec after the time of maximum compression, and this was interpreted as implying a slow outward radial motion of the plasma electrons of highest energy. Part of this radial motion, as well as the decay of the X-ray intensity, can be accounted for by the expansion of the plasma as the strength of the confining field decreases, but there still remans a slow drift which cannot be explained in this manner. A small electric field, fluctuating at a frequency comparable to or slower than the electron gyromagnetic (or cyclotron) frequency, would induce a randomly directed crossed electric and magnetic field drift which could account for the additional outward radial motion of the plasma.

MULTISTAGE HIGH COMPRESSION

9.143. In the work described above, a high compression factor was achieved by using a weak initial confining field. Since the radii of the particle orbits are then large, it is necessary to use a vacuum chamber of large diameter in order to avoid losses by the particles striking the walls. In this event, the energy expended in increasing the magnetic field strength, in order to produce the required radial compression of the plasma, is large. There is thus a practical limit to the compression factor that can be attained; in the Table Top II experiments, this was about 1000, as stated earlier. Stronger magnetic fields and higher compressions were possible in this work, in principle, but even a twofold increase would have been very costly.

9.144. One way of overcoming the difficulty of obtaining very high compressions is to carry out the process in stages. The plasma is introduced into a vacuum chamber of large diameter and is confined by a weak DC magnetic mirror field; here it is compressed radially by means of a pulsed field. Then, by moving one of the mirrors and decreasing the other, the plasma is transferred longitudinally into a chamber of smaller diameter with an appropriate DC field. The plasma is now further compressed and transferred, once again, into a chamber of still smaller diameter, and so on. In this manner, the strongest magnetic field is applied to a narrow tube and consequently does not require the expenditure of large amounts of energy. The radial compression factor is, as before, $B(t)/B(0)$, where $B(t)$ is the field strength at the end of the compression in the central region of the chamber of small diameter and $B(0)$ is the initial value in the wide tube.

9.145. In order that the flux tubes of the DC fields shall not strike the walls as they thread the entire system, it is necessary, in proceeding from one chamber to the next of smaller diameter, that the value of Br^2, where B is the strength of the confining DC field and r is the chamber radius, be at least equal to that in the preceding chamber. Thus, if B_1 and r_1 refer to one chamber and B_2 and r_2 to the next, where $r_1 > r_2$, the requirement is that $B_2 r_2{}^2 \geqslant B_1 r_1{}^2$, or

$$\frac{B_2}{B_1} \geqslant \left(\frac{r_1}{r_2}\right)^2.$$

9.146. An apparatus, known as Toy Top II $\phi\beta$, has been constructed to investigate the process of multiple compression [1, 42, 43]. It consists of three sections with diameters of approximately 46, 23, and 11.5 cm, respectively. Preliminary microwave and other measurements show that there is a steady increase in density as the plasma is transferred from one chamber to the next of smaller diameter.

PLASMA INJECTION STUDIES

9.147. The plasma sources used in most of the earlier work with magnetic mirror systems consisted of titanium electrodes loaded with deuterium, such as were described in Chapter 5. The maximum particle energy was then usually in the range from 1 to 10 ev. Injection has generally been made through one of the mirrors, trapping being achieved either by a time-rising field or as a result of collisions (or both). Attempts have also been made to use radial injection and it has been found that the extent of trapping can be greatly enhanced by taking advantage of the increased collision rate when two (or more) plasma pulses intercept one another at the center of the containment volume. Thus, by using two oppositely directed radial sources fired simultaneously, the trapped plasma density is about ten times as great as if only one source is used [43].

9.148. Experiments have also been made with three- and four-source radial injection. The sources are fired when the magnetic field is zero and a mirror field is energized when the plasma is roughly in the center of the vacuum chamber. In this way, it is hoped both to trap the plasma and to compress it, thus achieving a large increase in the energy component of the injected ions perpendicular to the field lines [44]. In a particular case, the plasma sources were inserted radially in a tube of 30 cm diameter. Probe measurements showed that the translational velocity of the injected plasma pulse was about 5×10^6 cm/sec. Hence, the magnetic mirror field was created at 3 to 6 microsec after the sources were fired.

9.149. The problems associated with heating a plasma by high compression could be largely avoided by the use of high-energy plasma sources. The sources referred to in §§5.80, 5.97, for example, are capable of producing ions of several kilo-electron volts energy, so that a relatively small compression should be sufficient to attain temperatures of thermonuclear interest. However, with this procedure the advantage of the great increase in density, resulting from the large compression factor, is lost and so it would be necessary to obtain a plasma of the required high ion density from the source alone.

9.150. The Felix apparatus was designed for the purposes of investigating the possibility of building up a plasma of fairly high energy by trapping pre-accelerated ions in a mirror system and of studying various loss mechanisms [45-47]. In order to increase the transit time between source and mirror and to accommodate fairly large particle orbits, a vacuum chamber of large dimensions is used; it has a length of 270 cm and is 46 cm in diameter. Since charge exchange of the injected ions with neutral particles of the residual gas is likely to be a serious source of loss, the vacuum chamber was so designed that background impurities could be kept to a minimum. The chamber itself is made of stainless steel and gaskets of gold wire are used throughout. The magnetic field coils (see Fig. 9.17) can be removed and replaced by an oven, so that the vacuum chamber and other parts of the system can be baked out at a temperature of 400°C. In this manner, base pressures of the order of 10^{-9} mm of mercury are attainable.

9.151. The magnetic mirror field is provided by two coils, at the ends of the chamber, which can be adjusted to give mirror ratios up to 3.0. The coils can be energized either by direct current, to produce a peak mirror field of 5000 gauss, or in a pulsed manner from a capacitor bank to give a field of 24,000 gauss maximum. In the pulsed operation, the field reaches its peak value in 4.7 millisec and decays with a L/R time constant of 200 millisec.

9.152. A monoenergetic beam of protons or deuterons, of energy ranging from 3 to 15 kev, is extracted radially from a movable plasma source through a narrow slot, as described in §5.81. The method of internal injection, with the source inside the mirror field, is used. The beam intensity can be varied from a few microamperes to more than 100 milliamp and the pulse durations from a few up to several hundred microseconds.

FIG. 9.17.   The Felix apparatus. (Courtesy Lawrence Radiation Laboratory, Livermore)

9.153. By intercepting a well-collimated beam of 1 milliamp intensity at various points along its path, by means of a conducting plate coated with a phosphor, the trajectory of the beam could be studied for different ion energies and source positions. The results were in general agreement with expectation and confirmed the existence of the precession effect referred to in §9.65. For suitably small particle orbits and source positions at some distance from the mirror peak, the beam is found to be completely reflected in front of the mirror, provided the source is not too far from the magnetic axis of the system. In some instances, the trajectory of a beam of ions was followed for as many as 40 transits between the mirrors. But as the ion source is moved nearer to the mirror or farther from the axis, deviations are observed from the trajectory expected on the basis of simple adiabatic theory, especially as the particle orbit increases. The effect of the size of the particle orbit is in agreement with expectation (§9.69).

9.154.   In attempting to increase the injected high-energy beam to a few hundredths of an ampere, however, serious difficulties were encountered due to space-charge effects. A small fraction of the beam is trapped, so that the charged particles are reflected back and forth many times between the mirrors, for a period of several milliseconds until they are lost in various ways, chiefly as a result of charge-exchange collisions with neutral particles in the residual gas. But most of the beam is lost either because the particles return to strike the source or because they escape through the farther mirror on the first transit. Since intense beams of ions are necessary, if full advantage is to be taken of their high energy, methods must be found for introducing sufficient numbers of electrons to neutralize the space charge.

CONFINEMENT OF BETA PARTICLES IN MIRROR GEOMETRY

9.155.   According to the theory of magnetic mirror systems, the reflection of charged particles should be independent of the nature of these particles. Experiments have consequently been made on the confinement of positive beta particles (positrons) produced in the decay of either tritium or neon-19. The advantage of using these gaseous source materials is that problems associated with the introduction of solid sources and of losses arising when the charged particles strike the source are eliminated.

9.156.   In the work with tritium, a known amount of the gas was introduced into the vacuum chamber in a mirror geometry by heating uranium tritide in an external container. The density of the positrons remaining within the chamber was calculated as a function of time by determining the number of ions they generated, as derived from measurements of the ion current. By working over a range of tritium pressures, down to about 3×10^{-6} mm of mercury, the effect of scattering collisions involving atoms of the background gas could be eliminated. It was then found that under suitable conditions a positron would be reflected more than 10^7 times by the mirror fields before being lost [48].

9.157.   When using neon-19 as the source of positrons, the gas, obtained by the (p,n) reaction of fast protons with fluorine-19, was injected radially at the midplane of the magnetic mirror system. After a short time, the gas was pumped out and observations were made with a scintillation counter of the rate of escape of the positrons through the mirror at one end, which was deliberately made somewhat weaker than the other. For positrons of small gyromagnetic radius, which are expected to behave adiabatically, the $1/e$-folding time for the confinement was determined to be about 6 sec. The losses which occurred were attributed mainly to scattering of the positrons by the atoms of gas remaining in the chamber. Two additional observations demonstrated clearly the confining effects of the mirror field. If the field at the end where the detector was located was made somewhat stronger than the other, the rate of particle escape was very greatly decreased. On the other hand, when this mirror was removed, there was a marked increase in the rate at which the positrons reached the counter [49].

THE DCX SYSTEM

9.158. The prime purpose of the direct-current experiment, known as DCX, is to test some of the concepts of thermonuclear ignition based on the injection of high-energy deuterons into a magnetic mirror configuration by dissociation of deuterium-molecule (D_2^+) ions in an arc (§9.89) [22, 50]. A diagrammatic representation of the apparatus is shown in Fig. 9.18. In order

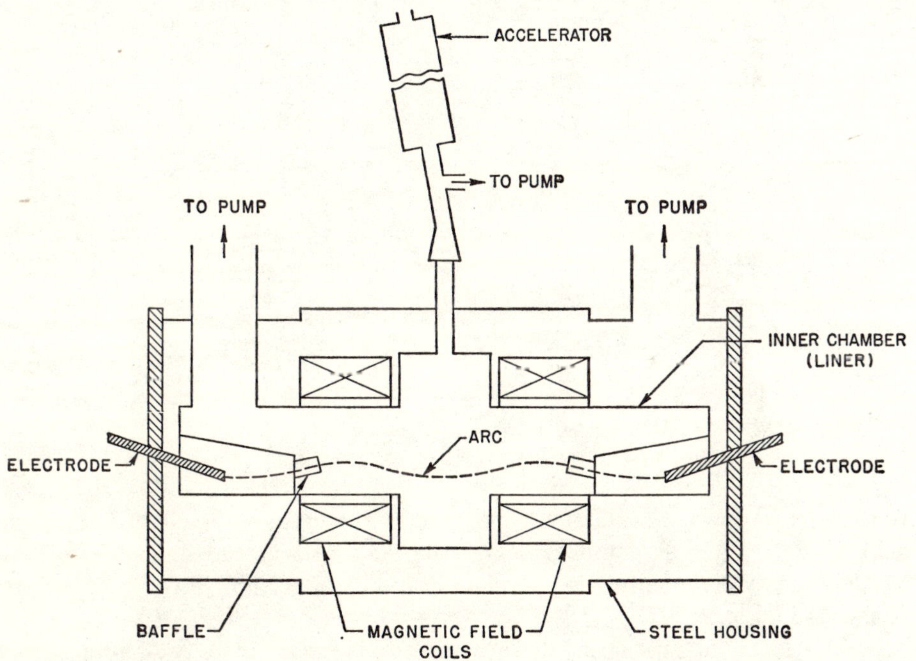

FIG. 9.18. Diagrammatic representation of the DCX system.

to obtain very low base pressures, thus decreasing the amount of neutral particles present, a dual vacuum system is employed. The inner copper chamber, 200 cm long and 50 cm diameter, called the liner, can be baked out for the removal of impurities by passing superheated steam at 400°C through pipes surrounding it, and base pressures of 3×10^{-8} mm of mercury have been attained. The liner and the coils which produce the magnetic field are enclosed in an evacuated housing of stainless steel about 250 cm long and 100 cm diameter. The steady (DC) magnetic field strength is 10,000 gauss at the midplane and the peak mirror ratio is 2.0.

9.159. Deuterium-molecule ions are accelerated to 600-kev energy by means of a vertically mounted DC cascade accelerator. The emerging ion beam is deflected magnetically through an angle of about 10°, partly to aid in directing the beam into the arc where dissociation is to take place (§5.104) and partly to remove unwanted species. The arc is struck between two carbon electrodes located in the outer vacuum chamber and it enters the liner through insulated graphite baffles. The electrodes are located off the axis of the magnetic field and photographs of the arc show a marked curvature corresponding to that of the magnetic field lines. The positions of the arc and of the entering beam of molecular deuterium ions are chosen so that the circular orbit of the dissociated atomic ion beam is concentric with the magnetic axis (Fig. 9.19).

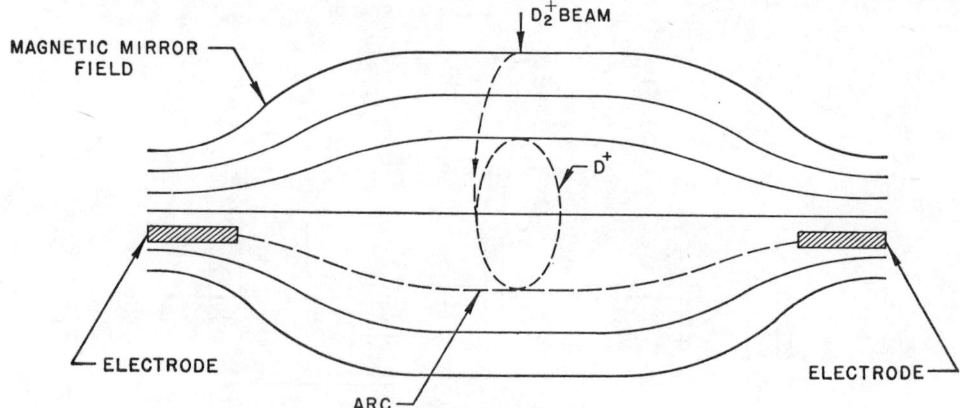

FIG. 9.19. Particle orbits in DCX magnetic field.

With a deuterium molecular-ion beam of 600-kev energy, the central magnetic field of 10,000 gauss should give an orbit of about 25 cm diameter for the dissociated 300-kev deuterons. A photograph of the complete DCX apparatus is shown in Fig. 9.20.

9.160. In usual operation, the arc voltage is approximately 150 volts and the current is in the range of 200 to 300 amp. It is of interest to mention that the arc helps to keep down the concentration of neutral particles in the liner by exhibiting a pumping action. Neutral atoms entering the arc are ionized and are thus constrained by the magnetic field to follow the arc column, through the baffle, into the outer vacuum chamber where they are removed by the pumps.

9.161. In the absence of the baffles, the trapped ring of deuterons, formed in the manner described in §5.22, is rendered visible by small carbon particles torn off the arc electrodes. These particles are normally invisible but they become incandescent upon passing through the ring. The baffling greatly reduces the number of carbon particles, which are undesirable because they tend to adsorb the deuterons. With the baffles in position, the trapped ring is faintly visible at higher pressures, but at a pressure of 5×10^{-7} mm of mercury a feeble glow fills a central region having a diameter of about 25 cm and an axial length of similar dimension.

9.162. A useful diagnostic procedure is the determination of the efflux of energetic neutral atoms from the deuteron ring as a result of charge ex-

FIG. 9.20.   The DCX apparatus.   (Courtesy Oak Ridge National Laboratory)

change therein. The neutral particle detector consists of a Faraday cup mounted inside a shielded box, with the aperture closed by a thin nickel foil. The thickness of the foil is such as to ensure an equilibrium charge distribution between ions and energetic neutral particles emerging from it, irrespective of the charge state of the incident particles.

9.163. An important measurement made by means of the energetic neutral detector is the $1/e$-decay time τ of the trapped ring of deuterons after the molecular-ion beam is switched off. With this information it is possible to determine the atomic ion density in the steady state. If this density is n_+ and V is the volume occupied by the circulating atomic ions, then n_+V is equal to $Ia\tau$, where I is the incoming molecular-ion current and a is the arc dissociation efficiency which can be measured by means of a Faraday cup temporarily inserted into the ion beam on its first turn. The value of V can be estimated from the positions at which the tip of a tungsten wire probe becomes incandescent; hence, all the information is available from which n_+ can be calculated.

9.164. In some experiments with the DCX apparatus using hydrogen, instead of deuterium, but with a magnetic field of decreased strength so as to maintain the same diameter of the trapped ring of atomic ions, the observed decay times were as long as 10 millisec at a liner pressure of 5×10^{-7} mm of mercury. These decay times were found to vary inversely as the pressure, so that charge exchange is probably the main cause of the loss of ions, at least within the experimental pressure range. The injected current of hydrogen-molecule ions was 5 milliamp and the measured atomic ion (proton) beam emerging from the arc was 0.25 milliamp, indicating that the dissociation efficiency of the arc was 5 per cent. During the 10-millisec decay time, the 300-kev protons will have made some 90,000 turns of the 25-cm-diameter trapped ring, and so the total ion current is roughly $90,000 \times 0.25$ milliamp or 22.5 amp. Under these conditions, the ion population is seen to spread longitudinally; the value of n_+ has been found to be 4×10^8 protons/cm³ [51].

9.165. At the pressure of 5×10^{-7} mm of mercury, burnout of the neutral particles, as defined in §9.91, should require an ion current some 300 times larger than the 0.25 milliamp proton current mentioned above. Efforts must therefore be made to increase the ion current and also to reduce the background pressure so as to decrease the density of neutral particles which must be burnt out.

9.166. An unexpected source of loss of charged particles was brought to light during the operation of the DCX system, namely, charge exchange occurring during the repeated transits of the ion beam through the arc. Unless this effect can be eliminated or at least greatly reduced, the burnout current will be larger than calculated and, hence, much more difficult to attain. Among the suggestions which have been made for decreasing the loss of ions by charge exchange in the arc, two may be mentioned.

9.167. One is to introduce a precession into the orbits of the dissociated ions by relocating the molecular-ion beam and the arc so that the orbits of the trapped atomic ions are no longer centered on the magnetic axis. Under the resulting conditions, the ions will frequently miss the arc as they revolve. This modification is expected to diminish the charge-exchange loss if the orbits are actually as ideal as they are thought to be. The other proposal is to replace the carbon arc with a deuterium arc, of the type described in §5.108.

This should eliminate the charge-exchange loss in the arc, if the deuterium is pure and completely ionized. Developments of the deuterium arc, however, indicate that the required gas input rate is about 5 cm³ (STP)/sec of deuterium gas for a 300-amp arc, and a severe pumping problem is posed if the pressure of 5×10^{-7} mm of mercury is to be maintained (or improved upon) in the face of such a gas influx.

THE OGRA DEVICE

9.168. The OGRA* device is a large magnetic mirror system in which a beam of high-energy, deuterium-molecule ions is injected between the mirrors in a direction perpendicular to the field lines, as in the DCX experiment. However, in the OGRA system dissociation of the molecular ions of long mean free path, and consequent trapping of the atomic ions, is to be brought about by collisions with molecules and ions, as mentioned in §9.102.

Fig. 9.21. The OGRA device. (Courtesy Journal of Nuclear Energy)

9.169. The vacuum chamber (Fig. 9.21) is a cylinder 12 meters in length and 1.4 meters internal diameter. The strength of the confining magnetic field is 5000 gauss at the center, increasing to 8000 gauss in the mirror regions at the ends. The chamber can be baked out at 450°C and by suitable pumping a final pressure of 10^{-8} mm of mercury should be attainable. Molecular

---

* Derived from the initial letters (reversed) of the names *Golovin* and *Artsimovich*.

ions from an arc source are accelerated to 200-kev energy; they are than focused and injected into the confining field between the mirrors by a system of electric and magnetic fields. It is hoped that the deuterium molecule-ion current entering the chamber will have a strength of several hundred milliamps, which should be sufficient to burn out the residual neutral particles [25, 26, 52].

SUMMARY AND CONCLUSIONS

9.170. A considerable number of experiments made in a variety of mirror systems have established the fact that, in agreement with theoretical expectation, plasmas can be confined for periods of several milliseconds. In one case, X-ray and electron signals persisted for 30 millisec, indicating that particles had been confined for that length of time. Although these confinement times are of interest, they are much less than those which would be required in a thermonuclear reactor.

9.171. It is of special significance that, in spite of early fears to the contrary, there has been no definite evidence so far of any gross instability of the plasma confined in a magnetic mirror geometry, such as are of common occurrence in pinched and stellarator systems. The stability has been ascribed to two circumstances, in particular: (1) the magnetic field lines are "tied down" outside the plasma by means of conductors, and (2) no electric currents flow parallel to the field lines. Whether plasma instabilities will develop at higher temperatures and plasma pressures than have been attained so far is, of course, uncertain.

9.172. Assuming, on the basis of the experimental results, that stable confinement by magnetic mirrors can be achieved, there remains the problem of building up a plasma of sufficiently high density and temperature to approach the conditions required for practical application. Three general lines of approach have been considered in this connection. One, which is based on the use of fast magnetic compression, will be described in Chapter 11, and the other two will be reviewed briefly here.

9.173. In the first of these, the plasma is injected at relatively low density, e.g., about 10^{11} or 10^{12} particles/cm^3, and at temperatures of a few electron volts. By magnetic compression, it should be possible to increase the density and temperature simultaneously. The feasibility of this procedure has been demonstrated in some work in which the magnetic field was increased by factors up to about 1000, so that the plasma was compressed (radially) to the same extent. Langmuir-probe and other observations indicated that the plasma had been compressed into a narrow filament at the axis of the containment tube. The density attained in this filament was probably in the vicinity of 10^{14} particles/cm^3 or more.

9.174. The measured energies of the electrons escaping from the plasma were in harmony with a Maxwellian distribution corresponding to a kinetic electron temperature of between 10 and 20 kev. The ion temperatures were not determined, but they were believed to be in the range from 1 to 3 kev, in different experiments, compared with an initial value at injection of about 1 to 2 ev.

9.175. One of the drawbacks to the use of magnetic compression of the plasma to achieve high densities and temperatures arises from the fact that the initial magnetic field will be relatively low, in order to attain as high a compression factor as possible, for a given value of the final (maximum) field value. Since the gyration radii of the charged particles in the initial weak field are large, it is necessary to contain the plasma in a vacuum chamber of large dimensions. Hence, the energy expended in producing the magnetic field, in order to compress and heat the plasma, will also be large.

9.176. One way of overcoming the problem is to design magnets of material of very high conductivity (§12.84). Another is to carry out the compression process in stages. After each stage, the plasma is transferred, by a moving mirror, to a vessel of smaller diameter, where it is compressed further by an increase in the field strength. The strongest fields are thus applied in a chamber of small size. An additional advantage claimed for this procedure is that the rapid transfer of the plasma from one vessel to the next tends to leave behind most of the neutral particles, which are not affected by the field, thus minimizing charge-exchange losses.

9.177. The second type of method for producing plasmas of high density and energy in a magnetic mirror geometry, which is attracting attention, is to inject large currents of particles of high energy which are trapped in a suitable manner. The function of the mirrors is then only to confine the plasma long enough for thermalization, as a result of collisions, and subsequent reaction to occur. Proposals have been made for the direct injection of deuterons, of neutral (deuterium) atoms, or of deuterium-molecule ions of high energy.

9.178. In some experimental studies using beams of high-energy atomic ions, serious difficulties due to space-charge effects were encountered when an attempt was made to increase the current to a few hundredths of an ampere. It appears, therefore, that if this procedure is to be successful, methods must be found for introducing a sufficient number of electrons together with the ions, in order to minimize the space charge.

9.179. Plans are being made for a device* in which a beam of neutral atoms of 10 to 20 kev energy will be injected into a magnetic mirror geometry [53]. The space-charge problem associated with the injection of ions will then not arise in as acute a form. However, since reliance would be placed on collisions with other atoms or with ions to produce ionization, it is neces-

---

* The system is called ALICE from the initial letters of Adiabatic Low-Energy Injection and Capture Experiment.

sary that the injected atoms should not have too high an energy, so that the ionization cross sections are reasonably large. Further heating will then be required e.g., by adiabatic compression, to obtain thermonuclear temperature plasmas.

9.180. The procedure for injecting molecular ions of high energy into a mirror system and employing an arc to bring about dissociation would appear to have many advantages. But experiments with this system have brought to light various unexpected channels whereby the ions are lost. Probably the outstanding problem in all procedures based on the injection of neutral particles or of molecular ions will be that of destroying (or burning out) the neutral particles which will inevitably be present. Unless this can be done effectively, energy losses due to charge exchange will make the methods useless for the practical release of thermonuclear energy.

REFERENCES FOR CHAPTER 9

1. R. F. Post, *Proc. Second U.N. Conf. on Peaceful Uses of Atomic Energy*, **32**, 245 (1958).
2. R. F. Post, USAEC Report UCRL-4231 (1954).
3. M. Bineau, T. Consoli, P. Hubert, F. Prévot, P. Ricateau, and A. Samain, *Nucl. Instr.*, **4**, 282 (1959).
4. M. Bineau, T. Consoli, C. Maissonnier, F. Prévot, and P. Ricateau, *Nucl. Instr.*, **4**, 290 (1959).
5. A. A. Garren, R. J. Riddell, L. Smith, G. Bing, L. R. Henrich, T. G. Northrop, and J. E. Roberts, *Proc. Second U.N. Conf. on Peaceful Uses of Atomic Energy*, **31**, 65 (1958).
6. R. F. Post, *Ann. Rev. Nuclear Sci.*, **9**, 367 (1959).
7. R. F. Post, R. E. Ellis, F. C. Ford, and M. N. Rosenbluth, *Phys. Rev. Letters*, **4**, 165 (1960).
8. J. Berkowitz, H. Grad, and H. Rubin, *Proc. Second U.N. Conf. on Peaceful Uses of Atomic Energy*, **31**, 177 (1958).
9. J. Berkowitz, K. O. Friedrichs, H. Goertzel, H. Grad, J. Killeen, and E. Rubin, *Proc. Second U.N. Conf. on Peaceful Uses of Atomic Energy*, **31**, 171 (1958).
10. G. Persico and J. G. Linhart, *Nuovo cimento*, **8**, 740 (1958).
11. H. Grad, unpublished.
12. A. C. Kolb, *Int. Symp. on Magneto-Fluid Dynamics* (1960).
13. D. Judd, W. McDonald, and M. N. Rosenbluth, USAEC Report WASH-289 (1955), p. 158.
14. A. N. Kaufman, USAEC Report TID-7520 (1956), p. 387.
15. R. F. Post, USAEC Report WASH-184 (1954), p. 18.
16. R. F. Post, USAEC Report WASH-289 (1955), p. 209.
17. R. F. Post, USAEC Report TID-7536 (1957), p. 135.
18. S. Chandrasekhar, *Principles of Stellar Dynamics*, University of Chicago Press, 1942, p. 48.
19. J. L. Tuck, *Phys. Rev. Letters*, **3**, 313 (1959).
20. M. S. Yoffe and V. G. Telkovski, quoted by E. E. Yushmanov, *Plasma Physics and the Problem of Controlled Thermonuclear Reactions*, Pergamon Press, Inc., Vol. IV, 1960.

21. G. Gibson, W. A. S. Lamb, and E. J. Lauer, *Proc. Second U.N. Conf. on Peaceful Uses of Atomic Energy*, **32**, 275 (1958); *Phys. Rev.*, **114**, 937 (1959).
22. C. F. Barnett, P. R. Bell, J. S. Luce, E. D. Shipley, and A. Simon, *Proc. Second U.N. Conf. on Peaceful Uses of Atomic Energy*, **31**, 298 (1958).
23. A. Simon and M. Rankin, USAEC Report ORNL-2354 (1957).
24. A. Simon, *Phys. Fluids*, **1**, 495 (1958); **2**, 336 (1959).
25. I. V. Kurchatov, *J. Nuclear Energy*, **8**, 168 (1958).
26. I. N. Golovin, *Proc. Inst. Elec. Eng.* (*London*), **106A,** Suppl. No. 2, 95 (1959).
27. A. Simon, USAEC Report ORNL-2831 (1959).
28. R. F. Post, with J. S. Steller, USAEC Report WASH-115 (1952), p. 88.
29. F. C. Ford, USAEC Report UCRL-4363 (1954).
30. F. H. Coensgen, USAEC Report WASH-184 (1954), p. 36.
31. R. F. Post, USAEC Report WASH-184 (1954), pp. 8, 20.
32. D. A. Edwards, USAEC Report WASH-289 (1955), p. 141.
33. R. F. Post, USAEC Report WASH-184 (1954), pp. 11, 20.
34. F. H. Coensgen, USAEC Report WASH-184 (1954), p. 30.
35. F. H. Coensgen, A. E. Sherman, and D. E. Showalter, USAEC Report TID-7503 (1955).
36. D. F. Martin and M. M. Hill, USAEC Reports TID-7520 (1956), p. 155; TID-7536, Part 2 (1957), p. 517.
37. S. G. Zizzo and L. S. Hall, USAEC Report WASH-289 (1955), p. 144.
38. A. L. Gardner, S. G. Zizzo, and L. S. Hall, USAEC Report TID-7503 (1955), p. 100.
39. F. C. Ford and S. G. Zizzo, USAEC Reports TID-7536, Part 2 (1957), p. 147; UCRL-5112 (1958).
40. F. H. Coensgen, F. C. Ford, and R. E. Ellis, *Proc. Second U.N. Conf. on Peaceful Uses of Atomic Energy*, **32**, 200 (1958).
41. R. E. Ellis and N. L. Parker, USAEC Report TID-7558 (1958), p. 195.
42. F. H. Coensgen, USAEC Report TID-7536, Part 2 (1957), p. 159.
43. F. H. Coensgen, F. W. Cummins, and A. E. Sherman, USAEC Report TID-7558 (1958), p. 211.
44. F. C. Ford and S. G. Zizzo, USAEC Report TID-7558 (1958), p. 184.
45. F. S. Eby, C. C. Damm, and E. C. Popp, USAEC Report TID-7536, Part 2 (1957), p. 168.
46. F. S. Eby and C. C. Damm, USAEC Report TID-7558 (1958), p. 168.
47. C. C. Damm and F. S. Eby, *Proc. Second U.N. Conf. on Peaceful Uses of Atomic Energy*, **32**, 273 (1958).
48. G. Gibson, W. C. Jordan, and E. J. Lauer, USAEC Reports UCRL-8775 (1959), p. 20; UCRL-8887 (1959), p. 26; UCRL-9002 (1959).
49. S. N. Rodionov, quoted in reference 26.
50. Members of the Oak Ridge National Laboratory, Thermonuclear Experiment Division, *The ORNL Thermonuclear Program*, USAEC Report ORNL-2457 (1958).
51. A. H. Snell, personal communication; cf. USAEC Report ORNL-2926 (1960).
52. L. A. Artsimovich, *Proc. Second U.N. Conf. on Peaceful Uses of Atomic Energy*, **31**, 6 (1958).
53. R. F. Post, *Fourth Int. Conf. on Ionization Phenomena in Gases*, p. 987 (1959).

Chapter 10

THE ASTRON SYSTEM

THE E-LAYER

FORMATION OF THE E-LAYER

10.1. The basic feature of the Astron concept for a thermonuclear reaction system is a cylindrical sheet of high energy, i.e., relativistic, gyrating electrons, called the *E-layer* [1]. The purpose of this layer of relativistic electrons is to produce and confine a plasma at high temperatures [2]. First, in conjunction with an externally applied magnetic field, it is expected that there will be formed a pattern of closed magnetic lines of force which can provide plasma confinement, since charged particles will be able to escape only by the relatively slow process of diffusion across the field lines (see Chapter 12). Second, interaction of the high-energy electrons of the E-layer with neutral atoms of deuterium and tritium will result in the formation of a plasma immediately as the system of closed magnetic lines is created. Subsequently, the electrons in the plasma gain energy from the E-layer electrons as a result of Coulomb scattering collisions, and part of this energy is then transferred to the ions. It is hoped that the temperature of the latter can thus be increased to the point where thermonuclear reactions take place at an appreciable rate.

10.2. In the Astron system it is proposed to establish the E-layer in a long, cylindrical vacuum chamber, in which an axial magnetic field is produced by means of a solenoid wound around the exterior of the cylinder in the usual manner. By increasing the field strength at the ends of the chamber, a magnetic mirror geometry is provided to minimize the loss of electrons that form the E-layer, but this mirror system plays no part in the confinement of the plasma. Electrons, accelerated externally to a high energy—probably 30 to 50 Mev in a hypothetical Astron reactor—are injected at one end of the vacuum chamber. As a result of the action of the magnetic field, the electrons will gyrate about the central axis while traveling back and forth in a longitudinal direction between the mirrors, thus forming the E-layer (Fig. 10.1) [1]. The electrostatic charge of the E-layer is assumed to be neutralized

404

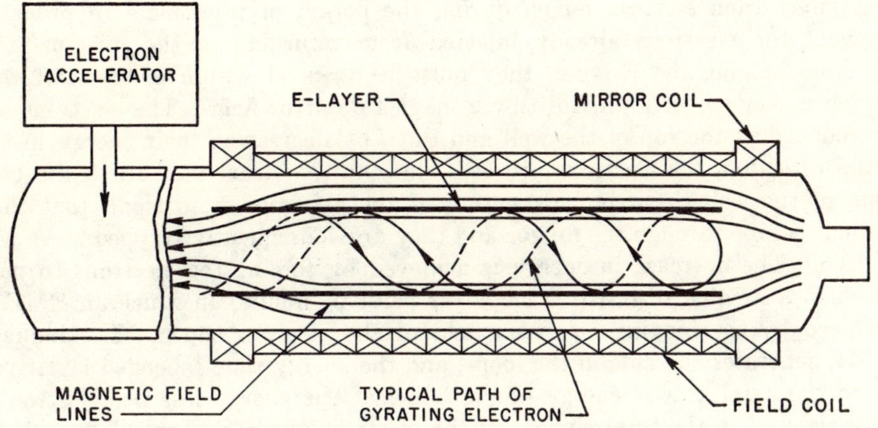

FIG. 10.1.   Gyration of electrons in E-layer.

in some manner by positive ions, so that there is effectively no space-charge potential.

10.3.   A theoretical treatment indicates that, if the electron orbits are to be stable, a requirement is

$$p_\theta{}^2 > \tfrac{1}{2}p^2,$$

where p_θ and p are the azimuthal momentum and total momentum, respectively, of the electrons far from the ends of the cylindrical chamber [3].   Furthermore, as a result of the method of injection, the electrons will always have a considerable axial momentum.   Consequently, a pressure will be exerted in the axial direction which will prevent collapse of the E-layer as a result of the contractive forces of its own magnetic field.

ELECTRON INJECTION

10.4.   Since the electrons in the E-layer are steadily losing energy, in the ionization of neutral atoms and in Coulomb scattering collisions with cold electrons in the plasma so produced, a continuous injection of high-energy electrons from outside is necessary in order to maintain the layer.   A problem arises in this connection from the fact that, when steady-state operation is attained, the magnetic field is constant in time.   A possible method by which this difficulty may be overcome is to use a traveling-wave system to carry a pulse of electrons over the barrier of the mirror field that confines the E-layer [4].

10.5.   A simpler procedure than the one using traveling waves for injecting electrons appears to be very promising [5].   The basic requirements of the injection system for the Astron device are that electron bunches be introduced into the E-layer region in an irreversible manner without disturbing the electrons already injected and in such a way that the presence of the E-layer will not affect each electron bunch during the period of injection. In order to prevent the electrons already injected from returning to the injector tube or from leaving the E-layer, they must be trapped within a potential well which is conveniently provided by a magnetic mirror field. The electrons are introduced at the top of the well and then, by decreasing their energy in the axial direction, the electrons proceed "downhill." When they reach the bottom of the well the axial momentum of the electrons is so small that they cannot escape through the mirror and they are consequently trapped.

10.6. The decrease in energy is achieved by forcing the electrons to pass through a number of resistive loops (or passive circuits) in which an EMF is induced by the moving magnetic field of the electron bunch. The induced EMF generates currents in the loops, and the energy thus expended is derived from the axial kinetic energy of the bunch. The energy loss per electron is proportional to the total current of the electron bunch, and calculations indicate that a circulating current of the order of a few thousand amperes will be required in order to obtain the desired result. It is expected that this requirement can be met by a current of 100 to 200 amp, from the electron gun producing the bunch, for a period of 0.1 to 0.3 microsec.

10.7. A sectional diagram of the proposed electron injection system is shown in Fig. 10.2 [5]. The resistive loops, which are high-frequency re-

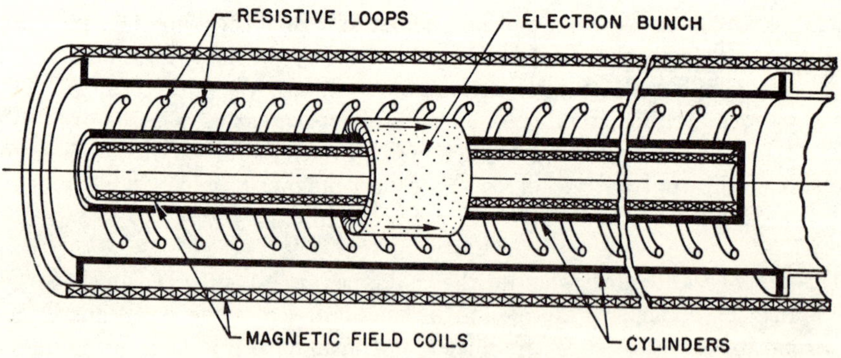

FIG. 10.2. Electron injection system for Astron.

sistors of coated glass, are located in the annular space between two metal coaxial cylinders. A bunch of high-energy electrons from an electron gun is injected at the left end in an approximately azimuthal direction, with a small velocity component in the axial direction to insure that the injected beam misses the injector after one turn. Thus the beam will follow a helical path along the tube. However, owing to the presence of the resistive loops, the head of the beam will be retarded, so that by the time the tail end of the beam is injected the axial velocity of the head will have been reduced considerably. As a result the length of the bunch in the axial direction will be limited, probably to about 30 or 40 cm. By means of a special arrangement of coaxial coils, one inside the inner cylinder and the other at the exterior of the outer cylinder, the radius of the electron bunch can be maintained constant as it moves through the mirror field.

10.8. An essential condition for the effective performance of the injection system described above is that the axial defocusing force at the head of the beam, due to the self-magnetic field of the electron bunch, must be smaller than the slowing-down force. If this were not the case, the bunch would tend to expand in the axial direction, with the result that the effective current of the bunch would be rapidly diminished and slowing down would not be achieved. Both the self-defocusing force and the slowing-down force are proportional to the current of the bunch, and so the ratio is independent of this current. But the defocusing force decreases with increasing energy of the electrons and it is estimated that an absolute minimum for the electron energy, for marginal performance of the injection system, is about 2 Mev. At an energy of 5 Mev, however, the self-defocusing force is only about one-tenth of the slowing-down force and satisfactory operation should be possible.

10.9. Upon entering the E-layer region, the electron bunch is freed from the constraints imposed on it by the resistive loops. Hence it will be spread along the E-layer, approximately 600 cm in length, which is confined by the mirror fields. The energy stored in the electromagnetic field of the bunch is gradually dissipated into electron kinetic energy as the bunch grows in the axial direction. The energy of the electrons in the E-layer is thus increased with each bunch that is injected, up to a limiting value determined by the rate of escape through the mirror fields.

10.10. The magnetic field of the E-layer would tend to attract the incoming electron pulses by creating a deeper potential well. This is not desirable, since it would affect the slowing-down field during the build-up of the E-layer. In order to avoid any modification of the magnetic mirror confining fields by the self-field of the E-layer, coils will be provided in the injection region to produce a magnetic field equal and opposite to that of the E-layer field. As a result, the magnetic field due to the E-layer electrons will be restricted to the E-layer region and will not exist in the injection region. However, an electron within the E-layer, moving from the end to the center, must cross all the flux created by the E-layer current, and this serves to maintain the radius of the layer constant during the build-up phase.

CONFINEMENT AND PROPERTIES OF PLASMA

FORMATION OF A MAGNETIC FIELD

10.11. The direction of the self-magnetic field of the E-layer, produced by the current due to the rotating electrons, is opposite to that of the external magnetic field. The net strength of the field within the E-layer region thus decreases continuously, at first, as the E-layer is built up and the charge per unit length increases. When the number of electrons per unit length of the layer reaches a critical value N_c, then the magnetic field inside the volume enclosed by the E-layer is reduced to zero, at points not too near the ends. The value of N_c is given by

$$N_c = \frac{m_e c^2}{e^2}\,\gamma, \tag{10.1}$$

where m_e is the rest mass of the electron and e is its charge, c is the velocity of light, and γ is the relativistic mass ratio of the electrons, i.e.,

$$\gamma = \left(1 - \frac{v^2}{c^2}\right)^{-\frac{1}{2}},$$

v being the electron velocity. By increasing the number of electrons in the E-layer somewhat beyond the critical value, the magnetic field within the enclosed volume is reversed. At this time, the combination of externally applied magnetic field and the E-layer field produces the pattern of closed lines

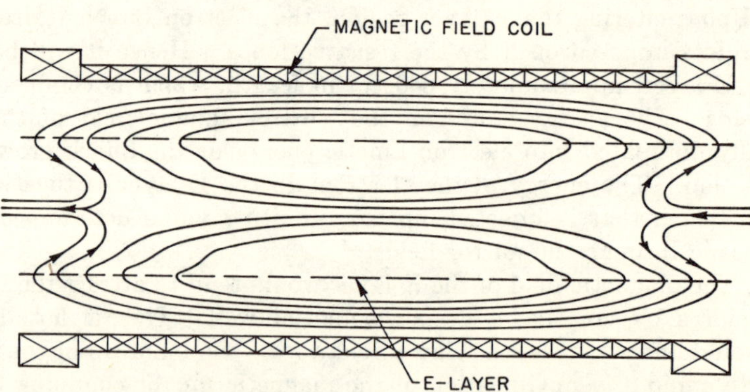

FIG. 10.3. Formation of pattern of closed magnetic field lines.

of force of cylindrical form, shown in Fig. 10.3, which serves to confine the plasma [1].

FORMATION AND HEATING OF PLASMA

10.12. As soon as this magnetic pattern is established, a plasma can be formed by injecting the thermonuclear fuel materials, i.e., deuterium and tritium, into the vacuum chamber. As they travel through the E-layer region, the atoms become ionized and form a plasma of low energy. Subsequently, the energy is increased, as stated earlier, as a result of Coulomb scattering interactions between plasma electrons and those of the E-layer. The rate of energy gain is proportional to the number of electrons in the E-layer and hence to the number per unit length. Since this number must be at least equal to N_c, as defined by equation (10.1), it is evident that the rate at which energy is transferred from the E-layer electrons to the plasma will increase as the relativistic mass ratio γ of the electrons increases. It is desirable, therefore, from this point of view, that the energy of the injected electrons should be high. As seen above, the characteristics of the resistive injection system require a minimum energy of 5 Mev. But if a higher value is selected, it will be determined by a balance between the energy expended in accelerating the electrons and the consequent gain in temperature of the plasma due to the accompanying increase in the relativistic mass ratio.

10.13. The temperature of the plasma will continue to rise as long as the rate of energy loss from the diffusion of charged particles across the magnetic field lines and to radiation is less than the rate of energy gain by the E-layer. A theoretical study of the pressure and magnetic field distributions during the plasma heating stage indicates that the diffusion loss passes through a maximum with increasing pressure (or temperature). It appears that, for the plasma to gain energy from the E-layer, the density of the former must be less than a certain value. This value varies as γ^4, so that it increases very rapidly with the relativistic mass ratio of the E-layer electrons. For γ equal to 100, which corresponds to electrons of 50-Mev energy, it has been estimated that, based on diffusion loss considerations, the limiting plasma density is about 10^{14} particles/cm^3.

10.14. Provided the concentration of impurities of high atomic number is kept within reasonable bounds, energy losses due to electron excitation followed by radiation emission will be negligible. Hence, the only radiation loss that has been considered is that due to bremsstrahlung.* In fact, at temperatures of thermonuclear interest, which are above that where the rate of diffusion is a maximum, bremsstrahlung losses greatly exceed those resulting from diffusion of charged particles to the walls. Assuming an electron temperature of 100 kev in the plasma, it appears that, provided the energy of the E-layer is high enough, e.g., about 50 Mev, it should be possible to maintain a plasma density of 10^{14} particles/cm^3 at any temperature required for thermonuclear reaction [1]. As seen from the general consideration in §2.46, this is the correct order of the plasma density that should be satisfactory for a power reactor.

PLASMA EQUILIBRIUM AND STABILITY

10.15. When, during the course of its formation, the plasma temperature rises, its pressure increases correspondingly. However, at the boundary of the plasma, corresponding to the last closed magnetic field lines in Fig. 10.3, the pressure is zero. There is consequently a pressure gradient within the plasma in the direction perpendicular to the field lines. Because of this gradi-

---

* If there is a magnetic field within the plasma, then allowance must also be made for cyclotron radiation (§2.77).

ent, the plasma starts to diffuse outward, but a current is created which, in conjunction with the magnetic field, balances the pressure gradient. The current, in turn, modifies the system of closed magnetic lines, with the result that the pattern becomes denser as the plasma pressure increases. The situation has been examined theoretically to determine if a self-consistent equilibrium of the plasma does exist. One of the possible solutions requires that a magnetic field discontinuity should be present in the plasma, and this can be realized by the presence of a sheath of rotating charged particles where, within the thickness of the sheath, the magnetic field passes through zero. This requirement is satisfied by the E-layer, as may be seen from Fig. 10.4, which

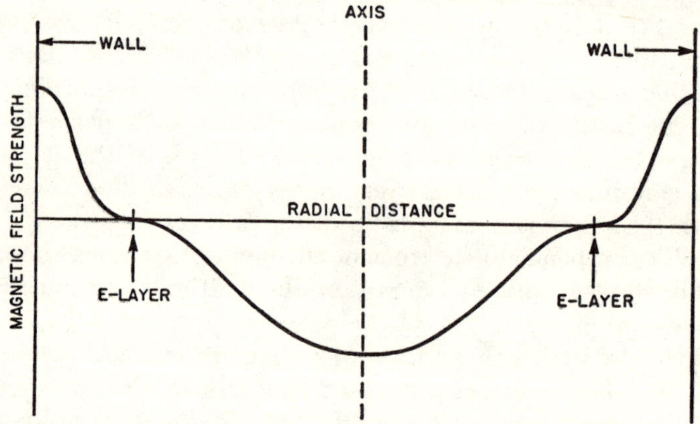

Fig. 10.4. Radial variation of magnetic field strength.

shows the radial variation of the magnetic field strength within the plasma in a plane, e.g., the midplane, normal to the axis of the vacuum chamber. Thus, in addition to providing a means for producing, confining, and heating a plasma, the E-layer apparently also permits the existence of a self-consistent equilibrium solution for the system.

10.16. Having ascertained that an equilibrium solution for the plasma is possible, it is necessary to examine the stability of this equilibrium against small perturbations. The possible perturbations can be divided into two classes depending upon whether they give rise to a propagating or nonpropagating disturbance along the magnetic field lines. The perturbations of the first type are difficult to study mathematically and no definite conclusions concerning them have yet been reached. However, there are reasons for believing that these perturbations are less catastrophic than those in the second category, which includes situations where the magnetic lines move parallel to themselves, e.g., flute (or interchange) instabilities.

10.17. The latter are susceptible to a relatively simple theoretical treatment and the results indicate that perturbations associated with a nonpropagating disturbance are either not possible at all in the Astron geometry, or if they are possible, then only stable modes are permissible. It has been concluded, therefore, that the plasma in an Astron system should be stable against what are expected to be the more destructive perturbations [1].

ASTRON OPERATING CONDITIONS

ENERGY BALANCE REQUIREMENTS

10.18. Because of the desirability of maintaining the E-layer once it has been established, a reactor based on the Astron concept would operate continuously, as far as energy production is concerned, rather than as a pulsed device. For this and other reasons, is has consequently been proposed that the feed should be an equimolar deuterium-tritium mixture, the temperature being raised above the ignition point by transfer of energy from the continuously renewed E-layer, without any dependence upon fusion energy [6]. In the operation of such a steady-state device, additional energy losses, by diffusion and as radiation, are necessary to remove the energy carried by the charged products of the thermonuclear reactions, i.e., mainly helium-4 nuclei in the case under consideration. This requirement leads to an operating temperature that is relatively low, so that the energy loss by charge exchange is too large. In order to minimize this loss, a higher temperature is desirable, but it appears that, as a result of the associated decrease in diffusion loss (§12.82), an energy balance in the plasma cannot be maintained except at very high temperatures. Even then, the deliberate introduction of impurities of moderately high atomic number might be necessary in order to increase the bremsstrahlung losses (§2.74).

10.19. There is, however, an upper limit to the operating temperature arising from the requirement that the helium nuclei produced in the thermonuclear D-T reaction be removed from the system by diffusion. Since the rate of diffusion of charged particles across the magnetic field lines in a highly ionized plasma decreases as the temperature is raised, this means that the temperature must not exceed a certain value. An examination of the various factors involved has led to the provisional choice of an operating temperature of 24 kev for an equimolar deuterium-tritium fuel. In the absence of any appreciable concentration of impurities, the energy loss by diffusion and as bremsstrahlung will then be less than that carried by the helium nuclei. An energy balance can be achieved by introducing the fuel material into the reacting volume largely in the form of deuterium and tritium cyclohexanes, either as a mixture of C_6D_{12} and C_6T_{12} or as the single compound $C_6D_6T_6$.

10.20. In addition to satisfying the energy balance requirement, the use of relatively heavy fuel molecules for the Astron system seems to be desirable from another point of view. The attainment of an equilibrium in the confined plasma is dependent upon a particular distribution of "sources" and this means that it should be feasible to control the distance beyond the plasma surface traveled by a fuel molecule before it is ionized. A rough calculation indicates that, in order to penetrate a distance of about 4 cm into the plasma, the fuel molecules must have an energy of roughly 1 kev per nucleon. To attain this energy, the molecules must be preionized and accelerated before injection into the plasma. But the penetration by these ions of the magnetic field barrier confining the plasma presents a difficulty. It could be overcome, in principle, by neutralizing the fuel ions after acceleration, but this would be a very inefficient procedure. An alternative possibility is to use a relatively heavy molecule of such magnetic rigidity that even when singly ionized it can penetrate some distance into the plasma. It has been estimated that a fuel molecule containing about 100 nucleons, i.e., with a molecular weight in the vicinity of 100, would be suitable for this purpose. The molecular weight of $C_6D_6T_6$ is 102. The total energy of the accelerated fuel molecule-ions prior to injection into the reaction volume would then have to be about 100 kev.

EXPERIMENTAL PROGRAM

10.21. The first stage in the program to test the Astron concept by experiment is the design of a model to demonstrate the establishment of the E-layer and the development, in association with an external magnetic field, of the closed pattern of magnetic lines of force. A pulsed electron gun capable of producing a beam of 1-Mev energy at a peak current of 100 amp, for pulses of about 1 microsec duration, has been constructed [7]. The repetition rate is variable, the maximum being 60 pulses per sec. With a modified accelerator, the electron energy will be increased to 2 Mev and the current to 200 amp. A further increase in energy to 5 Mev is planned in order to test the injection system described in §10.5 *et seq.* The electron current will be 200 amp and the injection period 0.15 microsec. The contemplated operating rate is 30 to 60 pulses/sec, and it is estimated that about 100 pulses will be required to establish the E-layer.

10.22. Some experiments have been made on the injection and trapping, in the form of an annulus, of 750-kev electrons in a long cylinder with magnetic mirrors at the ends. Scintillation detector signals were obtained at some distance from the point of injection for as long as 1.5 sec at a gas pressure of 3×10^{-7} mm of mercury. The confinement time decreases with increasing pressure and the rate of particle loss seems to be determined only by scattering through the mirrors. The preliminary results thus appear to be promising for the formation of the E-layer [8, 9].

REFERENCES FOR CHAPTER 10

1. N. C. Christofilos, *Proc. Second U.N. Conf. on Peaceful Uses of Atomic Energy,* **32,** 279 (1958); see also, USAEC Report TID-7558, Suppl. 1 (1958).

2. J. G. Linhart and A. Schoch, *Nucl. Instr.*, **4**, 332 (1959), contains a review of the uses of relativistic electron beam devices.
3. J. Killeen, USAEC Report TID-7558 (1958), p. 404; also in UCRL-5101 (1958).
4. N. C. Christofilos, USAEC Reports UCRL-5100 (1958); UCRL-5104 (1958).
5. N. C. Christofilos, USAEC Reports UCRL-8887 (1959), p. 40; UCRL-5617-T (1959).
6. N. C. Christofilos, USAEC Report UCRL-5095 (1958).
7. N. C. Christofilos, K. W. Ehlers, and F. Voelker, USAEC Report TID-7558 (1958), p. 444; also in UCRL-8151 (1958).
8. W. A. S. Lamb, USAEC Report UCRL-8887 (1959), p. 67.
9. W. A. S. Lamb and R. C. Johnson, USAEC Report UCRL-9002 (1959), p. 45.

Chapter 11

MISCELLANEOUS PROPOSALS FOR PLASMA CONFINEMENT AND HEATING

FAST MAGNETIC COMPRESSION EXPERIMENTS

GENERAL PRINCIPLES

11.1. The pinched discharge, stellarator, magnetic mirror, and Astron systems, to which major attention is being devoted in the effort toward the realization of thermonuclear power, have been discussed in the four preceding chapters. A number of other proposals have been made for the confinement and heating of plasmas and for other procedures of thermonuclear interest. Some of these will be considered in this chapter, namely, studies of fast magnetic compression, cusped geometries, rotating plasmas, confinement by radio-frequency fields, and inertial electrostatic confinement.

11.2. The magnetic compression studies to be described here are concerned with the confinement and heating of a plasma by an initial rapid compression phase followed by a slower adiabatic compression.* The procedure is to apply a rapidly increasing axial magnetic field which produces an azimuthal current, i.e., in the θ direction in the plasma [1-13]. It is because of the analogy with the conventional (or axial) pinch, described in Chapter 7, produced by axial (or *z direction*) currents that the magnetic compression process is sometimes referred to as the *theta pinch* (or θ *pinch*).† The heating which occurs in the first stage of the compression is considered to be similar to that in the conventional dynamic pinch (§7.19 *et seq.*). A thin, cylindrical sheath of plasma, formed near the walls, is driven inward toward the axis with a velocity greatly in excess of the speed of sound. The fast-moving sheath, acting as a magnetic piston, sweeps the plasma particles before it, thereby increasing their energy.

11.3. In the ordinary pinched discharge, the axial current density j_z gives rise to the azimuthal field B_θ; the plasma is then compressed by the $j_z B_\theta$ force (or pressure gradient) which, being perpendicular to both the j_z and B_θ

---

* The process described here as fast compression heating is sometimes referred to as "shock heating." However, as pointed out in §5.40, the disturbances differ from ordinary shocks in the respect that there is no strict mass-flow continuity across the front.

† It has also been called the *azimuthal pinch* and *orthogonal pinch*.

414

directions, acts radially inward.   In the theta pinch, on the other hand, passage of a discharge through a single-turn coil wound about the tube containing the gas induces an azimuthal current j_θ in the opposite direction.   As a result, the axial B_z field, which would normally have filled the whole cross section of the coil, does not exist in the interior of the current shell but only outside it (Fig. 11.1).   Thus, except for the 90° change in the current and field direc-

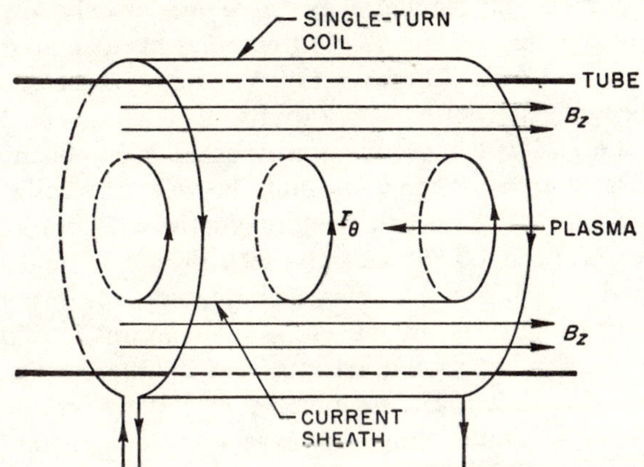

FIG. 11.1.   Formation of theta pinch.

tions, the conditions at the current sheath are identical with those for the conventional pinch.   It is now the $j_\theta B_z$ force which compresses the plasma cylinder.

11.4.   The original motivation for the theta-pinch work was the prospect of attaining high temperatures by irreversible, fast compression of a plasma. As will be seen shortly, it is probable that ion temperatures of 1 to 2 kev have been reached, as a result of such heating followed by adiabatic compression, and the only neutrons for which a thermonuclear origin seems to be a strong possibility have been obtained in this manner.   In the conventional stabilized pinch, most of the temperature increase during passage of the discharge current is due to ohmic heating, and relatively little to the (adiabatic) compression of the gas.   Because of the decrease in the resistivity of the plasma with increasing temperature, this method of heating becomes too slow to be useful at temperatures above about 0.5 kev.   When heating is produced by magnetic compression, there is no such limitation, in principle, although the technical difficulties, e.g., the requirements of large currents and a rapid rate of increase, become more serious with increasing temperature.

11.5.   Furthermore, in fast compression heating, the ions and electrons tend to acquire the same velocity and so most of the energy supplied should be initially transferred to the ions, rather than to the electrons as in ohmic heating.   These considerations apply also to the nonstabilized dynamic pinch, but it is probable that instability break-up is slower in the theta pinch (§11.7). It is largely for this reason, and also because experiments can be made in straight tubes without introducing electrodes into the plasma, that the study of fast magnetic compression has attracted world-wide interest [12, 13].

11.6.   In its simplest form, the theta-pinch system consists of a wide single-turn solenoidal coil wrapped around an insulating tube, usually of quartz or alumina, containing the experimental gas at a moderately low pressure, e.g., 0.01 to 0.1 mm of mercury.   The gas is first ionized, as stated in §5.50, in a convenient manner and then a high-voltage capacitor bank of low inductance is discharged through the coil.   The rapidly increasing current induces an electric field in the gas which produces azimuthal currents in the preionized plasma, giving rise to an axial magnetic field between the coil and the current sheath in the plasma.   The unbalanced magnetic pressure then drives the plasma rapidly toward the axis of the tube (see Fig. 11.1), and as a result, energy is transferred to the gas particles.   High temperatures, particularly of the ions, are attained due to the initial rapid compression followed by the slower adiabatic radial compression arising from the increasing magnetic field.

11.7.   It is probable that the theta pinch may have more intrinsic stability than a conventional dynamic pinch.   It was seen in Chapter 7 that, in the latter case, the growth of both the $m = 0$ (sausage-type) and $m = 1$ (kink) instability modes is favored by the azimuthal magnetic field.   However, in the theta pinch the situation appears to be different.   The analogy to the $m = 0$ stablity in the nonstabilized pinch would be the flute (or interchange) instability in the theta pinch.   Since the force driving the instability would arise from the curvature of the axial field lines around the plasma, it is to be expected that this mode would grow more slowly than the sausage-type instability because the line curvature is usually less.   There is evidence from streak photographs that macroscopic stability has been achieved in the theta pinch for periods up to 20 microsec, which is considerably longer than the stable lifetime of the ordinary dynamic pinch under similar conditions.

11.8.   The foregoing discussion has referred, in particular, to a cylindrical system.   The theta pinch may conceivably also be realized in a toroidal geometry.   The torus, of insulating material, would be completely (or largely) surrounded by a one-turn coil through which the primary current would flow in the azimuthal direction, rather than longitudinally as in the conventional toroidal pinch.   Since the confining field is longitudinal, however, a planar torus could not be used, because the plasma would then be driven to the walls (§8.1).   This difficulty could be overcome by introducing a rotational transform, just as in stellarator systems (Chapter 8).   The general name Thetatron has been given to devices in which the theta pinch is used for initial confinement and heating of a plasma [13].

11.9.   Experiments with the theta pinch have been made under a variety of conditions.   In some cases one single-turn coil has been used, whereas in others two such coils have been employed thus providing their own mirror geometry, since the axial magnetic field strength is weaker in the region between the coils. The two coils can either be connected in parallel with a single capacitor bank or be energized by the simultaneous discharge of two separate banks.   By suitable shaping (see Fig. 11.2) a single coil can provide a magnetic mirror

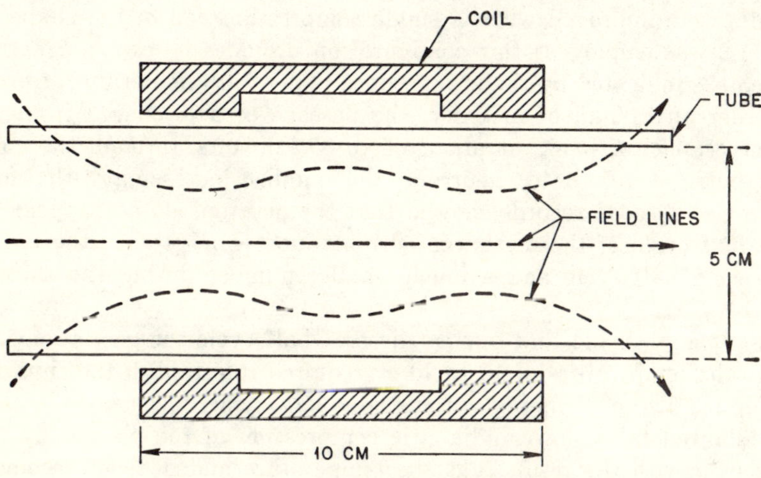

Fig. 11.2.   Magnetic field coil in Scylla.

configuration or a one-turn coil can be located between two mirrors which are independent of the discharge system.   The vacuum tube diameters have varied from 2 cm to 20 cm, and the peak magnetic fields from 2000 to over 100,000 gauss.   The length of the coils has ranged from about 0.64 cm to 30 cm, and longer ones have been considered.   The incentive to the use of longer coils is based on the view, outlined in §9.32, that the mirror confinement of the particles in a high-pressure plasma, in which the relaxation distance for large-angle scattering collisions is small, is determined by the distance between the mirrors.

11.10.   An early attempt at plasma heating by what has subsequently become known as the theta pinch were the so-called "collapse" experiments [3]. The initial ionization of the gas was produced by means of a P.I.G. discharge (§5.91) and this was followed by the application of an alternating axial magnetic field, referred to as "slow collapse."   The resulting currents in the partially ionized plasma served to heat and ionize it further.   After a few cycles of the alternating field, the plasma was sufficiently ionized to act as an almost perfect conductor. A sheath formed between the plasma and the magnetic field and was found to move away from the tube walls. At this point, the axial magnetic field was suddenly increased by discharging a high-frequency capacitor bank through a single-turn solenoid surrounding the tube in which the plasma was confined. The resulting fast collapse produced a strong shock in the plasma.

11.11. Neutrons which may possibly be of thermonuclear origin, that is, neutrons produced by reactions between deuterons which were essentially in thermal equilibrium, were obtained from the device known as Scylla. In its first form two single-turn coils were used to produce the magnetic field [4], but the results were improved when a single compression coil of the shape shown in Fig. 11.2 was employed; this configuration provides its own magnetic mirror system, as indicated by the broken lines [4]. With deuterium at pressures of the order of 0.1 mm of mercury, the passage of a discharge from a low-inductance capacitor bank, at about 65 to 85 kilovolts, through the coil produced a burst of 10^5 to 10^7 neutrons, the number increasing with the bank voltage. Oscilloscope records showed that the neutrons do not appear during the first half-cycle of the discharge current; most emerge near the maximum of the second half-cycle and a much smaller number during the third half-cycle.

11.12. It is probable that during the first half-cycle the gas is ionized and heated to the temperature of 10 to 20 ev required to establish the thin current sheath formed at the beginning of the second half-cycle. The fast compression is then followed by a slower adiabatic compression of the plasma by the increasing field with the result that the temperature and density become sufficiently high in the second half-cycle for neutrons to be produced at a significant rate. If the gas is preionized by a strong discharge parallel to the tube axis, neutrons are observed on the first half-cycle of the magnetic field.

11.13. Measurements made with a collimated detector show that nearly all the neutrons come from a central region roughly 2 cm long and 1.5 cm in diameter. This result suggests that the neutrons originate in the gas away from the tube walls and that the compression is three-dimensional, i.e., both axial and radial. Whether the total compression is alone sufficient to account for the observed temperature increase in the plasma is uncertain. Magnetic probe measurements, made at the time of maximum compression of the plasma, indicate that pressures of 0.8×10^8 dynes/cm^2 are attained near the axis of the cylindrical vacuum chamber.

11.14. Associated with neutron production, and apparently originating from the same limited plasma volume, are soft X-rays, with energies up to a few kiloelectron volts [14]. Assuming a Maxwellian distribution of electron velocities, a study of the absorption of the soft X-rays emitted in the axial direction leads to a kinetic electron temperature in the plasma of roughly 240 ev (§6.77), which is less than the estimated ion temperature of about 1.3 kev given below [15].   As stated earlier, it is to be expected that the ion temperature will be somewhat higher than the electron temperature.

11.15.   In addition to the soft X-rays, the emission of hard X-rays was observed, particularly at the lower gas pressures, e.g., 5 to 50×10^{-3} mm of mercury.   There is, however, no time correlation between the production of these X-rays and neutrons.   The latter appear at the magnetic field maximum and their intensity distribution is symmetrical; the hard X-ray pulses, on the other hand, have an asymmetric distribution and are observed when there is a high rate of change of magnetic flux, i.e., when there is a strong electric field. The hard X-rays appear to originate from the walls of the vacuum chamber at some distance outside the mirror fields, so that they may be thick-target bremsstrahlung produced by runaway electrons which have acquired high energies in the central region and then escaped through the mirrors by traveling along the field lines.   The absence of hard X-rays at the higher initial gas pressures is attributed to electron-ion collisions that prevent the electrons from attaining high energies.

11.16.   As the deuterium gas pressure is increased from a low value, the neutron yield first increases, then remains approximately constant, and finally decreases.   The small yield at low pressures is attributed to the delay in gas breakdown; this seems to be confirmed by the observation that neutron emission now occurs mainly in the third, rather than the second, magnetic compression.   At the higher pressures the decreased neutron yield may be due either to a lower efficiency in heating and ionization by the shock mechanism or to the distribution among a larger number of particles of the energy imparted to the plasma.

11.17.   The general nature of the results described above suggests that the neutrons may be of thermonuclear origin.   More striking evidence of the thermonuclear character of the reaction, however, has been obtained from a study, based on a large number of separate discharges, of the energy distribution of the protons and tritons emitted in the proton branch of the D-D reactions.   The experimental technique and the results for the protons were described in §6.144; similar observations were made with the tritons.   The data were found to agree well with a Maxwellian distribution of the plasma deuterons at a temperature of 1.3 kev [16, 17].

11.18.   In a different series of experiments, made in another laboratory, the behavior of the plasma during compression was observed with the aid of a streak camera [2].   By employing a radiofrequency field and a high-frequency voltage pulse to preionize and preheat the gas, it was possible to make effective use of the first half-cycle of the main discharge through two single-turn coils. As the magnetic field increases, the streak photographs show that a radial shock wave first forms.   This is reflected at the axis of the tube so that it begins to move outward, but the direction of motion is soon reversed as a result of the radial compression caused by the axial magnetic field.   Some "bouncing" then occurs, similar to that observed in conventional pinched plasmas (§7.59), until the axial shock waves, produced at the two ends of the system collide at the center of the vacuum chamber. The collision of the shock fronts increases the temperature of the plasma, so that it tends to expand, but it is soon compressed again by the increasing magnetic field. At the compression maximum, the volume of the gas is decreased by a factor of about 100. The plasma now begins to undergo large-amplitude radial oscillations, after which it remains quiescent for several microseconds. The spectrum shows the broadened Balmer lines of deuterium during the early stages, but after the axial shock waves collide at (or near) the midplane of the system there is only a continuum, implying complete ionization of the gas.

11.19. It appeared possible that the radial oscillations described above might arise from an axial magnetic field that was trapped within the plasma during the preheated discharge. Magnetic probe measurements showed that the axial field produced during the preheating stage, after initial radiofrequency breakdown, was indeed trapped in succeeding half-cycles. Furthermore, the axial field exterior to the plasma, arising from the main discharge, was in the direction opposite to the trapped field [5, 8, 18]. This phenomenon is referred to as *reversed-field trapping*.

11.20. With a single-turn solenoidal coil 10-cm long, a burst of about 10^5 neutrons, with an average emission time of 1 microsec, was observed when the reversed field situation existed. By increasing the length of the coil to 30 cm, the neutron yield rose to about 10^7, and the duration of the emission also increased. There thus seems to be some correlation between the neutron yield and the confinement time, which in the plasma of relatively high density was presumably determined by the distance between the mirror fields. From calculations based on the neutron yield and other considerations there is a possibility that a kinetic deuteron temperature of about 2 kev may have been attained. If the timing of the discharge was such that the trapped axial field was in the same direction as the external field, no neutrons were observed. Bremsstrahlung measurements (§6.75 *et seq.*) indicated an electron temperature of about 0.7 kev and a density of more than 10^{16} electrons/cm³ [19].

11.21. In later work with the Scylla apparatus, with the gas preionized so that neutrons were observed in the first compression half-cycle, it was found that the neutron output was greatly enhanced by the inclusion, prior to the discharge, of a small axial field in the direction opposite to the compression field. The significance of these and other results associated with trapped reverse fields is not yet clear. One possibility is that the deuterons obtain sufficient thermal energy to produce neutrons as a result of the heat liberated by the intermixing of the trapped fields (§5.8). If this is an ohmic process, then the electrons will acquire most of the energy, which must then be transferred to the ions. But the time before neutron emission occurs seems to be too short for the deuterons to be heated in this manner. It is not inconceivable that there is some other mechanism for energy dissipation arising from the mixing of the oppositely directed magnetic fields, but its nature is a matter for speculation.

CUSPED GEOMETRIES

INTRODUCTION

11.22.   In most systems in which a plasma is confined by a magnetic field that surrounds it smoothly, i.e., without a discontinuity, there will be a tendency toward instability.   Basically, the reason is that the lines of force, which may be thought of as being stretched around the plasma, can shorten themselves by burrowing into the gas and thus force it outward.   Theoretically, however, a confinement which is absolutely stable against arbitrary deformations, even of finite amplitude, can be achieved if the field lines everywhere curve away from a diamagnetic plasma, i.e., the field-plasma interface is everywhere convex on the side toward the plasma (§13.5 *et seq.*).   In order to satisfy this curvature requirement, the magnetic field configuration must possess cusps, i.e., points or lines (or both) through which the field lines pass radially outward from the center of the confinement region [21-26].   The general name *cusped geometry* is given to these configurations.

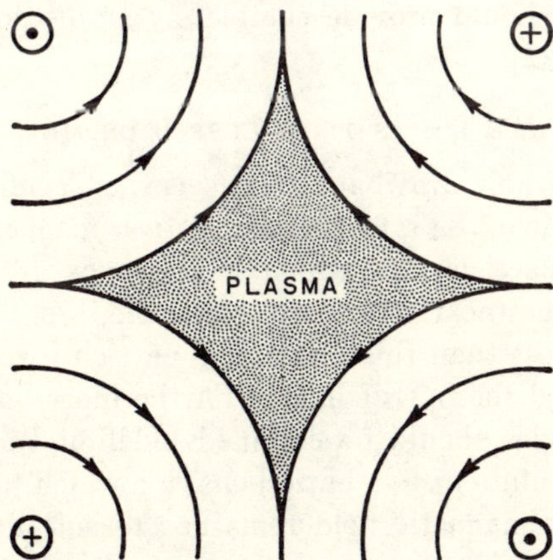

Fig. 11.3.   Simple cusped magnetic field.

11.23.   A simple example of a cusped geometry is that produced by an array of four line currents alternating in direction; a section through the field system in which it is supposed that a plasma is confined is shown in Fig. 11.3.*   It is

---

* A proposal for a confinement geometry of this kind, which became known as "picket fence," was made by J. L. Tuck, Jr. in 1954 [20]. It consisted of two layers of parallel wires carrying currents in alternating directions, so that the magnetic field had a series of cusps. The original purpose of the picket fence geometry was to generate magnetic fields only in the vicinity of the walls, thus decreasing the power expended in maintaining these fields. But it was soon realized that its real advantage would be that of providing stable confinement.

apparent that if there is a loss of plasma from the central volume, the lines of force must be stretched in order to fill the volume previously occupied by the plasma. Since this requires the expenditure of energy, the type of instability referred to above will probably not occur.* Another simple type of cusped geometry is the biconical cusp produced by a pair of field coils arranged as in a mirror system except that the currents flow in opposite directions (Fig. 11.4A). Various combinations of alternating line or coil currents can be en-

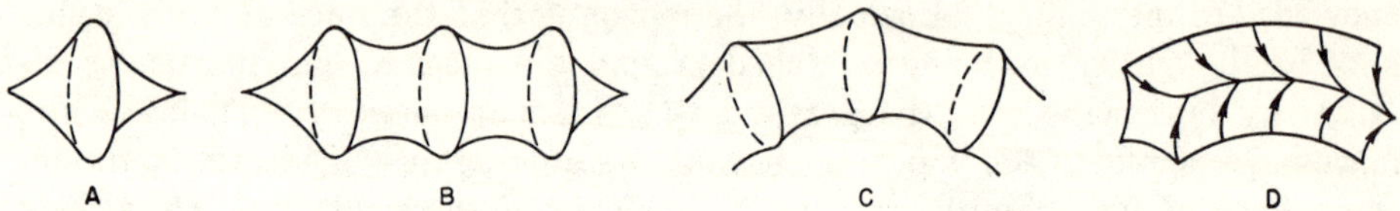

FIG. 11.4. Examples of cusped geometries.

visaged to give linear or toroidal cusped systems. For example, several coils placed alongside each other give the multiple cusp of Fig. 11.4B; these might be arranged so as to close upon themselves to form a torus, as in Fig. 11.4C. The four line current elements of Fig. 11.3 could also be bent into loops to produce a toroidal configuration of which part is shown in Fig. 11.4D. In principle, all of these should provide completely stable confinement of a diamagnetic plasma [21, 22].

PARTICLE LOSSES FROM CUSPED GEOMETRIES

11.24. The most serious drawback of the cusped confinement geometry is the expected high particle loss rate [21, 22]. In a magnetic mirror system, it was shown in §9.16, particles for which $W_\perp/W$ was greater than a certain value, namely, the reciprocal of the mirror ratio, were confined. In other words, if $W_\perp/W$ was less than this value, the particles would be lost immediately when they entered the mirror field. Furthermore, in the absence of collisions, all other particles should be confined indefinitely. These conclusions were based on the postulate of adiabatic behavior which involved the assumption that the confining magnetic field remained essentially constant over one gyromagnetic radius and for a gyromagnetic period of the particles. In these circumstances, provided scattering collisions are not significant, a confined particle of given energy will always be reflected at the same point in the magnetic mirror.

11.25. In a cusped configuration, however, a singular condition exists at a certain point, e.g., the center of Fig. 11.3 where the magnetic field goes exactly to zero. A particle passing in the vicinity of this point will experience a large

---

*It is of interest to note that in the central region between mirror fields the loss of plasma will result in a contraction of the field lines, so that the process should be favored and instability should result. As noted in §9.28, however, there appear to be factors operative which counteract the tendency to instability.

field change. This can easily occur in a time interval less than the gyromagnetic period, especially as this period is considerably lengthened by the decrease in the magnetic field strength. The magnetic moment is then no longer an adiabatic invariant of the system. In the extreme case of a particle actually passing through the point where the field is zero, the particle will momentarily travel in a straight line and its subsequent motion will bear little or no relationship to that prior to its passage through the point. Consequently, particles which do not escape through the cusps in the magnetic field at the first encounter may well do so in later encounters, even if there are no collisions.

11.26. Losses from a cusped geometry are thus quite similar to those from a mirror system with a high mirror ratio, the loss cone in velocity space being continuously refilled as a consequence of the nonadiabatic behavior of the particles in the center of the field. In addition, scattering collisions, as in a mirror system, can cause particles to enter the loss cone. It is seen, therefore, that losses from the cusps can be quite large.

11.27. The foregoing discussion of losses in a cusped geometry is a single-particle approximation, in which a vacuum field configuration, i.e., without plasma, has been assumed. Another approach is to treat the case of a completely diamagnetic plasma occupying the central region. There is then a sharp discontinuity separating field-free plasma from the vacuum containing the magnetic field.

11.28. From the standpoint of losses, two different approximations may be considered [21, 22]. The first supposes that the plasma-field interface is sharp and acts as a specular reflector from which particles bounce back with completely discontinuous velocity changes. In this event, plasma losses are negligible, since only particles aimed along the cusp axis (or median plane) will escape; the others are reflected back (Fig. 11.5A). On the other hand, if the

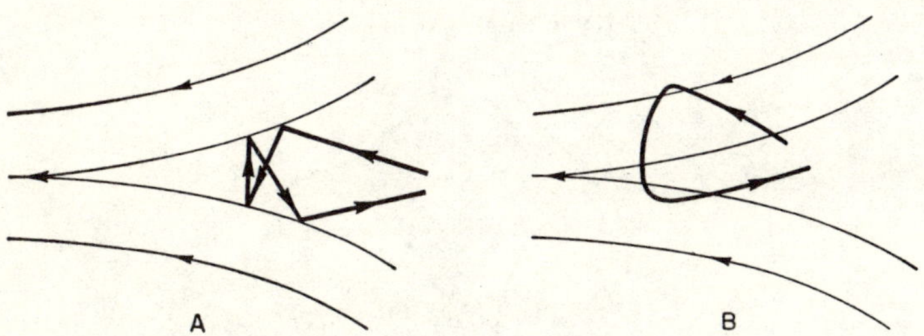

Fig. 11.5. Reflection of charged particle at a cusp: (A) sharp interface, (B) sheath of finite thickness.

interaction of the particles with the magnetic field is taken into consideration, it is seen that the particles are able to penetrate the field to some extent, thus leading to a sheath of finite thickness. A theoretical treatment indicates that this is roughly equal to the geometric mean of the gyromagnetic radii of the ions and electrons. A consequence of this finite thickness is that the plasma no longer comes to a point at the cusp, but extends to the outside through a "hole" of radius approximately equal to the sheath thickness (Fig. 11.5B). In this case losses may be considerable.

11.29. An approximate expression for the confinement time τ in the cusped system shown in Fig. 11.5B, when the particle pressure is approximately equal to that of the confining field, i.e., $\beta \approx 1$, has been found to be

$$\tau = \frac{R^2 B}{T} \times 10^{-10} \quad \text{sec,}$$

where R cm is the radius of the circular cusp, B is the corresponding value of the field in gauss, and T is the temperature in kilo-electron volts [21, 22]. It will be seen that the confinement time decreases with increasing temperature. For example, if R is 100 cm, then a field strength of 10,000 gauss will give a confinement time of 1 sec at a temperature of 10 ev, but this is decreased to 1 millisec at 10 kev. The confinement time can be increased by increasing the field strength and, in particular, the cusp radius, but there will be an accompanying increase in the total expenditure of magnetic field energy.

REDUCTION OF LOSSES

11.30. A number of proposals have been made for possible methods of partially stopping the escape of particles from a cusped geometry. One of these utilizes a subsidiary coil around the line cusp excited with a radiofrequency current [25-27]. The result is a rapid shifting back and forth of the effective position of the cusp at a rate which should produce a restoring force on any particle about to escape (Fig. 11.6). Confinement by radiation pressure at the

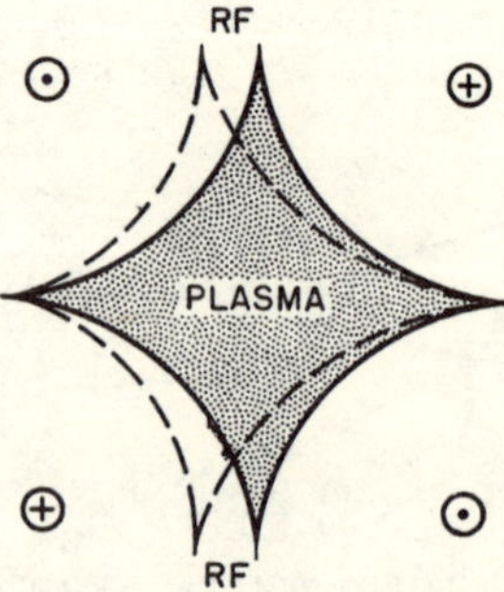

FIG. 11.6. Use of radiofrequency field to reduce cusp losses.

cusps is also being investigated (cf. §11.83). Another suggestion is to apply a crossed magnetic field at the cusp, provided by a solenoid, as shown in Fig. 11.7; since the magnetic field in the center of the cusp is never zero, losses due to nonadiabaticity are greatly reduced [22]. A third possibility is to strip

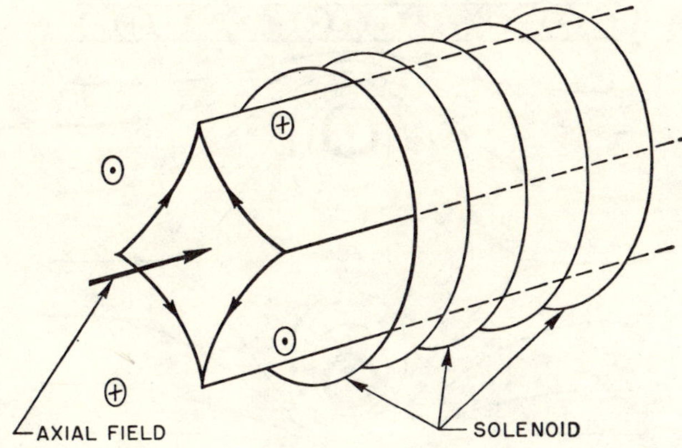

Fig. 11.7.   Cusp with crossed axial field.

the electrons from the outer edges of the boundary layer, as indicated schematically in Fig. 11.8.   This is expected to result in a decrease in the thickness
of the layer and hence in the area of the "holes" through which the particles can
escape [21, 22].

11.31.   A somewhat more complicated proposal is effectively to close the
cusps on themselves, in what has been called the "caulked picket fence."   A

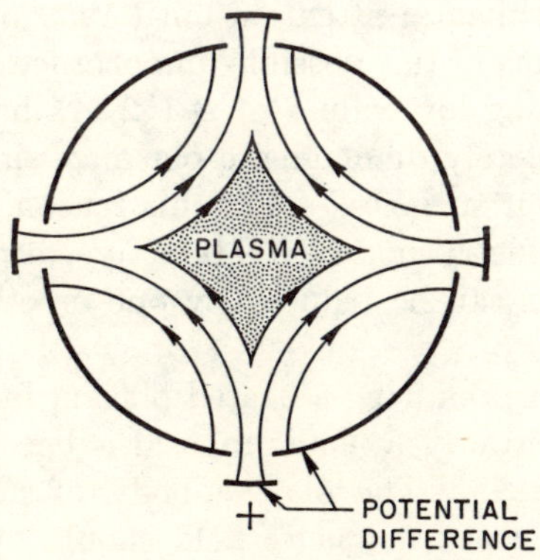

Fig. 11.8.   Reduction of cusp losses by stripping electrons from boundary.

highly conducting ring carries a current such as to produce a magnetic field
at its center which is opposite in direction to that formed by a solenoid wound
about the containing tube (Fig. 11.9).   As a result, the lines of force are convex toward the plasma except in the high-field region between the ring and
the tube wall.   In this configuration, plasma can reach the wall only by crossing magnetic lines, and the only region where the plasma could be hydromagnetically unstable with respect to the field is where, in pressure equilibrium,

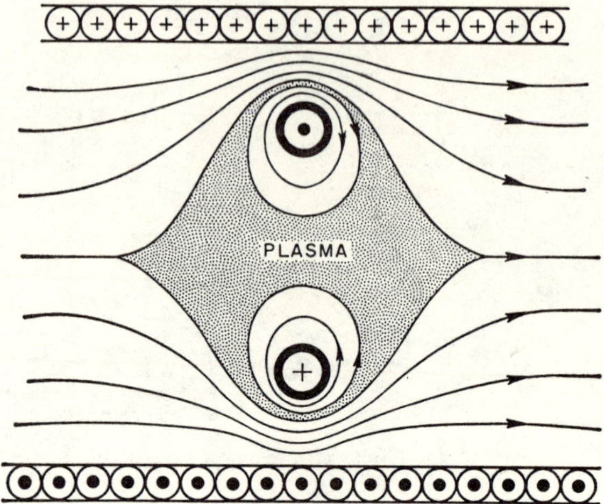

Fig. 11.9.   The caulked picket fence.

the plasma density should be very small.   The principle could be extended to
a torus with a levitated spiral winding [28, 29].

EXPERIMENTAL STUDIES

11.32.   Although the cusped geometry has been known for some years, it
received very little experimental attention until 1959 because of the high predicted loss rates.   However, the possible importance of cyclotron radiation
losses from plasmas having low values of β (§2.84) has directed attention to
the confinement of completely diamagnetic plasmas, since they would produce
this radiation only at their surfaces.*   For this reason, the great intrinsic stability of the cusped geometry makes it attractive, especially in view of the
possibility that the losses can be reduced by one or other of the methods described above.

11.33.   The problem of producing a heated plasma in a cusped geometry has
been approached in several ways.   One proposal is based on the use of entropy
trapping, described in §9.78.   The plasma body of high kinetic energy and
density injected into a cusped magnetic field should "pry apart" the field at
the cusp and enter the central region still in a completely diamagnetic state.
The expansion which occurs at the interior of the container should produce
a randomization of the directions of particle motion within the plasma and
thus make the entrance process irreversible [30, 30a].   Particle losses should
be appropriate to the thin-boundary condition, while cyclotron radiation should
be that for a high-β plasma, i.e., only from the surface.

11.34.   Another suggestion is to employ two single-turn solenoids, of the

---

* The term "completely diamagnetic plasma" is used here to describe a plasma which
excludes a magnetic field entirely, regardless of the cause; for such plasmas β is unity.

type shown in Fig. 11.2, side by side. After an initial preionization stage, fast capacitor banks are discharged simultaneously through the coils, producing two compressed and heated plasma threads at the center, in the manner described in §11.11 *et seq.* (Fig. 11.10A). An insert of higher resistivity at

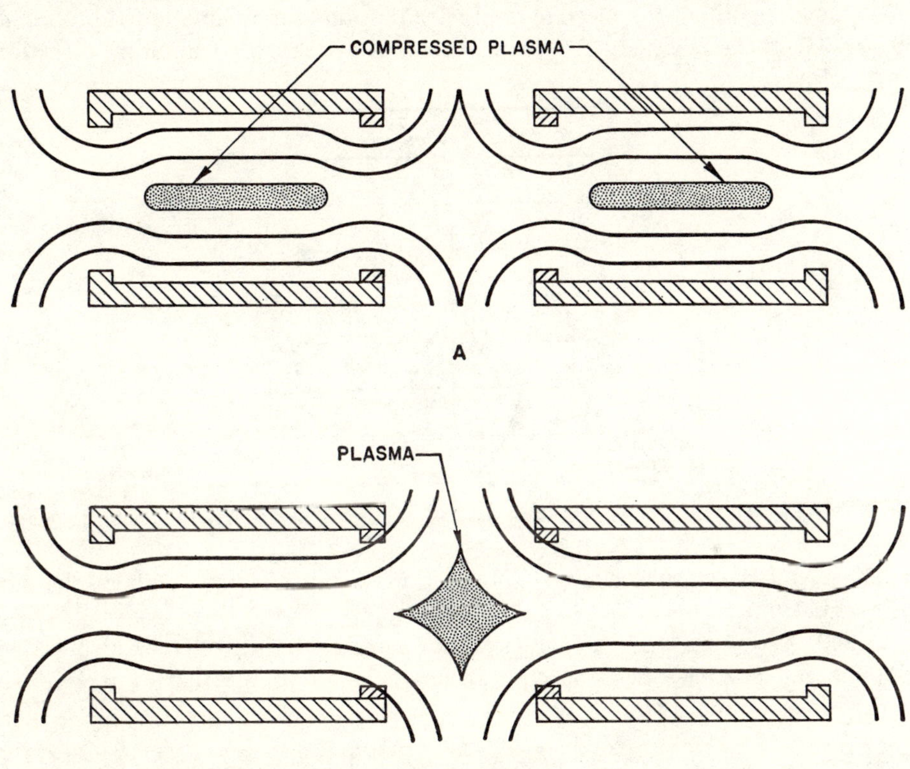

FIG. 11.10. Formation of plasma in cusped geometry (CHALICE).

the inner end of each coil causes the interior mirrors to disappear more rapidly than the remaining field. As a result, the two plasma threads are made to fuse, by a kind of axial compression, thus forming a spindle of plasma in a cusped geometry (Fig. 11.10B) [31].

ROTATING PLASMA STUDIES

INTRODUCTION

11.35. Rotating plasmas have attracted interest in thermonuclear research because, among other things, they offer possibilities in connection with both confinement and heating [32]. If subjected to the action of crossed electric and magnetic fields a charged particle will tend to drift in a direction perpendicular to both fields. As seen in §4.22 *et seq.*, the drift velocity, in a rectilinear system, is independent of both the charge sign and the mass of the particle, so that the plasma as a whole should drift in the same direction with a velocity given by equation (4.11). In a system having cylindrical symmetry, e.g., in which the electric field is radial and the magnetic field axial, as in Fig. 11.11, the charged particles will tend to drift in the azimuthal direction,

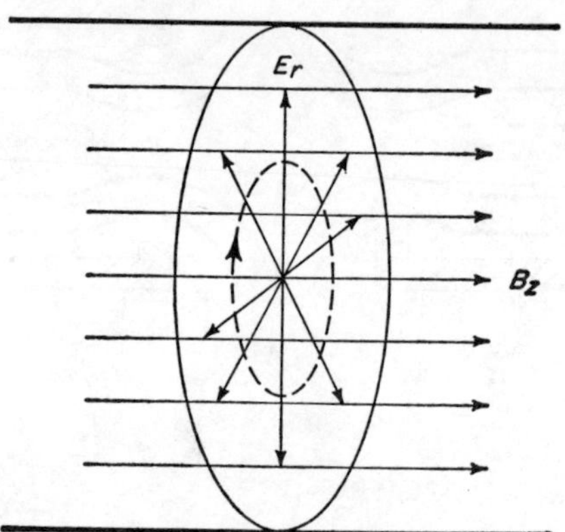

Fig. 11.11. Plasma rotation in crossed electric and magnetic fields.

as shown by the dashed curve. Hence, the plasma will rotate about the axis of symmetry.

11.30. As a result of the rotation, however, centrifugal forces also act on the charged particles and the equilibrium velocity of rotation v_θ is given in Gaussian-cgs units by

$$v_\theta = c \left(\frac{E}{B} + \frac{mv_\theta^2}{erB} \right), \tag{11.1}$$

where m and e are the mass and charge, respectively, of the charged particles. The first term on the right, which is identical with that in equation (4.11), is due to the crossed electric and magnetic fields whereas the second, equivalent to equation (4.19), arises from the centrifugal force. As a general rule this term is small, but the fact that it is different for ions and electrons leads to an azimuthal current, and the associated Lorentz force just balances that due to centrifugal pressure of the plasma.

11.37. A rotating plasma can store large amounts of kinetic energy of mass motion. If this could be converted into random (thermal) motion of the individual particles, heating of the plasma would result. There are certain natural processes, such as outward diffusion, viscous drag between adjacent layers which move at different velocities, and turbulent mixing, which tend to convert the nonthermal stored energy into thermal energy. In addition, artificial means may be used to enhance the conversion. For example, it has been suggested that by superimposing a rapidly oscillating component on the radial electric field, a standing hydromagnetic half-wave will be set up in the axial direction. The viscous drag arising from the associated velocity gradients along the field lines should result in a rapid conversion of the directed rotational kinetic energy into thermal energy. Perturbation of the cylindrical symmetry, after rotation has been established, e.g., by a local change in the magnetic or electric field, should also result in a randomization of the kinetic energy.

11.38. A particularly attractive manner of producing a rotating plasma of high internal energy, which could then be further thermalized in one of the ways described above, is to create ions in the crossed electric and magnetic fields. It can be shown that in this case the drift energy of the ions is equal to their gyromagnetic energy.

11.39. In a conventional discharge tube a space-charge sheath forms on the cathode because the electrons have a higher mobility than do the ions (§4.85). If, however, the surface of the anode is parallel to the lines of an applied magnetic field, as it is in some rotating plasma experiments, the sheath forms on the anode. The reason is that in order to travel in the direction of the electric field, the charged particles must cross the magnetic field lines, and the mobility of the ions is then greater than that of the electrons (cf. §12.20). As the ions move away from the anode surface, they leave behind a region with a net negative space charge. In this narrow sheath there may be large voltage drop which can accelerate the ions.

ROTATING PLASMA IN A MAGNETIC MIRROR FIELD

11.40. An interesting consequence of the rotation of a plasma is that it should lead to improved confinement in a mirror geometry [33, 34]. The reason is that the centrifugal force due to the rotational motion tends to keep the charged particles away from the axis of the containing tube. Since escape through a mirror requires a particle to approach this axis, the rotation of the plasma will make the escape more difficult.

11.41. If $W_{\theta(0)}$ is the magnitude of the rotational drift energy in the central region of the magnetic mirror system and $W_{\theta(m)}$ is the value at the mirror peak, then the conservation of energy requirement for motion along a given field line is now

$$W_{\perp(0)} + W_{\parallel(0)} + W_{\theta(0)} = W_{\perp(m)} + W_{\parallel(m)} + W_{\theta(m)}.$$

From the constancy of the magnetic moment it follows that

$$W_{\perp(m)} = W_{\perp(0)}R_m,$$

and since the condition for reflection is that $W_{\parallel(m)}$ should be zero or negative, this means that

$$W_{\parallel(0)} \lessgtr W_{\perp(0)}(R_m - 1) + W_{\theta(m)} - W_{\theta(0)}. \tag{11.2}$$

11.42. The drift energy W_θ can be expressed by

$$W_\theta = -\tfrac{1}{2}mv_\theta^2,$$

where m is the particle mass and v_θ is its velocity of rotation, so that

$$W_{\theta(m)} - W_{\theta(0)} = \tfrac{1}{2}mv_{\theta(0)}^2 \left[1 - \left(\frac{v_{\theta(m)}}{v_{\theta(0)}} \right)^2 \right]. \tag{11.3}$$

It was stated above that the second term on the right of equation (11.1) is small in comparison with the first, and so v_θ may be taken as being equal to cE/B. Furthermore, on a given flux surface, B is proportional to $1/r^2$, where r is the radial distance, since Br^2 is constant (§4.24), and E varies as $1/r$; hence, v_θ^2 is proportional to $1/B$ and equation (11.3) can be written as

$$W_{\theta(m)} - W_{\theta(0)} = \tfrac{1}{2}mv_{\theta(0)}^2 \left(\frac{R_m - 1}{R_m} \right).$$

Combination of this result with equation (11.2) gives the condition for trapping of charged particles in a rotating plasma confined by a magnetic mirror system as

$$W_{\parallel(0)} \leqslant W_{\perp(0)}(R_m - 1) + \tfrac{1}{2}mv_{\theta(0)}^2 \left(\frac{R_m - 1}{R_m} \right), \tag{11.4}$$

the last term being positive, since R_m must be greater than unity.

11.43. In the absence of the radial electric field, so that the plasma does not rotate, equation (9.26) shows that particles for which $W_{\parallel(0)}$ exceeds $W_{\perp(0)}(R_m - 1)$ will not be trapped. It follows, however, that, in a rotating plasma, particles having larger values of $W_{\parallel(0)}$ will still be confined by the given magnetic mirror system. The mirror enhancement due to rotation is represented by the last term in equation (11.4) and this is seen to increase with $v_{\theta(0)}$, that is, with the ratio of the field strengths E/B, and with the peak mirror ratio R_m.

11.44. In addition to enhancing the ability of a mirror system to confine charged particles, rotation may possibly exert a stabilizing influence on the plasma and also serve to increase its temperature (§11.37). There is, however, a drawback to the use of a rotating plasma in a thermonuclear device. In addition to the particle pressure, the confining magnetic field must now balance the centrifugal force on the charged particles. It is possible, therefore, that with a given magnetic field the density of the stable plasma which can be confined will be decreased.

THE HOMOPOLAR EXPERIMENT

11.45. The Homopolar device has been given this name because the plasma disk rotating in a magnetic field is similar, geometrically and electrically, to the flywheel of a conventional homopolar generator [34]. The apparatus is identical with that described in §7.158 and depicted in Fig. 7.26, which has been used for the formation of a disk-shaped pinch discharge, except that a magnetic field, up to 17,000 gauss, is applied in the axial direction, i.e., at right angles to the plates. The discharge space is bounded by a central electrode 7.5 cm in diameter and an outer electrode 25 cm in diameter; at the top and bottom are glass plates, about 1.5 cm apart, which lie against the pole faces of an electromagnet. The glass serves to separate the plasma discharge current from the return current that flows over the copper-plated pole faces. Most of the work on the Homopolar device has been done with argon gas at a pressure sufficiently high, of the order of 10^{-2} cm of mercury, for anode sheath effects to be negligible. Some observations have also been made with deuterium, helium, and neon.

11.46. When the discharge is first passed, the gas breaks down and the plasma is pinched away from the glass plates. The radial current produces a torque as it flows across the axial magnetic field lines and the plasma is accelerated into the rotational state. The pinch phase is accompanied by a rapid fall in the applied voltage, which rises again during the acceleration phase. The current increases during the pinch and early acceleration phases, but drops to a very small value as the equilibrium rotation state is reached. While the plasma is rotating, the high voltage decays slowly as a result of losses and internal heating.

11.47. Several different experimental observations provide evidence that the plasma is actually rotating. For example, the spectral lines show a Doppler shift which changes in direction as the magnetic field direction, and hence that of rotation, is reversed. From the shift of the 3888 A line in helium, with an applied potential of 8 kilovolts and a magnetic field of 17,000 gauss, the velocity of rotation of the plasma was calculated to be about 3×10^6 cm/sec. This is in agreement with the value expected according to equation (11.1).

11.48. Since the centrifugal force due to plasma rotation should give rise to an azimuthal current, as stated above, there should be an accompanying axial component of the magnetic field. The situation is physically equivalent to the rotating charged particles "leaning" on the magnetic field lines and bending them outward. This effect was observed by means of a magnetic probe coil oriented so as to measure the radial component of the magnetic field, which arises from the centrifugal distortion of the axial field. It was found, in agreement with expectation, that the value changed sign with the reversal of the applied (axial) magnetic field, but not with the reversal of the electric (radial) field.

11.49. An interesting aspect of the rotating plasma in the Homopolar device is that the system is able to store energy and so behaves like a capacitor. In fact, when the electrodes are connected to a charged capacitor bank, the current and voltage variations are similar to those observed when an uncharged capacitor is connected to a charged one through an intermediate resistance and inductance. The storage of energy in the Homopolar plasma has been demonstrated by the current obtained when it is crowbarred through a short circuit of low inductance.   Under favorable conditions more than half of the input energy can be recovered.   It is of interest to mention that the Homopolar device offers the possibility for use as a practical capacitor; its low inductance and high dielectric constant make it a very compact system of electrical energy storage [35].

11.50.   The effective capacitance of a rotating plasma system can be shown to be equal to its geometrical capacitance multiplied by a so-called "dielectric constant" K, given by equation (4.104).   Since, however, the stored energy is in the form of kinetic energy of rotation, rather than as energy of the electric field, it is clear that K is not a real dielectric constant but only a formally equivalent quantity.

11.51.   The basic ideas of the Homopolar system have been incorporated in a toroidal geometry, as shown in section on Fig. 11.12.   The electric field is

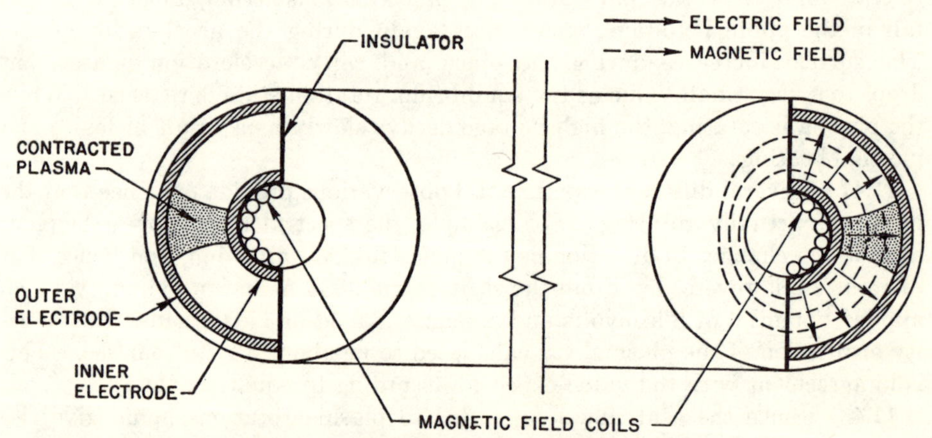

FIG. 11.12.   Rotating plasma in toroidal system.

radial, but the applied magnetic field is now azimuthal, as indicated at the right of the figure.   The plasma consequently rotates in the longitudinal direction, i.e., into and out of the plane of the paper.   It is anticipated that in this device the centrifugal force will be more effective in pulling the plasma away from the insulators than it is in the flat Homopolar system.   Preliminary experiments in the toroidal geometry have indicated that decay times of the rotating plasma approaching a millisecond can be realized.

IXION EXPERIMENTS

11.52.   The Ixion* experiments were designed to study the rotation of a plasma confined in a magnetic mirror system, using crossed electric and mag-

---

*Named after Ixion, the Thessalonian king who, according to Greek mythology, was bound to a revolving fiery wheel.

netic fields [33].   In the earlier observations, the radial electric field was applied between a central rod, at the axis of the tube, and a concentric outer cylinder, both made of stainless steel.   Because of the development of a low-impedance breakdown between these two electrodes, the central one was replaced, in the apparatus called Ixion III, by a 3-cm-diameter column of plasma injected down the axis by means of the device described in §5.89.   The outer cylindrical electrode (anode) was retained and contact with the central "plasma electrode" was made by means of tungsten rings located at the ends of the vacuum chamber (Fig. 11.13).   The latter consists of a steel cylinder 86 cm

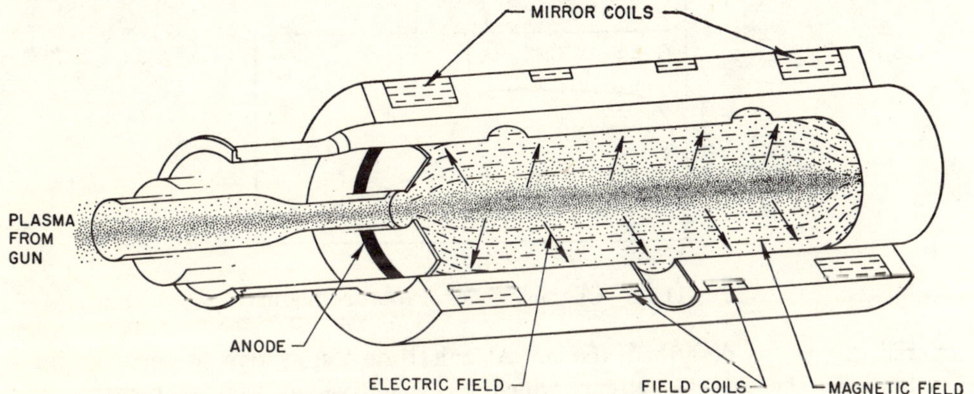

FIG. 11.13.   The Ixion III apparatus.

long and 24 cm in diameter which is normally evacuated to a pressure of 2×10^{-6} mm of mercury between discharges.   The magnetic mirror field is produced by pulsed coils, providing a field of roughly 9000 gauss at the midplane and a peak value of some 20,000 gauss at the mirrors.   A capacitor bank is used to supply the electric field, the potential of which is generally about 7.5 kilovolts.   Deuterium gas is admitted through the valve of the plasma injector, so as to give a pressure of about 10^{-3} mm of mercury in the vacuum chamber.

11.53.   Observations have been made with an oscilloscope of the time behavior of the voltage and current between the inner and outer electrodes.   Following a short period of delay, about 0.2 millisec after application of the voltage, during which negligible current was drawn, although the plasma from the injector had arrived promptly, a sudden discharge occurred.   Under appropriate conditions, i.e., moderately strong magnetic field but not too high electric field and gas pressure, the discharge was accompanied by (a) a sudden impulse of current, about 20,000 amp, lasting for roughly 25 microsec, which resulted in the transfer of a major portion of the capacitor bank charge to the plasma in the vacuum chamber, (b) a sharp drop in the field voltage to about half of its initial value, as the charge was shared between the plasma and the capacitor bank, and (c) intense emission of light due to atomic deuterium and various impurities. Following the impulsive discharge, the voltage decreased to zero in about 0.5 millisec.

11.54. The current impulse and decrease in voltage, illustrated in Fig. 11.14, have been attributed to the development of a rotating plasma containing

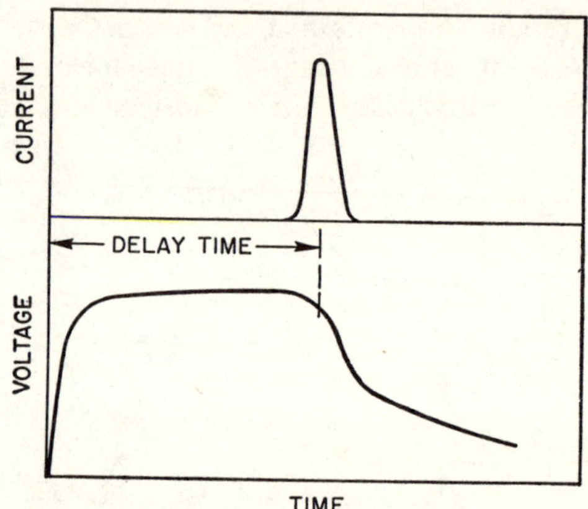

Fig. 11.14. Characteristics of rotating plasma.

stored energy, as described above. At this time the system presents an impedance to the external circuit which is capacitive, as well as resistive, so that the charge is shared between the capacitor bank and the plasma. Oscillations might be expected, as a result of the circuit inductance at the time of the voltage drop, but the failure to observe them may be due to the damping effects of plasma heating. The gradual decay of the voltage after the initial decrease is attributed to the loss of plasma to the walls.

11.55. The delay (or waiting) period, before the impulsive discharge occurs, is related to some extent to the fact that the bulk of the gas entering through the valve of the plasma injector reaches the vacuum chamber some time after the central plasma electrode. But the results of earlier experiments, with a solid central electrode, indicate that another factor is important in the initial high-impedance phase during which negligible current passes. This is believed to be the formation of an electron sheath around the anode (§11.39) which prevents the electric field from penetrating the main body of the plasma.

11.56. Very soon after the start of the current impulse, the spectral lines of various impurities, as well as of deuterium, appear in the plasma spectrum. The large amounts of hydrogen, carbon, oxygen, silicon, iron, etc. present must have a strong influence on the behavior of the plasma and undoubtedly prevent the attainment of high deuteron energies. Measurement of the intensities of the broadened inpurity lines, especially the C II line at 4267 A, as a function of wave length, shows that the maximum is displaced when the direction of the confining magnetic field is reversed. This is attributed to the Doppler effect of the plasma rotational drift and the observed shift indicates a drift velocity of about 4×10^6 cm/sec, a value consistent with calculations based on the known electric and magnetic field strengths. It would appear, therefore, that the impurities, at least, are rotating, and so it is probable that the plasma as a whole is also doing so. It may be mentioned that the Balmer lines of deuterium showed extreme broadening and some structure for which there is no obvious explanation.

11.57. If the magnetic field strength is decreased, the voltage raised, or the pressure increased excessively, the results depicted in Fig. 11.14 are not obtained. Following the delay period, an oscillatory discharge is observed and there are indications that in these circumstances the plasma is not confined. For the same field and pressure conditions, all the discharges were of this type, and magnetic probe coil measurements showed marked diamagnetic behavior of the plasma with almost complete removal of the magnetic field from the region around the midplane. The diamagnetic effect decreased, however, with increasing axial distance from the center and was essentially zero at the mirrors. The oscillatory discharge is generally accompanied by a burst of neutrons of uncertain—but probably not thermonuclear—origin and by the introduction of large amounts of impurities, presumably as a result of the discharge striking the walls.

11.58. If the plasma in the Ixion device is disconnected from the voltage source after being set into rotation, it becomes possible to make measurements of the equivalent capacitance and internal shunt resistance as a function of time. In such an experiment, in which the voltage decay at the terminals was observed with different values of the external shunt resistance, it was found that, while the internal capacitance decayed rapidly, the resistance increased at a rate which kept the time-constant for voltage decay unchanged over a wide range of applied magnetic fields. It has been suggested that this behavior can be accounted for by supposing that the energy loss process is charge exchange with neutral contaminants.

ION MAGNETRON EXPERIMENTS

11.59. The Berkeley ion magnetron device consists of a long solenoidal magnetic field with mirrors at the ends [36]. A radial electric field is applied between the outer walls and a central metal electrode (anode). The gas in the vacuum chamber is at low pressure, less than 10^{-3} mm of mercury, so that an anode sheath forms. In contrast to the Homopolar and Ixion systems, the Berkeley ion magnetron operates under continuous, instead of pulsed, conditions. The magnetic field is 8000 gauss at the center and 12,000 gauss at the mirrors, and the applied voltage is 10 kilovolts. In general, the operation appears to be comparable to that of Ixion during the initial delay (or waiting) period.

11.60.   The anode sheath has a considerable density of high-energy electrons circulating around the axis, so that there is a large probability that molecules (or atoms) entering the sheath will become ionized.   The ions thus formed will be rapidly accelerated out of the sheath by the electric field and will circulate about the axis in magnetronlike orbits.   Because of the relatively high strength of the axial confining field, the circulating ions are prevented from reaching the walls of the vacuum chamber.   Each time an ion returns toward the axis, it is reflected by the anode sheath potential, so that it is effectively trapped between the sheath and the confining field.   After several revolutions, however, an energetic ion will undergo a charge-exchange collision with a cold atom, so that it is replaced by a slow ion while the high-energy atom goes to the wall. This process limits the magnitude of the circulating ion current; it was estimated to be approximately 100 amp when the current passing through the gas was about 1 amp.

11.61.   An indication of the presence of appreciable ion currents was obtained from the observation of fairly violent oscillations of the central electrode, which was mechanically supported at one end only, so that it was free to vibrate.   It appeared that these oscillations are a result of the recoil of ions which are reflected by the anode sheath [36].

11.62.   With steady operation of the discharge tube containing deuterium, at a current of 1 amp and a potential of 10 kilovolts, it is found that neutrons are emitted at a rate between 10^4 and 10^5 per sec.   It is believed that they are produced by head-on collisions between ions entering the sheath and those leaving it.   By improving the vacuum system, so as to decrease the density of neutral particles and thus decrease the probability of charge exchange with energetic ions, the circulating ion current was increased.   At the same time, a marked increase in the neutron emission rate was observed.

11.63.   In the Moscow ion magnetron system a mirror geometry is used, but the central electrode consists of a beam of plasma emitted from a source located within one of the mirror regions [32, 33].   Ionization of the gas, e.g., deuterium introduced into the source, is brought about by high-energy electrons emitted from a hot cathode.   The potential of the plasma beam is up to 40 kilovolts positive with respect to the long metal containing vessel.   A "reflector" electrode, which is at the same potential as the source, located in the other mirror region serves to limit the central plasma beam.   The operating gas pressures have generally been quite low, in the vicinity of 5×10^{-5} mm of mercury.

11.64.   In the earlier measurements, with a device having a vacuum chamber 100 cm long and 46 cm in diameter, three modes of operation could be distinguished.   At high applied voltage, all the ions supplied by the plasma source are accelerated radially as soon as they leave the source, and so there is no central beam.   If the voltage is reduced somewhat, the beam extends beyond the source but not all the way to the reflector.   In these circumstances an anode sheath is formed. A stable radial potential distribution is observed, but most of the electric field is concentrated in the vicinity of the axis, i.e., across the sheath. At still lower voltage, the central electrode covers the whole distance between the source and the reflector electrode. The radial potential distribution then alternates rapidly between one having a steep slope near the axis, due to the presence of the sheath, and one with a gradual slope corresponding to the absence of a marked voltage drop at the anode.

11.65. Neutrons have been observed in a larger ion magnetron system (170 cm long and 50 cm diameter) of design similar to the one described above, with a deuterium plasma. When the high voltage is applied as a square-wave pulse, neutron emission occurs after the voltage has decreased to zero and continues for a few milliseconds. The confinement time thus appears to be longer than in either the Homopolar or Ixion devices; however, these latter systems are smaller in dimensions and the operating conditions are different. At 40 kilovolts anode potential and a tube current of 1 or 2 amp, the larger Moscow ion magnetron produces 10^7 to 10^8 neutrons/sec. The curve for neutron yield versus applied voltage appears to be directly related to the variation of the deuteron charge-exchange cross section with energy (cf. §12.66).

PLASMA CONFINEMENT BY RADIOFREQUENCY FIELDS

GENERAL PRINCIPLES

11.66. A number of different suggestions have been made for the application of radiofrequency fields to confine a plasma [41-52]. Provided the electromagnetic radiation frequency is less than the characteristic plasma frequency (§4.98), the radiation is to a great extent reflected by the plasma. In a suitable cavity, therefore, the surface of the plasma can be subjected to a radiation pressure which will serve to keep the plasma away from the walls. In other words, the plasma can probably be confined by means of the radiofrequency field.

11.67. A simple example is provided by a toroidal cavity with a filament of plasma forming a central core around the longitudinal axis of the torus. This arrangement resembles a coaxial cable closed upon itself to form a circle; the cavity walls represent the outer conductor and the plasma is the central conductor. To confine the plasma the space between its surface and the cavity walls is excited either in the transverse magnetic (TM_{01}) or the transverse electric (TE_{01}) mode having the lowest cut-off frequency.* The electromagnetic radiation energy in the former case (TM mode) then oscillates between a purely longitudinal electric field and a purely transverse (azimuthal) mag-

---

* For a discussion of the various types of waves which can be excited in resonant cavities or propagated in wave guides, one of the several available books on microwaves should be consulted.

netic field, whereas in the latter case (TE mode) the respective field directions are reversed, as shown in Fig. 11.15 [46].

11.68. In each situation, the electric and magnetic fields are out of phase by 90°. The electromagnetic energy supplied to the system, oscillating between these two fields, travels radially and is reflected back and forth between the plasma and the cavity wall. If the radiation intensity is sufficiently large, the pressure at the surface of the plasma will just balance the kinetic pressure of the particles in the interior. The plasma will then be confined in a region near the longitudinal axis of the toroidal cavity. The geometry of the system is similar to that of the pinched discharge, the confining forces in both cases being due to the excess electromagnetic energy density on the outside of the plasma. At the magnetic field maximum the TM mode is equivalent to a conventional pinch and the TE mode to a theta pinch (see Fig. 11.15).

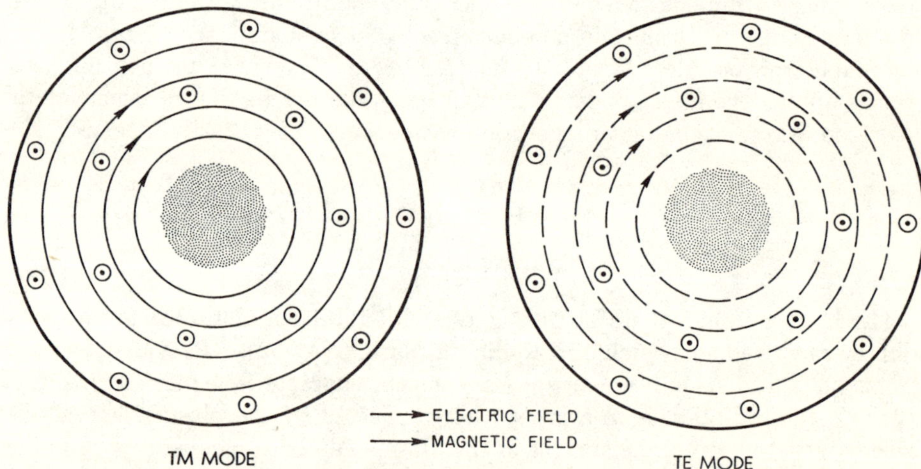

FIG. 11.15. Use of radiofrequency fields for plasma confinement.

11.69. It has been shown theoretically that the plasma and the radiation field penetrate each other to some extent. By investigating the radial motion of the charged particles in the electromagnetic field, it is found that the particles are confined in a potential well, with a minimum at the longitudinal axis of the toroidal cavity. For an oscillating field, the potential is proportional to $e^2/m\omega^2$, where e and m are the charge and mass, respectively, of the particles and ω is the angular frequency of the electromagnetic field. Since the potential depends upon e^2, both ions and electrons are attracted in the same direction, i.e., toward the bottom of the potential well at the axis of the torus. This tendency for particles of either sign to be driven toward the center of the cavity may be regarded as a manifestation of the radiation pressure. Because of the smaller mass of the electrons, the pressure is exerted mainly on these particles, since the potential varies inversely as the mass. The ions are then held in the confined plasma chiefly as a result of electrostatic attraction. A possible drawback to this method of confinement is that, at high densities, the plasma is expected to exhibit instabilities analogous to those of a dynamic pinched discharge [41].

11.70. In place of a toroidal cavity, it should be possible to confine a plasma at the center of a spherical cavity. In this event, the space between the plasma and the cavity wall could be excited by electromagnetic radiation at a frequency corresponding to resonance in the TE_{110} mode. However, in these circumstances it appears that there would be two points where the confining field is zero, so that the plasma could escape. The difficulty can perhaps be overcome if both the TE_{110} and TE_{111} modes were used, with a time phase difference of $90°$ between them. But it is probable that in this case the confined plasma would not be spherical, since the pressure due to the electromagnetic radiation would not be the same at all points on the surface. Whether such a plasma would be stable or not is uncertain.

11.71. Another possibility for plasma confinement in a spherical cavity is to make use of three different resonant frequencies, corresponding to the TE_{110} mode and two other modes with successively higher radial wave numbers (TE_{210} and TE_{310}). By establishing these three radiofrequency wave systems at right angles to each other in space, it should be possible to exert a uniform pressure over the surface of a spherical conductor, e.g., a plasma, at the center of the resonant cavity.

FREQUENCY OF CONFINING RADIATION

11.72. One of the anticipated advantages of the confinement of a plasma by by a radiofrequency field is the possible stabilizing effect of such a field [53, 54]. A simple physical argument in favor of this expectation is based on the fact that the organized plasma motions resulting from instabilities tend to be oriented by the confining magnetic field. If this field is caused to vary rapidly, particularly in direction, it is conceivable that the instability motions will lose coherence with the field structure and will not develop sufficiently to permit the plasma to escape from confinement.

11.73. A rough estimate of the frequency of the electromagnetic radiation required to maintain stability can be made in the following manner. It may be supposed that the growth of plasma perturbations of long wave length are suppressed by the finite size of the system and those of short wave length by the inertia of the plasma. It may then be inferred that there should exist a band of stable perturbation wave lengths bounded, approximately, by the dimensions of the plasma at the upper limit and by c_s/ω, where c_s is the sound (or Alfvén) speed in the plasma and ω is the angular frequency of the confining radiofrequency field. Thus, stability might be expected if the condition

$$l \gg \frac{c_s}{\omega}$$

was satisfied, where l is a linear dimension of the system. Assuming the kinetic value for the velocity of sound, it follows that the stability requirement is

$$\omega \gg \frac{1}{l}\left(\frac{\gamma kT}{m}\right)^{\frac{1}{2}},$$

where m is the average mass of the plasma particles, γ is the ratio of the specific heats, k is the Boltzmann constant, and T is the absolute temperature.

11.74. The average mass is essentially equal to half that of the ion, and taking l as being of the order of 100 cm, it is found that for temperatures up to 100 kev the frequency of the radiation must exceed 10^7 cycles/sec.* It is difficult to generate radiofrequency fields of high energy at frequencies far in excess of 1000 megacycles (10^9 cycles)/sec; hence, possible frequency values for stable plasma confinement are in the ultra-high-frequency range, e.g., 100 to 1000 megacycles/sec. The wave lengths are then of the order of magnitude of the dimensions of the system, i.e., 1 meter or so.

11.75. For the radiation to be mainly reflected at the surface, the frequency must be less than the characteristic frequency of the plasma. Assuming a particle density of 10^{14} electrons/cm^3, it is seen from equation (6.7) that the frequency of the electromagnetic radiation must be less than about 10^{11} cycles/sec, i.e., 10^5 megacycles/sec. The frequencies indicated above, on the basis of plasma stability, thus lie in the region where reflection of the radiation will occur from plasmas having densities similar to those that may be encountered in controlled thermonuclear research.

POWER DISSIPATION BY SKIN EFFECT

11.76. One of the main problems connected with the utilization of a radio-frequency field to confine a plasma is the large amount of energy dissipated at the interior surface of the cavity wall due to the skin effect. Although this energy is not completely lost, since it is converted into heat and so can be partially recovered, the skin effect dissipation means that a large amount of radiofrequency power must be supplied to the system. A rough calculation indicates that the power dissipation in the cavity walls, expressed in Gaussian-cgs units, is given by

$$\text{Power dissipation per unit area} \approx \alpha^2 cnkT \left(\frac{2\omega\eta}{\pi}\right)^{\frac{1}{2}}, \qquad (11.5)$$

where ω is the angular frequency of the confining field, η is the resistivity of the cavity wall material, n is the electron (or ion) number density in the plasma, α is the ratio of the magnetic field at the cavity wall to that in the plasma; k, T, and c have their usual significance [55].

* It will be seen in §11.79 that in a hypothetical thermonuclear reactor in which confinement of the plasma is achieved entirely by means of a radiofrequency field, the dimensions would probably be greater than 100 cm. In these circumstances fields of lower frequency would satisfy the stability requirement given above.

11.77.   Assuming the cavity wall to consist of a relatively good conductor of electricity, e.g., silver or copper, and taking n to be 10^{14} particles/cm^3 and the temperature as 10 kev, it can be calculated that, for the kind of radiofrequency field which might be used for confining a plasma, the power dissipation to the walls is of the order of 100 kilowatts/cm^2.   No means are at present available for supplying such large quantities of radiofrequency power to a system of practical dimensions.   In any case, even if this were feasible, the heat removal from the cavity walls would still represent a serious practical problem for which there is no simple solution.

11.78.   Independently of the frequency, the electric fields required for radiofrequency confinement of a plasma are larger than can be handled by present techniques.   The magnetic field required to confine the plasma referred to above, i.e., one having a density of 10^{14} particles/cm^3 at a temperature of 1 kev, is somewhat more than 6000 gauss.   Since an electromagnetic radiation field is being used, this means that the electric field strength would be 6000 esu, i.e., 1.8×10^6 volts/cm.   It seems impossible, with the means presently available, to support such a strong field over a system of appreciable dimensions.

11.79.   Suppose, however, that it were found feasible to design a thermonuclear reactor with a radiofrequency field used for confinement of the plasma. The power level at which the rate of energy production by thermonuclear reactions would be equal to the rate of dissipation in the cavity walls has been estimated to be in the vicinity of 10^8 megawatts.   This implies a minimum area of 10^9 cm^2 for the cavity and hence linear dimensions of the order of 100 meters.

11.80.   It is possible that the skin effect power dissipation could be reduced in some manner, e.g., by maintaining the cavity wall at a very low temperature to decrease the specific resistance or by special techniques such as are used to minimize losses in radiofrequency transmission lines [46].   The required power level and dimensions of the thermonuclear reactor would then be smaller than those given above.   The suggestion has also been made that a method might conceivably be devised whereby the charged particles of high energy resulting from the thermonuclear reactions could produce radiofrequency power directly.   While this would have no effect on the minimum operating power level, it would greatly decrease the overall losses.

11.81.   It may be concluded, therefore, that, even if stable confinement of a plasma by radiofrequency electromagnetic fields can be achieved, the design of a practical thermonuclear reactor utilizing this concept will be possible only if the problems arising from power dissipation in the cavity wall, due to the skin effect, can be solved.   There is a distinct prospect, however, that plasma confinement by means of radiofrequency radiation may be combined with ordinary confinement by magnetic fields.   The major part of the required pressure would then be supplied by the latter, whereas the superimposed radiofrequency field serves to decrease losses due to escape of particles, and possibly also to eliminate drifts and to stabilize the plasma. Arguments have been presented to show that, in these circumstances, the fields of very high frequency mentioned above and used in some experiments (§11.87) may be unnecessary [48]. In fact, frequencies as low as a few megacycles per second might be adequate. Among the advantages of such electromagnetic fields of lower frequency is a decrease in the energy losses due to the skin effect for a specified particle density in the plasma, or alternatively, an increase in the density for a given loss, as may be seen from equation (11.5).

11.82.   Among the proposed applications of combined stationary and radiofrequency magnetic fields, two may be mentioned. First, there is the use of radiofrequency "plugs" to prevent the escape of charged particles from the ends of a straight tube in which the plasma is confined by an externally applied axial magnetic field. The radiofrequency radiation pressure would thus act as a substitute for a magnetic mirror field. A preliminary experimental test of this idea is described below.

11.83.   Another suggestion is to use a radiofrequency field to minimize the serious escape of particles at the cusps in stable configurations, e.g., "picket fence," of a diamagnetic plasma confined by a magnetic field (§11.23). A possible arrangement is shown in Fig. 11.16 [48]. The cusped geometry is

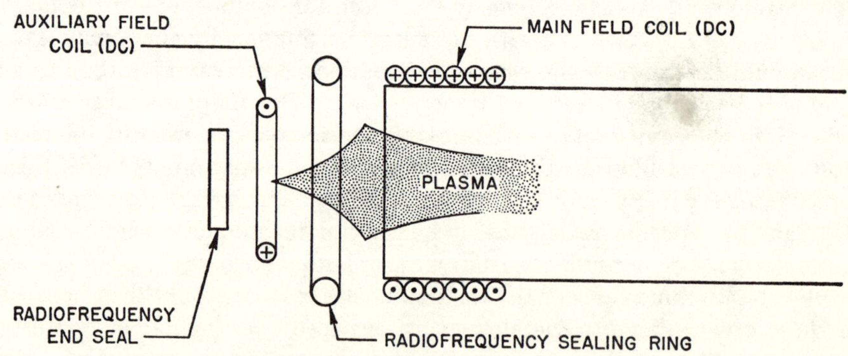

FIG. 11.16.   Use of radiofrequency field to decrease end losses.

produced by two magnetic field coils, one indicated as the main (DC) field coil and the other as the auxiliary (DC) field coil, with currents flowing in opposite directions. The field due to the auxiliary coil is several times stronger than the main field, thus producing a sharp cusp. The two possible paths of particle escape, namely, along the axis and radially between the main and auxiliary field coils, are closed by the application of appropriate radiofrequency fields as shown.

EXPERIMENTAL STUDIES

11.84.   A test of the confinement of a plasma by means of radiofrequency plugs at the end of a straight tube has been made with the apparatus represented in Fig. 11.17 [45]. The plasma was produced in a tube, 40 cm long and 1.5 cm in diameter, containing gas, such as hydrogen, nitrogen, or argon, at a pressure in the range from 5×10^{-5} to 5×10^{-3} mm of mercury. Confinement along the length of the tubes was achieved by means of an axial magnetic field, provided by the coils shown, up to 2000 gauss in strength. The two ends of the tube were set into resonant cavities in which electromagnetic oscillations of the TE_{101} type were excited from pulsed magnetron generators. The wave length of the radiofrequency radiation was in the 10-cm ($\sim$ 3000 megacycles/sec) range and the pulses lasted 120 microsec with a power up to 400 kw. The amplitude attained by the maximum high-frequency magnetic field in the resonators was 60 gauss.

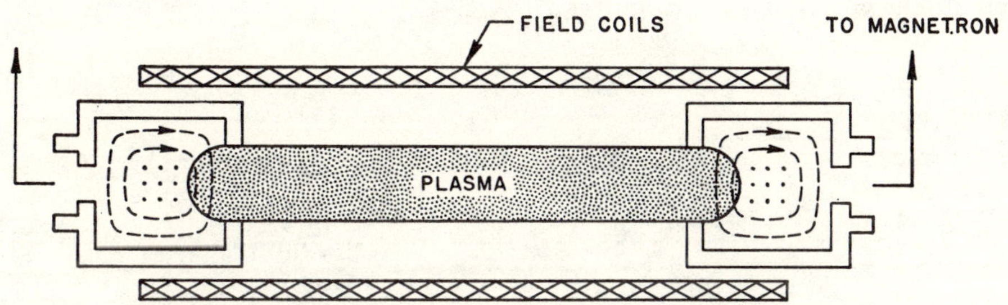

Fig. 11.17. Radiofrequency plugs for plasma confinement.

11.85. The behavior of the plasma under the influence of the radiofrequency field was studied by observing the change in the electron density, using microwave techniques (§6.16 *et seq.*). Evidence was obtained that the high-frequency electromagnetic field forced the plasma out of the ends of the tube located within the resonant cavities. It appears that, under the experimental conditions, the density of the plasma in the tube near to, but outside, the resonator corresponded to roughly 10^{13} electrons/cm³, whereas within the resonator region it was much less than 10^{10} electrons/cm³. The total kinetic pressure equivalent to the radiofrequency magnetic field B is equal to $B^2/8\pi$, and since the maximum value of B was 60 gauss, the pressure is about 140 dynes/cm². The maximum pressure of electrons within the plasma which can be supported was thus half of this amount. The electron temperature in the discharge was estimated to be approximately 5 ev, so that the density corresponding to a kinetic pressure of 70 dynes/cm² is close to 10^{13} electrons/cm³, in good agreement with the observed value.

11.86. Some other work designed to test the ability of a radiofrequency field to confine charged particles has been done, for simplicity, with an electron beam rather than with a neutral plasma [56]. The apparatus consists of a cylindrical wave guide 20 cm long and 4 cm in diameter, terminated at both ends by resonators of slightly larger diameter than the main section

(Fig. 11.18). It was designed so that the fields approximated that of the TE_{01} made of a circular wave guide at cut-off. Microwave radiation, at a frequency of about 9.3 kilomegacycles/sec, is fed into one of the end resonators from a magnetron which delivers 250 kev of power in 2 microsec pulses at a rate of 400 per sec. An electron beam of 0.3 milliamp is injected into the other resonator at 400 volts; at this low accelerating voltage the electrons interact with the field for some 170 radiofrequency periods. When no radiofrequency field is applied, the electron beam is found to spread out rapidly owing to Coulomb repulsion. But during the microwave pulses the beam is focused and about 90 per cent of the electrons are collected by an axial probe of 0.3-cm diameter located at the other end of the wave guide. It is clear, therefore, that the electrons are nearly all confined to the axis of the cylindrical cavity by the radiofrequency field.

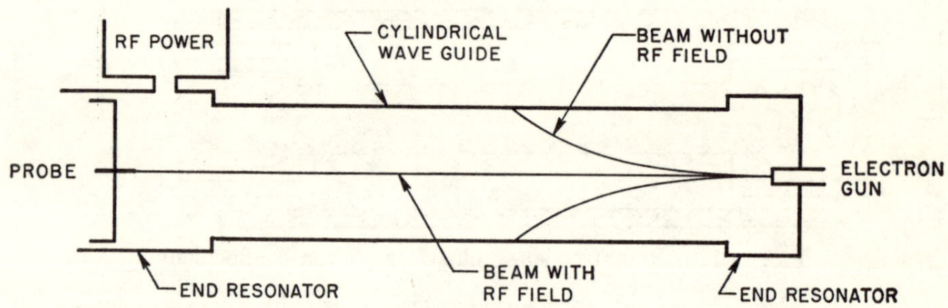

Fig. 11.18. Confinement of electron beam by radiofrequency field.

11.87. A design for an experimental facility to study plasma formation and confinement by radiofrequency fields makes use of a spherical cavity, 30 cm in radius, of copper-lined steel (Fig. 11.19) [46]. Since it is difficult to generate large amounts of radiofrequency power at frequencies greater than 1000 megacycles/sec, as mentioned earlier, an operating frequency of 800 megacycles/sec was chosen. For an assumed plasma density of 2×10^{12} electrons (or ions)/cm³ and a temperature of 1 kev, the power requirement, allowing for the skin effect dissipation, is estimated to be 500 kilowatts. It is proposed to supply this by means of eight beam power tetrodes operating as grid-driven oscillators; four of these would be used to drive the TE_{110} mode and the other four the TE_{111} mode with a time phase difference of 90° (§11.70). Pulsed operation is contemplated with a maximum pulse duration of 10 millisec.

11.88. Among the problems anticipated in the construction of the apparatus are those concerned with the band width and the synchronization of the two sets of drivers. The empty cavity will resonate at a frequency of 715 megacycles/sec, whereas with a fully developed plasma core of radius one-third that of the spherical cavity, the resonant frequency is expected to be 800 mega-

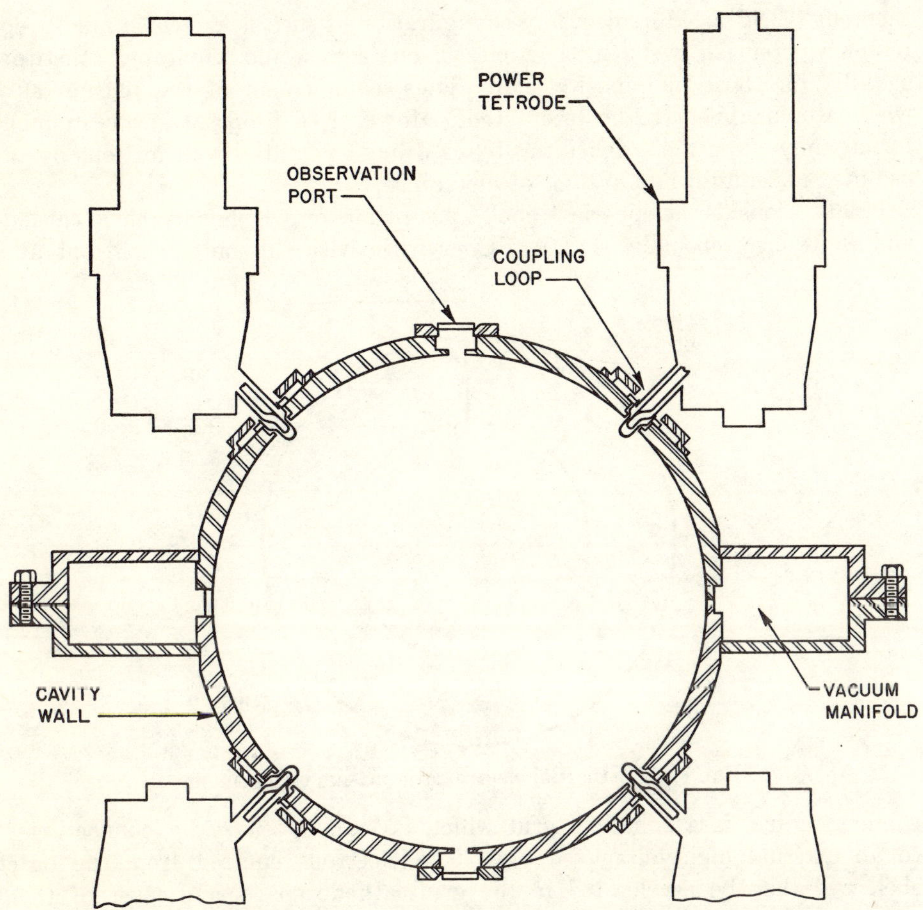

Fig. 11.19. Design of experimental facility for study of radiofrequency confinement.

cycles/sec. The radiofrequency oscillator circuit elements must be designed to follow this relatively large change in frequency. Moreover, in view of the mode coupling action of the plasma, it is uncertain whether it will be possible to maintain the desired relationship of equal stored energy at 90° phase difference in time between the two radiofrequency wave systems. If this proves to be difficult or impossible, it may be necessary to use three resonant modes of different frequencies, as described in §11.71.

INERTIAL-ELECTROSTATIC PLASMA CONFINEMENT

GENERAL PRINCIPLES

11.89. Schemes for equilibrium plasma confinement which involve only electrostatic fields may be expected to fail since they violate Earnshaw's theorem (§3.12). Moreover, any configuration of such fields which might represent a potential "well" for particles of one sign would inevitably constitute a "hill" for those of opposite sign. Thus, confinement of the plasma as a whole would appear to be prohibited. However, a proposal has been made for circumventing these difficulties by creating a potential well for ions by the use of a nonequilibrium system of electrons [57].

11.90. Consider a spherical shell, the inside of which acts as a cathode and emits electrons (Fig. 11.20). Concentric with this outer shell, but at a

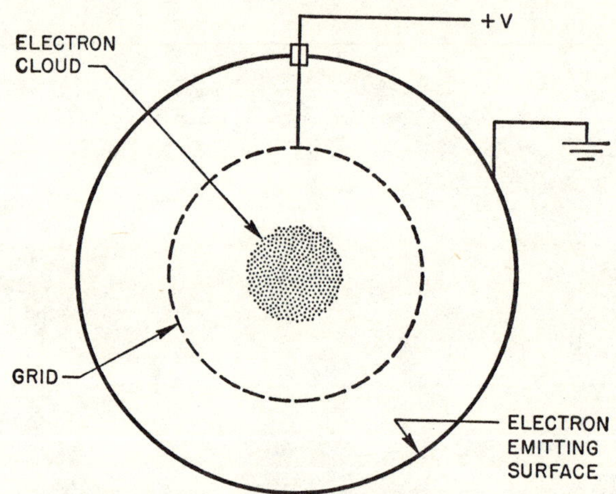

FIG. 11.20.   Inertial electrostatic plasma confinement.

smaller radius, is a spherical grid which is made positive by connecting it to an external high-voltage supply. The electrons emitted from the outer shell will then be accelerated in the grid-cathode space, and most of them will pass through the grid into the inner region around the center of the sphere. A space charge will consequently be built up in this central region, and incoming electrons will experience a strong repulsion. As a result they will be stopped and then turned back toward the grid. However, a single electron will, on the average, pass through the grid and oscillate between the cathode and central region several times before being intercepted and captured by the grid. Qualitatively, it is seen that a large potential hill for electrons develops in the central region, and this must represent a corresponding well for ions. Hence, if ions could be injected into the center of the sphere, e.g., by introducing neutral gas which will be ionized by the electrons, they would fall into the potential well and be trapped.

QUANTITATIVE TREATMENT

11.91. For the solution of the potential problem, five equations are used; two of these give the ion and electron currents, two express the requirement of energy conservation, and the final one is the Poisson equation. Suppose that, at a given radius r, all the electrons have a radial velocity v_e, half the total moving outward and the other half inward, whereas all the ions have the radial velocity v_i, again half moving outward and half inward. Let σ_e and σ_i represent the charge densities due to electrons and ions, respectively. Then, if it is assumed that the particles have only radial motion, the respective currents are given by

$$I_e = \tfrac{1}{2}(4\pi r^2 v_e \sigma_e) \quad \text{and} \quad I_i = \tfrac{1}{2}(4\pi r^2 v_i \sigma_i).$$

Poisson's equation for a spherically symmetrical system is

$$\frac{1}{r^2} \cdot \frac{\partial}{\partial r}\left(r^2 \frac{\partial V}{\partial r}\right) = -4\pi(\sigma_i - \sigma_e),$$

where V is the potential between the grid and the cathode, and the conservation of energy conditions are expressed by

$$\tfrac{1}{2}m_e v_e^2 - eV = \epsilon_e \quad \text{and} \quad \tfrac{1}{2}m_i v_i^2 + eV = \epsilon_i,$$

where m and ϵ represent the mass and total energy, respectively, of the indicated particles, and e is the electronic charge.

11.92.   By solving this set of equations, it has been found that, for a single energy ϵ_i of the ions, as well as ϵ_e for the electrons, the ions cannot be trapped, but if there is a distribution of ion energies, trapping is possible. Detailed calculations, assuming a Maxwellian distribution of the ion energies, make it clear, however, that densities sufficiently high for a practical thermonuclear reactor probably cannot be attained. It has also been found that the system becomes unstable at ion densities which would be required for such a reactor. Nevertheless, a system of the kind being considered might be useful for the production and confinement of hot plasmas for experimental purposes.

REFERENCES FOR CHAPTER 11

1. E. M. Little and D. B. Thomson, USAEC Report LA-2202 (1958).
2. A. C. Kolb, *Proc. Second U.N. Conf. on Peaceful Uses of Atomic Energy*, **31**, 328 (1958).
3. S. A. Colgate and R. E. Wright, *Proc. Second U.N. Conf. on Peaceful Uses of Atomic Energy*, **32**, 145 (1958).
4. W. C. Elmore, E. M. Little, and W. E. Quinn, *Proc. Second U.N. Conf. on Peaceful Uses of Atomic Energy*, **32**, 337 (1958).
5. A. C. Kolb, *Fourth Int. Conf. on Ionization Phenomena in Gases*, p. 1021 (1959).
6. V. H. Blackman and G. B. F. Niblett, in R. K. M. Landshof, Ed., *The Plasma in a Magnetic Field*, Stanford University Press, 1958, p. 87.
7. G. B. F. Niblett, *Proc. Inst. Elec. Eng.* (*London*), **106A**, Suppl. No. 2, 152 (1959).
8. A. C. Kolb and W. R. Faust, U.S. Naval Research Laboratory Report 5391 (1959).
9. L. M. Goldman, H. C. Pollock, J. Rouvina, and W. F. Westendorp, Division of Plasma Physics APS, Monterey Conf., B 5 (1959).
10. J. K. Wright, W. T. Eeles, and J. D. Herbert, *Nature*, 183, 1665 (1959).

11. H. G. Loos, *Phys. Rev. Letters,* **2,** 283 (1959).
12. H. L. Jordan, H. Kever, and K. Schindler, *Nucl. Instr.,* **4,** 322 (1958).
13. S. A. Colgate, *Nucleonics,* **17,** No. 10, 82 (1959); see *Fourth Int. Conf. on Ionization Phenomena in Gases,* pp. 1021-1080 (1959).
14. K. Boyer, E. M. Little, W. E. Quinn, G. A. Sawyer, and T. F. Stratton, *Phys. Rev. Letters,* **2,** 279 (1959).
15. F. C. Jahoda, E. M. Little, W. E. Quinn, G. A. Sawyer, and T. F. Stratton, *Phys. Rev.,* **119** (1960).
16. D. L. Nagle, W. E. Quinn, W. B. Riesenfeld, and W. Leland, *Phys. Rev. Letters,* **3,** 318 (1959).
17. D. L. Nagle, W. E. Quinn, F. L. Ribe, and W. B. Riesenfeld, *Phys. Rev.,* **119** (1960).
18. A. C. Kolb, C. B. Dobbie, and H. R. Griem, *Phys. Rev. Letters,* **3,** 5 (1959).
19. H. R. Griem, A. C. Kolb, and W. R. Faust, *Phys. Rev. Letters,* **2,** 281 (1959).
20. J. L. Tuck, USAEC Report WASH-184 (1955), p. 77.
21. H. Grad, USAEC Reports WASH-289 (1955), p. 115; NYO-7969 (1957).
22. J. Berkowitz, K. O. Friedrichs, H. Goertzel, H. Grad, J. Killeen, and E. Rubin, *Proc. Second U.N. Conf. on Peaceful Uses of Atomic Energy,* **31,** 171 (1958).
23. B. B. Kadmotsev and S. I. Braginsky, *Proc. Second U.N. Conf. on Peaceful Uses of Atomic Energy,* **32,** 233 (1958).
24. G. Persico and J. G. Linhart, *Nuovo cimento,* **8,** 740 (1958).
25. C. L. Longmire, USAEC Report WASH-289 (1955), p. 11.
26. J. L. Tuck, USAEC Report WASH-289 (1955), p. 7.
27. C. M. Braams and H. C. Brinkman, unpublished.
28. C. L. Longmire and J. L. Tuck, unpublished.
29. J. L. Tuck, *Inatom Review* (1960).
30. J. L. Tuck, *Phys. Rev. Letters,* **3,** 313 (1959).
30a. H. Grad, *Phys. Rev. Letters,* **4,** 222 (1960).
31. G. Schmidt, S. Koslov, D. Finkelstein, and K. Rogers, *Phys. Fluids,* **3** (1960).
32. J. M. Wilcox, USAEC Report UCRL-8584 (1959); *Rev. Mod. Phys.,* **31,** 1045 (1959).
33. K. Boyer, J. E. Hammel, C. L. Longmire, D. Nagle, F. L. Ribe, and W. B. Riesenfeld, *Proc. Second U.N. Conf. on Peaceful Uses of Atomic Energy,* **31,** 319 (1958).
34. O. A. Anderson, W. R. Baker, A. Bratenahl, H. P. Furth, J. Ise, W. B. Kunkel, and J. M. Stone, *Proc. Second U.N. Conf. on Peaceful Uses of Atomic Energy,* **32,** 155 (1958); USAEC Report UCRL-8062 (1958).
35. O. A. Anderson et al., *J. Appl. Phys.,* **30,** 188 (1959).
36. J. Wilcox, J. D. Gow, and L. Smith, USAEC Report UCRL-8579 (1958).
37. E. E. Yushmanov, *Plasma Physics and the Problem of Controlled Thermonuclear Reactions,* Pergamon Press, Inc., Vol. IV, 1960.
38. M. S. Yoffe, quoted in reference 32.
39. M. L. Good, USAEC Report WASH-146 (1953).
40. F. B. Knox, *Australian J. Phys.,* **10,** 221, 565 (1957).
41. E. S. Weibel, Ramo-Wooldridge Corp. Report ARL-57-1026 (1957); *J. Electronics and Control,* **5,** 435 (1958); *Phys. Rev.,* **114,** 18 (1959).
42. E. S. Weibel, in R. K. M. Landshoff, Ed., *The Plasma in a Magnetic Field,* Stanford University Press, 1958, p. 60.
43. H. A. H. Boot, S. A. Self, and R. B. R. Shersby-Harvie, *J. Electronics and Control,* **4,** 434 (1958); S. A. Self, *Phys. Fluids,* **3,** 488 (1960).
44. M. U. Clauser and E. S. Weibel, *Proc. Second U.N. Conf. on Peaceful Uses of Atomic Energy,* **32,** 161 (1958).

45. A. A. Vedenov et al., *Proc. Second U.N. Conf. on Peaceful Uses of Atomic Energy*, **32**, 239 (1958).

46. J. W. Butler, A. J. Hatch, and A. T. Ulrich, *Proc. Second U.N. Conf. on Peaceful Uses of Atomic Energy*, **32**, 324 (1958).

47. R. F. Wuerker, H. Shelton, and R. V. Langmuir, *J. Appl. Phys.*, **30**, 242 (1959).

48. D. G. Dow and R. C. Knechtli, *J. Electronics and Control*, 7, 316 (1959).

49. C. M. Braams, W. J. Schwader, and J. C. Terlouw, *Nucl. Instr.*, **4**, 327 (1959).

50. V. Cushing and M. S. Sodha, *Phys. Fluids*, **2**, 494 (1959), **3**, 142 (1960).

51. J. W. Butler, *Fourth Int. Conf. on Ionization Phenomena in Gases*, p. 620 (1959).

52. R. T. P. Whipple, UKAEA Report AERE-R-2917 (1959).

53. J. Berkowitz, H. Grad, and H. Rubin, *Proc. Second U.N. Conf. on Peaceful Uses of Atomic Energy*, **31**, 177 (1958).

54. R. J. Tayler, USAEC Report TID-7536, Part 2 (1957), p. 202.

55. E. S. Weibel and G. L. Clark, Space Technology Laboratories Report TR-59-0000-00838 (1959).

57. W. C. Elmore, J. L. Tuck, and K. M. Watson, *Phys. Fluids*, **2**, 239 (1959).

Chapter 12

ENERGY LOSSES AND SCALING LAWS

SOURCES OF ENERGY LOSS

DIRECT AND INDIRECT LOSSES

12.1.   Energy losses from a controlled thermonuclear reaction system may be divided into two broad categories—direct losses and indirect losses.   Among direct losses are those due to plasma instabilities, radiation, charge exchange, thermal conduction, and diffusion.   These are all energy losses that take place directly from the plasma itself.   The indirect losses are those arising from external circumstances, such as the losses resulting from joule (or ohmic) heating in the magnetic field coils and from the conversion of the thermonuclear reaction energy into useful power.

12.2.   Unless the problems of plasma instability are largely overcome, a practical thermonuclear reactor would appear to be out of the question.   It is evident that considerable effort will have to be made in order to find satisfactory solutions to these problems.   Since various proposals for stabilizing plasmas in different situations have been mentioned in earlier sections of this book, and the general subject of plasma stability and instability will be treated from the theoretical standpoint in Chapter 13, the matter will not be considered here.

12.3.   There are certain inevitable radiation losses from a high-temperature hydrogen isotope plasma; as seen in Chapter 2, these arise from bremsstrahlung, i.e., free-free electronic transitions, and from cyclotron radiation emitted by the electrons in a magnetic field.   Although the cyclotron radiation would not be seriously affected by impurities, except insofar as they increase the electron density, the bremsstrahlung radiation could be very greatly increased. The rate of emission of this radiation increases with the electron temperature but below about 500 ev, which is the range for the majority of laboratory plasmas, another source of radiation loss becomes much more significant. This consists of the excitation (line) radiation that accompanies bound-bound electronic transitions and the recombination (continuum) radiation from free-bound transitions in incompletely stripped atoms of impurity elements.   These

450

radiations become less and less important, in comparison with bremsstrahlung, as the temperature of the plasma increases, but they nevertheless play a role that cannot be overlooked in experimental studies. This source of energy loss will therefore be considered in this chapter.

12.4. Charge exchange (or charge transfer) can represent a serious loss of energy from a plasma, especially in the early stages of its production when the temperature is low and neutral particles are still present. If charge exchange occurs in a collision between a high-energy (or hot) ion and a low-energy (or cold) neutral particle, a cold ion and a hot neutral are formed. The neutral particle is not confined by the magnetic field and so it can escape to the walls and there give up its energy. Since charge-exchange cross sections generally decrease rapidly with increasing energy of the ions, especially at high energies, the losses due to charge exchange become less significant with increasing plasma temperature. Because of the importance of charge transfer reactions in connection with the processes occurring in ionized gases in general, the matter has been the subject of much theoretical and experimental study. Some of the results obtained will be described later (§12.61 *et seq.*).

12.5. Most of the present chapter will be devoted to a consideration of diffusion and heat conduction, since these are direct sources of energy loss that can perhaps be reduced to some extent by suitable adjustment of the operating conditions. Subsequently, reference will be made to the manner in which other losses, both direct and indirect, can be decreased, e.g., by a choice of temperature, magnetic field strength, and dimensions of the thermonuclear reaction system.

DIFFUSION OF PLASMA IN A MAGNETIC FIELD

INTRODUCTION

12.6. In a confined plasma there will generally be a density (or pressure) gradient, from the interior to the outside, which will cause the charged particles to diffuse and eventually escape to the walls of the containing vessel. This represents a source of energy loss to a thermonuclear reacting system. If there were no collisions among the particles in a plasma confined by a magnetic field, then, provided there are no electric fields, the guiding center of each particle would remain fixed on a given field line and diffusion across the magnetic field would not take place. However, as a result of collisions, the guiding centers can move from one field line to another in a random manner, so that a net diffusion in a direction transverse to the magnetic field becomes possible if a density gradient exists. The rate of diffusion of charged particles across the lines of force is dependent on the strength of the magnetic field. But since there is no restraint upon the particle motion in the direction parallel to the field lines, the diffusion rate in this direction is essentially the same as in the complete absence of the magnetic field.

12.7. The discussion given below of the diffusion of a plasma in a magnetic field falls into two main parts: one deals with weakly ionized plasmas and the other with plasmas that are highly (or fully) ionized. First, the behavior of the weakly ionized plasma will be considered in the absence of a magnetic field, and then the effect of a magnetic field will be derived. In a later section, the diffusion of a fully ionized plasma will be considered.

WEAKLY IONIZED PLASMA: NO MAGNETIC FIELD

12.8. In a weekly ionized plasma the great majority of the collisions responsible for diffusion of the plasma are those between a neutral particle, on the one hand, and either an ion or an electron, on the other hand. Because of the relatively small density of charged particles, electron-electron and ion-ion collisions occur much less frequently and make a negligible contribution to the plasma diffusion. Hence, no electrical forces are involved in the collisions and the plasma may be treated as an ordinary gas by familiar kinetic theory methods. Several procedures can be used for relating the diffusion coefficients of the charged particles to other properties [1]. The one adopted here was chosen because it is simple, even if not completely rigorous, and can be readily modified to allow for the effect of a magnetic field [2].

12.9. Suppose that the charged particles in the plasma are continually colliding with a lattice of fixed collision centers. It will be assumed that the charged particles are monoenergetic and that the density varies only in the

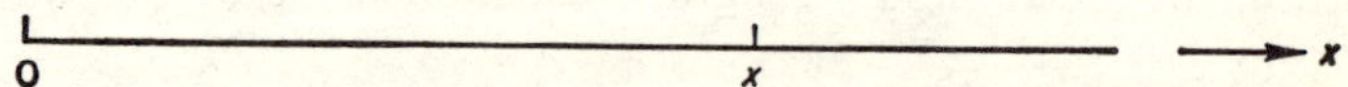

FIG. 12.1.   Motion of particle along x axis.

x direction. Consider a point x on the x axis (Fig. 12.1); the diffusion collision rate per unit volume at this point is given by

$$\text{Collisions per cm}^3 \text{ per sec} = n(x)\nu,$$

where $n(x)$ is the particle density at x and ν is the frequency of collisions of a single particle. Let v be the particle velocity and λ the mean free path for collisions; then ν is equal to v/λ, so that

$$\text{Collisions per cm}^3 \text{ per sec} = \frac{v}{\lambda} n(x).$$

12.10. If the velocity distribution of the charged particles after collision with the fixed centers is isotropic, then on the average one-third of the particles will move in the direction of the x axis following collisions. Of these, one half, on the average, will travel to the left and the other half to the right. Thus, only one-sixth of the charged particles colliding with another particle at x will move along the x axis toward the point 0 in Fig. 12.1. This same result applies irrespective of whether x is to the right or to the left of the origin at 0. Hence, the rate at which charged particles move toward 0 after collision will be

$$\text{Collisions per cm}^3 \text{ per sec toward } 0 = \frac{v}{6\lambda}\, n(x).$$

12.11. The probability that a particle colliding at x will reach 0 without undergoing a further collision is $e^{-x/\lambda}$, where x here represents the absolute distance, i.e., irrespective of sign, of the point x from the origin 0. It follows, therefore, that the charged particle current dI at 0, i.e., number of particles per cm² per sec passing in the x direction through a small area normal to the x axis, due to collisions occurring between x and $x + dx$ is given by

$$dI = \frac{v}{6\lambda}\, n(x)e^{-x/\lambda}\, dx. \tag{12.1}$$

12.12. The net current at 0 is then obtained by integrating over all values of x from minus infinity to zero in one direction and from zero to infinity in the other direction and subtracting the integrals; thus,

$$I = \frac{v}{6\lambda}\left[\int_{-\infty}^{0} n(x)e^{-x/\lambda}\, dx - \int_{0}^{\infty} n(x)e^{-x/\lambda}\, dx\right]. \tag{12.2}$$

If $n(x)$ is expanded in a Maclaurin series and all terms beyond the second are neglected, then

$$n(x) \approx n(0) + x\,\frac{dn}{dx},$$

and substitution into equation (12.2) gives

$$I = \frac{v}{6\lambda}\cdot\frac{dn}{dx}\left[\int_{-\infty}^{0} xe^{-x/\lambda}\, dx - \int_{0}^{\infty} xe^{-x/\lambda}\, dx\right]$$

$$= -\frac{\lambda v}{3}\cdot\frac{dn}{dx}.$$

The diffusion coefficient D is defined quite generally by Fick's law, i.e.,

$$I = -D\,\frac{dn}{dx}, \tag{12.3}$$

so that

$$D = \tfrac{1}{3}\lambda v. \tag{12.4}$$

12.13. Since the sign of the charged particle has not been specified in the foregoing derivation, equation (12.4) is applicable to both ions and electrons; hence, the individual particle diffusion coefficients may be represented by

$$D_i = \tfrac{1}{3}\lambda_i v_i \quad \text{and} \quad D_e = \tfrac{1}{3}\lambda_e v_e, \tag{12.5}$$

where λ_i and λ_e are the mean free paths for ion-neutral and electron-neutral collisions, and v_i and v_e are the ion and electron velocities, respectively. As a general rule, the two mean free paths will not be very different, but at the same (or approximately the same) temperature, the average velocity of the electrons will be much higher than that of the ions. Hence, in a given plasma the diffusion coefficient of the electrons will be greater than that of the ions.

12.14. The electron and ion density gradients in the bulk of the plasma must be identical, because of space-charge considerations. It follows, therefore, that the electrons will tend to diffuse out of a plasma, in the absence of a magnetic field, at a faster rate than the ions. As a result of the consequent separation of charges, however, a space-charge electric field will be established which has the effect of retarding the electrons and accelerating the ions, so that the (ambipolar) diffusion losses are equalized. The situation will be treated more fully below (§12.21), but consideration will first be given to the individual diffusion behavior of charged particles in a magnetic field.

WEAKLY IONIZED PLASMA IN A MAGNETIC FIELD

12.15. In a magnetic field, the charged particles will move in circular paths of radius r_g, equal to the gyromagnetic radius, which has a definite value for a particle of velocity v and mass m moving in a magnetic field B. If, after collision with another particle, a charged particle is to pass in the x direction through the small area, at the origin, normal to the x axis, it must travel along

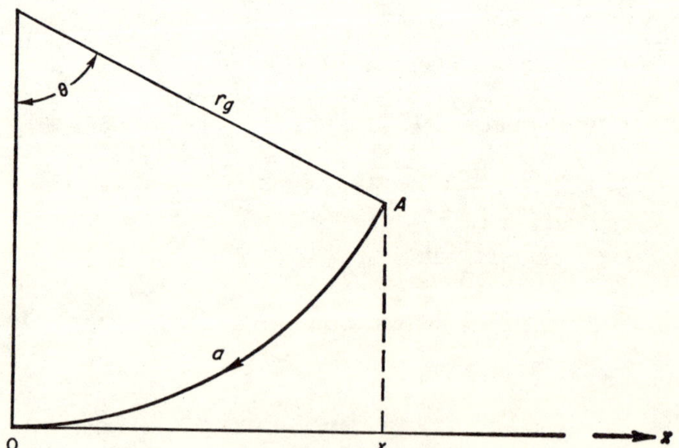

FIG. 12.2. Charged particle in a magnetic field.

the arc of a circle of radius r_g, e.g., from the point A to 0 in Fig. 12.2. Let a be the length of this arc; then it can be shown by arguments similar to those used in deriving equation (12.2) that

$$I = \frac{v}{6\lambda}\left[\int_{-\infty}^{0} n(a)e^{-a/\lambda}\,da - \int_{0}^{\infty} n(a)e^{-a/\lambda}\,da\right]. \tag{12.6}$$

12.16.   From Fig. 12.2 it is apparent that

$$a = r_g\theta$$

and

$$x = r_g \sin\theta,$$

where x is the x coordinate of the point A.   Further, if it is postulated, as above, that the particle density varies only in the x direction, then

$$n(a) \approx n(0) + x\,\frac{dn}{dx}.$$

Upon making the appropriate substitutions and recalling that r_g is constant for a given charged particle, equation (12.6) reduces to

$$I = -\frac{vr_g^2}{3\lambda}\cdot\frac{dn}{dx}\int_0^\infty \sin\theta\, e^{-r_g\theta/\lambda}\,d\theta$$

$$= -\frac{vr_g^2}{3\lambda}\cdot\frac{dn}{dx}\cdot\frac{1}{1+r_g^2/\lambda^2}$$

$$= -\frac{\lambda v}{3}\cdot\frac{1}{1+\lambda^2/r_g^2}\cdot\frac{dn}{dx}.$$

The diffusion coefficient D_{mag} in the magnetic field is thus

$$D_{\text{mag}} = \frac{\lambda v}{3}\cdot\frac{1}{1+\lambda^2/r_g^2}$$

$$= \frac{D}{1+\lambda^2/r_g^2}, \qquad (12.7)$$

where D is the diffusion coefficient in the absence of a magnetic field.

12.17.   If the magnetic field direction is perpendicular to the x axis, so that v is the velocity of the charged particle normal to the field, the gyromagnetic frequency ω is equal to v/r_g (cf. §4.12); that is,

$$\omega = \frac{v}{r_g}. \qquad (12.8)$$

Further, the average time τ between collisions is given by

$$\tau = \frac{\lambda}{v}. \qquad (12.9)$$

Hence, equation (12.7) may be written in the form*

$$D_\perp = \frac{D}{1+\omega^2\tau^2}, \qquad (12.10)$$

---

* It will be understood in this and the subsequent treatments that symbols such as D, D_i, D_e, or D_{amb}, without the subscript $\perp$ or $\parallel$, refer to diffusion coefficients in the absence of a magnetic field.

where $D_\perp$ is the diffusion coefficient of either ions or electrons, respectively, in the direction perpendicular to the magnetic field lines [2].

12.18. It is seen from equations (12.8) and (12.9) that the product $\omega\tau$ is equal to λ/r_g, and since $2\pi r_g$ is the circumference of the circle of gyration in the magnetic field, $\omega\tau$ is equal to 2π times the number of gyrations an electron or an ion, respectively, makes before it collides with a neutral particle. In most cases of interest, the charged particle density is relatively low and the magnetic field strength is large, so that the particles perform many gyrations between collisions; in other words, the radius of gyration is very much less than the collision mean free path. Hence, $\omega\tau$ is large in comparison with unity and then equation (12.10) reduces to

$$D_\perp \approx \frac{D}{\omega^2\tau^2}. \tag{12.11}$$

12.19. The influence of magnetic field strength and temperature on the diffusion coefficient may be determined by replacing D by $\tfrac{1}{3}\lambda v$, the frequency ω by its equivalent given by equation (4.3), and τ by λ/v; the result can be written as

$$D_\perp = \text{constant} \times \frac{m^2 v^3}{\lambda B^2}.$$

Assuming a Maxwellian energy distribution, so that the kinetic energy $\tfrac{1}{2}mv^2$ is proportional to the absolute temperature T, and if the mean free path, like that for collision between two neutral particles, varies directly with T, it is seen that

$$D_\perp = \text{constant} \times \frac{m^{1/2} T^{1/2}}{B^2}. \tag{12.12}$$

The diffusion coefficient of a charged particle in the direction perpendicular to the magnetic field may thus be expected to vary directly as the square root of the temperature and inversely as the square of the magnetic field strength.

12.20. Attention may be called to an interesting difference between the result just derived and the behavior in the absence of a magnetic field or for diffusion parallel to the field direction. For the latter cases, equation (12.5) is applicable and, as already seen, there is a tendency for the electrons to diffuse more rapidly than the ions. However, for diffusion perpendicular to the field lines, equation (12.12) shows that the diffusion coefficient increases with the particle mass. The ions will thus tend to diffuse more rapidly than the electrons across the magnetic field lines but more slowly in the direction of the field.

AMBIPOLAR DIFFUSION

12.21. Whenever charged particles of one sign in a plasma tend to diffuse at a different rate from that of the particles of the other sign, a space-charge electric field is established which acts in such a manner as to equalize the diffusion rates. The resultant coefficient of diffusion of the particles of both signs, i.e., of the plasma as a whole, is called the *ambipolar diffusion coefficient.* Its relationship to the single-particle diffusion coefficients considered above may be derived by introducing the effect of an electric field on the motion of the charged particles [2].

12.22. If, under the influence of an electric field E, a charged particle moves in the field direction with a velocity v, then the *mobility* μ of the charged particle is defined by

$$v = \mu E. \tag{12.13}$$

The force Ee exerted by the field on a particle of charge e may be set equal to the rate of change of momentum between collisions; the time between collisions is λ/v, and consequently

$$Ee = mv \times \frac{v}{\lambda} = \frac{mv^2}{\lambda}. \tag{12.14}$$

If E is eliminated from equations (12.13) and (12.14), it is found that

$$\mu = \frac{e\lambda}{mv}.$$

By equation (12.4), λ may be replaced by $3D/v$ and, assuming a Maxwellian distribution, the particle kinetic energy is equal to $\tfrac{3}{2}kT$; hence,

$$\mu = \frac{eD}{kT}. \tag{12.15}$$

12.23. In expressing the charged-particle current, a term which allows for the effect of the electric field must be added to the general diffusion equation (12.3). For diffusion in the x direction, the ion current I_i in the absence of a magnetic field is now given by

$$I_i = -D_i \frac{dn_i}{dx} + \mu_i E n_i, \tag{12.16}$$

assuming that the electric field acts in the x direction only. If v_i is the ion velocity in this direction, then I_i is also equal to $n_i v_i$ and upon introducing this result into equation (12.16) it is seen that

$$v_i = -\frac{D_i}{n_i} \cdot \frac{dn_i}{dx} + \mu_i E. \tag{12.17}$$

For the negatively charged electrons, on the other hand, the corresponding expression is

$$v_e = -\frac{D_e}{n_e} \cdot \frac{dn_e}{dx} - \mu_e E, \tag{12.18}$$

with a negative sign preceding the last term because the electric field affects the velocity of the electrons in the opposite sense to that of the ions. The effect of the space-charge electric field on diffusion is to equalize the ion and electron velocities, so that the common (ambipolar) velocity v is

$$v = - \frac{D_i}{n_i} \cdot \frac{dn_i}{dx} + \mu_i E = - \frac{D_e}{n_e} \cdot \frac{dn_e}{dx} - \mu_e E. \tag{12.19}$$

12.24. Since the plasma may be supposed to remain electrically neutral and the charged particle gradients are approximately equal, it is possible to write

$$n_i \approx n_e = n \tag{12.20}$$

and

$$\frac{dn_i}{dx} \approx \frac{dn_e}{dx} = \frac{dn}{dx}, \tag{12.21}$$

where n refers to the individual particle density of the plasma, regardless of sign. Hence, equation (12.19) may be written as

$$v = - \frac{D_i}{n} \cdot \frac{dn}{dx} + \mu_i E = - \frac{D_e}{n} \cdot \frac{dn}{dx} - \mu_e E.$$

Upon eliminating E, the result is

$$I = nv = - \left(\frac{\mu_e D_i + \mu_i D_e}{\mu_i + \mu_e} \right) \frac{dn}{dx}.$$

In the absence of a magnetic field, the ambipolar diffusion coefficient, which is the effective diffusion coefficient of both ions and electrons, is thus seen to be

$$D_{\text{amb}} = \frac{\mu_e D_i + \mu_i D_e}{\mu_i + \mu_e}. \tag{12.22}$$

12.25. If the expression for μ given by equation (12.15) is used in equation (12.22), it is found, assuming the electron and ion temperatures to be approximately equal, that

$$D_{\text{amb}} \approx \frac{2 D_i D_e}{D_i + D_e}. \tag{12.23}$$

As noted earlier, D_e is much greater than D_i, in the absence of a magnetic field, and so

$$D_{\text{amb}} \approx 2 D_i. \tag{12.24}$$

The ambipolar (or effective) diffusion coefficient of both ions and electrons in the weakly ionized plasma is thus twice that calculated for the ions, i.e., for the slower-moving particles, when the effect of the space-charge electric field is not taken into consideration.

12.26. In the presence of a magnetic field, it is found, by a method similar to that used in §12.15, that the charged particle mobility $\mu_\perp$ normal to the field is

$$\mu_\perp = \frac{\mu}{1 + \omega^2 \tau^2}, \tag{12.25}$$

where μ is the mobility of the given particle in the absence of a field or in the direction parallel to the field lines. Thus, comparison with equation (12.10) indicates that the diffusion coefficient and mobility are changed in the same proportion by the magnetic field.

12.27. It can then be shown that equation (12.23) is still applicable with D_i and D_e, actually $D_{e\perp}$ and $D_{i\perp}$, defined by equation (12.10) or (12.11). However, in this case $D_{i\perp}$ greatly exceeds $D_{e\perp}$ as stated in §12.20, so that the ambipolar diffusion coefficient across the magnetic field lines is now

$$D_{\text{amb}\,\perp} \approx 2D_{e\perp}, \tag{12.26}$$

provided the ion and electron temperatures are not very different. Here again, the effective diffusion coefficient of both ions and electrons is twice the value calculated for the slower moving particles when the space-charge field is neglected. The ambipolar diffusion coefficient in the direction of the magnetic field is the same as in the absence of a field; hence, by equation (12.24),

$$D_{\text{amb}\,\|} \approx 2D_i, \tag{12.27}$$

the ions being the particles which move more slowly in this direction.

DIFFUSION IN MAGNETIC FIELD OF FINITE DIMENSIONS

12.28. The foregoing treatment has been based on the tacit assumption that the plasma has essentially infinite dimensions, so that the individual ion and electron currents balance out in each direction in order to maintain space-charge neutrality. However, in a field of finite dimensions, space is no longer isotropic, and all that is necessary is that the total ion and electron currents in all directions should balance one another. The diffusion coefficients in these circumstances may be quite different from those for an infinite plasma derived above.

12.29. As a simple illustration, consider a two-dimensional plasma, with finite dimensions in the x and y directions, enclosed by a vessel with conducting walls, as shown in Fig. 12.3. The magnetic field, as before, is taken to be parallel to the y axis, so that diffusion in the x direction is perpendicular to the field lines. Since the ions tend to diffuse more rapidly than the electrons in this direction, there will be a transient charge separation with an excess of negative charge as indicated at A and of positive charge at B. The electric field which immediately develops then acts so as to neutralize the space charge resulting from the separation. Normally, this would be achieved by the electric field, acting perpendicular to the magnetic field direction, accelerating the electrons from A to B while retarding the motion of the ions. However, if the distance from A to one of the conducting walls in the x direction is not large, neutralization of the space charge can be achieved preferentially by electrons traveling from A, along the magnetic field lines, to the end wall, along the conducting wall, and then to B, also along the field lines, as shown by the heavy arrows in Fig. 12.3. The resulting electron "short-circuit" current thus obviates the necessity for ambipolar diffusion and the observed diffusion rate across the field lines should be essentially that of the ions in the x direction [3, 4].

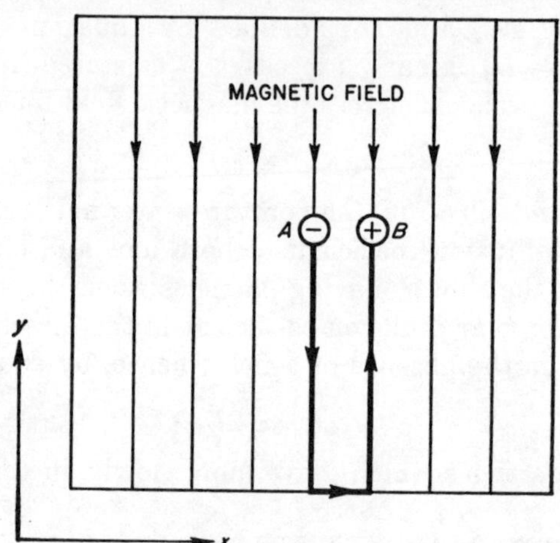

FIG. 12.3. Illustration of "short-circuit" effect.

12.30. The situation may be treated mathematically in the following manner. Consider, first, a general finite two-dimensional system, such as that represented in Fig. 12.3, where charge separation arising from different rates of diffusion of ions and electrons results in the development of electric fields E_x and E_y, in the x and in the y direction, respectively. A different approach is now required to evaluate the effective diffusion constants in a magnetic field, and use is made of the principle of particle conservation (Fick's second law). In the present case this takes the form

$$\frac{\partial n}{\partial t} = D_{\text{eff}\perp} \frac{\partial^2 n}{\partial x^2} + D_{\text{eff}\parallel} \frac{\partial^2 n}{\partial y^2} \tag{12.28}$$

or, using equation (12.3),

$$\frac{\partial n}{\partial t} = -\frac{\partial I_\perp}{\partial x} - \frac{\partial I_\parallel}{\partial y},$$

for both ions and electrons. Making the simplifications permitted by equations (12.20) and (12.21), it is found, based upon the arguments in §12.23, that for the ions

$$\frac{\partial n}{\partial t} = D_{i\perp} \frac{\partial^2 n}{\partial x^2} - \mu_{i\perp} \frac{\partial}{\partial x}(E_x n) + D_i \frac{\partial^2 n}{\partial y^2} - \mu_i \frac{\partial}{\partial y}(E_y n) \tag{12.29a}$$

and for the electrons

$$\frac{\partial n}{\partial t} = D_{e\perp} \frac{\partial^2 n}{\partial x^2} + \mu_{e\perp} \frac{\partial}{\partial x}(E_x n) + D_e \frac{\partial^2 n}{\partial y^2} + \mu_e \frac{\partial}{\partial y}(E_y n), \qquad (12.29b)$$

where, as in Fig. 12.3, the directions x and y are perpendicular and parallel, respectively, to the magnetic field lines. Since the values of $D_{\parallel}$ and $\mu_{\parallel}$ are identical with those in the absence of a magnetic field, the subscript $_{\parallel}$ has been omitted from these symbols in equations (12.29a) and (12.29b).

12.31. For a magnetic field of appreciable strength, it is seen from equation (12.25) that $\mu_{\perp}$ will be considerably less than μ, and since the spatial variations of E_x and E_y will generally not be very different, it follows that the second terms on the right of equations (12.29a) and (12.29b) can be neglected in comparison with the fourth terms. Upon omitting the second terms and setting the right sides of these equations equal to one another, it is readily found that

$$\frac{\partial n}{\partial t} = \frac{\mu_e D_{i\perp} + \mu_i D_{e\perp}}{\mu_i + \mu_e} \cdot \frac{\partial^2 n}{\partial x^2} + \frac{\mu_e D_i + \mu_i D_e}{\mu_i + \mu_e} \cdot \frac{\partial^2 n}{\partial y^2}. \qquad (12.30)$$

According to equation (12.28) the effective diffusion coefficient perpendicular to the field lines is equal to the coefficient of $\partial^2 n/\partial x^2$ in equation (12.30); hence,

$$D_{\text{eff}\perp} = \frac{\mu_e D_{i\perp} + \mu_i D_{e\perp}}{\mu_i + \mu_e}$$

$$\approx D_{i\perp}, \qquad (12.31)$$

since $D_{i\perp} \gg D_{e\perp}$ and $\mu_e \gg \mu_i$. It is seen, therefore, that in the case under consideration, i.e., when the short-circuit effect exists, the diffusion coefficient of the charged particles across the magnetic field is essentially equal to that of the ions alone in the absence of a space-charge field and short-circuit current. This is the same conclusion as was reached earlier from a general consideration of the physical situation.

12.32. It should be noted that, if the conditions are such that there is no short-circuit effect, the observed diffusion coefficient of the charged particles across the field lines would be given by equation (12.26) as $2D_{e\perp}$. Since $D_{i\perp}$ is much greater than $D_{e\perp}$ (or $2D_{e\perp}$), it is apparent that, as a consequence of the short-circuit effect, there is a marked increase in the rate of diffusion of the plasma particles in the direction perpendicular to the magnetic field.

12.33. The effective diffusion coefficient in the y direction, i.e., parallel to the field lines, is given by the coefficient of $\partial^2 n/\partial y^2$ in equation (12.30); thus,

$$D_{\text{eff}\parallel} = \frac{\mu_e D_i + \mu_i D_e}{\mu_i + \mu_e}. \qquad (12.32)$$

Comparison of this result with equation (12.22) shows that the diffusion coefficient $D_{\text{eff}\parallel}$ is identical with the ambipolar diffusion coefficient in the absence of a magnetic field. It is therefore possible to write

$$D_{\text{eff}\parallel} = D_{\text{amb}} \approx 2D_i. \tag{12.33}$$

The short-circuit effect thus has essentially no influence on the diffusion of the plasma particles in the direction parallel to the magnetic field lines.

12.34. In the design of chambers for containing plasmas to be confined by a magnetic field, for use in thermonuclear studies, the field direction invariably corresponds with that of the longer dimension of the chamber. Thus, the conditions are not those for which the short-circuit effect is to be expected. Moreover, under conditions of thermonuclear interest, the plasmas are fully ionized and the diffusion mechanism is different from that considered here for weakly ionized plasmas. Nevertheless, the treatment of the diffusion situation in which there is a short-circuit current of electrons has played an important role in the experimental establishment of the variation with field strength of the diffusion coefficient across the magnetic field lines. This aspect of the subject will be considered below.

EXPERIMENTAL DIFFUSION STUDIES

12.35. Prior to 1955 the only experimental information concerning the relationship between the diffusion coefficient and magnetic field strength was based on some measurements made during World War II studies of weakly ionized arc plasmas [5]. Two general conclusions, of considerable significance for controlled thermonuclear research, were drawn from this work. First, that the diffusion coefficient of the plasma across the field lines was much larger than that calculated from the collision theory given above, and second, that the diffusion coefficient varied inversely as the magnetic field strength rather than inversely as its square. In order to account for these discrepancies from the expected behavior, it was suggested that the principal mechanism of diffusion was not collisions but plasma oscillations. On the basis of this postulate, a theoretical treatment, the details of which have not been published, was reported to have led to an expression for the effective diffusion coefficient of the charged particles that can be written in the form

$$D_{\perp} \approx \frac{T_e}{16B} \times 10^{11} \quad \text{cm}^2/\text{sec}, \tag{12.34}$$

where T_e is the electron temperature in kilo-electron volts and B is the magnetic field strength in gauss [6].

12.36. The method used to determine the diffusion coefficient was to strike an arc along the axis of a graphite chamber of rectangular cross section. By means of a collimating slot, a narrow discharge was transmitted symmetrically through the chamber containing argon gas at a pressure of about 10^{-3} mm of mercury. The entire chamber was placed in a magnetic field, so that the lines of force were parallel to the discharge axis. The arc ionized the gas in its vicinity and the resulting plasma diffused outward across the field lines because of the particle density gradient in this direction. By means of cold, shielded probes of the Langmuir type (§6.82), the ion density was determined at various distances from the arc in the x direction, i.e., perpendicular to the field lines. The density was found to decrease exponentially, i.e.,

$$n(x) = n(0)e^{-x/q},$$

and a theoretical treatment showed that the $1/e$-folding distance q, i.e., the distance from the arc in which the density decreased by a factor of $1/e$, is given by

$$q \approx \left(\frac{lD_{e\perp}}{u}\right)^{\frac{1}{2}}, \tag{12.35}$$

where l is the length of the arc, u is the velocity of streaming of the electrons to the end walls, and $D_{e\perp}$ has the same significance as above. It appears that, to a good approximation, u may be set equal to the thermal velocity of the ions at the temperature of the plasma.

12.37. A typical value of q, in a magnetic field of about 4000 gauss, was found to be 0.3 cm in a plasma having an estimated temperature of 2 ev. The length of the arc was 12 cm and u was roughly 4×10^5 cm/sec, so that $D_{e\perp}$ was determined to be about 3×10^3 cm²/sec. The value calculated from collision theory was much smaller, namely 20 cm²/sec, although even this result, which depended on a choice of the collision mean free path of the electrons, was apparently too high. The correct calculated value of $D_{e\perp}$ is probably in the vicinity of 5 cm²/sec, so that it is smaller than the observed diffusion coefficient by a factor of about 600. On the other hand, if equation (12.34), presumably based on the concept of plasma oscillations as the cause of diffusion, is used, the diffusion coefficient, for a field strength of 4000 gauss, is found to be almost exactly equal to the observed value. It was as a result of this agreement that the general conclusions given at the beginning of the present section were reached.

12.38. There are two matters in connection with the foregoing work to which attention should be drawn. In the first place, the accepted variation of the diffusion coefficient as $1/B$, according to equation (12.34), was based on measurements at only one magnetic field strength, namely, about 4000 gauss. Further, it is probable that the short-circuit effect would have been operative under the conditions of the experiment. Consequently, the diffusion coefficient determined was not $D_{e\perp}$, but something much closer to $D_{i\perp}$. As seen from equation (12.12), at the same particle temperature, $D_{i\perp}/D_{e\perp}$ is equal to the square root of the ratio of the respective masses. Hence, in argon gas, $D_{i\perp}$ might be expected to be about 300 times $D_{e\perp}$. Taking the latter to be 5 cm²/sec, as given above, it appears that $D_{i\perp}$ should be roughly 1.5×10^3 cm²/sec, which is reasonably close to the observed diffusion coefficient across the field lines. In view of the uncertainties in the experimental determination and in the calculation of $D_{e\perp}$, the agreement is probably as good as could be expected.

12.39. The supposed dependence of the plasma diffusion coefficient perpendicular to the magnetic field on the reciprocal of the field strength, i.e., equation (12.34), rather than on its square, i.e., equation (12.12), was based on the apparent failure of collision theory to account for the observed diffusion coefficient. But, as just seen, if the short-circuit effect had been in operation in the experimental chamber, the discrepancy would not exist. It would then not be necessary to use equation (12.34) to account for the results and the diffusion coefficient would not necessarily vary as $1/B$. It was obviously desirable, therefore, to make plasma diffusion measurements over a range of magnetic field strengths in order to determine directly the manner in which the plasma diffusion coefficient perpendicular to the lines of force varies with the field strength.

12.40. Studies of this type have been made with an apparatus similar to that described above except that the graphite chamber was a cylinder 18 cm long and with a 5-cm radius (Fig. 12.4). The chamber was placed between

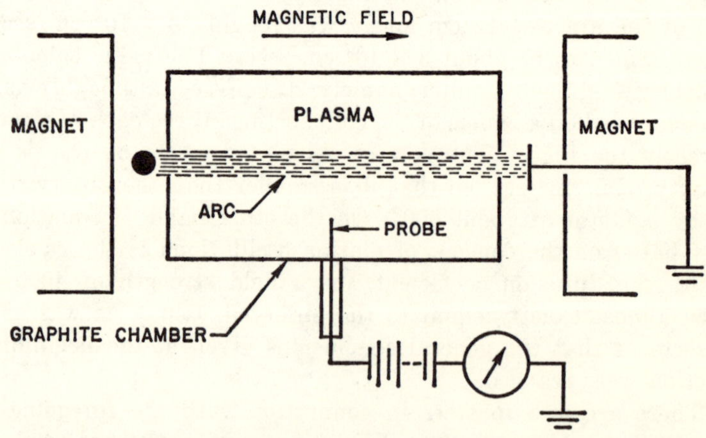

Fig. 12.4. Measurement of plasma diffusion in magnetic field.

the pole pieces of a magnet, so as to produce a magnetic field parallel to the axis of the chamber and to the direction of the arc. The field strength could be varied from 2500 to 14,000 gauss [7, 8]. The gas in the chamber was nitrogen at a pressure of 1.5×10^{-3} mm of mercury. The temperature in the plasma formed around the arc was estimated to be about 2 ev and the degree of ionization of the gas was probably less than 10 per cent. The plasma diffusion coefficient was determined in the direction perpendicular to the magnetic field from probe measurements in the manner described in §12.36. The result obtained, for conditions similar to those in the earlier experiments, was 2.7×10^3 cm²/sec, which is of the same order as that reported from the aforementioned work and in good agreement with the value to be expected if the short-circuit effect were operative.

12.41. According to collision theory, e.g., equation (12.12), the diffusion coefficient $D_\perp$ should be approximately proportional to $1/B^2$, irrespective of whether it applies to ions or electrons. Hence, by equation (12.35), the $1/e$-folding length, q, should vary as $1/B$. The results obtained over the range of magnetic field strengths mentioned above completely satisfy this requirement, as may be seen from Fig. 12.5, which is a plot of $1/q$ versus

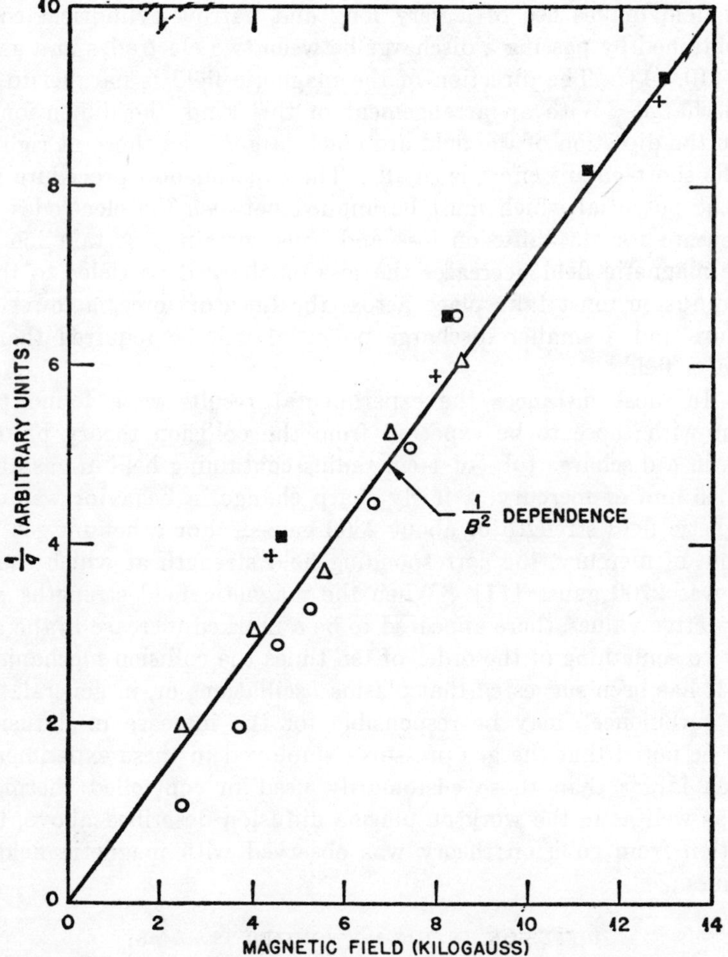

FIG. 12.5. Diffusion parameter and magnetic field strength.

B, obtained from observations with two different magnets and four different arc currents.

12.42. Consequently, whether the short-circuit effect exists or not, there is no reason to doubt the fact that the diffusion coefficient is dependent upon $1/B^2$, as required by collision theory. Hence, equation (12.34) cannot be correct, and there seems to be no reason to suppose that, under the conditions of the experiment, plasma oscillations have any significant effect on diffusion. It is of interest to note, too, that the rate of diffusion of electrons across a magnetic field, as determined from the spreading of a beam, has also been found to vary as $1/B^2$, so that the collision mechanism of diffusion is presumably applicable in this case also [9].

12.43. An indirect method of studying diffusion of charged particles in a magnetic field makes use of a very long and narrow cylindrical column of plasma obtained by passing a discharge between two electrodes in a gas at low pressure [10, 11]. The direction of the magnetic field is parallel to the axis of the discharge. With an arrangement of this kind, the dimensions of the plasma in the direction of the field are much larger than those at right angles, so that the short-circuit effect is small. The experimental procedure is to determine the potential which must be applied between the electrodes in order to compensate for the diffusion loss and thus sustain a certain ion density. Since the magnetic field decreases the loss of charged particles to the walls, because diffusion must take place across the lines of force, a lower electron temperature and a smaller discharge potential will be required than in the absence of a field.

12.44. In most instances the experimental results were found to be in agreement with those to be expected from the collision theory of diffusion. However, in a discharge tube of 1-cm radius containing helium gas at a pressure of 1.46 mm of mercury, a fairly sharp change in behavior was observed at a magnetic field strength of about 2300 gauss. For a helium gas pressure of 3.40 mm of mercury, the corresponding field strength at which the change occurred was 2700 gauss [11]. When the magnetic field strengths exceeded these respective values, there appeared to be a marked increase in the diffusion coefficient to something of the order of ten times the collision mechanism magnitude. It has been suggested that plasma oscillations or, in general, "electromagnetic turbulence" may be responsible for the increase in diffusion rate. It should be noted that the gas pressures employed in these experiments were very much larger than those customarily used in controlled thermonuclear research as well as in the work on plasma diffusion described above, in which no deviation from collision theory was observed with magnetic fields up to 14,000 gauss.

DIFFUSION IN HIGHLY IONIZED PLASMAS

12.45. In a highly ionized plasma, in which the proportion of neutral particles is negligible, diffusion will result from ion-electron, electron-electron, and ion-ion collisions. It might be expected at first sight that ion-ion collisions would be important because the exchange of momentum is greater in such collisions than it is in the other two types of charged particle collisions. However, theoretical treatment indicates that collisions between like-charged particles produces no diffusion in the first order, i.e., proportional to $- \, dn/dx$, because of a peculiar cancellation of terms. The first nonvanishing term for the charged particle current due to like-particle collisions is the one involving the third derivative of the particle density, i.e., $- \, d^3n/dx^3$. It appears from these results that, if the particle density does not vary greatly over a distance equal to the gyromagnetic radius, as is usually the case, diffusion due to collisions of like particles is negligible in comparison with that arising from ion-electron collisions (see also §12.50) [12, 13].

12.46. For the treatment of diffusion in a highly ionized plasma due to ion-electron interactions, use is made of two hydromagnetic equations of motion of the system [2, 14]. In Gaussian-cgs units, the first, derived in §3.16 as equation (3.12), is

$$\frac{1}{c} \, (\mathbf{j} \times \mathbf{B}) = \nabla p, \tag{12.36}$$

and the second, which has been called the generalized Ohm's law, is

$$\mathbf{E} + \frac{1}{c} \, (\mathbf{v} \times \mathbf{B}) = \eta \mathbf{j} + \frac{1}{en_e} \, \nabla p_i, \tag{12.37}$$

were $\mathbf{E}$ and $\mathbf{B}$ are the electric and magnetic field vectors, $\mathbf{j}$ is the current density vector, $\mathbf{v}$ is the macroscopic (or mass) velocity of the plasma, η is its specific resistance, and ∇p and ∇p_i are the gradients of the total particle pressure and the ion pressure, respectively; c is the velocity of light, e the electronic charge, and n_e the electron particle density. In the present situation, it will be postulated that the electric field can be neglected; furthermore, since the velocity of the ions is much less than that of the electrons within the moving plasma, the second term on the right of equation (12.37) is small, so that this equation reduces to

$$\frac{1}{c} \, (\mathbf{v} \times \mathbf{B}) \approx \eta \mathbf{j}. \tag{12.38}$$

12.47. By combining equations (12.36) and (12.38) so as to eliminate $\mathbf{j}$, the result is

$$\nabla p = \frac{1}{\eta c^2} \, [(\mathbf{v} \cdot \mathbf{B})\mathbf{B} - B^2 \mathbf{v}].$$

The component of the mass velocity of the plasma in the direction perpendicular to the magnetic field is therefore determined, from the second term in the brackets, to be

$$\mathbf{v}_\perp = - \, \frac{\eta c^2}{B^2} \, \nabla p. \tag{12.39}$$

Assuming an isothermal plasma, ∇p may be replaced by $kT\nabla n$, where ∇n is the particle density gradient, and so equation (12.39) yields

$$n\mathbf{v}_\perp = -\frac{nkT\eta c^2}{B^2}\nabla n. \tag{12.40}$$

The effective diffusion coefficient of the plasma across the field lines is thus given by

$$D_\perp = \frac{nkTc^2}{B^2}\eta. \tag{12.41}$$

12.48. In estimating the effect of temperature on the diffusion coefficient, it should be noted that, according to equation (12.41), $D_\perp$ is somewhat unusual in the respect that the particle density n is involved. Since the kinetic pressure p of the particles is equal to nkT, it follows from equation (3.17) which defines the ratio β that, for small β,

$$\frac{nkT}{B^2} \approx \frac{\beta}{8\pi}, \tag{12.42}$$

and hence equation (12.41) becomes

$$D_\perp = \frac{\beta c^2}{8\pi}\eta. \tag{12.43}$$

Since the resistivity η varies as $1/T_e^{3/2}$, as may be seen from equation (4.73), it follows that, provided β does not change appreciably with temperature, $D_\perp$ will be roughly proportional to $1/T_e^{3/2}$. The behavior in this respect is different from that of a weakly ionized plasma where the diffusion coefficient increases with temperature in proportion to $T^{1/2}$. Another difference between the behavior in weakly and highly ionized plasma is in connection with the effect of the field strength. At constant particle density, $D_\perp$ in the latter case is also proportional to $1/B^2$, but at constant β it is independent of the field strength.

12.49. An alternative expression for $D_\perp$ in a highly ionized plasma, which is of interest, can be obtained by utilizing the approximate equation (4.70) for the resistivity; thus,

$$\eta \approx \frac{m_e}{ne^2\tau_c},$$

where m_e is the electron mass and τ_c represents the mean time between electron-ion collisions; the electron density n_e in equation (4.70) has been replaced here by the particle density n. Upon substituting into equation (12.41), it is found that

$$D_\perp \approx \frac{m_e kT}{\tau_c}\left(\frac{c}{eB}\right)^2.$$

For a Maxwellian energy distribution, $\tfrac{3}{2}kT$ is equal to the mean kinetic energy of the electrons, i.e., $\tfrac{1}{2}m_e v_e^2$, and so

$$D_\perp \approx \left(\frac{m_e c}{eB}\right)^2 \frac{v_e^2}{3\tau_c}. \tag{12.44}$$

It is seen from equation (4.3) that the gyromagnetic frequency of the electron is $\omega_e = eB/m_e c$, and since τ_c is equal to v_e/λ, where λ is the mean free path (or relaxation length) for electron-ion collisions, equation (12.44) can be simplified to

$$D_\perp \approx \frac{\lambda v_e}{3} \cdot \frac{1}{\omega^2 \tau_c^2}$$

$$\approx \frac{D_e}{\omega^2 \tau_c^2}, \tag{12.45}$$

which is of the same form as equation (12.11) for the diffusion coefficient of individual charged particles in a weakly ionized plasma.

12.50.   In spite of the similarity in form of equations (12.11) and (12.45), there is an important difference in the diffusion behavior of weakly and highly ionized plasmas.   In the former case the charged particles tend to diffuse at different rates, and it is the establishment of a space-charge electric field or, as an alternative, the short-circuit effect which results in an equalization of the diffusion rates.   In a highly ionized plasma a different situation exists because of the different nature of the collisions responsible for diffusion. As the result of a collision between an ion and an electron in a magnetic field, the guiding centers are shifted by equal amounts in the same direction.   The diffusion coefficients are thus the same for both electrons and ions.   It should be noted, however, that this condition would not be applicable in circumstances such that ion-ion or electron-electron collisions were important.   To a first approximation, the guiding centers of the two colliding particles shift in opposite directions by equal amounts, so that the net (first-order) diffusion is zero, as mentioned in §12.45 [15].

THERMAL CONDUCTION LOSSES

INTRODUCTION

12.51.   In addition to the energy loss by diffusion, due to the presence of a density (or pressure) gradient in a plasma, there can be a loss resulting from the flow of heat energy across a temperature gradient.   Such a gradient will generally exist in a thermonuclear reaction system because the temperature in the interior will usually be greater than at the exterior of the plasma. Considering, for simplicity, the transport of heat in one (x) direction, the thermal flux ϕ, expressed, for example, in ergs/(cm²)(sec), is given by

$$\phi = -K \frac{dT}{dx}, \tag{12.46}$$

where K is the thermal conductivity of the medium.

12.52. In the absence of a magnetic field, the charged particles in a plasma will behave like neutral atoms, to a first approximation, as far as thermal transport is concerned. Hence, an expression for K can be derived in a relatively simple manner by means of the kinetic theory of gases. For the present purpose, however, it is required to know the effect of a magnetic field on the thermal conductivity of a plasma. Consequently, the treatment to be used here will follow that employed in the earlier part of this chapter in connection with diffusion, since this lends itself more readily than does the conventional kinetic theory approach to the modifications resulting from the presence of a magnetic field.

THERMAL CONDUCTION IN THE ABSENCE OF A MAGNETIC FIELD

12.53. The element dI of charged-particle current flowing through the origin in the x direction, due to collisions occurring between x and $x + dx$, is given by equation (12.1) as

$$dI = \frac{v(x)}{6\lambda}\, ne^{-x/\lambda}\, dx.$$

It should be noted that n is written here in place of $n(x)$, since the particle density is assumed to be constant, i.e., n does not depend on x. On the other hand, the velocity v is now a function of x, since it is associated with the particle energy and, hence, its temperature which is supposed to have a spatial gradient. The corresponding flow of heat energy $d\phi$ can be obtained upon multiplying dI by the average particle energy in the x direction in the region between x and $x + dx$ from the origin, i.e., $\frac{1}{2}m[v(x)]^2$; hence,

$$d\phi = \frac{mn}{12\lambda}\, [v(x)]^3 e^{-x/\lambda}\, dx.$$

The net heat energy flow past the origin in the x direction is then obtained by the method used in §12.12, that is, by taking the difference in the flows in the $+x$ and $-x$ directions; thus,

$$\phi = \frac{mn}{12\lambda}\left[\int_{-\infty}^{0} [v(x)]^3 e^{-x/\lambda}\, dx - \int_{0}^{\infty} [v(x)]^3 e^{-x/\lambda}\, dx\right]. \tag{12.47}$$

12.54. A Maclaurin series expansion of $[v(x)]^3$ to the first order gives

$$[v(x)]^3 \approx v_0^3 + 3xv_0^2\left(\frac{dv}{dx}\right)_0,$$

where the zero subscript indicates that the quantity is to be taken at the origin. Insertion of this result into equation (12.47) then yields

$$\phi = \frac{mnv_0^2}{4\lambda}\left(\frac{dv}{dx}\right)_0\left[\int_{-\infty}^{0} xe^{-x/\lambda}\, dx - \int_{0}^{\infty} xe^{-x/\lambda}\, dx\right]$$

$$= -\frac{mnv_0^2\lambda}{2}\left(\frac{dv}{dx}\right)_0. \tag{12.48}$$

12.55. If there is a Maxwellian distribution among the particles, the average kinetic energy $\frac{1}{2}mv^2$ in the x direction is equal to $\frac{1}{2}kT$, where k is the Boltzmann constant.   Consequently,

$$mv_0 \left(\frac{dv}{dx}\right)_0 = \frac{d}{dx}\left(\frac{1}{2}mv^2\right)_0 = \frac{d}{dx}\left(\frac{1}{2}kT\right)_0,$$

so that equation (12.48) becomes

$$\phi = -\frac{nv\lambda k}{2}\cdot\frac{dT}{dx},$$

where v and dT/dx are determined at the same point.   Upon comparison of this expression with equation (12.46), it follows that

$$K = \tfrac{1}{2}nv\lambda k,$$

which is identical with the result obtained from kinetic theory.   It should be noted that K refers to the thermal conduction in a particular direction and v is the velocity component in that direction.

12.56. As seen in §2.36, $n\lambda$ is equal to $1/\sigma$, where σ is the cross section for the scattering process, so that

$$K = \frac{vk}{2\sigma}. \tag{12.49}$$

In a moderately (or highly) ionized plasma, the transfer of thermal energy will be due mainly to collisions among charged particles.   Hence, σ is approximately equal to the long-range Coulomb scattering cross section, which is of the same order for any pair of singly charged particles, provided the energies are the same.   It is seen, therefore, from equation (12.49) that K is roughly proportional to v, and since the velocity is much greater for electrons than for ions at the same temperature, it follows that only the electrons need be considered as the carriers of thermal energy.   According to equation (4.27), σ can thus be replaced by $\pi e^4/4W^2$, and since W may be expressed as $\frac{1}{2}m_e v_e^2$, equation (12.49) takes the form

$$K = \frac{m_e^2 v^5 k}{2\pi e^4}$$

or, since $\frac{1}{2}m_e v_e^2$ is equal to $\frac{1}{2}kT$, as seen above,

$$K = \frac{1}{2\pi}\cdot\frac{k(kT)^{5/2}}{m_e^{1/2}e^4}. \tag{12.50}$$

12.57. A more exact derivation, that treats the plasma as a fully ionized gas in which the electrons do not interact with one another and all the ions are at rest, leads to

$$K = \left(\frac{2}{\pi}\right)^{3/2}\frac{20}{\ln\Lambda}\cdot\frac{k(kT)^{5/2}}{m_e^{1/2}e^4}, \tag{12.51}$$

where Λ is defined in §4.57 [16]. This result differs from the value given by equation (12.50) by only a small numerical factor. However, it should be noted that, whereas the expression for K obtained by the approximate treatment is independent of the plasma density, equation (12.51) includes the density in the $\ln \Lambda$ term. However, since this does not vary greatly, as can be seen from the values in Table 4.1, the dependence of K on the electron density is not very significant.

THERMAL CONDUCTION IN MAGNETIC FIELD

12.58. Thermal transport by the charged particles in a plasma in the direction parallel to the magnetic field lines is not affected by the field, and so the thermal conductivity K is given by equation (12.51). For the direction perpendicular to the field lines, the simple derivation given above can be modified in the same manner as that used in §12.15 for the treatment of diffusion in a magnetic field. The result obtained is exactly analogous to equation (12.10), namely,

$$K_\perp = \frac{K}{1 + \omega^2 \tau^2}.$$

In most cases of interest, $\omega^2 \tau^2$ is much greater than unity, so that

$$K_\perp \approx \frac{K}{\omega^2 \tau^2}. \tag{12.52}$$

The thermal conductivity across the field lines will thus generally be considerably less than in the direction parallel to the magnetic field. Since the gyromagnetic frequency ω is proportional to the field strength B, the thermal conductivity $K_\perp$ varies as $1/B^2$. Increasing the field strength will thus decrease the thermal conduction perpendicular to the field lines.

12.59. It is seen from equation (12.51) that the thermal conductivity in the field direction varies as $T^{5/2}$ and so it would be very large in a thermonuclear reactor operating at a temperature of several kilo-electron volts. In the direction perpendicular to the field lines, however, not only will the thermal conductivity be less because of the ω^2 factor in the denominator of equation (12.50), but the temperature dependence will also be smaller because the collision time τ between charged particles increases with temperature. As a general rule, these times vary as $W^{3/2}$, and hence as $T^{3/2}$ (cf. §4.68 *et seq.*), and so τ^2 will be proportional to T^3, at constant particle density. In this event, $K_\perp$ will vary as $1/T^{1/2}$, and so it appears that the thermal conductivity across the magnetic field lines may actually decrease with increasing temperature.

12.60. In general, it can be concluded that thermal conduction perpendicular to the field direction would probably not be a serious source of energy loss in a thermonuclear reactor. Furthermore, in a toroidal (or similar) geometry, thermal conduction parallel to the field direction would either be nonexistent, because of the absence of a temperature gradient, or if it did exist it would not be associated with any net loss of energy from the system. In a reactor having a cylindrical vacuum chamber with its axis parallel to the confining field lines, losses by thermal conduction through the ends would be a serious matter at very high temperatures.

LOSSES DUE TO CHARGE EXCHANGE

MEASUREMENT OF CROSS SECTIONS

12.61. The possibility of energy loss from energetic deuterium ions as a result of charge-exchange (or charge-transfer) interaction with neutral particles of low energy has been mentioned a number of times in earlier chapters. There are two aspects of this problem, namely, charge exchange of the deuterons or tritons (a) with deuterium or tritium atoms (or molecules), and (b) with impurity atoms (or molecules). These situations could arise either in an incompletely ionized deuterium or deuterium-tritium plasma, i.e., at relatively low ion energies, or when energetic deuterons are injected into a magnetic mirror (or other) geometry. A considerable amount of work, both theoretical and experimental, has been done in connection with the evaluation of charge-exchange cross sections [17, 18]. All that is necessary here is to indicate the general principles involved in the measurement of these cross sections and to review the results of interest for thermonuclear research.

12.62. If a beam of energetic deuterons, at a density n_+, passes through a gas target x cm in thickness, the rate of change of ion density with distance traversed is

$$\frac{dn_+}{dx} = -n_+ n \sigma_{+0} + n_0 n \sigma_{0+}, \tag{12.53}$$

where n is the density of the gas target, n_0 is the density of deuterium atoms in the beam, and σ_{+0} and σ_{0+} represent the cross sections for electron capture and electron loss by the deuterium ions and atoms, respectively. The first term on the right gives the decrease in deuteron density due to capture of electrons from the target gas atoms (or molecules), and the second term is the corresponding increase resulting from loss of electrons by deuterium atoms. After passage through a sufficient length of the gas target, an equilibrium will be attained in which n_+ does not change further. In these circumstances, dn_+/dx is zero, so that, by equation (12.53)

$$\frac{n_+}{n_0} = \frac{\sigma_{0+}}{\sigma_{+0}}, \tag{12.54}$$

where n_+/n_0 is the ratio of the densities of deuterons to deuterium atoms at equilibrium.

12.63. By means of equation (12.54) it is possible to determine the ratio of the two cross sections; it is required, therefore, for only one to be measured for the other to be known.   The simplest procedure is to determine σ_{0+} in the following manner.   If the number of D^- ions present, which would form deuterium atoms by electron loss, is small, as is undoubtedly the case, the rate of change of n_0 with distance is given by equation (12.53) with the sign reversed; that is,

$$\frac{dn_0}{dx} = n_+ n\sigma_{+0} - n_0 n\sigma_{0+}. \tag{12.55}$$

Suppose the beam of high-energy particles consists initially of deuterium atoms only and that the deuterons formed by electron loss are removed by an electric field as fast as they are formed, so that they cannot return to the neutral state. The first term on the right of equation (12.55) is then zero, so that

$$\frac{dn_0}{dx} = -n_0 n\sigma_{0+}.$$

Upon integration, this gives

$$n_{0(x)} = n_0 e^{-n\sigma_{0+}x}, \tag{12.56}$$

where n_0 is the initial density of neutral deuterium atoms in the beam and $n_{0(x)}$ is the value after traversal of a distance x through the gas target.   If these densities can be measured, σ_{0+} can be calculated, n being determined from the temperature and pressure of the gas target.

12.64.   The experimental procedure which can be used to determine the electron capture and loss cross sections is to pass a beam of high-energy ions, obtained from an accelerator, through the gas target.   The densities of ions and atoms in the emergent beam are measured by means of a Faraday cage (for ions only) and a combination of thermocouple and secondary-electron emission detector (for the ions and atoms).   By increasing the gas pressure of the target gas until the ratio n_+/n_0 in the beam becomes constant, the equilibrium condition is attained, so that σ_{0+}/σ_{+0} can be determined, in accordance with equation (12.54).

12.65.   The same apparatus can be used to measure σ_{0+} with the aid of equation (12.56).   The accelerated ion beam is first passed through a chamber, called a "neutralizer," where considerable charge exchange occurs; the emergent beam, after passage through an electrostatic field to divert the ions, then consists entirely of high-energy atoms.   By observing the density (or relative density) after passage through various thicknesses of target, with the ions being removed continuously, the cross section σ_{0+} can be obtained.

CROSS SECTION RESULTS

12.66.   Measurements of the cross sections σ_{+0} and σ_{0+} for hydrogen atoms and ions have been made with various gas targets, including hydrogen, helium, oxygen, nitrogen, neon, and argon.   Although no experiments have been reported for deuterium or tritium, it is probable that the cross sections will be

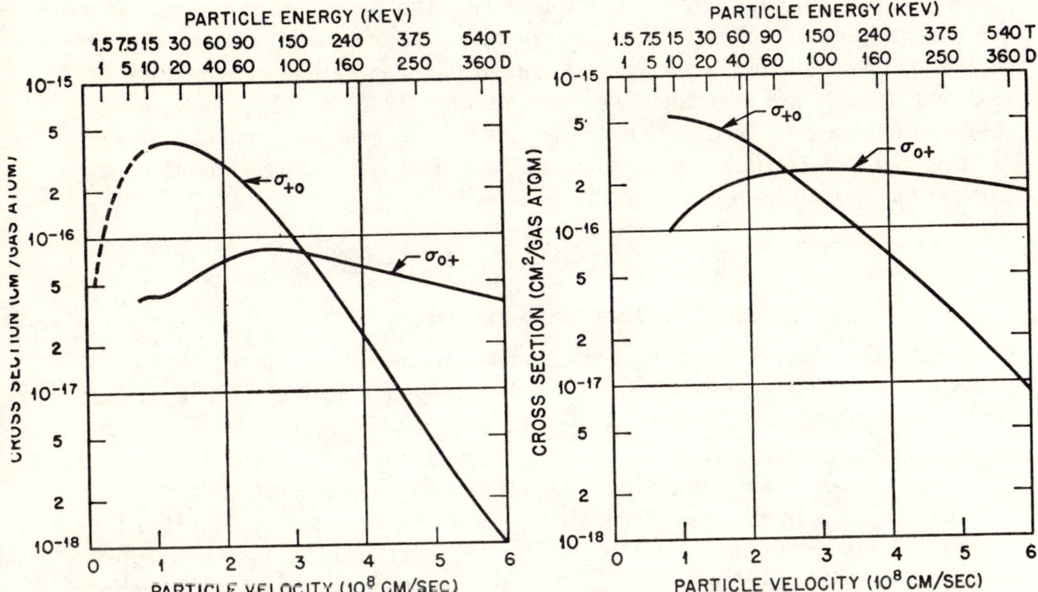

FIG. 12.6. Neutralization and ionization cross sections of D (and T) in deuterium gas.

FIG. 12.7. Neutralization and ionization cross sections of D (and T) in nitrogen gas.

equal to those for hydrogen ions and atoms at the same velocity. The cross sections are given in Figs. 12.6 and 12.7, for the electron capture (or neutralization) cross section (σ_{+0}) of deuterons or tritons and the electron loss (or ionization) cross section (σ_{0+}) of deuterium or tritium atoms in their passage through deuterium gas and nitrogen gas, respectively [17]. The dashed portion of Fig. 12.6, for σ_{+0} at low energies, is based on calculated values; the other curves are plotted from experimental data.*

12.67. It will be noted, in the first place, that the cross sections are higher in nitrogen than in deuterium or tritium. In general, the cross sections increase with the atomic number of the target gas, and the results in Fig. 12.7 may be regarded as applying to the common impurities, namely, carbon, oxygen, and nitrogen, which have been detected in discharges in deuterium. The cross sections for silicon would probably be higher than those observed with nitrogen.

12.68. In spite of the fact that the neutralization cross section σ_{+0} for deuterium ions increases sharply with energy below about 10 kev, it is probable that, in stellarator and pinched discharge systems, charge-exchange losses will be significant only at energies less than about 1 kev. This is because ohmic (and other) heating in these devices should result in essentially com-

*Somewhat different values have also been reported [18].

plete ionization of both deuterium and impurity atoms, so that charge exchange becomes negligible.   Where deuterons are injected into a mirror system, charge-exchange losses due to the presence of neutral particles can be quite serious unless the energies are very high.   As seen in Figs. 12.6 and 12.7, the σ_{+0} cross sections decrease markedly at deuteron energies in excess of roughly 20 kev, and at 100 kev (in deuterium) they are less than the σ_{0+} cross sections for ionization of deuterium atoms.

OPTICAL RADIATION LOSSES

EFFECT OF IMPURITIES

12.69.   The presence in a plasma of neutral or partially ionized atoms will inevitably lead to the loss of energy from the system in the form of optical radiation.   "Optical radiation" is used here in a very general sense for any kind of electromagnetic radiation, no matter what its wave length, that is due to energy transitions involving the outer electrons of any incompletely stripped atomic species.   In the first place, absorption of energy can result in the formation of an electronically excited state; the excitation energy will subsequently be emitted as line radiation, called *excitation radiation*, which represents a loss of energy from the plasma.   Since the excited electron remains attached to the atom at all times, the transition leading to the emission of energy is referred to as "bound-bound."   In addition, a partially stripped atom can capture a free electron, with the emission of radiation in the form of a continuum.   This *recombination radiation*, as it is termed, is then due to "free-bound" electronic transitions [20, 21].

12.70.   Above a temperature of about 50 ev, the hydrogen isotopes are essentially completely ionized and the optical radiation losses are then negligible. However, if impurities are present, as is inevitably the case under normal operating conditions, partially stripped atoms will exist up to quite high temperatures, because of their higher atomic numbers.   Consequently, radiation losses can be considerable.   The very large energy loss rates from a stabilized pinch discharge in deuterium, for example, has been found to be mainly due to ultraviolet line radiation from incompletely stripped oxygen (and other) atoms, as mentioned in §7.124.

12.71.   The impurities in a plasma originate from two main sources.   First, there are the contaminants which are adsorbed on the chamber walls and driven out during the passage of the discharge.   These are chiefly oxygen, nitrogen, and carbon, which are typical impurities in experimental plasmas.   Of greater importance are the contaminants which arise from the bombardment of the wall material itself by energetic charged particles and radiations.   These may be silicon, oxygen, and aluminum from ceramic tubes, or metallic elements, such as aluminum or iron from metal-walled chambers.   Improved vacuum techniques and baking at high temperatures (cf. §8.78) can perhaps help in minimizing adsorbed impurities, but it seems that little can be done to prevent erosion of the chamber walls by particles and radiations from the discharge. Possibly recourse will have to be had to such devices as the divertor which was designed for use in stellarator systems (§8.56).

EXTENT OF IONIZATION

12.72. In estimating the effect of an impurity on the energy loss as optical radiation, it should be noted that the degree of ionization of a particular species in a laboratory plasma is much less than what it would be in the interior of a star, for example, at the same temperature and density. The reason is that, in an optically thick body such as a star, there is thermodynamic equilibrium between the plasma and the black-body radiation field. This radiation field will generally be intense enough to make a significant contribution to the ionization of the atoms by the photoionization process, i.e., removal of electrons by interaction with the radiation photons. When the body of plasma becomes smaller than the mean free path of the photons, however, photoionization becomes insignificant, leaving collisions between atoms (or ions) and electrons as the principal means of ionization. It is to be expected, therefore, that the extent of stripping of the atoms in a terrestrial plasma, which must be of moderate size, will be less than in the interior of a star under the same conditions [20, 22].

12.73. By considering the steady-state situation, in which the degree of ionization of an impurity atom is determined by a balance between ionization by collision with electrons and recombination of ions with electrons, it has

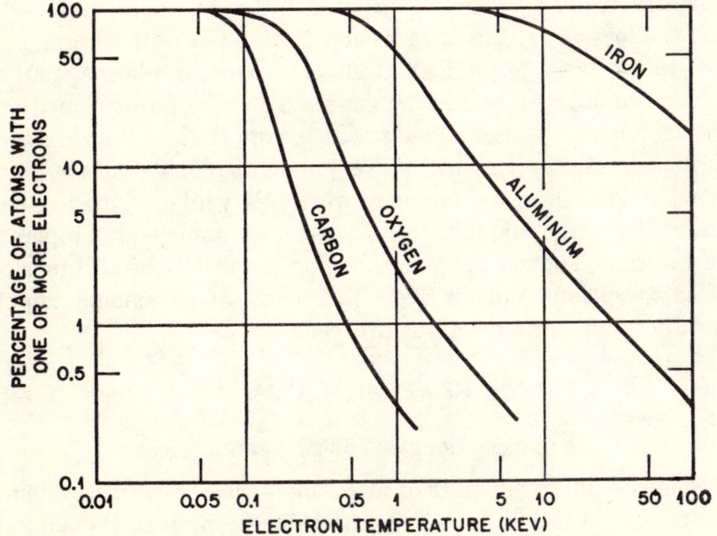

Fig 12.8.   Dependence of electron stripping on temperature.

been possible to estimate the degree of stripping of various atoms over a range of temperatures. At low densities the processes of ionization and recombination are dependent in the same manner on the density, and so the results obtained are independent of density. A case of special interest is the determination of the proportion of the atoms which still retain one or more bound electrons as a function of electron temperature. The results of such calculations for carbon, oxygen, aluminum, and iron are shown in Fig. 12.8, which is based on the assumption of a Maxwellian distribution of electron velocities [21].

12.74. It is seen that even at a temperature of 2 kev, about 1 per cent of oxygen atoms are still not completely stripped. With increasing atomic number, so that the atom has more electrons, the situation becomes worse. At a temperature as high as 100 kev, even traces of iron in the plasma could have serious consequences as far as optical radiation losses are concerned. Of course, the impurity losses increase with decreasing temperatures, and so it can be understood why they play such an important role in laboratory plasmas with temperatures in the vicinity of 500 ev or less. However, as the temperatures approach those appropriate to the operation of a thermonuclear reactor, which may well be as high as 100 kev, optical radiation losses due to impurities become less significant while those from bremsstrahlung become much more so.

MAGNITUDE OF RADIATION LOSS

12.75. Some indication of the magnitude of the excitation radiation energy loss may be obtained from the following data. At an electron temperature of 100 ev, the rate of emission of line radiation from aluminum is about 10^5 times as great as the bremsstrahlung loss from a hydrogen isotope plasma. For an electron density of 10^{15} particles/cm^3, the latter is about 0.15 watt/cm^3 at 100 ev, and the loss as excitation radiation from 1 per cent aluminum impurity will be 10^3 times as large, namely, 150 watts/cm^3 of plasma [20]. This is roughly of the same order as may be expected for the power density of an operating thermonuclear reactor (§2.42). It is true that at the reactor temperature the optical radiation loss will be very much smaller, but the plasma must obviously pass through a low temperature range while it is being heated. If the loss rates are large enough, the desired thermonuclear temperature may never be attained. The necessity for keeping impurities of all kinds out of the hydrogen isotope plasma is thus likely to become of paramount importance as research on thermonuclear reactions proceeds.

SCALING LAWS

THERMONUCLEAR POWER PRODUCTION

12.76. Some of the parameters of a thermonuclear reacting system, notably the dimensions of the reaction chamber and the strength of the confining magnetic field, can be varied in such a manner as to increase the power output and minimize losses. The broad rules which are applicable are referred to as *scaling laws* [23]. Some aspects of the subject have been treated in earlier chapters in connection with special plasma confinement devices. The present discussion is, however, more general in nature.

For a reaction chamber having a circular cross section of radius r, the thermonuclear power *per unit length* of the chamber is given by equation (2.18) as

$$P_{\text{th}} = \tfrac{1}{2}n_{\text{D}}{}^2\overline{\sigma v}Q \times \pi r^2 \tag{12.57}$$

for the D-D reactions, where the terms have been defined in §2.40. Let B be the strength of the confining field, irrespective of whether it is an external field produced by means of a current through coils surrounding the reaction chamber or whether it is a self-magnetic field associated with passage of a current through the gas, as in the pinch effect. If the system is in a steady state, the particle density is related to the field strength and the temperature; thus, according to equation (12.42), $n_{\text{D}}{}^2$ is proportional to $\beta^2 B^4/T^2$. If the ratio β is assumed to be the same in different systems, in which the magnetic field and the radius of the reaction chamber are different, then it follows from equation (12.57) that

$$P_{\text{th}} \approx \text{constant} \times \frac{\overline{\sigma v}}{T^2} \times r^2 B^4. \tag{12.58}$$

12.77. It may be concluded, therefore, that the maximum thermonuclear power output, for a reaction chamber of specified size and a given confining magnetic field strength, should be obtained at the temperature for which $\overline{\sigma v}/T^2$ is a maximum. It can be shown from the data in Table 2.1 that for both the D-D and D-T reactions this maximum is in the temperature region of 10 to 20 kev. Since the ideal ignition temperature for the D-D reaction system is about 36 kev, as indicated in §2.68, it would not be possible to operate a thermonuclear reactor in this temperature range if deuterium alone were the fuel material. But where a deuterium-tritium mixture is used, advantage could be taken of the expected maximum in the thermonuclear power output in the vicinity of 10 kev, since the ideal ignition temperature is about 4 kev.

12.78. The dependence of the thermonuclear power per unit length on r^2 is obvious because the total power is proportional to the volume of the reactor chamber, for a given fuel particle density. The variation of the power output as the fourth power of the magnetic field strength, as implied by equation (12.58), is of considerable importance. It shows that the use of very strong magnetic fields for confinement of the plasma would be desirable, even if not absolutely necessary, in a thermonuclear reactor. The relevant problem of the power which must be expended to maintain the magnetic field will be considered below (§12.83).

SCALING OF DIRECT ENERGY LOSSES

12.79. One of the most significant sources, and perhaps the main source, of energy loss at high temperatures is bremsstrahlung. This is expected to be independent of both the magnetic field strength and the dimensions of the reacting system. As seen in Chapter 2, the thermonuclear power and the rate of energy loss as bremsstrahlung vary in the same manner with the density of the reacting nuclei. Hence, the ratio of bremsstrahlung to thermonuclear power production is independent of the dimensions and of the magnetic field strength, since the latter determines the particle density of the plasma that can be confined. The effect of temperature is apparent from Fig. 2.9 which shows that, as the temperature increases, the ratio of the power dissipated in the form of bremsstrahlung to that produced by thermonuclear reaction decreases.

12.80. Another manner in which energy may possibly be lost from a high-temperature plasma confined by (and containing) a magnetic field is as cyclotron radiation, which was discussed in Chapter 2. As pointed out in §2.84, perhaps the most significant factor in determining the loss of energy in the form of cyclotron radiation is the value of the ratio of the kinetic particle pressure to the magnetic field pressure within the plasma. If this is large, the losses will be small. In the extreme case of a completely diamagnetic plasma, which excludes the confining field entirely, the cyclotron radiation losses should be small.

12.81. The loss of energy by diffusion of charged particles to the walls of the reaction chamber, in the direction perpendicular to the magnetic field, will be determined by the particle current in this direction, the mean energy per particle, and the area of the chamber walls. Since the plasma in a thermonuclear reaction system would be completely ionized, the particle current may be expressed by means of equation (12.40). The mean energy per particle, assuming an approximately Maxwellian distribution, should be proportional to the temperature, whereas the wall area per unit length will vary as the radius r of a chamber of circular cross section.

12.82. It follows, therefore, that the power loss by diffusion per unit length, represented by P_{dif}, is given by

$$P_{\text{dif}} = \text{constant} \times \frac{nT\eta}{B^2}\, \nabla n \times Tr. \tag{12.59}$$

The particle gradient ∇n will be roughly proportional to n/r, and η varies as $1/T^{3/2}$, according to equation (4.73); furthermore, nT is proportional to B^2, by equation (12.42), assuming β to remain approximately constant; hence, equation (12.59) becomes

$$P_{\text{dif}} = \text{constant} \times \frac{B^2}{T^{3/2}}.$$

It follows, therefore, upon introducing equation (12.58), that the fraction of the thermonuclear power lost as a result of diffusion can be represented by

$$\frac{P_{\text{dif}}}{P_{\text{th}}} \approx \text{constant} \times \frac{T^{1/2}}{\overline{\sigma v}} \cdot \frac{1}{r^2 B^2}. \tag{12.60}$$

The function $\overline{\sigma v}/T^{1/2}$ does not exhibit a maximum, for the D-D and D-T reactions, at temperatures of thermonuclear interest, but increases, so that $T^{1/2}/\overline{\sigma v}$ decreases, monotonically up to about 60 kev, at least. It is to be concluded that the relative loss of thermonuclear energy by diffusion of charged particles to the chamber walls, across the magnetic field lines, will decrease with increasing temperature. Increasing the dimensions of the system and the magnetic field strength should also serve to decrease the energy lost by this mechanism [23].

SCALING OF MAGNET LOSSES

12.83. The power required to maintain the axial magnetic field used to confine the plasma, in such systems, for example, as the stellarator and magnetic mirror geometries, can be related to the dimensions of the magnet [23-25]. The strength of the magnetic field produced by the current I flowing in a solenoidal conductor is proportional to NI, where N is the number of coil terms per unit length of the confinement chamber, so that

$$B = \text{constant} \times NI. \tag{12.61}$$

The ohmic power dissipated by the current in the magnet coils per unit length of the chamber, i.e., P_{mag}, is given by

$$P_{\text{mag}} = I^2 \times \frac{2\pi RN}{A}\,\eta, \tag{12.62}$$

where R is the coil radius, A is the cross sectional area of the coil material, and η is its resistivity. If the actual diameter (or thickness) of the coil material is small relative to that of the chamber which the solenoid surrounds, then R may be taken to be equal to the radius r of the chamber. By equation (12.61), I is proportional to B/N and so equation (12.62) yields

$$P_{\text{mag}} \approx \text{constant} \times \frac{B^2 r\eta}{NA}.$$

Upon introducing the expression for the thermonuclear power per unit length given above, it follows that

$$\frac{P_{\text{mag}}}{P_{\text{th}}} \approx \text{constant} \times \frac{T^2}{\overline{\sigma v}} \cdot \frac{1}{rB^2} \cdot \frac{\eta}{NA}. \tag{12.63}$$

12.84. If the diameter (or thickness) of the coil material may be assumed to increase in proportion to the chamber radius, then it can be readily shown that NA will vary as r, provided the coil is always wound as closely as possible so as to maximize N. In these circumstances, equation (12.63) takes the form

$$\frac{P_{\text{mag}}}{P_{\text{th}}} \approx \text{constant} \times \frac{T^2}{\overline{\sigma v}} \cdot \frac{\eta}{r^2 B^2}.$$

It is evident, therefore, that the proportion of thermonuclear energy that is lost as joule heat in the coils of the magnet used to produce the confining field can be decreased by increasing the chamber radius and the field strength, as well as by decreasing the resistivity of the coil material.

12.85. Since the electrical resistance of metals generally decreases with decreasing temperature and becomes very small at low temperatures, investigations are proceeding on the possibility of constructing "cryogenic magnets" [26, 27]. Two materials which appear to be promising for operation in the vicinity of 10 to 20°K are sodium and aluminum. The net power requirements are estimated to be very much less than for magnets with copper coils operating at ordinary temperatures [28, 29].

CONCLUSION

12.86. The general conclusion to be drawn from the foregoing discussion, as well as from various points of view considered in earlier chapters, is that a thermonuclear reaction chamber should have as large an internal diameter as is practical. In addition, the strength of the confining magnetic field should be as high as possible. As regards the operating temperature, each case must be evaluated upon its own merits. While the temperature must, of necessity, exceed the appropriate ideal ignition temperature, consideration must be given to the decrease in the thermonuclear power production resulting from the decrease in $\overline{\sigma v}/T^2$ above 10 to 15 kev, on the one hand, and the relative decrease in bremsstrahlung and other losses, on the other hand, as the temperature is increased.

REFERENCES FOR CHAPTER 12

1. S. Chapman and T. G. Cowling, *The Mathematical Theory of Non-Uniform Gases*, Cambridge University Press, 1953, Chapter 18.
2. A. Simon, *An Introduction to Thermonuclear Research*, Pergamon Press, Inc., 1959, Chapter IX; also in USAEC Report ORNL-2285 (1957), p. 186.
3. A. Simon, *Phys. Rev.*, **98**, 317 (1955).
4. A. Simon, *Proc. Second U.N. Conf. on Peaceful Uses of Atomic Energy*, **32**, 343 (1958).
5. D. Bohm, E. H. S. Burhop, H. S. W. Massey, and R. M. Williams, in A. Guthrie and R. K. Wakerling, Eds., *Characteristics of Electrical Discharges in Magnetic Fields*, NNES Div. 1, Vol. 5, McGraw-Hill Book Co., Inc., 1949, pp. 197 ff.
6. D. Bohm, E. H. S. Burhop, H. S. W. Massey, Reference 5, pp. 62 ff.
7. A. Simon and R. V. Neidigh, USAEC Report ORNL-1890 (1955).

8. A. Simon, *Proc. Second U.N. Conf. on Peaceful Uses of Atomic Energy,* **32,** 343 (1958).
9. R. J. Bickerton, *Proc. Phys. Soc. (London),* **B70,** 305 (1957).
10. R. J. Bickerton and A. von Engel, *Proc. Phys. Soc. (London),* **B69,** 468 (1956).
11. B. Lehnert, *Proc. Second U.N. Conf. on Peaceful Uses of Atomic Energy,* **32,** 349 (1958).
12. A. Simon, *Phys. Rev.,* **100,** 1557 (1955).
13. C. L. Longmire and M. Rosenbluth, *Phys. Rev.,* **103,** 57 (1956).
14. L. Spitzer, *Physics of Fully Ionized Gases,* Interscience Publishers, Inc., 1956, pp. 23, 38.
15. Reference 14, p. 39.
16. Reference 14, p. 87.
17. P. M. Stier and C. F. Barnett, *Phys. Rev.,* **103,** 896 (1956).
18. C. F. Barnett and H. K. Reynolds, *Phys. Rev.,* **109,** 355 (1958).
19. I. V. Kurchatov, *J. Nuclear Energy,* **8,** 168 (1958).
20. R. F. Post, *Ann. Rev. Nuclear Sci.,* **9,** 367 (1958).
21. G. Knorr, *Z. Naturforsch.,* **13a,** 941 (1958).
22. G. Elwert, *Z. Naturforsch.,* **7a,** 432 (1952).
23. R. F. Post, *Rev. Mod. Phys.,* **28,** 338 (1956).
24. Reference 2, p. 79.
25. L. Geller, R. W. Rupp, and S. Koslov, *Proc. Second U.N. Conf. on Peaceful Uses of Atomic Energy,* **31,** 257 (1958).
26. H. L. Laquer and E. F. Hammel, *Rev. Sci. Instr.,* **28,** 875 (1957).
27. H. L. Laquer, *Proc. Symp. on Mag. Field Design,* USAEC Report ORNL-2745, p. 44 (1959).
28. R. G. Mallon, USAEC Report UCRL-5792-T (1959).
29. R. J. Corruccini, NBS Boulder Lab. Mem. Report CM-2 (1959).

Chapter 13

PLASMA STABILITY THEORY *

INTRODUCTION

STABLE AND UNSTABLE EQUILIBRIA

13.1.   Nearly all the proposals for confining plasmas at high temperatures, with a view to the realization of thermonuclear reactions, involve the use of magnetic fields in one form or another. Hence, the problem of the stability of the plasma in a magnetic field is of paramount importance. Early experimental and theoretical work showed that the plasmas in simple pinch and stellarator systems were basically unstable. Consequently, efforts have been made toward the development of a theory of plasma stability in the hope that the instabilities might be understood and methods developed for eliminating them or at least inhibiting their rate of growth.

13.2.   Since the whole field of plasma confinement, and particularly of stability during confinement, is relatively new, and since there are undoubtedly several fundamentally different types of instability, no really satisfactory theory of plasma stability exists at the present time. Some progress has been made in connection with hydromagnetic instabilities, in which the plasma is treated as a simple hydrodynamic fluid interacting with a magnetic field, and a few significant conclusions capable of experimental test have been reached. However, it must be admitted that, except for one special case, the requirements for the stabilization of a plasma under conditions of practical interest have not yet been elucidated. Because the stability problem is basic to all aspects of research in controlled thermonuclear reactions, various methods of solving it which have been tried will be outlined and their limitations discussed. Although the treatments are highly mathematical, emphasis will be placed as far as possible on their physical significance.

13.3.   In an unperturbed plasma in a state of equilibrium and at rest, the isotropic (scalar) pressure balance equation is

$$\nabla p = \frac{1}{c}(\mathbf{j} \times \mathbf{B}),$$

*Based on a draft prepared by Bergen R. Suydam.

which has three components corresponding to the three directions of the vectors. However, these three component equations relate four functions, namely, the pressure and the three components of the magnetic field. Consequently, equilibrium can be realized, in principle, with a wide variety of magnetic fields. The condition required is that $\nabla \times (\mathbf{B} \times \nabla \times \mathbf{B}) = 0$; for such fields, and only for such, is confinement of a scalar pressure possible. The central problem of plasma confinement is to determine which of the equilibria, of the many possible, are stable. In a stable system, a small perturbation will lead to an oscillation about the equilibrium state, but if the system is unstable, the perturbation will grow indefinitely. It will be seen that the treatment of the stability of a plasma confined by a magnetic field has many features in common with that of wave propagation. The question that will be asked is: Do perturbations remain at the same level or do they continue to grow?

13.4. Before discussing the various approaches to the subject of hydromagnetic plasma stability, it should again be emphasized that no really satisfactory theory has yet been developed. The present chapter should therefore be regarded as being in the nature of a progress report rather than a definitive account; it is included in this book because of the essential importance of the problem with which it is concerned.

SIMPLE EQUILIBRIUM SYSTEMS

13.5. A simple situation, for which it appears that an unequivocal solution to the stability problem is possible, may be treated in the following manner. Consider a portion of plasma having a uniform kinetic pressure p_0. The plasma is assumed to be perfectly diamagnetic, so that it contains no internal magnetic field. Suppose this plasma is confined by an external field, the value of which is B_0 at the surface of the plasma. The condition of equilibrium is then

$$p_0 = \frac{B_0{}^2}{8\pi},$$

so that the (uniform) pressure of the plasma is exactly balanced by the magnetic pressure of the confining field.

13.6. In order to determine whether this equilibrium is stable or not, imagine the surface of the plasma to be disturbed slightly by forming a wave (or "flute") with its crest and trough running parallel to the magnetic field lines. If the trough and crest have the proper dimensions, there will be no change in the volume of the plasma and hence no change in the pressure. Suppose, in the first place, that the confining field lines are curved so that they are concave toward the plasma. This is represented in Fig. 13.1A, where I shows the side view and II the end view, in which the wavelike disturbance is indicated. It follows from Maxwell's equation $\nabla \times \mathbf{B} = 0$ that the magnetic field strength will diminish with increasing distance from the center of curvature of the field lines.   Consequently, in the case under consideration, the magnetic field on the crest of the wave is such that $B^2 < B_0{}^2$, whereas in the trough $B^2 > B_0{}^2$.   As a result, a net force acts in such a sense as to heighten the crest and deepen the trough, that is to say, the disturbance will tend to grow in amplitude.   It follows, therefore, that the particular equilibrium is unstable.   On the other hand, if the magnetic field lines are convex toward the plasma, as represented in Fig. 13.1B, the net magnetic force is such as to depress the crest and raise the

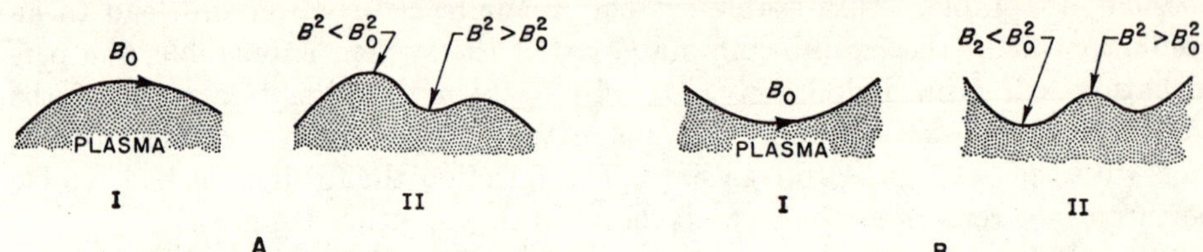

FIG. 13.1.   Unstable and stable magnetic field-plasma configurations.

trough of the disturbance under consideration.   In other words, the force acts in such a manner as to counteract the disturbance and the equilibrium is stable.

13.7.   The foregoing discussion suggests and rigorous analysis (see Appendix A) confirms that a perfectly diamagnetic plasma, i.e., one with no internal magnetic field, is stable or not according to the curvature of the lines of the confining field at its surface [1-3].   If every such line is *everywhere* convex toward the plasma or, in other words, if the center of curvature of the magnetic lines on the plasma-field interface always lies in the direction away from the plasma, then the system is stable.   On the other hand, if the lines are *anywhere* concave toward the plasma, so that the center of curvature is within (or toward) the plasma, then the system must be unstable.   Thus, a pinched discharge consisting of a very thin sheath of essentially infinite conductivity, with no trapped axial field, will be fundamentally unstable.

13.8.   In order for a completely diamagnetic plasma to be stable when confined by a magnetic field, it cannot form a convex body, but must have cusps. A number of different configurations satisfying this requirement, some of which are being studied experimentally, were described in §11.22 *et seq.*

HYDROMAGNETIC THEORY

NORMAL MODE ANALYSIS

13.9.   The foregoing discussion has been essentially qualitative in nature and has dealt with the restricted situation of a perfectly diamagnetic plasma. For practical purposes, information is required concerning the rate of growth of various instabilities that might develop and also on how the growth may be inhibited.   Furthermore, in many cases of interest, there is a magnetic field inside as well as outside the plasma; the situation is then more realistic, but also more complicated, than that considered above.   The displacement of the plasma surface now results not only in a shift of the magnetic lines in the vacuum field outside, but will also cause the lines within the plasma to bend, stretch, or be distorted in other ways. This leads to additional forces which require more detailed analysis. The mathematical treatment of such a system is very complex and so it is necessary to make approximations in the model of the plasma in order to simplify the results to the point where they can be applied to a practical geometry.

13.10. The simplest theory to be described here is based on the treatment of a plasma as a simple (ideal) hydrodynamic fluid which is subjected to the action of electromagnetic forces. In other words, it involves a combination of the standard equations of hydrodynamics with Maxwell's equations of electromagnetism, leading to a complete set of differential equations determining the motion of the plasma. This hydromagnetic theory is of interest both because it has led to concrete results and because the domain of its validity has been shown to be much wider than might have been anticipated from the approximations made in its development (§13.46). Essentially, the theory of plasma stability has features in common with that of plasma waves, since instabilities may be regarded merely as being waves which grow at an ever-increasing rate.

13.11. In order to apply the hydromagnetic equations to a plasma, a number of simplifying postulates are made. First, collisions between particles are assumed to occur so frequently that the velocity distribution of the particles is isotropic, or very nearly so. Actually, this means that the plasma can be characterized at each point by a mass density ρ, a scalar pressure p, and a single temperature. Incidentally, the frequent collisions provide a mechanism for keeping the neighboring particles together so that they form a coherent element of fluid which can be treated by simple hydrodynamics. The second postulate is that, as a result of the high rate of particle collisions, heat flow is inhibited within the plasma, so that the system is adiabatic. Finally, it is supposed that the plasma temperature is so high that it is essentially a perfect conductor. This means that such currents as flow produce no ohmic heating nor do they require electric fields to drive them.

13.12. With these simplifications, the plasma behaves like the perfect fluid of ordinary hydrodynamics, except that it is subject to electromagnetic forces. Exactly as in fluid mechanics, there are three conservation equations, as follows:

Conservation of Mass

$$\frac{d\rho}{dt} + \rho \boldsymbol{\nabla} \cdot \mathbf{v} = 0, \tag{13.1}$$

where d/dt represents the hydrodynamic operator, i.e.,

$$\frac{d}{dt} \equiv \frac{\partial}{\partial t} + \mathbf{v} \cdot \boldsymbol{\nabla},$$

$\mathbf{v}$ being the vector velocity of the plasma and t the time. Equation (13.1) is also known as the equation of continuity.

Conservation of Momentum

$$\rho \frac{d\mathbf{v}}{dt} = -\boldsymbol{\nabla} p + \frac{1}{c} (\mathbf{j} \times \mathbf{B}), \tag{13.2}$$

where d/dt is the hydrodynamic operator, and $\mathbf{j}$ and $\mathbf{B}$ are, as before, the current density and magnetic field vectors. Equation (13.2), which neglects the effect of gravity and of the electric field, since the latter is assumed to be zero, is sometimes known as the force (or motion) equation (cf. §4.116). It describes the motion of the plasma under the combined action of a pressure gradient and a magnetic body force.

Conservation of Energy

$$\frac{dp}{dt} + \gamma p (\boldsymbol{\nabla} \cdot \mathbf{v}) = 0, \tag{13.3}$$

where d/dt is the hydrodynamic operator defined above and γ is the ratio of the specific heats, which may generally be taken as the monatomic gas value, i.e., 5/3, for a plasma. This equation implies adiabatic behavior, as may be seen by combining it with equation (13.1) to give

$$\frac{1}{p} \cdot \frac{dp}{dt} = \frac{\gamma}{\rho} \cdot \frac{d\rho}{dt}$$

and integrating; the result is

$$p = \text{constant} \times \rho^{\gamma},$$

which is the familiar adiabatic law for a gas.

13.13. In addition to the three hydrodynamic equations given above, there are the Maxwell equations. As the plasma is assumed to be perfectly conducting, an observer moving with the local velocity of the plasma would experience no electric field; hence,

$$\mathbf{E} + \frac{1}{c} (\mathbf{v} \times \mathbf{B}) = 0.$$

When this is set into Maxwell's equation

$$\boldsymbol{\nabla} \times \mathbf{E} = -\frac{1}{c} \cdot \frac{\partial \mathbf{B}}{\partial t},$$

there is obtained

$$\frac{\partial \mathbf{B}}{\partial t} = \boldsymbol{\nabla} \times (\mathbf{v} \times \mathbf{B}) \tag{13.4}$$

as the equation of motion of the magnetic field. Physically, this result implies that the plasma and the magnetic field move together.

13.14. Use will also be made of the Maxwell relationship, neglecting the displacement current (cf. §3.11),

$$\nabla \times \mathbf{B} = \frac{4\pi}{c}\,\mathbf{j}, \tag{13.5}$$

which is essentially a definition of the current density. Finally, the particles move in such a way that the charge density σ is given by

$$\nabla \cdot \mathbf{E} = 4\pi\sigma,$$

and the divergence of the magnetic field is zero, i.e.,

$$\nabla \cdot \mathbf{B} = 0.$$

13.15. Consider an equilibrium system at rest, so that $\mathbf{v} = 0$, characterized by the state variables ρ_0, p_0, and $\mathbf{B}_0$. These variables are connected by equation (13.2) which now takes the form

$$0 = -\nabla p_0 + \frac{1}{c}\,(\mathbf{j}_0 \times \mathbf{B}_0).$$

Suppose that the state of equilibrium is perturbed by a very small amount, so that the pressure becomes $p_0 + p'$, the magnetic field becomes $\mathbf{B}_0 + \mathbf{B}'$, and so on. The system will now have a small velocity $\mathbf{v}$, and the quantities p', $\mathbf{B}'$, and $\mathbf{v}$ will develop with time in accordance with the equations of motion given above. It is essential that the perturbations p', $\mathbf{B}'$, etc., are considered to be small, so that their products, both with themselves and with each other, are negligible. In this way, it is possible to achieve *linear* equations of motion.

13.16. It is convenient to introduce a displacement variable vector $\boldsymbol{\xi}$ defined by

$$\frac{\partial \boldsymbol{\xi}}{\partial t} \equiv \mathbf{v} \quad \text{or} \quad \boldsymbol{\xi} \equiv \int \mathbf{v}\,dt,$$

so that equation (13.4) leads to

$$\frac{\partial \mathbf{B}'}{\partial t} = \nabla \times \left(\frac{\partial \boldsymbol{\xi}}{\partial t} \times \mathbf{B}_0 \right)$$

which, upon integration, yields

$$\mathbf{B}' = \nabla \times (\boldsymbol{\xi} \times \mathbf{B}_0). \tag{13.6}$$

13.17. Writing $p_0 + p'$ for p in equation (13.3) and introducing the definition of $\boldsymbol{\xi}$, this expression takes the linearized form

$$\frac{\partial p'}{\partial t} + \left(\frac{\partial \boldsymbol{\xi}}{\partial t} \right) \cdot \nabla p_0 + \gamma p_0 \nabla \cdot \left(\frac{\partial \boldsymbol{\xi}}{\partial t} \right) = 0$$

and integration then gives

$$p' = -\gamma p_0 \nabla \cdot \boldsymbol{\xi} - \boldsymbol{\xi} \cdot \nabla p_0. \tag{13.7}$$

From equation (13.2), with $\mathbf{j}$ equal to $\mathbf{j}_0 + \mathbf{j}'$, it follows that

$$\rho_0 \frac{\partial^2 \xi}{\partial t^2} = -\nabla p' + \frac{1}{c}\left[(\mathbf{j}' \times \mathbf{B}_0) + (\mathbf{j}_0 \times \mathbf{B}') \right].$$

Upon substituting equations (13.6) and (13.7) for $\mathbf{B}'$ and p', respectively, and utilizing equation (13.5), there results

$$\rho_0 \frac{\partial^2 \xi}{\partial t^2} = \nabla[\gamma p_0 \nabla \cdot \xi + \xi \cdot \nabla p_0]$$

$$+ \frac{1}{4\pi}\left[\nabla \times \nabla \times (\xi \times \mathbf{B}_0) \times \mathbf{B}_0 + \nabla \times \mathbf{B}_0 \times \nabla \times (\xi \times \mathbf{B}_0)\right], \quad (13.8)$$

where ρ_0, p_0, and $\mathbf{B}_0$ are determined by the initial equilibrium distribution.

13.18. Because equation (13.8) is linear in the unknown vector function ξ, it can, without loss of generality, be simplified by analyzing it into its Fourier components. Thus, assuming the time dependence to be of the form

$$\xi(\mathbf{x}, t) = \xi(\mathbf{x}) \exp (i\Omega t),$$

where $\mathbf{x}$ represents the space coordinate vector, equation (13.8) becomes

$$-\rho_0 \Omega^2 \xi = \nabla[\gamma p_0 \nabla \cdot \xi + \xi \cdot \nabla p_0] + \frac{1}{4\pi}\left[(\nabla \times \mathbf{Q}) \times \mathbf{B}_0 + (\nabla \times \mathbf{B}_0) \times \mathbf{Q}\right], \quad (13.9)$$

where $\mathbf{Q}$ is defined by

$$\mathbf{Q} \equiv \nabla \times (\xi \times \mathbf{B}_0).$$

It should be noted that the ξ in equation (13.9) and subsequently is really the spatial component $\xi(\mathbf{x})$, as is apparent from the presence of Ω^2 on the left side of the equation. The argument $(\mathbf{x})$ is omitted here to simplify the representation. Equation (13.9), called the *normal mode equation*, combined with the boundary conditions on ξ, namely, (*a*) ξ is finite everywhere, and (*b*) the normal component of ξ vanishes on any metallic conductors present, e.g., magnetic field coils or external conductor, forms an eigenvalue problem for the determination of Ω^2. If every eigenvalue is positive, then every mode is periodic in time and is a simple (traveling or standing) wave. If, on the other hand, one of the eigenvalues of Ω^2 is negative, then one of the two associated modes grows exponentially with time and the system is unstable.

13.19. The normal mode equation (13.9) is analogous to that used in the study of acoustical waves and is identical with that which is applicable to Alfvén (plane) waves in a plasma having a finite pressure. When the magnetic field is uniform, a further Fourier analysis, with respect to the space variable $\mathbf{x}$, is possible; thus, ξ may be taken to be of the form

$$\xi = (\xi_1, \xi_2, \xi_3) \exp (i\mathbf{k} \cdot \mathbf{x} + i\Omega t), \quad (13.10)$$

and then equation (13.9) reduces to a system of linear algebraic equations which determine the so-called dispersion relationship between Ω^2 and $\mathbf{k}$. In many cases, the situation is further simplified by taking p to be constant, so that ∇p is zero. An example of the application of the normal mode method to the stability of a pinched discharge is given in Appendix B to this chapter.

VARIATIONAL METHOD

13.20. Although solution of the normal mode equation is possible in the simpler cases, with more complicated geometries equation (13.9) remains as a system of three coupled partial differential equations and is quite unmanageable. In these circumstances, another procedure can be used which replaces the normal mode problem of equation (13.9) by a simple variational problem [4]. By sacrificing detailed knowledge of the normal modes, it is possible, in the following manner, to determine merely whether a given system is stable or not. First, as a result of scalar multiplication of equation (13.9) by ξ followed by integration over the volume of the plasma, there is obtained

$$\Omega^2 \int \rho(\xi \cdot \xi)\, d^3x = \int \xi \cdot F\xi\, d^3x, \tag{13.11}$$

where F represents the negative of the complicated differential operator (on ξ) on the right side of equation (13.9). Because the normal component of ξ vanishes on the boundary, i.e., at infinity or at a conductor, the right side of equation (13.11) can be simplified by partial integration; thus,

$$\Omega^2 = \frac{\delta W}{\int \rho(\xi \cdot \xi)\, d^3x}, \tag{13.12}$$

where δW is defined by

$$\delta W \equiv \int \xi \cdot F\xi\, d^3x$$

$$= \int \left[\frac{1}{4\pi} (Q \cdot Q - \nabla \times B \cdot Q \times \xi) + \gamma p (\nabla \cdot \xi)^2 + (\nabla \cdot \xi)(\xi \cdot \nabla p) \right] d^3x. \tag{13.13}$$

13.21. The operator F can be shown to be self-adjoint, and so it follows that any vector function ξ which makes equation (13.12) stationary is an eigenmode and vice versa; furthermore, this stationary value of Ω^2 is an eigenvalue. Thus, the eigenvalue problem is exactly equivalent to minimizing Ω^2 as given by equation (13.12). As seen earlier, stability of the plasma system to a perturbation depends upon whether Ω^2 is always positive (stable) or whether any of the Ω^2 values can be negative (unstable). But the denominator of equation (13.12) is positive, and so the sign of Ω^2 depends only on that of the numerator, i.e., δW. Thus, the normal mode stability condition can be replaced by the following variational principle: the necessary and sufficient condition that the system be stable is that δW, as given by equation (13.13), be not negative for every ξ which satisfies the boundary conditions.

13.22. The variational principle developed above is also called an energy principle, for it turns out, as might have been expected, that δW is equal to the change in potential energy of the system resulting from the displacement ξ. Consequently, only those equilibrium systems are stable in which every conceivable small perturbation is accompanied by an increase of potential energy, i.e., by doing work on the system.

13.23. The energy principle described above has been used to provide a rigorous confirmation of the general arguments presented earlier which indicate that stable confinement of a perfectly diamagnetic plasma by a magnetic field is possible only when the center of curvature of the lines of force always lies on the field side of the plasma-field boundary. The details of the argument are given in Appendix A to this chapter.

APPLICATIONS OF STABILITY THEORY

STABILITY OF THE PINCHED DISCHARGE

13.24. An important application of stability theory has been in the study of the straight cylindrical pinched discharge [5]. Here the simple nature of the geometry permits some generality in the consideration of pressure distributions. In the earliest models, the plasma was treated as a cylindrical rod in which the current was confined to an infinitely thin layer at the surface (cf. §7.35). As ∇p_0 and $\mathbf{j}_0$ vanish, the normal mode equation (13.9) then reduces to the simpler form

$$-\rho\Omega^2\xi = \nabla(\gamma p_0 \nabla\cdot\xi) - \frac{1}{4\pi}[\mathbf{B}_0 \times (\nabla \times \mathbf{Q})]. \tag{13.14}$$

The spatial component of the displacement variable vector ξ or, in fact, of any perturbation represented by the primed quantities given above, may be written in the form of equation (13.10), so that for cylindrical geometry

$$\xi = \xi(r) \exp(im\theta + ikz), \tag{13.15}$$

where m is an integer, assumed to be positive, and k is real. As seen in §7.34, the value of m determines the nature of the plasma perturbation, thus, $m = 0$ represents a necking-off, $m = 1$ is a kink or spiral, etc. The wave length of the perturbation is equal to $2\pi/|k|$. If equation (13.15) is substituted into equation (13.14), there is obtained a set of three coupled, differential equations in the three space coordinates. Upon applying the boundary conditions appropriate to the problem, namely, that (a) the magnetic field is tangent to the plasma-field interface and (b) the total pressure $p + B^2/8\pi$ is continuous across the interface, the required dispersion relation can be found. From this the condition for stability, i.e., Ω^2 is always positive, is derived. Except for the simplest cases, the actual determination of which configurations are stable and which are not, for different values of m, involves laborious numerical calculations.

13.25. The first application of the normal mode analysis to a pinched discharge was made for a cylindrical plasma in which the current was confined to an extremely thin sheath at the surface, and there was no axial magnetic field [6].   For the case of $m = 1$, it was found that the system was unstable.   The calculations were extended to other values of m and it was shown that instability would also arise from $m = 0$, and, to a lesser extent, for $m \geqslant 2$. For sufficiently short wave lengths, the growth of the unstable perturbations, for all m values, was found to have an e-folding time of $(2|k|p_0/\rho_0 r_0)^{\frac{1}{2}}$, where r_0 is the tube radius [7].   Since the velocity of sound is equal to $(\gamma p_0/\rho_0)^{\frac{1}{2}}$, it has become the common practice to state that the instabilities in a pinched discharge grow, roughly, with a velocity equal to that of sound in the given plasma.   At high temperatures, this can be very large.

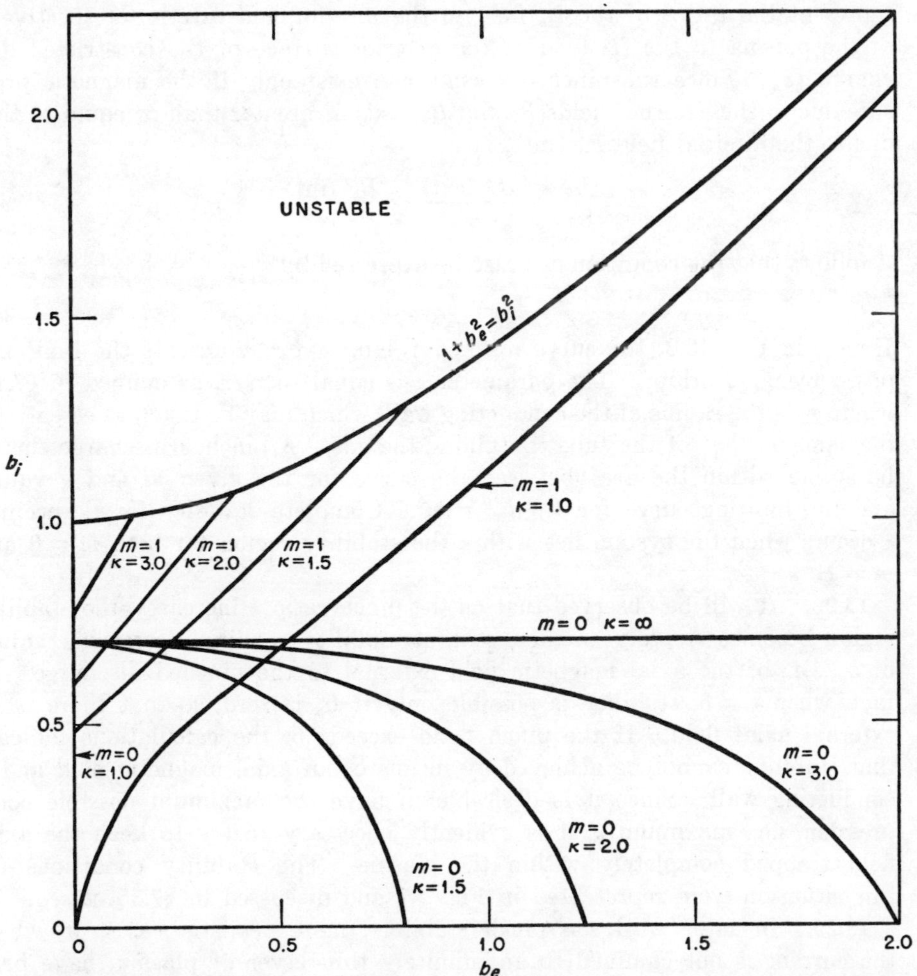

FIG. 13.2.   Conditions of stability to $m = 0$ and $m = 1$ perturbations in pinched discharge.

13.26. The normal mode method has also been applied to the case in which there is an axial magnetic field inside as well as outside the plasma, and also with conducting walls surrounding the system [8-10]. The stability conditions derived in this manner were found to be essentially identical with those obtained by a procedure similar in principle to the variational method described above [11].

13.27. The results of the computations are summarized in Fig. 13.2, for the two cases $m = 0$ and $m = 1$; the quantities b_i and b_e, defined by

$$b_i \equiv \frac{B_z \text{ (int)}}{B_\theta} \quad \text{and} \quad b_e \equiv \frac{B_z \text{ (ext)}}{B_\theta},$$

represent the ratios of the B_z field in the interior and exterior, respectively, of the plasma to the B_θ field at the exterior surface of the constricted discharge [8]. Since the pinched plasma can exist only if the magnetic pressure due to the external fields B_θ and B_z (ext) is greater than or equal to that due to the internal field B_z (int), i.e.,

$$\frac{B_\theta^2}{8\pi} + \frac{B_z^2 \text{ (ext)}}{8\pi} \geqslant \frac{B_z^2 \text{ (int)}}{8\pi},$$

it follows that the condition can also be expressed by

$$1 + b_e^2 \geqslant b_i^2.$$

Hence, in Fig. 13.2, the curve marked $1 + b_e^2 = b_i^2$ represents the limit imposed by equilibrium. The parameter κ is equal to r_0/r, as defined in §7.17, where r_0 is the radius of the conducting wall, which may be taken as essentially the same as that of the tube containing the gas. A pinched discharge should be stable within the area between the curve for the given m and κ values and the limiting curve for $1 + b_e^2 = b_i^2$. Complete stability for a specified κ occurs when the system lies within the stability region for both $m = 0$ and $m = 1$.

13.28. It will be observed that as the pinch ratio κ increases, the stability region becomes steadily smaller, so that stability requires decreasing values of b_e, i.e., of the axial magnetic field external to the pinched discharge. In fact, when $\kappa = 5$, stability is possible only if b_e is zero, so that there is no external axial field. If the pinch ratio exceeds 5, the calculations indicate that stability cannot be achieved by means of an axial magnetic field and a conducting wall. Since it is desirable to have the maximum possible compression, i.e., maximum κ, it is evidently necessary to try to keep the axial field trapped completely within the plasma. The stability conditions for this situation were represented in Fig. 7.7 and discussed in §7.35 *et seq.*

13.29. In later work on pinch stability, more realistic cases, in which the current is not confined to an infinitely thin layer of plasma, have been investigated. By allowing the current layer to have structure, more complicated plasma modes become possible and the requirements for stability become correspondingly more severe. The general nature of the results given above remains unchanged, but an analysis now indicates that stability is not possible if there is an axial magnetic field outside the plasma in the same direction as the field within the plasma. It appears that, when the current layer is thin but finite, an axial field in the vacuum outside the plasma is required for stability but it must be in a sense *opposite* to that inside the plasma. However, the reversed field alone is not sufficient for stability, since a special surface distribution is required in addition [12].

13.30. In the case of very thick current layers, it is a relatively simple matter to utilize the variational principle to derive a necessary, although not sufficient, condition for stability which relates the pressure gradient in the plasma to the torsion of the magnetic field lines [13]. The general conclusion was given in §7.97, and will be repeated here for completeness. The lines of the combined axial and azimuthal fields form spirals and their pitch, μ, may be defined by

$$\mu \equiv \frac{B_\theta}{rB_z},$$

where r is the radius of the pinched discharge; the quantity $(1/\mu)(\partial\mu/\partial r)$ measures the torsion (or "shear") of the field lines. The necessary stability condition alluded to above is then

$$\frac{r}{4}\left(\frac{1}{\mu}\cdot\frac{\partial\mu}{\partial r}\right)^2 + \frac{8\pi}{B_z^2}\left(\frac{\partial p}{\partial r}\right) > 0.$$

As seen in §7.99, one possible way of realizing this condition is to apply an axial magnetic field outside the plasma in a direction opposite to that of the field within the plasma.

13.31. It may be mentioned that stability conditions for the pinched discharge, which are both necessary and sufficient, have been derived [5, 14, 15]. However, these involve complex mathematical expressions and their application requires lengthy calculations. It would appear to be almost as simple to perform the variational problem directly for each case with the aid of a computing machine.

STABILITY IN STELLARATOR SYSTEMS

13.32. As indicated in Chapter 8, some use of hydromagnetic stability theory has been made in connection with stellarator systems, particularly of the kink ($m = 1$) and interchange instabilities. The situation in the stellarator differs from that of the pinched discharge in the respect that there is essentially only one type of magnetic field, namely the axial field, in the former case and this is largely within the plasma. Any azimuthal field that arises from current passing through the plasma, e.g., for ohmic heating, is negligible.

13.33. The normal mode treatment has been applied to the situation while the plasma is being heated ohmically [16]. It has been found that, in a stellarator in which the rotational transform arises from the large-scale geometry of the tube rather than from small local perturbations of the magnetic field produced by helical windings, the $m = 1$ mode is the only one which can cause serious instability. This occurs when the rotational transform due to the longitudinal current just compensates for that caused by the geometry of the stellarator tube. The theoretical treatment indicates that, in this case, the effect of an external conductor is negligible, although such a conductor tends to suppress the kink instability in a pinched discharge (§7.27).

13.34. By means of the variational principle, it has been shown that, when the rotational transform arises from the geometry of the stellarator, as considered above, the plasma is always subject to the interchange (or flute) instability. Utilizing the same procedure it has been found, however, that this can be stabilized by means of a transverse, helical magnetic field, which can itself produce a rotational transform, as explained in Chapter 8 [17]. Consider a plasma with an axial magnetic field inside it; this field must be so disposed that any conceivable displacement requires work to be done. The plasma may be visualized as a set of "onionskin" layers, each layer being a surface made up of magnetic field lines. If now the field lines are given torsion (or shear), that is, if the directions of the lines are always different in adjacent layers, then no displacement of the plasma will be possible which does not bend the lines. Even if there is a wavelike displacement in which the crests and troughs are aligned with the magnetic field in one particular layer, the torsion of the field line will spoil the alignment in neighboring layers. In such a system work must be done on the magnetic field for any conceivable displacement, and stability will be determined by whether the heat energy of the plasma can supply this work or not.

13.35. With the aid of the variational (or energy) principle it has been found that torsion of the magnetic field lines, such as would be produced by means of a transverse, helical field, as described in §8.19, has a stabilizing effect. The system is stable against interchange so long as the heat energy available is not too large or, in other words, provided the plasma pressure is small. These limiting pressures appear to be so low as to require β values of the order of 0.01. Furthermore, provided the rotational transform angle produced by the helical field is greater than and has the same sign as that caused by the ohmic heating current, it is expected that the plasma will be stable to perturbations of all m values.

COLLISIONLESS (NONEQUILIBRIUM) THEORIES

THE BOLTZMANN EQUATION

13.36. Although the hydromagnetic model of a plasma, based on the hydrodynamic equations (13.1), and (13.2), and (13.3), has been treated at some length, it is not because it is by any means a final or definitive theory. It is rather because it has led to results which have provided a significant insight into the stability of a plasma confined by a magnetic field. There are several defects in the theory, and so efforts are being made to develop other theories which are more rigorous but, on the other hand, more complicated; some of these alternative models will be discussed below.

13.37. The main objection to the hydrodynamic theory is concerned with the postulate of frequent collisions. As the temperature increases, the cross sections for Coulomb scattering collisions decrease (§4.48 *et seq.*) and at temperatures of thermonuclear interest they become so small that the mean free path of a particle may be comparable with or, depending on the density, may even be considerably larger than the dimensions of the containing vessel. Thus, although collisions may be frequent at the densities and temperatures of most laboratory experiments, they certainly will be much less common in an actual thermonuclear reactor. The absence of collisions has several important consequences.

13.38. In the first place, there is now no mechanism for partitioning the kinetic energy of the particles equally among their three degrees of freedom; thus, $p_\parallel$, the pressure parallel to the magnetic field lines, will not necessarily be the same as $p_\perp$, the pressure perpendicular to the field. Second, there is nothing to inhibit the motion of particles along the lines of force, and so heat flow parallel to these lines is likely to be very rapid. Third, in a reacting plasma, at least, electrons and ions will undoubtedly be at different temperatures. In brief, therefore, it may be concluded that the plasma will not be even in local thermodynamic equilibrium. If by any chance it is, then this local equilibrium will be upset if the plasma is in any way disturbed. It should be pointed out, too, that electrons and ions do not have to move together, and local deviations of charge density from zero may occur.

13.39. For the foregoing reasons, the nonequilibrium alternatives to the hydromagnetic theory start from the opposite assumption to that in §13.11, namely, it is supposed that there are no collisions or, if there are any, their effects are neglected. This postulate alone is not enough to bring about a sufficient simplification of the problem and another assumption is needed. The one made is to say that the magnetic field is so strong that the radii and periods of the gyromagnetic orbits of all particles are much smaller than any distances and times, respectively, of interest. In place of the three hydrodynamic equations (§13.12), the nonequilibrium theories either utilize directly a form of the Boltzmann distribution equation applicable to charged particles in which the collision term is disregarded [19-21], or they take averages over individual orbits which lead to results essentially equivalent to the Boltzmann equation [22, 23].

13.40. The so-called collisionless Boltzmann equation* may be written as

$$\frac{\partial f}{\partial t} + \mathbf{v}\cdot\nabla f + \frac{e}{m}\left[\mathbf{E} + \frac{1}{c}(\mathbf{v}\times\mathbf{B})\right]\cdot\nabla_v f = 0, \qquad (13.16)$$

*Some workers in the field refer to this as the Liouville equation, since the latter has no collision term. However, it seems preferable to use the expression "collisionless Boltzmann equation" to imply that the Boltzmann equation is "correct," but collisions are being ignored as an approximation in order to simplify the treatment.

where the distribution function f, which is really $f(\mathbf{x}, \mathbf{v}, t)$ is defined so that $f\delta^3\mathbf{x}\delta^3\mathbf{v}$ is the number of particles in the volume element $\delta^3\mathbf{x}$ and in the velocity range $\delta^3\mathbf{v}$ at position $\mathbf{x}$ and velocity $\mathbf{v}$ in the six-dimensional (position-velocity) phase space at time t; the symbols e and m represent the charge and mass of the particles and ∇_v is the gradient operator for differentiation with respect to the components of the velocity vector rather than with respect to the space coordinates [24, 25].

13.41. In order to obtain equations for the mass and momentum balance, the zero- and first-order moments of the Boltzmann equation are derived by multiplying equation (13.16) by $md\mathbf{v}$ and $m\mathbf{v}d\mathbf{v}$, respectively, and integrating over all velocity space. Utilizing the Maxwell relationships, as in §13.13, the results are found to resemble formally equations (13.1) and (13.2) of hydrodynamics, except that the gradient of a scalar pressure, i.e., ∇p, in the latter is replaced by the divergence of a pressure tensor.

13.42. A difficulty arises in connection with the development of the second-order moment of equation (13.16) to yield an expression for the energy balance. This introduces third moments of the distribution function which are the components of the heat flow vector. In ordinary hydrodynamics, use is made of the fact that, as a result of collisions, the distribution function is Maxwellian (or nearly so) in form. The third moments can then be related to the second moments and the equations can be closed. In addition, the heat flow is often so small, because of frequent collisions, that it can be neglected; in this event, the result reduces to the same form as the adiabatic equation (13.3).

13.43. In applying the situation to a plasma, however, the neglect of collisions makes it impossible to obtain closed equations because the heat flow terms cannot be evaluated. The simplest way of dealing with the situation is arbitrarily to cut off the energy balance equation at the third moment, which is equivalent to assuming no heat flow. The equations now form a closed system. The postulate of a strong magnetic field (§13.39) means that it is necessary to distinguish only between the components $p_\parallel$ and $p_\perp$ of the pressure tensor. As a result, the single adiabatic law, which can be written as $d(p/\rho^\gamma)/dt = 0$, is replaced by

$$\frac{d}{dt}\left(\frac{p_\parallel B^2}{\rho^3}\right) = 0$$

and

$$\frac{d}{dt}\left(\frac{p_\perp}{\rho B}\right) = 0,$$

where the d/dt is the hydrodynamic operator defined in §13.12.

13.44. The nonequilibrium theory outlined above has a serious defect, since it artificially suppresses heat flow under collisionless conditions when thermal transport along the magnetic lines is expected to be considerable.

Nevertheless, the theory has led to some new results of interest. One of these, which is typical of nonequilibrium theories, arises from a consideration of the simple case of a plasma having initially uniform pressures $p_\parallel$ and $p_\perp$ and density ρ_0 in a uniform field $\mathbf{B}_0$. By solving the problem of plane waves, it is found that exponentially growing waves are possible. In other words, instability may occur even in a plasma with a uniform distribution. The condition for stability is

$$\frac{p_\perp^2}{6(p_\perp + B^2/8\pi)} < p_\parallel < p_\perp + \frac{B^2}{4\pi},$$

where $p_\perp$ and $p_\parallel$ are the respective undisturbed values; hence, for stability of the plasma, $p_\parallel$ must lie within the range indicated [26]. A slightly more restrictive stability condition, which takes heat flow into consideration, may be written in the analogous form [27]

$$\frac{p_\perp^2}{p_\perp + B^2/8\pi} < p_\parallel < p_\perp + \frac{B^2}{4\pi}.$$

13.45. An interesting new general result derived from a more elaborate nonequilibrium theory is that the Maxwellian velocity distribution leads to the most stable state and that deviations from this distribution, if sufficiently great, can in themselves cause instability. A particular case arises when $p_\parallel$ and $p_\perp$ are unequal.

13.46. In addition to the foregoing conclusions, some other interesting features of nonequilibrium theories will be noted. First, it may be mentioned that it has been found possible to develop an energy principle, analogous to that described in §13.21, based directly on the collisionless Boltzmann equation [28]. An essentially equivalent principle has also been derived by summing over the orbits of individual particles [23]. These theories lead to the important result that, if $p_\parallel$ and $p_\perp$ are initially equal, then the δW predicted is at least as large as that obtained from the simple hydromagnetic theory, as given by equation (13.13). This means that, when stability is predicted on the basis of the energy (or variational) principle in hydromagnetic theory, the conclusion can be accepted with confidence in many cases. The utility of the simple theory would thus appear to be greater than its numerous deficiencies would suggest.

13.47. All three energy principles mentioned so far are applicable only to states of static equilibrium of the system as a whole. If the plasma is in a steady state of motion, e.g., if it is rotating, the kinetic energy of this motion is an additional source from which an instability might be driven. What this means mathematically is that, in linearizing equation (13.2), or its equivalent, it is not permissible to neglect terms arising from $\rho(\mathbf{v}\cdot\nabla)\mathbf{v}$.

13.48. A method has been developed for handling the problem arising when the plasma is in motion, by assuming a strong magnetic field and no collisions

[20, 29, 30]. The results lead to linearized equations of motion, i.e., to normal mode equations, rather than to a variational principle. This means that the treatment is very difficult to apply to anything but the simplest geometry, although it may be amenable to machine calculations in other cases.

APPENDIX A: APPLICATION OF THE VARIATIONAL PRINCIPLE TO A PERFECTLY DIAMAGNETIC PLASMA

13.49. In the following treatment it will be shown how the variation principle (or δW-formalism) has been applied to determine how the stability of a perfectly diamagnetic plasma confined by a magnetic field depends on the direction of the curvature of the vectors of the field lines [4]. Starting with the basic equation (13.12), with δW defined by equation (13.13), some general results will first be derived, and then they will be applied to the case of immediate interest.

13.50. The integral for δW can be transformed by partial integration utilizing the relationships

$$-\boldsymbol{\xi}\cdot\boldsymbol{\nabla}\times\mathbf{Q}\times\mathbf{B}_0 = Q^2 + \boldsymbol{\nabla}\cdot[\mathbf{Q}\times(\boldsymbol{\xi}\times\mathbf{B}_0)]$$

and

$$-\boldsymbol{\xi}\cdot\boldsymbol{\nabla}[\gamma p\boldsymbol{\nabla}\cdot\boldsymbol{\xi}+\boldsymbol{\xi}\cdot\boldsymbol{\nabla}p_0] = \gamma p_0(\boldsymbol{\nabla}\cdot\boldsymbol{\xi})^2 + (\boldsymbol{\nabla}\cdot\boldsymbol{\xi})(\boldsymbol{\xi}\cdot\boldsymbol{\nabla}p_0) - \boldsymbol{\nabla}\cdot[\gamma p_0\boldsymbol{\xi}\boldsymbol{\nabla}\cdot\boldsymbol{\xi}+\boldsymbol{\xi}(\boldsymbol{\xi}\cdot\boldsymbol{\nabla}p_0)].$$

It follows, therefore, that

$$\delta W = \delta W_f - \frac{1}{2}\int_S (n\cdot\boldsymbol{\xi})[\gamma p_0\boldsymbol{\nabla}\cdot\boldsymbol{\xi}+\boldsymbol{\xi}\cdot\boldsymbol{\nabla}p_0 - \frac{1}{4\pi}\mathbf{B}_0\cdot\mathbf{Q}]\,ds \qquad (13.17)$$

where δW_f is defined by

$$\delta W_f \equiv \int_V \left[\frac{Q^2}{4\pi} - \mathbf{j}\cdot\mathbf{Q}\times\boldsymbol{\xi} + \gamma p_0(\boldsymbol{\nabla}\cdot\boldsymbol{\xi})^2 + (\boldsymbol{\nabla}\cdot\boldsymbol{\xi})(\boldsymbol{\xi}\cdot\boldsymbol{\nabla}p)\right]d^3x, \qquad (13.18)$$

and ds is an element of surface; the integral in equation (13.17) is taken over the surface of the plasma whereas that in equation (13.18), as in equation (13.13), is over the volume.

13.51. In normal laboratory experiments, the "vacuum" outside the plasma usually contains enough particles to conduct current, even if it will not produce a pressure; hence, the plasma boundary lies on the conducting walls, where $(n\cdot\boldsymbol{\xi}) = 0$, the symbol n representing the unit vector normal to the surface; the surface integral then contributes nothing. In idealized cases, however, where a perfect vacuum exists or when the walls are nonconducting, the surface integral must be evaluated. This is best done by using boundary conditions to simplify the integral.

13.52. The first step is to calculate the vacuum fields; these are indicated by a circumflex. The electric field is given by

$$\hat{\mathbf{E}} = \hat{\mathbf{E}}_0 + \hat{\mathbf{E}}' = \hat{\mathbf{E}}'.$$

since $\hat{\mathbf{E}}_0 = 0$, and the magnetic field is

$$\hat{\mathbf{B}} = \hat{\mathbf{B}}_0 + \hat{\mathbf{B}}'.$$

The perturbations $\hat{\mathbf{E}}'$, $\hat{\mathbf{B}}'$ together form an electromagnetic field and are therefore derivable from a potential. Thus, if $\mathbf{A}$ represents a vector potential, then

$$\hat{\mathbf{B}}' = \nabla \times \mathbf{A} \tag{13.19}$$

and

$$\hat{\mathbf{E}}' = -\frac{1}{c} \cdot \frac{\partial \mathbf{A}}{\partial t} = -\frac{i\Omega}{c} \mathbf{A}, \tag{13.20}$$

and $\mathbf{A}$ satisfies the equation

$$\nabla \times \nabla \times \mathbf{A} = 0. \tag{13.21}$$

No scalar potential is necessary as no charge accumulations are assumed.

13.53. Next, the boundary conditions will be derived. In the first place, the magnetic field must be tangent to the plasma-vacuum interface. That this is true for the interior field follows directly from equation (13.6). The condition on the vacuum field is

$$\mathbf{n} \cdot \hat{\mathbf{B}}' = \mathbf{n} \cdot \nabla \times (\boldsymbol{\xi} \times \hat{\mathbf{B}}_0). \tag{13.22}$$

The second condition follows from the assumption of perfect conductivity: to an observer moving with the interface, the tangential component of the electrical field is zero. Transcribed into the laboratory frame, this condition is

$$\mathbf{n} \times \left[\mathbf{E}' + \frac{1}{c}(\mathbf{v} \times \mathbf{B}_0) \right] = \mathbf{n} \times \left[\hat{\mathbf{E}}' + \frac{1}{c}(\mathbf{v} \times \hat{\mathbf{B}}_0) \right].$$

But

$$\mathbf{E}' + \frac{1}{c}(\mathbf{v} \times \mathbf{B}_0) = 0,$$

therefore

$$\mathbf{n} \times \mathbf{A} = -(\mathbf{n} \cdot \boldsymbol{\xi})\hat{\mathbf{B}}_0. \tag{13.23}$$

The third boundary condition is that the total pressure, $p + B^2/8\pi$, must be continuous across the boundary, as otherwise infinite acceleration would result. This condition can be written

$$p_0 + \delta p + \frac{(\mathbf{B}_0 + \delta\mathbf{B})^2}{8\pi} = \frac{(\hat{\mathbf{B}}_0 + \delta\hat{\mathbf{B}})^2}{8\pi},$$

together with the equilibrium equations

$$p_0 + \frac{B_0^2}{8\pi} = \frac{\hat{B}_0^2}{8\pi}.$$

Upon subtraction, the result is

$$\delta p + \frac{1}{4\pi} \mathbf{B}_0 \cdot \delta\mathbf{B} = \frac{1}{4\pi} \hat{\mathbf{B}}_0 \cdot \delta\hat{\mathbf{B}}.$$

In these equations δp means the change in p resulting from the displacement over $\boldsymbol{\xi}$; the significance of $\delta \mathbf{B}$ is similar. In each case the change is computed moving with the displacement. It follows then that

$$\delta p = p' + \boldsymbol{\xi} \cdot \boldsymbol{\nabla} p$$
$$\delta B = \mathbf{B}' + (\boldsymbol{\xi} \cdot \boldsymbol{\nabla})\mathbf{B},$$

and when these are used together with equation (13.7) for p' and equation (13.6) or (13.19), for $\mathbf{B}'$ or $\hat{\mathbf{B}}'$, respectively, there is obtained the condition

$$-\gamma p_0 \boldsymbol{\nabla} \cdot \boldsymbol{\xi} + \frac{1}{4\pi} \mathbf{B}_0 \cdot \mathbf{Q} = \frac{1}{4\pi} \mathbf{B}_0 \cdot \boldsymbol{\nabla} \times \mathbf{A} + \frac{1}{8\pi} (\boldsymbol{\xi} \cdot \boldsymbol{\nabla})[\hat{B}_0{}^2 - B_0{}^2]. \quad (13.24)$$

13.54. Returning now to equation (13.17) and setting equation (13.24) into the surface integral, it is found that

$$\delta W - \delta W_f = -\frac{1}{2} \int_S (\mathbf{n} \cdot \boldsymbol{\xi}) \left\{ (\boldsymbol{\xi} \cdot \boldsymbol{\nabla}) \left[p_0 + \frac{B_0{}^2 - \hat{B}_0{}^2}{8\pi} \right] - \frac{1}{4\pi} \hat{\mathbf{B}}_0 \cdot \boldsymbol{\nabla} \times \mathbf{A} \right\} ds.$$

Since both $\mathbf{B}_0$ and $\mathbf{j}_0$ are parallel to the interface, the tangential derivative of $p_0 + \dfrac{B_0{}^2 - \hat{B}_0{}^2}{8\pi}$ vanishes, and so

$$\boldsymbol{\xi} \cdot \boldsymbol{\nabla} \left[p_0 + \frac{B_0{}^2 - \hat{B}_0{}^2}{8\pi} \right] = (\boldsymbol{\xi} \cdot \mathbf{n})\mathbf{n} \cdot \boldsymbol{\nabla} \left[p_0 + \frac{B_0{}^2 - \hat{B}_0{}^2}{8\pi} \right].$$

Furthermore, from the boundary condition equation (13.23),

$$\int_S \frac{1}{4\pi} (\mathbf{n} \cdot \boldsymbol{\xi}) \hat{\mathbf{B}}_0 \cdot \boldsymbol{\nabla} \times \mathbf{A} \, ds = -\int_S \frac{1}{4\pi} (\mathbf{n} \times \mathbf{A}) \cdot \boldsymbol{\nabla} \times \mathbf{A} \, ds$$

$$= -\int_S \frac{1}{4\pi} \mathbf{n} \cdot (\mathbf{A} \times \boldsymbol{\nabla} \times \mathbf{A}) \, ds.$$

Finally, this may be transformed into a volume integral taken over the vacuum region; thus,

$$\int \frac{1}{4\pi} (\mathbf{n} \cdot \boldsymbol{\xi}) \hat{\mathbf{B}}_0 \cdot \boldsymbol{\nabla} \times \mathbf{A} \, ds = \int_{\hat{V}} \frac{1}{4\pi} \boldsymbol{\nabla} \cdot [\mathbf{A} \times \boldsymbol{\nabla} \times \mathbf{A}] \, d^3x$$

$$= \int_{\hat{V}} \frac{1}{4\pi} [(\boldsymbol{\nabla} \times \mathbf{A})^2 - \mathbf{A} \times \boldsymbol{\nabla} \times \boldsymbol{\nabla} \times \mathbf{A}] \, d^3x.$$

Combining these results, and making use of equation (13.21), it follows that

$$\delta W = \delta W_f + \delta W_S + \delta W_{\hat{V}},$$

where δW_f is defined by equation (13.18) and

$$\delta W_S = -\frac{1}{2} \int_S (\mathbf{n} \cdot \boldsymbol{\xi})^2 \mathbf{n} \cdot \boldsymbol{\nabla} \left[p_0 + \frac{B_0{}^2 - \hat{B}_0{}^2}{8\pi} \right] ds \quad (13.25)$$

and

$$\delta W_{\hat{V}} = \int_{\hat{V}} \frac{1}{8\pi} (\boldsymbol{\nabla} \times \mathbf{A})^2 \, d^3x = \int_{\hat{V}} \left[\frac{(\hat{B}')^2}{8\pi} \right] d^3x. \quad (13.26)$$

It will be recalled from §13.21, that the condition for stability of the plasma is that δW be not negative for every ξ which satisfies the boundary conditions.

13.55.   By means of the results derived above, it is now possible to determine the conditions of stability of the perfectly diamagnetic plasma confined by a magnetic field.   First, it should be noted that if $\mathbf{B}_0 = 0$, then $\mathbf{Q} = 0$ and p_0 is a constant, so that equation (13.18) becomes

$$\delta W_f = \tfrac{1}{2} \int_{\hat{v}} \gamma p_0 (\nabla \cdot \xi)^2 \, d^3x \geqslant 0. \tag{13.27}$$

In fact, δW_f may be minimized to zero, by choosing ξ divergenceless, without compromising the freedom to choose $(\xi \cdot \mathbf{n})$ arbitrarily.

13.56.   Pressure balance requires $\hat{B}_0{}^2$ to be constant on the surface; thus, writing

$$\hat{\mathbf{B}}_0 = \hat{B}_0 \tau,$$

where τ is a unit vector in the $\hat{B}$ direction, so that

$$\nabla(\hat{B}_0{}^2) = 2\hat{B}_0{}^2 (\tau \cdot \nabla)\tau.$$

Consequently, equation (13.25) for δW_S may be written as

$$\delta W_S = \frac{\hat{B}_0{}^2}{8\pi} \int_S (\mathbf{n} \cdot \xi)^2 (\mathbf{n} \cdot \mathbf{K}) \, ds, \tag{13.28}$$

where $\mathbf{K}$ is the principal curvature vector* defined by

$$\mathbf{K} \equiv (\tau \cdot \nabla)\tau.$$

Because $\hat{B}_0{}^2$ is constant over the surface, it follows that $\mathbf{K}$ is normal to the surface, and

$$\mathbf{n} \cdot \mathbf{K} = \pm \frac{1}{R},$$

where R is the principal radius of curvature of the $\hat{B}$ line through the point in question, and the sign depends upon whether $\mathbf{K}$ points outward or inward from the surface.   It will be apparent that if $\mathbf{K}$ points outward, the integrand of equation (13.28) is everywhere nonnegative.   Since $\delta W_{\hat{v}}$ is certainly not negative, the system is stable.

13.57.   Suppose, now, that over some region $\mathbf{K}$ points inward.   The value of $(\xi \cdot \mathbf{n})$ will be chosen so as to vanish outside the given region.   In order to treat this region, it is convenient to introduce orthogonal, curvilinear coordinates (u^1, u^2, u^3), so chosen that $\hat{B}_1 = 0$ everywhere and, further, that $\hat{B}_3 = 0$ on the surface.   Thus the surface is a u^1 surface and on it $\hat{B}$ points in the u^2 direction.   Over the region being considered it is postulated that

$$(\mathbf{n} \cdot \xi) = \sin \alpha \chi_2 \cdot \sin \beta \chi_3, \tag{13.29}$$

---

*This is one of the Frenet formulae [31].

where χ_i is defined by

$$\chi_i \equiv \int h_i \, du^i,$$

so that the χ_i are arc lengths along the i curves.

13.58.   In order to calculate $\delta W_{\hat{v}}$, it should be noted that, since $\hat{B}'$ is curl-free, it is possible to write

$$\hat{B}' = -\nabla\Phi$$

and, therefore,

$$\delta W_{\hat{v}} = \frac{1}{8\pi} \int_{\hat{v}} (\nabla\Phi)^2 \, ds.$$

This is minimized by choosing Φ to satisfy the equation

$$\nabla^2\Phi = 0 \tag{13.30}$$

and as a result of this choice it is possible to perform a partial integration which yields

$$\delta W_{\hat{v}} = -\frac{1}{8\pi} \int_S (\mathbf{n}\cdot\nabla\Phi)\Phi \, ds. \tag{13.31}$$

13.59.   In the postulated coordinate system, the boundary condition equation (13.22) becomes

$$-(\mathbf{n}\cdot\nabla\Phi) = \frac{h_1}{h_2} \hat{B}_0 \partial_2 \left(\frac{\xi_1}{h_1}\right) = -\frac{1}{h_1} \partial_1\Phi,$$

so that on the surface, Φ has the form

$$\frac{1}{h_1} \partial_1\Phi = -B_0\alpha \cos \alpha\chi_2 \cdot \sin \beta\chi_3 + \Theta. \tag{13.32}$$

Solving the Laplace equation with this boundary condition yields

$$\Phi = B_0 \frac{\alpha}{\gamma} \cos \alpha\chi_2 \cdot \sin \beta\chi_3 \cdot e^{-\gamma\chi_1} + \Theta \tag{13.33}$$

where

$$\gamma^2 = \alpha^2 + \beta^2. \tag{13.34}$$

Here the symbol Θ represents terms of the order of curvature divided by α or β.

13.60.   The use of equation (13.29) now yields

$$\delta W_S = -\frac{A \hat{B}^2}{32\pi\overline{R}},$$

where A is the area over which $(\mathbf{n}\cdot\boldsymbol{\xi})$ is not equal to zero and $\overline{R}$ is the average radius of curvature over this area. Similarly, equations (13.32) and (13.33) give

$$\delta W_{\hat{v}} = \frac{A \hat{B}^2}{32\pi} \cdot \frac{\alpha^2}{\gamma} + \Theta,$$

so that

$$\delta W = \frac{A\hat{B}^2}{32\pi}\left[\frac{\alpha^2}{\gamma} - \frac{1}{R} + \Theta'\right].$$

Choosing α as small as possible, that is, of the order of the reciprocal length of the region A, and β very large, the terms $\frac{\alpha^2}{\gamma} + \Theta'$ may be made as small as desired. Hence, δW can be made negative and the system is unstable.

APPENDIX B: NORMAL MODE ANALYSIS OF THE PINCH WITH SHARP BOUNDARY

13.61. As an illustration of the application of the normal mode procedure for determining the conditions of stability of a plasma, an idealized model of the cylindrical pinch will be considered. The plasma is assumed to be in the form of a cylindrical rod in the interior of which no current flows. There is, however, a magnetic field in the plasma; hence, in cylindrical coordinates, the interior field has the form

$$\mathbf{B}_0 = (0, 0, B_s); \quad B_s = \text{constant}, \tag{13.35}$$

where, as before, $\mathbf{B}_0$ is the equilibrium field vector. Outside the plasma, the most general field possible is assumed, namely,

$$\hat{\mathbf{B}}_0 = (0, \hat{B}_\theta, \hat{B}_s) \tag{13.36}$$

$$\hat{B}_s = \text{constant} \tag{13.37}$$

$$\hat{B}_\theta = \hat{b}/r, \quad \hat{b} = \text{constant}. \tag{13.38}$$

As in Appendix A, a circumflex over a symbol indicates the value in the vacuum outside the plasma. The plasma pressure is constant, and so the normal mode equation (13.9) reduces to

$$-\rho\Omega^2\xi = \nabla(\gamma p\nabla\cdot\xi) - \frac{1}{4\pi}\mathbf{B}\times(\nabla\times\mathbf{Q}), \tag{13.39}$$

the zero subscripts, indicating equilibrium quantities, having been omitted since this will cause no confusion. The three components of equation (13.39) represent the equations of motion which, together with the appropriate boundary conditions, permit a solution of the problem.

13.62. Because of the linear nature of the equations, the spatial component of the displacement vector may be analyzed into its Fourier components by writing

$$\xi = [\xi_r(r)\xi_\theta(r)\xi_s(r)]\exp(im\theta + ikz). \tag{13.40}$$

When this is done it follows, from the definition of $\mathbf{Q}$ in §13.18, that

$$Q_r = ikB_s\xi_r \tag{13.41}$$

$$Q_\theta = ikB_s\xi_\theta \tag{13.42}$$

$$Q_s = -B_s\left[\frac{1}{r}\partial_r(r\xi_r) + \frac{im}{r}\xi_\theta\right]. \tag{13.43}$$

The individual components of equation (13.39) then give the following set of normal mode equations:

$$-\rho\Omega^2\xi_r = \partial_r(\gamma p\nabla\cdot\xi) - \frac{B_z}{4\pi}(\partial_r Q_z - ikQ_r) \tag{13.44}$$

$$-\rho\Omega^2\xi_\theta = \frac{im}{r}(\gamma p\nabla\cdot\xi) - \frac{B_z}{4\pi}\left(\frac{im}{r}Q_z - ikQ_\theta\right) \tag{13.45}$$

$$-\rho\Omega^2\xi_z = ik(\gamma p\nabla\cdot\xi). \tag{13.46}$$

The form of these equations suggests that the dependent variables be changed by introducing the new function Ψ defined by*

$$\Psi \equiv \gamma p\nabla\cdot\xi - \frac{1}{4\pi}\mathbf{B}\cdot\mathbf{Q}$$

$$= \left(\gamma p + \frac{B_z^2}{4\pi}\right)\left[\frac{1}{r}\partial_r(r\xi_r) + \frac{im}{r}\xi_\theta\right] + ik\gamma p\xi_z. \tag{13.47}$$

In terms of Ψ, equations (13.44-46) may be rewritten, respectively, as

$$\left(\frac{k^2 B_z^2}{4\pi} - \rho\Omega^2\right)\xi_r = \partial_r\Psi \tag{13.48}$$

$$\left(\frac{k^2 B_z^2}{4\pi} - \rho\Omega^2\right)\xi_\theta = \frac{im}{r}\Psi \tag{13.49}$$

$$\left[\frac{k^2 B_z^2}{4\pi} - \rho\Omega^2\left(1 + \frac{B_z^2}{4\pi\gamma p}\right)\right]\xi_r = ik\Psi. \tag{13.50}$$

13.63. The values of ξ_r, ξ_θ, ξ_z given by the equations (13.48-50) may now be set into equation (13.47). The result is a differential equation for Ψ; thus,

$$\frac{\partial^2\Psi}{\partial r^2} + \frac{1}{r}\cdot\frac{\partial\Psi}{\partial r} - \left(K^2 + \frac{m^2}{r^2}\right)\Psi = 0 \tag{13.51}$$

where K^2 is defined by

$$K^2 \equiv \frac{\dfrac{k^2 B_z^2}{4\pi} - \rho\Omega^2}{\gamma p + \dfrac{B_z^2}{4\pi}}\left\{1 + \frac{k^2\gamma p}{\dfrac{k^2 B_z^2}{4\pi} - \rho\Omega^2\left(1 + \dfrac{B_z^2}{4\pi\gamma p}\right)}\right\}. \tag{13.52}$$

This, together with equations (13.48-50), which now define the ξ_i, is equivalent to the original equations of motion. Equation (13.51) is a form of Bessel's equation for an imaginary argument [32] and the solution is

$$\Psi = I_m(Kr). \tag{13.53}$$

---

*The use of this procedure for solving equations 13.44, 13.45, and 13.46 is based on a suggestion made in connection with a similar problem by Reimar Lüst (private communication). It would seem to be applicable in many cases where there is no current flow in the interior of a plasma.

The constant K^2, and therefore Ω^2, is determined by the boundary conditions. For the vacuum field it is possible to write

$$\hat{\mathbf{B}}' = -\nabla\Phi,$$

and, representing Φ by

$$\Phi = \Phi(r)\exp(im\theta + ikz),$$

Laplace's equation takes the form

$$\frac{d^2\Phi}{dr^2} + \frac{1}{r}\cdot\frac{d\Phi}{dr} - \left(k^2 + \frac{m^2}{r^2}\right)\Phi = 0,$$

so that

$$\Phi = CI_m(kr) + DK_m(kr), \tag{13.54}$$

where C and D are constants, to be evaluated below.

13.64.   It will now be supposed that there is a perfectly conducting rigid wall of radius R_0 located in the vacuum.*   At this wall

$$CI_m'(kR_0) + DK_m'(kR_0) = 0, \tag{13.55}$$

where the prime implies the derivative with respect to the argument.   At the plasma-vacuum interface, where $r = R$, the boundary condition given by equation (13.22) becomes

$$k[CI_m'(kR) + DK_m'(kR)] = i\left(k\hat{B}_z + \frac{m}{r}\hat{B}_\theta\right)\xi_r$$

or, using equation (13.48),

$$CI_m'(kR) + DK_m'(kR) = \frac{iK\left(k\hat{B}_z + \dfrac{m}{r}\hat{B}_\theta\right)}{k\left(\dfrac{k^2 B_z^2}{4\pi} - \rho\Omega^2\right)}I_m'(kR). \tag{13.56}$$

From equations (13.55) and (13.56), it follows that

$$C = \frac{K_m'(kR_0)}{I_m'(kR)K_m'(kR_0) - K_m'(kR)I_m'(kR_0)}\cdot\frac{iK\left(k\hat{B}_z + \dfrac{m}{r}\hat{B}_\theta\right)I_m'(KR)}{k\left(\dfrac{k^2 B_z^2}{4\pi} - \rho\Omega^2\right)}$$

$$D = \frac{-I_m'(kR_0)}{I_m'(kR)K_m'(kR_0) - K_m'(kR)I_m'(kR_0)}\cdot\frac{iK\left(k\hat{B}_z + \dfrac{m}{r}\hat{B}_\theta\right)I_m'(KR)}{k\left(\dfrac{k^2 B_z^2}{4\pi} - \rho\Omega^2\right)}$$

13.65.   The boundary condition represented by equation (13.24) can be written as

---

*In order to avoid possible confusion with the general coordinate r, the radii of the wall, and of the plasma will here be represented by R_0 and R, respectively, instead of the corresponding lower-case symbols used in the main text.

$$-\Psi = \frac{i}{4\pi}\left(kB_z + \frac{m}{r}B_\theta\right)\Phi + \frac{1}{8\pi}\xi_r\partial_r(B_z{}^2 + B_\theta{}^2),$$

and upon setting in the values for Ψ, Φ, ξ_r, B_θ, and the expressions for the constants C and D given above, the result is

$$4\pi\frac{\left(\dfrac{k^2B_z{}^2}{4\pi} - \rho\Omega^2\right)I_m(KR)}{KI_m{}'(KR)} = \frac{B_\theta{}^2(R)}{R}$$

$$+ \frac{1}{k}\left(kB_z + \frac{m}{r}B_\theta\right)^2\left[\frac{K_m{}'(kR_0)I_m(kR) - I_m{}'(kR_0)K_m(kR)}{K_m{}'(kR_0)I_m{}'(kR) - I_m{}'(kR_0)K_m{}'(kR)}\right]. \qquad (13.57)$$

This is seen to be a dispersion equation which relates the frequency Ω of the waves to the propagation vector defined by

$$\mathbf{k} \equiv \left(0, \frac{m}{r}, k\right).$$

13.66. The dispersion equation is transcendental and can be solved only by numerical means. If the values of Ω^2 are positive then the system is stable, but if they are negative then it must be unstable. The actual determination of which configurations are stable and which are not is laborious, but it has been carried through and the results obtained were given in §13.27 *et seq.*

REFERENCES FOR CHAPTER 13

1. J. Berkowitz, H. Grad, and H. Rubin, *Proc. Second U.N. Conf. on Peaceful Uses of Atomic Energy*, **31**, 177 (1958).
2. B. B. Kadmotsev and S. I. Braginsky, *Proc. Second U.N. Conf. on Peaceful Uses of Atomic Energy*, **32**, 233 (1958).
3. G. Persico and J. G. Linhart, *Nuovo cimento*, **8**, 740 (1958).
4. I. B. Bernstein, E. A. Frieman, M. D. Kruskal, and R. M. Kulsrud, *Proc. Roy. Soc.*, **A244**, 17 (1958).
5. R. J. Tayler, *Proc. Second U.N. Conf. on Peaceful Uses of Atomic Energy*, **31**, 160 (1958).
6. M. D. Kruskal and M. Schwarzschild, *Proc. Roy. Soc.*, **A233**, 348 (1954).
7. R. J. Tayler, *Proc. Phys. Soc. (London)*, **B70**, 31 (1957).
8. R. J. Tayler, *Proc. Phys. Soc. (London)*, **B70**, 1049 (1957).
9. M. Kruskal and J. L. Tuck, *Proc. Roy. Soc.*, **A245**, 222 (1958).
10. V. D. Shafranov, *J. Nuclear Energy*, **5**, 86 (1957).
11. M. N. Rosenbluth, USAEC Report LA-2030 (1956).
12. M. N. Rosenbluth, *Proc. Second U.N. Conf. on Peaceful Uses of Atomic Energy*, **31**, 85 (1958).
13. B. R. Suydam, *Proc. Second U.N. Conf. on Peaceful Uses of Atomic Energy*, **31**, 157 (1958).
14. W. A. Newcomb, USAEC Report UCRL-5447 (1959).
15. B. R. Suydam, USAEC Report LAMS-2381 (1960).
16. M. D. Kruskal, J. L. Johnson, M. B. Gottlieb, and L. Goldman, *Proc. Second U.N. Conf. on Peaceful Uses of Atomic Energy*, **32**, 217 (1958).

17. J. L. Johnson, C. R. Oberman, R. M. Kulsrud, and E. A. Frieman, *Proc. Second U.N. Conf. on Peaceful Uses of Atomic Energy*, 31, 198 (1958).
18. K. M. Watson, *Phys. Rev.*, 102, 12 (1956).
19. K. A. Brueckner and K. M. Watson, *Phys. Rev.*, 102, 19 (1956).
20. S. Chandrasekhar, A. N. Kaufman, and K. M. Watson, *Ann. Physics*, 2, 435 (1957).
21. G. F. Chew, M. L. Goldberger, and F. E. Low, *Proc. Roy. Soc.*, A236, 112 (1956).
22. M. N. Rosenbluth and C. L. Longmire, *Ann. Physics*, 1, 120 (1957).
23. M. N. Rosenbluth and N. Rostoker, *Proc. Second U.N. Conf. on Peaceful Uses of Atomic Energy*, 31, 144 (1958).
24. S. Chapman and T. G. Cowling, *The Mathematical Theory of Non-Uniform Gases*. Cambridge University Press, 1953, Chapter 18.
25. L. Spitzer, *Physics of Fully Ionized Gases*, Interscience Publishers, Inc., 1956, Appendix.
26. B. R. Suydam, unpublished.
27. M. N. Rosenbluth, *Proc. Second U.N. Conf. on Peaceful Uses of Atomic Energy*, 31, 89 (1958).
28. M. D. Kruskal and C. R. Oberman, *Proc. Second U.N. Conf. on Peaceful Uses of Atomic Energy*, 31, 137 (1958).
29. S. Chandrasekhar, A. N. Kaufman, and K. M. Watson, *Proc. Roy. Soc.*, A245, 435 (1958).
30. S. Chandrasekhar, A. N. Kaufman, and K. M. Watson, *Ann. Physics*, 5, 1 (1958).
31. A. P. Wills, *Vector Analysis*, Prentice-Hall, Inc., 1940, pp. 56-60.
32. E. T. Whittaker and G. N. Watson, *A Course of Modern Analysis*, The Macmillan Co., 1943, pp. 372-374.

GENERAL REFERENCES

H. Alfvén, *Cosmical Electrodynamics*, Clarendon Press, 1939

W. P. Allis, Editor, *Nuclear Fusion*, D. Van Nostrand Co., Inc., 1960

A. S. Bishop, *Project Sherwood*, Addison-Wesley Publishing Co., Inc., 1958

S. Chandrasekhar, *Principles of Stellar Dynamics*, University of Chicago Press, 1942

S. Chandrasekhar, *Plasma Physics*, University of Chicago Press, 1960

F. H. Clauser, General Editor, *Symposium of Plasma Dynamics*, Addison-Wesley Publishing Co., Inc., 1960

T. G. Cowling, *Magnetohydrodynamics*, Interscience Publishers, Inc., 1957

J. L. Delcroix, *Introduction to the Theory of Ionized Gases*, Interscience Publishers, Inc., 1960

G. Francis, *Ionization Phenomena in Gases*, Academic Press, Inc., 1959

R. K. M. Landshoff, Editor, *Magnetohydrodynamics*, Stanford University Press, 1957

R. K. M. Landshoff, Editor, *The Plasma in a Magnetic Field*, Stanford University Press, 1958

J. G. Linhart, *Plasma Physics*, Interscience Publishers, Inc., 1960

C. Longmire, J. L. Tuck, and W. B. Thompson, Editors, Progress in Nuclear Energy, Series XI, Vol. I, *Plasma Physics and Thermonuclear Research*, Pergamon Press, Inc., 1959

J. D. Ramey, Compiler, *Bibliography on Plasma Physics and Magnetohydrodynamics*, Engineering and Physics Sciences Library, University of Maryland, 1959

A. Simon, *An Introduction to Thermonuclear Research*, Pergamon Press, Inc., 1959

B. A. Spence, Compiler, *Bibliography on Magnetohydrodynamics, Plasma Physics, and Controlled Thermonuclear Processes*, Report AERL-AMP-36 (1959), Avco-Everett Research Laboratory

L. Spitzer, Jr., *Physics of Fully Ionized Gases*, Interscience Publishers, Inc., 1956

Nuclear Instruments, **4**, 245-390 (1960)

Plasma Physics and the Problem of Controlled Thermonuclear Reactions, Vols. I, II, III, and IV (M. A. Leontovich, Responsible Editor), English Translation (J. Turkevich and D. Hicks, Translation Editors), Pergamon Press, Inc., 1959-1960

Proceedings of the Institution of Electrical Engineers (London), **106**, Part A, Supplement No. 2 (Convention on Thermonuclear Processes), 1959

Proceedings of the Second United Nations Conference on the Peaceful Uses of Atomic Energy, Volumes 31 and 32 (1959)

Proceedings of the Conference on Controlled Thermonuclear Reactions, Harwell, England, United Kingdom Atomic Energy Authority Report AERE GP/R-2371 (1957), Parts I and II

Proceedings of the Project Sherwood Conferences, United States Atomic Energy Commission Reports WASH-115 (1952); WASH-146 (1953); WASH-184 (1954); WASH-289 (1955); TID-7503 (1955); TID-7520 (1956); TID-7536 (1957); TID-7558 (1958)

Proceedings of the Fourth International Conference on Ionization Phenomena in Gases, Uppsala, Sweden (1959), North-Holland Publishing Co., 1960, Vol. II

Index

References are to paragraph numbers rather than to pages.